Handbook of Exact Solutions to the Nonlinear Schrödinger Equations

Handbook of Exact Solutions to the Nonlinear Schrödinger Equations

Usama Al Khawaja and Laila Al Sakkaf
Department of Physics, UAE University, Al-Ain, UAE

IOP Publishing, Bristol, UK

Multimedia to accompany this book can be downloaded from https://iopscience.iop.org/book/978-0-7503-2428-1.

ISBN 978-0-7503-2428-1 (ebook)
ISBN 978-0-7503-2426-7 (print)
ISBN 978-0-7503-2427-4 (mobi)

DOI 10.1088/978-0-7503-2428-1

Version: 20191101

IOP ebooks

British Library Cataloguing-in-Publication Data: A catalogue record for this book is available from the British Library.

Published by IOP Publishing, wholly owned by The Institute of Physics, London

IOP Publishing, Temple Circus, Temple Way, Bristol, BS1 6HG, UK

US Office: IOP Publishing, Inc., 190 North Independence Mall West, Suite 601, Philadelphia, PA 19106, USA

Contents

Preface x

Acknowledgments xii

Author Biographies xiii

Notation xiv

1 Introduction 1-1

References 1-6

2 Fundamental Nonlinear Schrödinger Equation 2-1

2.1 NLSE with Cubic Nonlinearity 2-1

2.1.1 Real Dispersion and Nonlinearity Coefficients 2-2

2.2 Summary of Subsection 2.1.1 2-33

2.2.1 Complex Dispersion and Nonlinearity Coefficients 2-40

2.3 Summary of Subsection 2.2.1 2-43

References 2-45

3 Nonlinear Schrödinger Equation with Power Law and Dual Power Law Nonlinearities 3-1

3.1 NLSE with Power Law Nonlinearity 3-1

3.1.1 Reduction to the Fundamental NLSE 3-2

3.2 Summary of Section 3.1 3-6

3.3 NLSE with Dual Power Law Nonlinearity 3-8

3.4 Summary of Section 3.3 3-14

References 3-17

4 Nonlinear Schrödinger Equation with Higher Order Terms 4-1

4.1 NLSE with Third Order Dispersion, Self-Steepening, and Self-Frequency Shift 4-3

4.2 Summary of Section 4.1 4-9

4.3 Special Cases of Equation (4.1) 4-13

4.3.1 Case I: Hirota Equation (HE) 4-13

4.3.2 Case II: Sasa–Satsuma Equation (SSE) 4-13

4.4 NLSE with First and Third Order Dispersions, Self-Steepening, Self-Frequency Shift, and Potential 4-13

4.5 Summary of Section 4.4 4-16
4.6 NLSE with Fourth Order Dispersion 4-17
4.7 Summary of Section 4.6 4-19
4.8 NLSE with Fourth Order Dispersion and Power Law Nonlinearity 4-20
4.9 Summary of Section 4.8 4-22
4.10 NLSE with Third and Fourth Order Dispersions and Cubic and Quintic Nonlinearities 4-24
4.11 Summary of Section 4.10 4-29
4.12 NLSE with Third and Fourth Order Dispersions, Self-Steepening, Self-Frequency Shift, and Cubic and Quintic Nonlinearities 4-32
4.13 Summary of Section 4.12 4-36
4.14 NLSE with $|\psi|^2$-Dependent Dispersion 4-39
4.15 Infinite Hierarchy of Integrable NLSEs with Higher Order Terms 4-40
4.15.1 Constant Coefficients 4-40
4.15.2 Function Coefficients 4-43
4.16 Summary of Section 4.15 4-46
References 4-49

5 Scaling Transformations 5-1

5.1 Fundamental NLSE to Fundamental NLSE with Different Constant Coefficients 5-4
5.2 Defocusing (Focusing) NLSE to Focusing (Defocusing) NLSE 5-5
5.3 Galilean Transformation (Movable Solutions) 5-6
5.4 Function Coefficients 5-10
5.4.1 Constant Dispersion and Complex Potential 5-10
5.4.2 Constant Dispersion and Real Quadratic Potential 5-11
5.4.3 Constant Dispersion and Real Linear Potential 5-18
5.4.4 Constant Nonlinearity and Complex Potential 5-24
5.4.5 Constant Nonlinearity and Real Quadratic Potential 5-25
5.4.6 Constant Nonlinearity and Real Linear Potential 5-25
5.5 Solution-Dependent Transformation 5-26
5.5.1 Special Case I: Stationary Solution, Constant Dispersion and Nonlinearity Coefficients 5-27
5.5.2 Special Case II: PT-Symmetric Potential 5-28
5.5.3 Special Case III: Stationary Solution, Constant Dispersion and Nonlinearity Coefficients, and Real Potential 5-29
5.6 Summary of Sections 5.1–5.5 5-30
5.7 Other Equations: NLSE with Periodic Potentials 5-38

5.7.1 General Case: $\text{sn}^2(x, m)$ Potential 5-38
5.7.2 Specific Case: $\sin^2(x)$ Potential 5-39
5.8 Summary of Section 5.7 5-40
Reference 5-40

6 Nonlinear Schrödinger Equation in (N + 1)-Dimensions 6-1

6.1 (N + 1)-Dimensional NLSE with Cubic Nonlinearity 6-4
6.2 (N + 1)-Dimensional NLSE with Power Law Nonlinearity 6-11
6.3 (N + 1)-Dimensional NLSE with Dual Power Law Nonlinearity 6-12
6.4 Galilean Transformation in (N + 1)-Dimensions (Movable Solutions) 6-16
6.5 NLSE in (2 + 1)-Dimensions with $\Phi x_1 x_2$ Term 6-22
6.6 Summary of Sections 6.1–6.5 6-24
6.7 (N + 1)-Dimensional Isotropic NLSE with Cubic Nonlinearity in Polar Coordinate System 6-33
6.7.1 Angular Dependence 6-34
6.7.2 Constant Dispersion and Real Potential 6-35
6.8 Summary of Section 6.7 6-38
6.9 Power Series Solutions to (2 + 1)-Dimensional NLSE with Cubic Nonlinearity in a Polar Coordinate System 6-41
6.9.1 Family of Infinite Number of Localized Solutions 6-42
References 6-42

7 Coupled Nonlinear Schrödinger Equations 7-1

7.1 Fundamental Coupled NLSE *Manakov System* 7-4
7.2 Summary of Section 7.1 7-13
7.3 Symmetry Reductions 7-17
7.3.1 Symmetry Reduction I *From Manakov System to Fundamental NLSE* 7-17
7.3.2 Symmetry Reduction II *From Manakov System to Fundamental NLSE* 7-17
7.3.3 Symmetry Reduction III *From Vector NLSE to Fundamental NLSE* 7-18
7.3.4 Symmetry Reduction IV *From Three Coupled NLSEs to Manakov System* 7-19
7.3.5 Symmetry Reduction V *From Vector NLSE to Manakov System* 7-22
7.4 Scaling Transformations 7-22
7.4.1 Linear and Nonlinear Coupling 7-22
7.4.2 Complex Coupling 7-25

7.4.3 Function Coefficients 7-26

7.5 Summary of Sections 7.3–7.4 7-30

7.6 (N + 1)-Dimensional Coupled NLSE (N + 1)-*Dimensional Manakov System* 7-32

7.6.1 Reduction to 1D Manakov System 7-32

7.7 Symmetry Reductions of (N + 1)-Dimensional CNLSE to Scalar NLSE 7-34

7.7.1 Symmetry Reduction I *From (N + 1)-Dimensional Manakov System to (N + 1)-Dimensional Fundamental NLSE* 7-34

7.7.2 Symmetry Reduction II *From (N + 1)-Dimensional Manakov System to (N + 1)-Dimensional Fundamental NLSE* 7-35

7.7.3 Symmetry Reduction III *From (N + 1)-Dimensional Vector NLSE to (N + 1)-Dimensional Fundamental NLSE* 7-36

7.8 (N + 1)-Dimensional Scaling Transformations 7-37

7.8.1 Linear and Nonlinear Coupling 7-37

7.8.2 Complex Coupling 7-39

7.9 Summary of Sections 7.7–7.8 7-40

References 7-42

8 Discrete Nonlinear Schrödinger Equation 8-1

8.1 Discrete NLSE with Saturable Nonlinearity 8-2

8.1.1 Nonstaggered Solutions 8-2

8.1.2 Staggered Solutions 8-9

8.2 Summary of Section 8.1 8-16

8.3 Short-period Solutions with General, Kerr, and Saturable Nonlinearities 8-22

8.4 Ablowitz–Ladik Equation 8-22

8.5 Summary of Section 8.4 8-30

8.6 Cubic-quintic Discrete NLSE 8-33

8.7 Summary of Section 8.6 8-37

8.8 Generalized Discrete NLSE 8-39

8.9 Summary of Section 8.8 8-47

8.10 Coupled Salerno Equations 8-48

8.11 Summary of Section 8.10 8-55

8.12 Coupled Ablowitz–Ladik Equation 8-58

8.13 Summary of Section 8.12 8-67

8.14 Coupled Saturable Discrete NLSE 8-71

8.15 Summary of Section 8.14 8-73

References 8-74

9 Nonlocal Nonlinear Schrödinger Equation 9-1

9.1 Nonlocal NLSE 9-3

9.2 Nonlocal Coupled NLSE 9-4

9.3 Symmetry Reductions to Scalar Nonlocal NLSE 9-7

9.3.1 Symmetry Reduction I *From Nonlocal Manakov System to Scalar Nonlocal NLSE* 9-7

9.3.2 Symmetry Reduction II *From Nonlocal Manakov System to Scalar Nonlocal NLSE* 9-8

9.3.3 Symmetry Reduction III *From Nonlocal Vector NLSE to Scalar Nonlocal NLSE* 9-9

9.4 Scaling Transformations 9-10

9.4.1 Linear and Nonlinear Coupling 9-10

9.4.2 Complex Coupling 9-12

9.5 Nonlocal Discrete NLSE with Saturable Nonlinearity 9-13

9.5.1 Nonstaggered Solutions 9-13

9.5.2 Staggered Solutions 9-15

9.6 Nonlocal Ablowitz–Ladik Equation 9-15

9.7 Nonlocal Cubic-Quintic Discrete NLSE 9-17

9.8 Summary of Chapter 9 9-21

Appendices

A Derivation of Some Solutions of Chapters 2 and 3 A-1

B Darboux Transformation Single Soliton and Breather Solutions B-1

C Derivation of the Similarity Transformations in Chapter 5 C-1

Preface

We have been involved for the past two decades with the nonlinear Schrödinger equation (NLSE) and its many variations including NLSE with higher-order terms, two- and three-dimensional NLSE, discrete NLSE, nonlocal NLSE, and coupled NLSEs. We noticed that, throughout the long history of NLSE, a large number of exact analytical solutions have been found and the number is still increasing as new solutions are being sought and discovered. As the basis for theoretical models of various research fields such as Bose–Einstein condensates of ultracold gases, nonlinear optics, and deep water waves, this equation is a subject of interest by the scientific communities in all three areas. Its known solutions are scattered in the literature of the different fields. For a beginner as well as for an expert researcher, it is difficult to keep track of the large number of known solutions. It is important for a researcher or a reviewer to know if a certain solution is a new solution, belongs to an existing class of solutions, or can be trivially obtained from another solution by a transformation.

This book is the result of our effort towards serving the research communities involved with the NLSE in the different fields by collecting all known solutions in one document. In addition, the book organizes the solutions by classifying and grouping them based on the aspects and symmetries they have. Although most of the solutions presented in this book have been derived elsewhere using various methods, we attempt here to present a systematic derivation of many solutions and even have derived some new ones. We have also presented symmetries and reductions that connect different solutions through transformations and enable classifying new solutions into known classes.

For the user to verify that the presented solutions do satisfy the NLSE, we provide Mathematica note books (available online at https://iopscience.iop.org/book/978-0-7503-2428-1) containing all solutions in a one-to-one correspondence with the solutions in the text. The reader can run the Mathematica cell and see for themselves that it indeed satisfies the NLSE. This is also an efficient method for detecting and avoiding possible typo mistakes in the text. A large number of figures and animations are included to help visualize solutions and their dynamics.

We have applied the following rules while collecting the solutions from the literature: (1) It has to be NLSE-related. As a result of this restriction, some interesting equations have been excluded such as the nonlinear Dirac equation, which we may consider in a future edition. (2) It has to satisfy the NLSE. We attempted to fix typo errors whenever it was obvious, but normally we did not invest much time in discovering what is wrong in the solution that does not satisfy the NLSE. (3) It has to be analytical. This excludes all numerical solutions. Some NLSEs do not admit analytical solutions but do have important stable numerical solutions such as in the case of two-dimensional and discrete NLSEs. Nonetheless, and in order to set a well-defined scope, we restrict our book to analytical solutions.

We made our best effort to cite the reference where the solution appeared first. Quite often, however, solutions were either rederived in a later reference and put in a different form, or were derived under different conditions. We cite the reference that we copied the solution from although a different variation of the solution may have appeared earlier. We do not claim any credit for the solutions collected, and we apologize for any citation mistake and for missing any solution. It should be mentioned that the solutions presented in this book have been the subject of study by a very large number of references. However, and as mentioned above, we cite only references that we copied the solutions from giving priority to any reference containing a large number of solutions. Due to the fact that we employed scaling transformations and symmetry reductions, many solutions published in the literature were not presented; the reader can reproduce those solutions using the transformations and symmetry reductions presented here. We give numerous examples on such cases. We should be grateful, however, if readers would draw our attention to any missing solutions. We also welcome criticism and comments hoping that they will lead to an enhanced second edition.

Acknowledgments

We are grateful for the substantial support of UAE University, including internal grant funding that supported this work. Fruitful discussions with our colleagues and collaborators, Lincoln Carr, Abdulaziz Alhaidari, Bakhtiyor Baizakov, Hocine Bahlouli, Saeed Al-Marzoug, Yuri Kivshar, Nail Akhmedeiv, Andrey Sokhorokov, Fathulla Abdullaev and Majid Taki are acknowledged.

The process of collecting and deriving solutions and writing this book took about two years. We are grateful to our families for their support and patience during this time.

Author Biographies

Usama Al Khawaja

Usama Al Khawaja obtained his Bachelor's degree in Physics from the University of Jordan in 1992 and Master's degree in Physics with thesis research on *two-dimensional neutral Fermi systems* from the University of Jordan in 1996. He earned his PhD degree in theoretical Physics with dissertation research on *Bose–Einstein condensation* from the University of Copenhagen in 1999. Afterwards, he spent three years of postdoctoral research at Utrecht University in the Netherlands before joining the United Arab Emirates University in 2002 as an assistant professor. He is currently a full professor and Chairman of the Physics department at the United Arab Emirates University. His main areas of research are Bose–Einstein condensation, nonlinear and quantum optics, integrability, and exact solutions. His main achievements in integrability and exact solutions include developing a systematic search method of finding Lax pairs of a given nonlinear partial differential equation. He also developed a convergent power series method for solving nonlinear differential equations. He has authored more than 70 papers and obtained one patent on applying discrete solitons in all-optical operations.

Laila Al Sakkaf

Laila Al Sakkaf obtained her Bachelor's degree in Physics from the United Arab Emirates University in 2015. Then she obtained her Master's degree in Physics from the United Arab Emirates University in 2018 with thesis research on the *iterative power series method for solving nonlinear differential equations*. She is currently a research assistant and a PhD student at the Physics department of the United Arab Emirates University. Her current research focus is on integrability and exact solutions of differential equations modeling nonlinear physical phenomena.

Notation

NLSE	Nonlinear Schrödinger equation
DNLSE	Discrete nonlinear Schrödinger equation
CNLSE	Coupled nonlinear Schrödinger equation
HONLSE	Nonlinear Schrödinger equation with higher-order terms
1D	One-dimensional/dimension
2D	Two-dimensional/dimensions
3D	Three-dimensional/dimensions
ND	N-dimensional/dimensions
$(N + 1)$-D	N dimensions in space, 1 refers to time
CW	Continuous wave
DW	Decaying wave
SW	Solitary wave
GN	General nonlinearity
SN	Saturable nonlinearity
KN	Kerr nonlinearity
IPS	Iterative power series
IST	Inverse scattering transform
PT	Parity-Time
AL	Ablowitz–Ladik
LP	Lax-Pair
DT	Darboux transformation
HE	Hirota equation
SSE	Sasa–Satsuma equation
sn, cn, dn, nd, cd, sd, cs, ds, dc, ns	Jacobi elliptic functions

IOP Publishing

Handbook of Exact Solutions to the Nonlinear Schrödinger Equations

Usama Al Khawaja and Laila Al Sakkaf

Chapter 1

Introduction

The nonlinear Schrödinger equation is known in the literature, most commonly, with the following dimensionless form

$$i\psi_t + \frac{1}{2}\psi_{xx} + \sigma|\psi|^2\psi = 0, \tag{1.1}$$

where σ is a real constant, $\psi = \psi(x, t)$ is a complex function, and the subscripts are partial derivatives in terms of its two independent variables, x and t. It is the basis for theoretical models describing three major fields, namely: Bose–Einstein condensates of ultracold gases [1], nonlinear optics in fibers and waveguide arrays [2], and deep water waves [3].

In Bose–Einstein condensates, which is a quantum system, the nonlinear Schrödinger equation is the classical field limit of the analogous quantized field equation. The function $\psi(x, t)$ is the wave function of the macroscopic many-particle system. To realize Bose–Einstein condensation, a confining (trapping) magnetic and optical potential is needed. This is accounted for by adding a potential term, $V(x)\psi(x, t)$, to the NLSE that then becomes the *Gross–Pitaevskii equation* [4, 5]. The nonlinear term corresponds to the interatomic interaction known as the Hartree–Fock energy with σ being proportional to the s-wave scattering length. The sign of σ can be both positive and negative, corresponding to attractive or repulsive interatomic interactions, respectively. The dispersion term corresponds to the kinetic energy pressure [1].

In nonlinear optics, the NLSE describes the propagation of pulses in nonlinear media such as optical fibers, photonic crystals, or waveguide arrays. It can be derived from Maxwell's equations with $\psi(x, t)$ corresponding to the envelope of modulated electrical (or magnetic) field strength of the propagating pulse [6, 7]. The nonlinear term corresponds to the modulation of the refractive index of the medium as a response to the propagating light pulse, which is known in the nonlinear optics

doi:10.1088/978-0-7503-2428-1ch1

community as the Kerr nonlinearity. The constant σ represents, in this case, the strength of the Kerr nonlinearity, which can also be positive or negative, leading to the *focusing* or *defocusing* NLSE, respectively. Here, the term ψ_{xx} corresponds to the dispersion of the pulse [2].

The NLSE describes also surface water waves where $\psi(x, t)$ corresponds to the intensity and phase of the waves. This description is restricted to deep water waves with a wavelength much smaller than the water depth. Shallow water waves are not described by the NLSE. The nonlinearity originates from the Bernoulli equation, its strength depends on the water depth, and it is always negative for deep water waves [3].

The above three examples suggest that the NLSE is a universal equation that describes the propagation of wave modulations in media with dispersion and nonlinearity.

The NLSE is integrable and admits, in principle, an infinite number of independent solutions [8]. It was first solved by Zakharov and Shabat using the Inverse Scattering Transform (IST), which relies on associating the NLSE to a linear system of differential equations [9]. The system has been known since then as the Zakharov–Shabat system and the method was adopted to find other solutions of the NLSE and its variations. In general, the linear system is given in terms of a pair of matrices, known as the Lax pair, acting on an *auxiliary* field. The existence of a Lax pair establishes the integrability of a differential equation, at least within the Lax pair sense [10]. The IST is a powerful method for finding solutions of nonlinear differential equations [11]. It is distinguished among other methods by generating classes of an infinite hierarchy of solutions. It can also be used to exactly solve the nonlinear initial value problem for a given nonlinear differential equation. Many other methods of solving nonlinear differential equations have been applied to the NLSE. For the systematic derivations of solutions we present in this book, we use mainly the IST and separation of variables methods.

There are many variations of the NLSE including NLSE with higher-order terms, NLSE in higher dimensions, NLSE with function coefficients and potential terms, coupled NLSEs, discrete NLSE, and nonlocal NLSE. It should be noted that we often refer to the NLSE and its variations simply by NLSE. Many of these variations turn out to be integrable and many others turn out to be related to the fundamental NLSE via some *scaling* transformations. All of these variations will be considered in this book.

The book begins in chapter 2 with the fundamental NLSE. In this chapter we prefer to present solutions of NLSE with arbitrary constant coefficients a_1 and a_2 for dispersion and nonlinearity, respectively (see section 2.1). One may argue that this is not the 'fundamental' NLSE since with a simple scaling transformation, as shown in chapter 5, it transforms into another NLSE with no coefficients ($i\psi_t + \psi_{xx} + |\psi|^2\psi = 0$), which may be more accurately denoted as the fundamental NLSE. This is indeed the case when a_2 is real, but the scaling transformation does not work when a_2 is complex. Therefore, by keeping the coefficients a_1 and a_2 explicitly in the NLSE, we will be able to consider solutions when the coefficients are complex. This chapter contains the largest number of solutions collected. The solutions of this chapter can be used as a *seed* for transformations generating many solutions of other NLSEs in the subsequent chapters. The solutions can be categorized as: (i) stationary solutions of the form $\psi(x, t) = u(x)e^{i\phi t}$, where $u(x)$ is a real function that can be localized or oscillatory and ϕ is a real constant, (ii) a class of breathers family, (iii) class of N-bright solitons, (iv) rational solutions that are fundamentally different from the breathers class.

In chapter 3, we consider the NLSE with power law and dual power law nonlinearities. Here, the cubic nonlinearity of the fundamental NLSE is replaced first by a nonlinearity with general power, n, that is not restricted to integers. Then we consider an NLSE with two nonlinear terms; one with power n and another with power m, where again n and m are arbitrary real constants that do not have to be integers. In the first case we show, at the beginning of the chapter, that with a scaling transformation applied to stationary solutions, the NLSE with power law non-linearity reduces to the fundamental NLSE, and thus all stationary solutions of chapter 2 lead to stationary solutions to the NLSE with power law nonlinearity. Many examples have been worked out explicitly. We could not find a similar transformation for the dual power law nonlinearity, but we have found 14 solutions for this case.

In chapter 4, we consider the NLSE with higher-order terms. These include, following the nonlinear optics terminology: third-order dispersion, fourth-order dispersion, self-steepening, self-frequency shift, and power law nonlinearity. At the end, we consider an infinite hierarchy of integrable NLSEs. The first member of the hierarchy being the fundamental NLSE, the higher-order members turn out to comprise most of the known higher-order variations of the NLSE. Solutions to all member equations of the infinite hierarchy have been presented.

In chapter 5, we present scaling transformations that reduce many variations of the NLSE to the fundamental one. These include transforming the NLSE with arbitrary constant coefficients to the one with no coefficients, transforming the NLSE with focusing (defocusing) nonlinearity to the NLSE with defocusing (focusing) nonlinearity, Galilean transformation to obtain movable solutions from static ones, transforming the NLSE with function coefficients and complex potential to the fundamental NLSE, and finally introducing a solution-dependent transformation where a seed solution is used to construct the transformation operator. The latter allowed for more possibilities including transforming an NLSE with constant coefficients and PT-symmetric potential to the fundamental NLSE.

In chapter 6, we consider NLSE in higher dimensions. We start with a scaling transformation showing that an $(N + 1)$-dimensional NLSE, with N denoting the spatial dimensions and 1 denoting the temporal dimension, can be reduced to the one-dimensional NLSE in terms of a reduced spatial variable. A Galilean transformation is then shown to apply where movable solutions of the $(N + 1)$-dimensional NLSE are obtained from the static solutions of the one-dimensional NLSE. An NLSE with mixed derivatives is also considered. Then, solutions of the $(N + 1)$-dimensional NLSE with power law and dual power law nonlinearities are presented. A scaling transformation of the $(N + 1)$-dimensional NLSE in polar coordinates is also worked out allowing one to consider solutions with cylindrical and spherical symmetries in two- and three-dimensional geometries, respectively. At the end, we present our iterative power series method of obtaining convergent power series representations of the solutions. The method is applied here for nonintegrable cases such as some two- and three-dimensional NLSEs.

In chapter 7, we consider the coupled NLSE. At first, we consider the fundamental coupled NLSE, known as the Manakov system. Being an integrable system, many solutions have been found and presented. Then, we show three simple symmetry reductions that transform the coupled NLSE to the scalar NLSE and symmetry reduction that transforms the vector NLSE (N-coupled NLSEs) to the Manakov system. Some interesting examples have been shown explicitly. We consider also a coupled system with additional linear coupling terms. A scaling transformation is performed to reduce this system to the fundamental Manakov system. Here, we found that, as a special case, one may obtain the solutions of a Manakov system from those of another Manakov system that differs in the values of the constant coefficients. Furthermore, one may obtain a solution of the Manakov system from another solution of the same system, which invokes the superposition principle known for linear differential equations. It is interesting to find such a principle applying for nonlinear differential equations. We have worked out explicitly a nontrivial example in this case. We have also considered a coupled system with additional complex coupling terms, which again with a scaling transformation was reduced to the fundamental Manakov system. Then, we consider the $(N + 1)$-dimensional coupled NLSEs. First, we show that, with a proper scaling transformation, this system reduces to the one-dimensional Manakov system. Then, we consider scaling transformations that reduce this system to the $(N + 1)$-dimensional scalar NLSE. Scaling transformations were then found for linear and nonlinear coupling, as mentioned above, but here generalized for $(N + 1)$ dimensions.

In chapter 8, we consider the discrete NLSE. The chapter starts with the discrete NLSE with saturable nonlinearity, which is integrable. Here, the solutions are classified into staggered and nonstaggered solutions. We show the transformation that links the two kinds of solutions. Then, we consider other types of nonlinearity including a general form of nonlinearity. Then, we considered the Ablowitz–Ladik equation, which is also integrable. Then, we consider an integrable discrete NLSE with cubic and quintic nonlinearities. A generalized discrete NLSE was then considered with many solutions satisfying the equation under some integrability conditions. Then, we consider coupled discrete NLSEs including the coupled Salerno equations and the coupled Ablowitz–Ladik equations.

In chapter 9, we consider the nonlocal NLSE. We start with transformations that reduce the nonlocal NLSE to the fundamental NLSE. This is possible only for even or odd solutions in the variable x. Then, we consider coupled nonlocal NLSEs and reduce them to the local Manakov system. Then, we consider nonlocal coupled NLSEs with linear, nonlinear, and complex coupling. With scaling transformation, they are reduced to their local counterparts. Then, we consider the nonlocal discrete NLSE, nonlocal discrete coupled NLSEs, nonlocal Ablowitz–Ladik coupled system, and nonlocal NLSE with cubic and quintic nonlinearities.

Chapters 2 and 3 are supplemented by appendix A and appendix B, where we lay out detailed derivations of most of the solutions presented in these chapters. In appendix A, we follow the standard methods of solving differential equations in a systematic manner in order to account for all possible solutions. Appendix B explains the Lax pair and Darboux transformation method and gives a detailed derivation of the bright soliton and breather solutions. Chapter 5 is supplemented with appendix C where we show the detailed derivations of the scaling transformation. All chapters are started by a 'glance'; a summary that helps the reader to easily navigate through the text.

All of the solutions presented in the text are rewritten together with the NLSE they satisfy in a Mathematica notebook (available online at https://iopscience.iop.org/book/978-0-7503-2428-1). The reader can run the cells in the Mathematica notebook in order to verify that the solutions do indeed satisfy the NLSE. This minimizes any possible typo errors in the text. It is also convenient to have the solution typed in for those readers who want to use the solutions in their calculations. Verification of the solutions was possible at different levels of accuracy. Some solutions satisfy the NLSE with all variables and parameters unspecified. For some other solutions, Mathematica could not verify the solution in a reasonable time, therefore we set arbitrary values for the variables and parameters. Here we used only integers or ratios of integers. The verification is then obtained with the integer '0' as a result of substituting the solution in the NLSE. This result is numerical but with infinite accuracy. For the rest of the solutions, Mathematica could not verify the solutions with infinite accuracy. In this case, we set numerical values for the parameters and variables with a chosen number of digits. The verification leads to a numerical zero with accuracy increasing when the number of digits is increased. Some readers may still want to verify the last two cases with unspecified values of the variables. In such cases they will need to wait longer. For example, with a typical personal computer one may have to wait 15 minutes to verify the two-bright-solitons solution with unspecified variables.

There are many solutions that can be obtained from other solutions for some specific values of the parameters. For the reader to have quick and easy access to the solutions that they might be interested in, we present such solutions explicitly.

This book can be considered as the nucleus of a growing collection of solutions to the NLSE and its various variations. The search for new solutions is an ongoing effort by many researchers. We believe that new solutions will continue to appear. Our book will be a suitable host and help one to keep track of the new solutions and classify them into their proper classes, if any. It will be also a useful reference to judge what is claimed a new solution. In addition, the scaling transformations presented in this book will be an efficient tool to reveal whether a new solution can be obtained from an existing solution with a simple transformation. They will be helpful to derive new solutions for some specific setups. We aim at monitoring the literature for the appearance of new solutions of the NLSE and plan to update the book frequently with future editions. Researchers and users of this book are invited to contribute by suggesting and pointing out new or even missing solutions so that we can incorporate them into future editions.

References

[1] Pethick C J and Smith H 2001 *Bose-Einstein Condensation in Dilute Gases* (Cambridge: Cambridge University Press)

[2] See for instance: Hasegawa A and Kodama Y 1995 *Solitons in Optical Communications* (New York: Oxford University Press)

Mollenauer L F and Gordon J P 2006 *Solitons in Optical Fibers* (Boston: Academic Press)

Agrawal G P 2001 *Nonlinear Fiber Optics* 3rd ed (San Diego: Academic)

Akhmediev N N and Ankiexicz A 1997 *Solitons: Nonlinear Pulses and Beams* (London: Chapman and Hall)

Akhmediev N N and Ankiewicz A 2008 *Dissipative Solitons: From Optics to Biology and Medicine* (Berlin: Springer)

Taylor J (ed) 1992 *Optical Solitons: Theory and Experiment* Cambridge Studies in Modern Optics pp I–Vi (Cambridge: Cambridge University Press)

Kivshar Y S and Agrawal G P 2003 *Optical Solitons* (Burlington: Academic)

Butcher P and Cotter D 1990 *The Elements of Nonlinear Optics* (Cambridge Studies in Modern Optics) (Cambridge: Cambridge University Press)

Newell A C and Moloney J V 1992 *Nonlinear Optics* (Redwood City: Addison-Wesley)

Taylor J R (ed) 1992 *Optical Solitons-Theory and Experiment* (Cambridge: Cambridge University Press)

[3] Kharif C, Pelinovsky E and Slunyaev A 2009 Rogue waves in the ocean *Advances in Geophysical and Environmental Mechanics and Mathematics* (Berlin: Springer)

[4] Gross E P 1961 Structure of a quantized vortex in boson systems *Il Nuovo Cimento (1955–1965)* **20** 454–77

[5] Pitaevskii L P 1961 Vortex lines in an imperfect Bose gas *Sov. Phys. JETP* **13** 451–4

[6] Hasegawa A and Tappert F 1973 Transmission of stationary nonlinear optical pulses in dispersive dielectric fibers. I. Anomalous dispersion *Appl. Phys. Lett.* **23** 142–4

[7] Hasegawa A and Tappert F 1973 Transmission of stationary nonlinear optical pulses in dispersive dielectric fibers. II. Normal dispersion. *Appl. Phys. Lett.* **23** 171–2

[8] Zakharov V E E and Manakov S V 1974 On the complete integrability of a nonlinear Schrödinger equation *Theor. Math. Phys.* **19** 332–43

[9] Shabat A and Zakharov V 1972 Exact theory of two-dimensional self-focusing and one-dimensional self-modulation of waves in nonlinear media *Sov. Phys. JETP* **34** 62–9

[10] Al Khawaja U 2010 A comparative analysis of Painlevé, Lax pair, and similarity transformation methods in obtaining the integrability conditions of nonlinear Schrödinger equations *J. Math. Phys.* **51** 053506–11

[11] Ablowitz M J and Clarkson P A 1991 *Solitons, Nonlinear Evolution Equations and Inverse Scattering* **149** (Cambridge: Cambridge University Press)

Chapter 2

Fundamental Nonlinear Schrödinger Equation

A Glance at Chapter 2

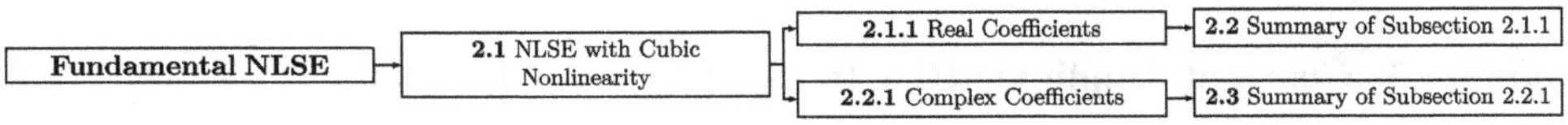

A Statistical View of Chapter 2

	Equation	Solutions
1	$i\,\psi_t + a_1\,\psi_{xx} + a_2\,\lvert\psi\rvert^2\,\psi = 0$	49
2	$i\,\psi_t + (a_{1r} + i\,a_{1i})\,\psi_{xx} + (a_{2r} + i\,a_{2i})\,\lvert\psi\rvert^2\,\psi = 0$	10
Total	2	59

2.1 NLSE with Cubic Nonlinearity

Equation:

$$i\,\psi_t + a_1\,\psi_{xx} + a_2\,\lvert\psi\rvert^2\,\psi = 0, \tag{2.1}$$

where

$\psi = \psi(x, t)$ is the complex function profile,
x and t are its two independent variables,
a_1 and a_2 are arbitrary constants.

doi:10.1088/978-0-7503-2428-1ch2

Solutions:

2.1.1 Real Dispersion and Nonlinearity Coefficients

***Solution* 1. Constant Amplitude I** *continuous wave (CW), t-dependent phase*

$$\psi(x,\,t) = A_0\, e^{i\left[a_2\, A_0^2\,(t-t_0)+\phi_0\right]}, \tag{2.2}$$

where A_0, t_0, and ϕ_0 are arbitrary real constants.

- *Reference*: [1].

***Solution* 2. Constant Amplitude II** *CW, x-dependent phase*

$$\psi(x,\,t) = A_0\, e^{i\left[\pm A_0\sqrt{\frac{a_2}{a_1}}\,(x-x_0)+\phi_0\right]}, \tag{2.3}$$

where
$a_1\, a_2 > 0$,
A_0, x_0, and ϕ_0 are arbitrary real constants.

- *Derived in appendix* A.

***Solution* 3. Constant Amplitude III** *CW, t- and x-dependent phase*

$$\psi(x,\,t) = A_0\, e^{i\left[A_1\,(x-x_0)+\left(A_0^2\,a_2-A_1^2\,a_1\right)(t-t_0)+\phi_0\right]}, \tag{2.4}$$

where
A_0, A_1, x_0, t_0, and ϕ_0 are arbitrary real constants.

- *Reference*: [1].

***Solution* 4. Rational Solution I** *decaying wave (DW)*

$$\psi(x,\,t) = \frac{A_0}{\sqrt{A_1+t-t_0}}\, e^{i\left\{\frac{[a_1\,A_2+x-x_0]^2}{4\,a_1\,[A_1+t-t_0]}+a_2\,A_0^2\,\ln[A_1+t-t_0]+\phi_0\right\}}, \tag{2.5}$$

where
A_0, A_1, A_2, t_0, x_0, and ϕ_0 are arbitrary real constants.

- *Reference*: [1].

***Solution* 5. Rational Solution II**

$$\psi(x,\,t) = \sqrt{\frac{-2\,a_1}{a_2}}\,\frac{1}{x-x_0}\,e^{i\,\phi_0}, \tag{2.6}$$

where
$a_1 a_2 < 0$,
x_0 and ϕ_0 are arbitrary real constants.

- *Reference*: [2].

***Solution* 6. Rational Solution III** *higher order of* (2.6)

$$\psi(x, t) = \frac{1}{\sqrt{-a_2}} \frac{q_1(x, t)}{q_2(x, t)}, \tag{2.7}$$

where

$$q_1(x, t) = 16 x_1^2 + 2\left[2 A_1 x_0 \left(\sqrt{\frac{2}{a_1}} x_1 x + \lambda - \lambda^*\right) - A_0 \lambda^2 e^{2\lambda^*\left(\frac{x}{\sqrt{2 a_1}} - i \lambda^* t\right)}\right]$$
$$\times e^{2\lambda\left(\frac{x}{\sqrt{2 a_1}} + i \lambda t\right)},$$
$$q_2(x, t) = 4 x_0 \frac{\lambda^2 A_0}{A_1} e^{2\lambda^*\left(\frac{x}{\sqrt{2 a_1}} - i \lambda^* t\right)}$$
$$+ e^{2\lambda\left(\frac{x}{\sqrt{2 a_1}} + i \lambda t\right)}\left[2 x_0 A_1 + A_0 \left(\sqrt{\frac{2}{a_1}} x_1 x - x_0\right) e^{2\lambda^*\left(\frac{x}{\sqrt{2 a_1}} - i \lambda^* t\right)}\right]$$
$$- 8 \lambda^2 \left(\sqrt{\frac{2}{a_1}} x_1 x + x_0\right),$$

$\lambda = \lambda_r + i \lambda_i$,
$c_1 = c_{1r} + i c_{1i}$,
$c_2 = c_{2r} + i c_{2i}$,
$x_0 = \lambda + \lambda^*$,
$x_1 = |\lambda|^2$,
$A_0 = \frac{|c_2|^2}{2 \lambda^{*2} |c_1|^2}$,
$A_1 = \frac{c_2}{c_1}$,
$a_1 > 0$,
$a_2 < 0$,
λ_r, λ_i, c_{1r}, c_{1i}, c_{2r}, c_{2i}, and ϕ_0 are arbitrary real constants.

The * denotes the complex conjugate.

- *Derived by Khelifa Elhadj and U Al Khawaja using the Darboux transformation, unpublished.*

***Solution* 7. sec(x)**

$$\psi(x,\, t) = A_0 \sqrt{\frac{-2\, a_1}{a_2}}\, \sec[A_0\, (x - x_0)]\, e^{-i\left[a_1\, A_0^2\, (t - t_0) + \phi_0\right]}, \tag{2.8}$$

where

$a_1\, a_2 < 0$,

A_0, x_0, t_0, and ϕ_0 are arbitrary real constants.

- *Reference*: [3] *with* $m = 1$ *and* $\alpha = \gamma = \lambda = \nu = 0$.

***Solution* 8. csc(x)**

$$\psi(x,\, t) = A_0 \sqrt{\frac{-2\, a_1}{a_2}}\, \csc[A_0\, (x - x_0)]\, e^{-i\left[a_1\, A_0^2\, (t - t_0) + \phi_0\right]}, \tag{2.9}$$

where

$a_1\, a_2 < 0$,

A_0, x_0, t_0, and ϕ_0 are arbitrary real constants.

- *Reference*: [3] *with* $m = 1$ *and* $\alpha = \gamma = \lambda = \nu = 0$.

***Solution* 9. tan(x)**

$$\psi(x,\, t) = A_0 \sqrt{\frac{-2\, a_1}{a_2}}\, \tan[A_0\, (x - x_0)]\, e^{i\left[2\, a_1\, A_0^2\, (t - t_0) + \phi_0\right]}, \tag{2.10}$$

where

$a_1\, a_2 < 0$,

A_0, x_0, t_0, and ϕ_0 are arbitrary real constants.

- *Reference*: [3] *with* $m = 1$ *and* $\alpha = \gamma = \lambda = \nu = 0$.

***Solution* 10. cot(x)**

$$\psi(x,\, t) = A_0 \sqrt{\frac{-2\, a_1}{a_2}}\, \cot[A_0\, (x - x_0)]\, e^{i\left[2\, a_1\, A_0^2\, (t - t_0) + \phi_0\right]}, \tag{2.11}$$

where

$a_1\, a_2 < 0$,

A_0, x_0, t_0, and ϕ_0 are arbitrary real constants.

- *Reference*: [3] *with* $m = 1$ *and* $\alpha = \gamma = \lambda = \nu = 0$.

***Solution* 11. sech(x)** *bright soliton*
(Figure 2.1)

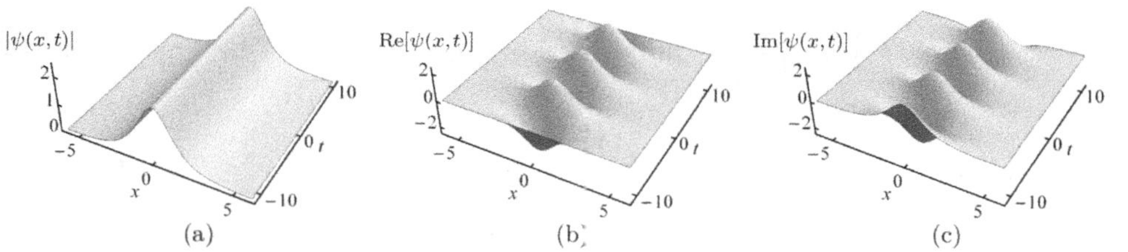

Figure 2.1. Bright soliton (2.12), with $a_1 = 1$, $a_2 = 1/2$, $A_0 = 1$, and $x_0 = t_0 = \phi_0 = 0$. (a) Absolute value of (2.12), (b) real part of (2.12), and (c) imaginary part of (2.12).

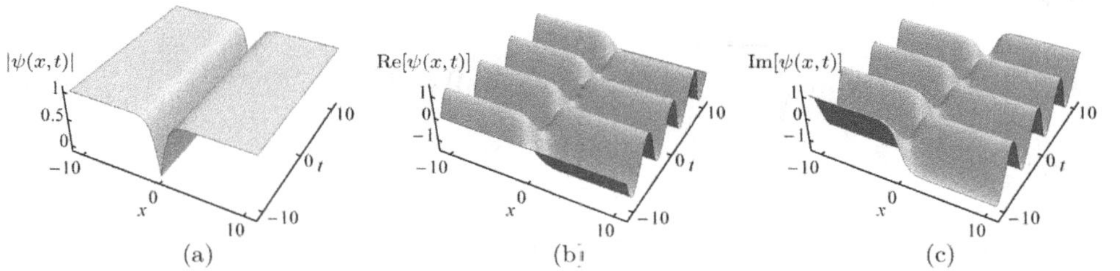

Figure 2.2. Dark soliton (2.14), with $a_1 = 1/2$, $a_2 = -1$, $A_0 = 1$, and $x_0 = t_0 = \phi_0 = 0$. (a) Absolute value of (2.14), (b) real part of (2.14), and (c) imaginary part of (2.14).

$$\psi(x,\, t) = A_0 \sqrt{\frac{2\, a_1}{a_2}}\ \mathrm{sech}[A_0\,(x - x_0)]\, e^{i\left[a_1\, A_0^2\,(t-t_0)+\phi_0\right]}, \tag{2.12}$$

where

$a_1\, a_2 > 0$,

A_0, x_0, t_0, and ϕ_0 are arbitrary real constants.

- *Reference*: [1].

***Solution* 12. csch(x)**

$$\psi(x,\, t) = A_0 \sqrt{\frac{-2\, a_1}{a_2}}\ \mathrm{csch}[A_0\,(x - x_0)]\, e^{i\left[a_1\, A_0^2\,(t-t_0)+\phi_0\right]}, \tag{2.13}$$

where

$a_1\, a_2 < 0$,

A_0, x_0, t_0, and ϕ_0 are arbitrary real constants.

- *Reference*: [3] *with* $m = 1$ *and* $\alpha = \gamma = \lambda = \nu = 0$.

***Solution* 13. tanh(x)** *dark soliton*

(Figure 2.2)

$$\psi(x,\, t) = A_0 \sqrt{\frac{-2\, a_1}{a_2}}\; \tanh[A_0\,(x - x_0)]\; e^{-i\left[2\, a_1\, A_0^2\,(t-t_0)+\phi_0\right]}, \tag{2.14}$$

where
$a_1\, a_2 < 0$,
A_0, x_0, t_0, and ϕ_0 are arbitrary real constants.

- *Reference*: [3] *with* $m = 1$ *and* $\alpha = \gamma = \lambda = \nu = 0$.

***Solution* 14. coth(x)**

$$\psi(x,\, t) = A_0 \sqrt{\frac{-2\, a_1}{a_2}}\; \coth[A_0\,(x - x_0)]\; e^{-i\left[2\, a_1\, A_0^2\,(t-t_0)+\phi_0\right]}, \tag{2.15}$$

where
$a_1\, a_2 < 0$,
A_0, x_0, t_0, and ϕ_0 are arbitrary real constants.

- *Reference*: [3] with $m = 1$ and $\alpha = \gamma = \lambda = \nu = 0$.

***Solution* 15.**
(Figure 2.3)

$$\begin{aligned}\psi(x,\, t) = &\left(A_0 + i\, A_1 \tan\left\{A_1\left[\sqrt{\frac{-a_2}{2\, a_1}}\,(x - x_0) + a_2\, A_0\,(t - t_0)\right]\right\}\right)\\ &\times e^{i\left[a_2\left(A_0^2 - A_1^2\right)(t-t_0)+\phi_0\right]},\end{aligned} \tag{2.16}$$

where
$a_1 a_2 < 0$,
A_0, A_1, x_0, t_0, and ϕ_0 are arbitrary real constants.

- *Reference*: [4].

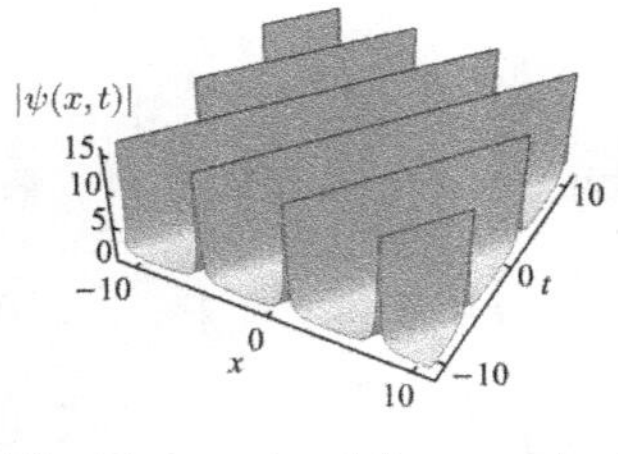

Figure 2.3. Plot of solution (2.16) with $a_1 = A_1 = 1/2$, $a_2 = -1$, $A_3 = 3/4$, and $x_0 = t_0 = \phi_0 = 0$.

***Solution* 16.**

$$\psi(x, t) = A_0 \sqrt{\frac{-a_1}{2\, a_2}} \{\sec[A_0 (x - x_0)] + \tan[A_0 (x - x_0)]\} \times e^{i\left[\frac{a_1 A_0^2}{2} (t - t_0) + \phi_0\right]}, \tag{2.17}$$

where

$a_1\, a_2 < 0$,

A_0, x_0, t_0, and ϕ_0 are arbitrary real constants.

- *Reference*: [3].

***Solution* 17.**

$$\psi(x, t) = A_0 \sqrt{\frac{-2}{a_2}} \left(\cot\left\{A_0 \left[\frac{x - x_0}{\sqrt{a_1}} - 2\, A_1 (t - t_0)\right]\right\} - \tan\left\{A_0 \left[\frac{x - x_0}{\sqrt{a_1}} - 2\, A_1 (t - t_0)\right]\right\}\right) \times e^{i\left[\frac{A_1}{\sqrt{a_1}} (x - x_0) - \left(A_1^2 - 8\, A_0^2\right)(t - t_0) + \phi_0\right]}, \tag{2.18}$$

where

$a_1 > 0$,

$a_2 < 0$,

A_0, A_1, x_0, t_0, and ϕ_0 are arbitrary real constants.

- *Reference*: [4].

***Solution* 18.**

$$\psi(x, t) = A_0 \sqrt{\frac{-a_1}{2\, a_2}} \left\{\frac{\cos[A_0 (x - x_0)]}{1 - \sin[A_0 (x - x_0)]}\right\} e^{i\left[\frac{a_1 A_0^2}{2} (t - t_0) + \phi_0\right]}, \tag{2.19}$$

where

$a_1\, a_2 < 0$,

A_0, x_0, t_0, and ϕ_0 are arbitrary real constants.

- *Reference*: [3].

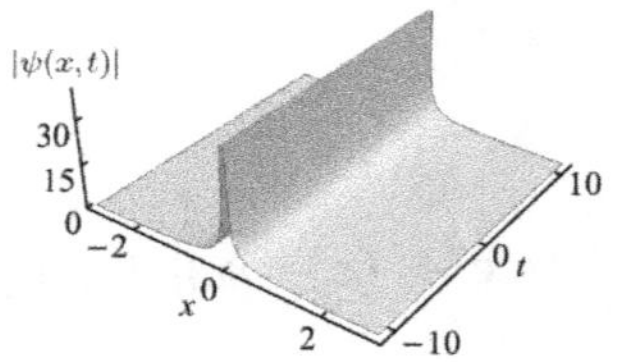

Figure 2.4. Plot of solution (2.20) with $a_1 = -1/2$, $a_2 = 1/2$, $A_0 = 1/20$, and $x_0 = t_0 = \phi_0 = 0$.

***Solution* 19.**
(Figure 2.4)

$$\psi(x,\, t) = A_0 \sqrt{\frac{-2\, a_1}{a_2}} \left\{ \frac{1 + \tanh^2[A_0\,(x - x_0)]}{\tanh[A_0\,(x - x_0)]} \right\} e^{-i\left[8\, a_1\, A_0^2\,(t - t_0) + \phi_0\right]}, \tag{2.20}$$

where
$a_1\, a_2 < 0$,
A_0, x_0, t_0, and ϕ_0 are arbitrary real constants.

- *Reference*: [3].

***Solution* 20.**

$$\psi(x,\, t) = A_0 \sqrt{\frac{-2}{a_2}} \left(\frac{b_0 \tan\left\{A_0\left[\frac{x - x_0}{\sqrt{a_1}} - 2\, A_1\,(t - t_0)\right]\right\} - b_1}{b_0 + b_1 \tan\left\{A_0\left[\frac{x - x_0}{\sqrt{a_1}} - 2\, A_1\,(t - t_0)\right]\right\}} \right) \times e^{i\left[\frac{A_1\,(x - x_0)}{\sqrt{a_1}} - \left(A_1^2 - 2\, A_0^2\right)(t - t_0) + \phi_0\right]}, \tag{2.21}$$

where
$a_1 > 0$,
$a_2 < 0$,
A_0, A_1, b_0, b_1, x_0, t_0, and ϕ_0 are arbitrary real constants.

- *Reference*: [4].

***Solution* 21.**
(Figure 2.5)

$$\psi(x,\, t) = \frac{q_1(x,\, t)}{q_2(x,\, t)}\, e^{i\left[\frac{A_0}{\sqrt{a_1}}\,(x - x_0) - A_0^2\,(t - t_0) + \phi_0\right]}, \tag{2.22}$$

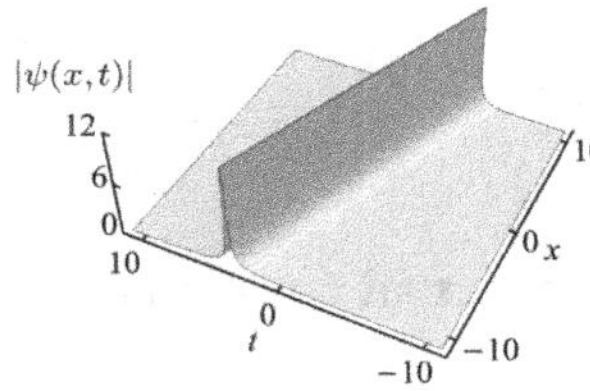

Figure 2.5. Plot of solution (2.22) with $a_1 = 1/2$, $a_2 = -1/4$, $A_0 = -2$, $b_0 = b_1 = -1$, and $x_0 = t_0 = \phi_0 = 0$.

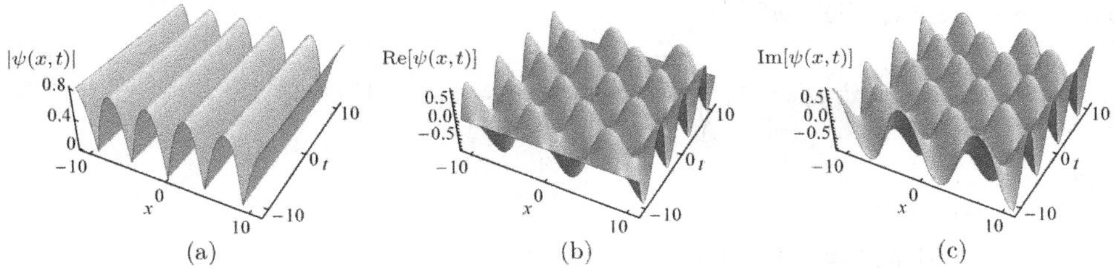

Figure 2.6. Solitary wave (2.23), with $a_1 = 1$, $a_2 = -1$, $A_0 = 1$, $x_0 = t_0 = \phi_0 = 0$, and $m = 1/2$. (a) Absolute value of (2.23), (b) real part of (2.23), and (c) imaginary part of (2.23).

where

$$q_1(x, t) = -2\, b_0\, A_0 \sqrt{-2\, a_2} + 2\, b_1 - \sqrt{\frac{-2\, a_2}{a_1}}\, (x - x_0) + 2\, A_0 \sqrt{-2\, a_2}\, (t - t_0),$$

$$q_2(x, t) = 4\, b_0^2\, a_2\, A_0^2 + 2\, b_1^2 + \frac{4\, b_0\, a_2\, A_0}{\sqrt{a_1}}\, (x - x_0) - 8\, b_0\, a_2\, A_0^2\, (t - t_0) + \frac{a_2}{a_1}\, (x - x_0)^2 - \frac{4\, a_2\, A_0}{\sqrt{a_1}}\, (x - x_0)\, (t - t_0) + 4\, a_2\, A_0^2\, (t - t_0)^2,$$

$a_1 > 0$,
$a_2 < 0$,
A_0, b_0, b_1, x_0, t_0, and ϕ_0 are arbitrary real constants.

- *Reference*: [4].

***Solution 22.* sn(*x*, *m*)** *solitary wave (SW)*
(Figure 2.6)

$$\psi(x, t) = A_0 \sqrt{\frac{2\, m}{-a_2\, (1 + m)}}\ \mathrm{sn}\left[\frac{A_0}{\sqrt{a_1\, (1 + m)}}\, (x - x_0),\, m\right] e^{-i\left[A_0^2\, (t - t_0) + \phi_0\right]}, \quad (2.23)$$

where
$a_1\,(1+m)>0$,
$a_2\,m\,(1+m)<0$,
$m\neq -1$,
A_0, t_0, x_0, and ϕ_0 are arbitrary real constants.

- *Reference*: [5].

***Solution* 23. sn(*x*, −1)** *SW*

$$\psi(x,\,t)=A_0\sqrt{\frac{2\,a_1}{a_2}}\;\mathrm{sn}\,[A_0\,(x-x_0),\,-1]\,e^{i\,\phi_0}, \tag{2.24}$$

where
$a_1\,a_2>0$,
A_0, x_0, and ϕ_0 are arbitrary real constants.

- *Derived in appendix* A.

***Solution* 24. cn(*x*, *m*)** *SW*
(Figure 2.7)

$$\psi(x,\,t)=A_0\sqrt{\frac{2\,m}{a_2\,(2\,m-1)}}\;\mathrm{cn}\left[\frac{A_0}{\sqrt{a_1\,(2\,m-1)}}\,(x-x_0),\,m\right]\times e^{i\left[A_0^2\,(t-t_0)+\phi_0\right]}, \tag{2.25}$$

where
$a_1\,(2\,m-1)>0$,
$a_2\,m\,(2\,m-1)>0$,
$m\neq 1/2$,
A_0, t_0, x_0, and ϕ_0 are arbitrary real constants.

- *Reference*: [5].

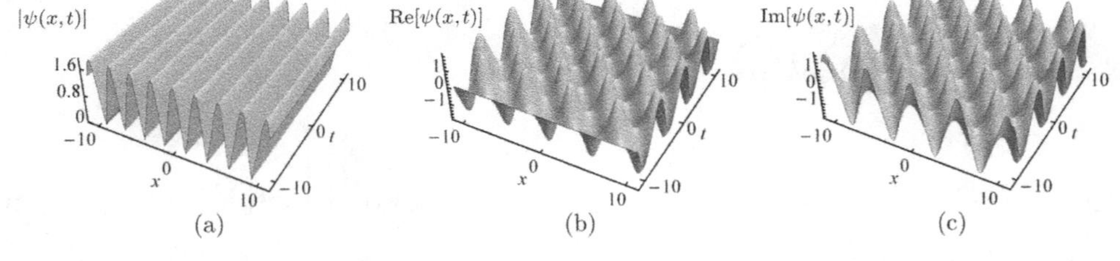

Figure 2.7. Solitary wave (2.25), with $a_1=1$, $a_2=1$, $A_0=1$, $x_0=t_0=\phi_0=0$, and $m=7/10$. (a) Absolute value of (2.25), (b) real part of (2.25), and (c) imaginary part of (2.25).

***Solution* 25. cn(x, 1/2)** *SW*

$$\psi(x,\,t) = A_0\,\sqrt{\frac{a_1}{a_2}}\;\mathrm{cn}\left[A_0\,(x - x_0),\,\frac{1}{2}\right]e^{i\,\phi_0}, \tag{2.26}$$

where
$a_1\,a_2 > 0$,
$m = 1/2$,
A_0, x_0, and ϕ_0 are arbitrary real constants.

- *Derived in appendix* A.

***Solution* 26. dn(x, m)** *SW*
(Figure 2.8)

$$\psi(x,\,t) = A_0\,\sqrt{\frac{2}{a_2\,(2-m)}}\;\mathrm{dn}\left[\frac{A_0}{\sqrt{a_1\,(2-m)}}\,(x - x_0),\,m\right]e^{i\left[A_0^2\,(t-t_0)+\phi_0\right]}, \tag{2.27}$$

where
$a_1\,(2 - m) > 0$,
$a_2\,(2 - m) > 0$,
$m \neq 2$,
A_0, t_0, x_0, and ϕ_0 are arbitrary real constants.

- *Reference*: [5].

***Solution* 27. dn(x, 2)** *SW*

$$\psi(x,\,t) = A_0\,\sqrt{\frac{2\,a_1}{a_2}}\;\mathrm{dn}\left[A_0\,(x - x_0),\,2\right]e^{i\,\phi_0}, \tag{2.28}$$

where
$a_1\,a_2 > 0$,
A_0, x_0, and ϕ_0 are arbitrary real constants.

- *Derived in appendix* A.

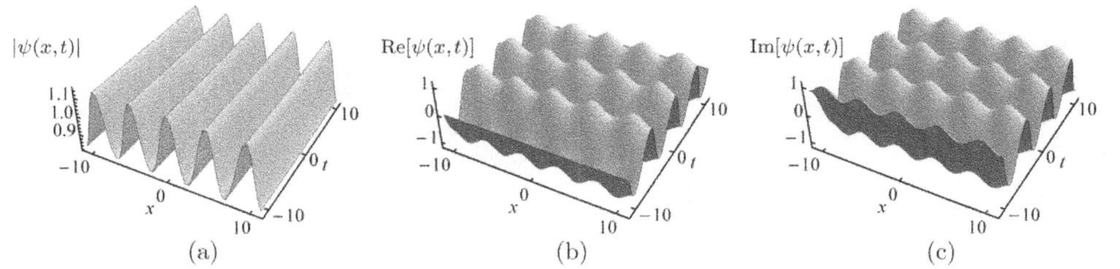

Figure 2.8. Solitary wave (2.27), with $a_1 = 1$, $a_2 = 1$, $A_0 = 1$, $x_0 = t_0 = \phi_0 = 0$, and $m = 1/2$. (a) Absolute value of (2.27), (b) real part of (2.27), and (c) imaginary part of (2.27).

Solution* 28. nd(*x*, *m*) *SW

$$\psi(x,\ t) = A_0\ \sqrt{1-m}\ \text{nd}\left[\frac{A_1}{\sqrt{a_1}}\ (x - x_0),\ m\right] e^{-i\,[A_2\,(t-t_0)+\phi_0]}, \tag{2.29}$$

where

$A_2 = (m-2)\,A_1^2$,
$a_1 > 0$,
$a_2 = \frac{2\,A_1^2}{A_0^2}$,
$0 < m \leqslant 1$,
A_0, A_1, x_0, t_0, and ϕ_0 are arbitrary real constants.

- *Reference*: [6], *taken from the nonlocal case.*

Solution* 29. sd(*x*, *m*) *SW

$$\psi(x,\ t) = A_0\ \sqrt{m\,(1-m)}\ \text{sd}\left[\frac{A_1}{\sqrt{a_1}}\ (x - x_0),\ m\right] e^{-i\,[A_2\,(t-t_0)+\phi_0]}, \tag{2.30}$$

where

$A_2 = (1-2\,m)\,A_1^2$,
$a_1 > 0$,
$a_2 = \frac{2\,A_1^2}{A_0^2}$,
$0 < m \leqslant 1$,
A_0, A_1, x_0, t_0, and ϕ_0 are arbitrary real constants.

- *Reference*: [6], *taken from the nonlocal case.*

Solution* 30. cd(*x*, *m*) *SW

$$\psi(x,\ t) = A_0\ \sqrt{m}\ \text{cd}\left[\frac{A_1}{\sqrt{a_1}}\ (x - x_0),\ m\right] e^{-i\,[A_2\,(t-t_0)+\phi_0]}, \tag{2.31}$$

where

$A_2 = (m+1)\,A_1^2$,
$a_1 > 0$,
$a_2 = \frac{-2\,A_1^2}{A_0^2}$,
$m > 0$,
A_0, A_1, x_0, t_0, and ϕ_0 are arbitrary real constants.

- *Reference*: [6], *taken from the nonlocal case.*

***Solution* 31. $\mathbf{sn}^2(x, m)$** *solitary wave on a finite background* (Figure 2.9)

$$\psi(x,\, t) = A(x)\, e^{i\,[\phi(x,t)+\phi_0]}, \tag{2.32}$$

where

$A(x) = \sqrt{R_3 + (R_2 - R_3)\, \mathrm{sn}^2[A_0\,(x - x_0),\, m]}$,

$\phi(x,\, t) = \dfrac{\sqrt{2}\, \lambda_0\, \Pi\left\{\frac{R_3 - R_2}{R_3},\ \mathrm{am}[A_0\,(x - x_0), m], m\right\}\, \mathrm{dn}[A_0\,(x - x_0), m]}{\sqrt{2}\, R_3\, A_0\sqrt{1 - m\, \mathrm{sn}^2[A_0\,(x - x_0), m]}} + \lambda_2\,(t - t_0)$,

$m = \frac{R_2 - R_3}{R_1 - R_3}$,

$A_0 = \sqrt{\frac{a_2\,(R_3 - R_1)}{2\, a_1}}$,

$a_1\, a_2\,(R_3 - R_1) > 0$,

R_j, $j = 1, 2, 3$ are the three roots of $Y(x) = 2\, a_1\, \lambda_0^2 - 2\, a_1\, \lambda_1\, x - 2\, \lambda_2\, x^2 + a_2\, x^3$,

Π is the incomplete elliptic integral,

am is the amplitude for Jacobi elliptic functions,

x_0, t_0, ϕ_0, λ_0, λ_1, and λ_2 are arbitrary real constants.

- *Derived in appendix* A.
- *This solution can be written in terms of* cn^2 or dn^2 *by using*: $\mathrm{sn}^2(x, m) = 1 - \mathrm{cn}^2(x, m) = [1 - \mathrm{dn}^2(x, m)]/m$.

***Solution* 32. $\mathbf{dn}(x, m) + \mathbf{cn}(x, m)$** *SW*

$$\psi(x,\, t) = \left\{\frac{A_0}{2}\, \mathrm{dn}\left[\frac{A_1}{\sqrt{a_1}}\,(x - x_0),\, m\right] + \frac{B_0\,\sqrt{m}}{2}\, \mathrm{cn}\left[\frac{A_1}{\sqrt{a_1}}\,(x - x_0),\, m\right]\right\} \times e^{-i\,[A_2\,(t - t_0)+\phi_0]}, \tag{2.33}$$

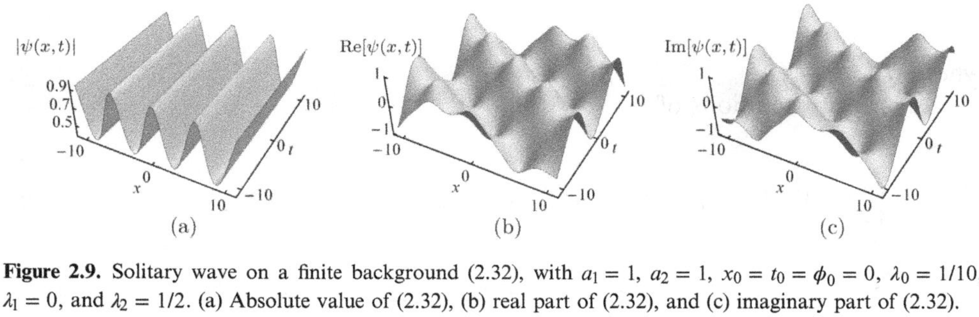

Figure 2.9. Solitary wave on a finite background (2.32), with $a_1 = 1$, $a_2 = 1$, $x_0 = t_0 = \phi_0 = 0$, $\lambda_0 = 1/10$, $\lambda_1 = 0$, and $\lambda_2 = 1/2$. (a) Absolute value of (2.32), (b) real part of (2.32), and (c) imaginary part of (2.32).

where

$A_2 = \frac{-(m+1)\,A_1^2}{2}$,
$B_0 = \pm A_0$,
$a_1 > 0$,
$a_2 = \frac{2\,A_1^2}{A_0^2}$,
$0 < m \leqslant 1$,
A_0, A_1, x_0, t_0, and ϕ_0 are arbitrary real constants.

- *Reference*: [6], *taken from the nonlocal case.*

***Solution* 33.** *SW*

$$\psi(x,\, t) = A_0\, m\, \frac{\mathrm{cn}\left[\frac{A_1}{\sqrt{a_1}}\,(x - x_0),\, m\right] \mathrm{sn}\left[\frac{A_1}{\sqrt{a_1}}\,(x - x_0),\, m\right]}{\mathrm{dn}\left[\frac{A_1}{\sqrt{a_1}}\,(x - x_0),\, m\right]}\, e^{-i\,[A_2\,(t-t_0)+\phi_0]}, \quad (2.34)$$

where

$A_2 = 2\,(2 - m)\,A_1^2$,
$a_1 > 0$,
$a_2 = -\frac{2\,A_1^2}{A_0^2}$,
$m > 0$,
A_0, A_1, x_0, t_0, and ϕ_0 are arbitrary real constants.

- *Reference*: [6], *taken from the nonlocal case; we corrected the denominator.*

***Solution* 34. *N*-Bright Solitons**

$$\psi(x,\, t) = \frac{1}{\sqrt{a_2}}\, \sum_{j=1}^{N} \psi_j(x,\, t), \quad (2.35)$$

where

$\psi_j(x,\, t)$ are solutions of

$$\sum_{k=1}^{N} M_{j\,k}\left[\gamma_j^{-1}(x,\, t) + \gamma_k^{*}(x,\, t)\right]\psi_k(x,\, t) = 1, \quad j = 0,\, 1,\, 2,\, \ldots,\, N, \quad (2.36)$$

$a_1 > 0,$
$a_2 > 0,$
$\lambda_j = \alpha_j + i\,\nu_j,$
$M_{jk} = 1/(\lambda_j + \lambda_k^*),$
$\gamma_j(x,\,t) = e^{\frac{\lambda_j}{\sqrt{2a_1}}(x-x_{0j})+i\,(\lambda_j^2\,(t-t_0)/2+\phi_{0j})},$
$\gamma_k^*(x,\,t)$ and $\gamma_j^{-1}(x,\,t)$ are the complex conjugate and the inverse of γ_k and γ_j, respectively,
λ_k^* is the complex conjugate of λ_k,
α_j, ν_j, x_{0j}, t_0, and ϕ_{0j} are arbitrary real constants.

- *Reference*: [7].

Solution 35. Two Bright Solitons

(Figure 2.10)

$$\psi(x,\,t) = \frac{1}{\sqrt{a_2}}\,[\psi_1(x,\,t) + \psi_2(x,\,t)], \tag{2.37}$$

where

$$\psi_1(x,\,t) = \frac{M_{12}\left[\gamma_1^{-1}(x,t)+\gamma_2^*(x,t)\right] - M_{22}\left[\gamma_2^{-1}(x,t)+\gamma_2^*(x,t)\right]}{M_{12}\,M_{21}\left[\gamma_1^*(x,t)+\gamma_2^{-1}(x,t)\right]\left[\gamma_1^{-1}(x,t)+\gamma_2^*(x,t)\right] - M_{11}\,M_{22}\left[\gamma_1^{-1}(x,t)+\gamma_1^*(x,t)\right]\left[\gamma_2^{-1}(x,t)+\gamma_2^*(x,t)\right]},$$

$$\psi_2(x,\,t) = \frac{-M_{11}\left[\gamma_1^{-1}(x,t)+\gamma_1^*(x,t)\right] + M_{21}\left[\gamma_1^*(x,t)+\gamma_2^{-1}(x,t)\right]}{M_{12}\,M_{21}\left[\gamma_1^*(x,t)+\gamma_2^{-1}(x,t)\right]\left[\gamma_1^{-1}(x,t)+\gamma_2^*(x,t)\right] - M_{11}\,M_{22}\left[\gamma_1^{-1}(x,t)+\gamma_1^*(x,t)\right]\left[\gamma_2^{-1}(x,t)+\gamma_2^*(x,t)\right]},$$

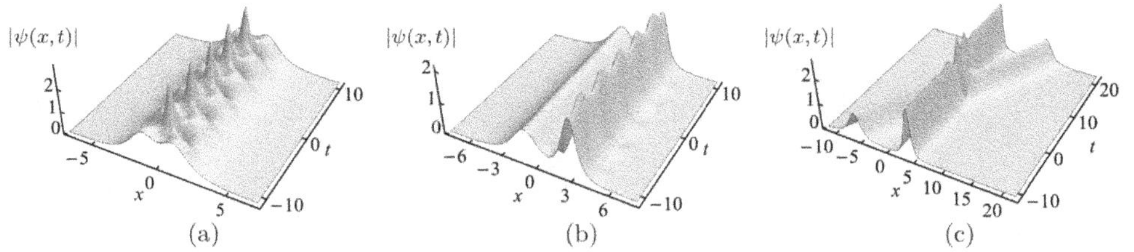

Figure 2.10. Two bright solitons (2.37), with $a_1 = 1/2$, $a_2 = 1$, $\alpha_1 = 1$, $\alpha_2 = 2$, and $x_{01} = \nu_2 = \phi_{01} = \phi_{02} = 0$. (a) $x_{02} = \nu_1 = 0$, (b) $x_{02} = 2$ and $\nu_1 = 0$, and (c) $x_{02} = 3$ and $\nu_1 = 1/2$. Animation available online at https://iopscience.iop.org/book/978-0-7503-2428-1.

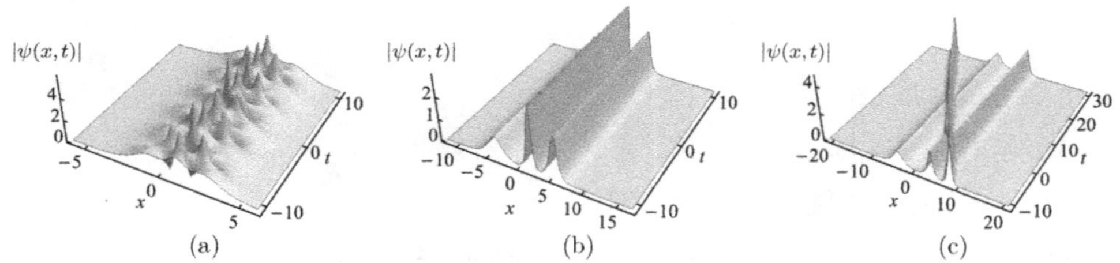

Figure 2.11. Three bright solitons (2.38), with $a_1 = 1/2$, $a_2 = 1$, $\alpha_1 = 2$, $\alpha_2 = 3$, $\alpha_3 = 1$, and $\phi_{01} = \phi_{02} = \phi_{03} = t_0 = 0$. (a) $x_{01} = x_{02} = x_{03} = 1$ and $\nu_1 = \nu_2 = \nu_3 = 0$, (b) $x_{01} = 4$, $x_{02} = 2$, $x_{03} = -2$, and $\nu_1 = \nu_2 = \nu_3 = 0$, and (c) $x_{01} = 4$, $x_{02} = 2$, $x_{03} = -2$, $\nu_1 = \nu_3 = 0$, and $\nu_2 = -1/2$. Animation available online at https://iopscience.iop.org/book/978-0-7503-2428-1.

$a_1 > 0,$
$a_2 > 0,$
$M_{jk} = 1/(\lambda_j + \lambda_k^*),$
$\gamma_j(x, t) = e^{\frac{\lambda_j}{\sqrt{2 a_1}}(x - x_{0j}) + i\,[\lambda_j^2\,(t - t_0)/2 + \phi_{0j}]},$
$\lambda_j = \alpha_j + i\,\nu_j,$
α_j, ν_j, x_{0j}, t_0, and ϕ_{0j} are arbitrary real constants,
$N = 2$ in (2.35).

- *Reference*: [7].

Solution **36. Three Bright Solitons**
(Figure 2.11)

$$\psi(x, t) = \frac{1}{\sqrt{a_2}}\,[\psi_1(x, t) + \psi_2(x, t) + \psi_3(x, t)], \tag{2.38}$$

where

$$
\begin{aligned}
\psi_1(x,t) = -\Big[\Big(\Big\{ & M_{13} M_{22} \big[\gamma_2^{-1}(x,t) + \gamma_2^{*}(x,t)\big]\big[\gamma_1^{-1}(x,t) + \gamma_3^{*}(x,t)\big] \\
& -M_{12} M_{23} \big[\gamma_1^{-1}(x,t) + \gamma_2^{*}(x,t)\big]\big[\gamma_2^{-1}(x,t) + \gamma_3^{*}(x,t)\big]\Big\} \\
& \times \Big\{-M_{13} \big[\gamma_1^{-1}(x,t) + \gamma_3^{*}(x,t) + M_{33} \big[\gamma_3^{-1}(x,t) + \gamma_3^{*}(x,t)\big]\big]\Big\} \\
& - \Big\{-M_{13} \big[\gamma_1^{-1}(x,t) + \gamma_3^{*}(x,t)\big] + M_{23} \big[\gamma_2^{-1}(x,t) + \gamma_3^{*}(x,t)\big]\Big\} \\
& \times \Big\{M_{13} M_{32} \big[\gamma_3^{-1}(x,t) + \gamma_2^{*}(x,t)\big]\big[\gamma_1^{-1}(x,t) + \gamma_3^{*}(x,t)\big] \\
& - M_{12} M_{33} \big[\gamma_1^{-1}(x,t) + \gamma_2^{*}(x,t)\big]\big[\gamma_3^{-1}(x,t) + \gamma_3^{*}(x,t)\big]\Big\}\Big) \\
& \div \Big(\Big\{M_{13} M_{22} \big[\gamma_2^{-1}(x,t) + \gamma_2^{*}(x,t)\big]\big[\gamma_1^{-1}(x,t) + \gamma_3^{*}(x,t)\big] \\
& - M_{12} M_{23} \big[\gamma_1^{-1}(x,t) + \gamma_2^{*}(x,t)\big]\big[\gamma_2^{-1}(x,t) + \gamma_3^{*}(x,t)\big]\Big\} \\
& \times \Big\{M_{13} M_{31} \big[\gamma_3^{-1}(x,t) + \gamma_1^{*}(x,t)\big]\big[\gamma_1^{-1}(x,t) + \gamma_3^{*}(x,t)\big] \\
& - M_{11} M_{33} \big[\gamma_1^{-1}(x,t) + \gamma_1^{*}(x,t)\big]\big[\gamma_3^{-1}(x,t) + \gamma_3^{*}(x,t)\big]\Big\} \\
& - \Big\{M_{13} M_{21} \big[\gamma_2^{-1}(x,t) + \gamma_1^{*}(x,t)\big]\big[\gamma_1^{-1}(x,t) + \gamma_3^{*}(x,t)\big] \\
& -M_{11} M_{23} \big[\gamma_1^{-1}(x,t) + \gamma_1^{*}(x,t)\big]\big[\gamma_2^{-1}(x,t) + \gamma_3^{*}(x,t)\big]\Big\} \\
& \times \Big\{M_{13} M_{32} \big[\gamma_3^{-1}(x,t) + \gamma_2^{*}(x,t)\big]\big[\gamma_1^{-1}(x,t) + \gamma_3^{*}(x,t)\big] \\
& - M_{12} M_{33}\big[\gamma_1^{-1}(x,t) + \gamma_2^{*}(x,t)\big]\big[\gamma_3^{-1}(x,t) + \gamma_3^{*}(x,t)\big]\Big\}\Big)\Big],
\end{aligned}
$$

$$
\begin{aligned}
\psi_2(x,\,t) = \Big\{ & (M_{23}\,M_{31} - M_{21}\,M_{33})\,\gamma_2^{-1}(x,\,t)\,\gamma_3^{-1}(x,\,t) + \Big[M_{21}\,(M_{13} - M_{33})\,\gamma_2^{-1}(x,\,t) \\
& + (-M_{13} + M_{23})\,M_{31}\,\gamma_3^{-1}(x,\,t)\Big]\,\gamma_3^{*}(x,\,t) + \gamma_1^{-1}(x,\,t)\Big[M_{13}\,(M_{21} - M_{31}) \\
& \times \gamma_1^{*}(x,\,t) + (M_{13}\,M_{21} - M_{11}\,M_{23})\,\gamma_2^{-1}(x,\,t) + (-M_{13}\,M_{31} + M_{11}\,M_{33}) \\
& \times \gamma_3^{-1},(x,\,t) + M_{11}\,(-M_{23} + M_{33})\,\gamma_3^{*}(x,\,t)\Big] + \gamma_1^{*}(x,\,t)\,[M_{23}\,(-M_{11} + M_{31}) \\
& \times \gamma_2^{-1}(x,\,t) + (M_{11} - M_{21})\,M_{33}\,\gamma_3^{-1}(x,\,t) + (M_{13}\,M_{21} - M_{11}\,M_{23} \\
& - M_{13}\,M_{31} + M_{23}\,M_{31} + M_{11}\,M_{33} - M_{21}\,M_{33})\,\gamma_3^{*}(x,\,t)\Big]\Big\} \\
& \div \Big(M_{12}\,(M_{23}\,M_{31} - M_{21}\,M_{33})\gamma_2^{-1}(x,\,t)\,\gamma_2^{*}(x,\,t)\,\gamma_3^{-1}(x,\,t) + \Big\{(-M_{13}\,M_{22} \\
& + M_{12}\,M_{23})\,M_{31}\,\gamma_2^{*}(x,\,t)\,\gamma_3^{-1}(x,\,t) + \gamma_2^{-1}(x,\,t)\Big[M_{21}\,(M_{13}\,M_{32} - M_{12}\,M_{33}) \\
& \times\ \gamma_2^{*}(x,\,t) + M_{13}\,(-M_{22}\,M_{31} + M_{21}\,M_{32})\,\gamma_3^{-1}(x,\,t)\Big]\Big\}\ \gamma_3^{*}(x,\,t) + \gamma_1^{-1}(x,\,t) \\
& \times \Big\{M_{22}\,(-M_{13}\,M_{31} + M_{11}\,M_{33})\,\gamma_2^{*}(x,\,t)\,\gamma_3^{-1}(x,\,t) + \Big[M_{11}\,(-M_{23}\,M_{32} \\
& + M_{22}\,M_{33})\,\gamma_2^{*}(x,\,t) + M_{23}\,(M_{12}\,M_{31} - M_{11}\,M_{32})\,\gamma_3^{-1}(x,\,t)\Big]\gamma_3^{*}(x,\,t) \\
& + \gamma_2^{-1}(x,\,t)\Big[(M_{13}\,M_{21} - M_{11}\,M_{23})\,M_{32}\,\gamma_2^{*}(x,\,t) + (-M_{13}\,M_{22}\,M_{31} \\
& + M_{12}\,M_{23}\,M_{31} + M_{13}\,M_{21}\,M_{32} - M_{11}\,M_{23}\,M_{32} - M_{12}\,M_{21}\,M_{33} \\
& + M_{11}\,M_{22}\,M_{33})\,\gamma_3^{-1}(x,\,t) + (-M_{12}\,M_{21} + M_{11}\,M_{22})\,M_{33}\,\gamma_3^{*}(x,\,t)\Big] \\
& + \gamma_1^{*}(x,\,t)\Big[(-M_{13}\,M_{22} + M_{12}\,M_{23})\,M_{31}\,\gamma_2^{-1}(x,\,t) + M_{13}\,(-M_{22}\,M_{31} \\
& + M_{21}\,M_{32})\,\gamma_2^{*}(x,\,t) + M_{21}\,(M_{13}\,M_{32} - M_{12}\,M_{33})\,\gamma_3^{-1}(x,\,t) \\
& + M_{12}\,(M_{23}\,M_{31} - M_{21}\,M_{33})\,\gamma_3^{*}(x,\,t)\Big]\Big\} \\
& \times \gamma_1^{*}(x,\,t)\,\Big\{(M_{13}\,M_{21} - M_{11}\,M_{23})\,M_{32}\,\gamma_3^{-1}(x,\,t)\,\gamma_3^{*}(x,\,t) \\
& + \gamma_2^{-1}(x,\,t)\Big[M_{23}\,(M_{12}\,M_{31} - M_{11}\,M_{32})\,\gamma_2^{*}(x,\,t) + M_{11}\,(-M_{23}\,M_{32} \\
& + M_{22}\,M_{33})\,\gamma_3^{-1}(x,\,t) + M_{22}\,(-M_{13}\,M_{31} + M_{11}\,M_{33})\,\gamma_3^{*}(x,\,t)\Big] \\
& + \gamma_2^{*}(x,\,t)\Big[(-M_{12}\,M_{21} + M_{11}\,M_{22})\,M_{33}\,\gamma_3^{-1}(x,\,t) + \Big(-M_{13}\,M_{22}\,M_{31} \\
& + M_{12}\,M_{23}\,M_{31} + M_{13}\,M_{21}\,M_{32} - M_{11}\,M_{23}\,M_{32} - M_{12}\,M_{21}\,M_{33} \\
& +\ M_{11}\,M_{22}\,M_{33})\,\gamma_3^{*}(x,\,t)\Big]\Big\}\Big),
\end{aligned}
$$

$$
\begin{aligned}
\psi_3(x, t) = {} & \{ M_{21}(M_{12} - M_{32})\gamma_2^{-1}(x, t)\gamma_2^*(x, t) + [(M_{22}M_{31} - M_{21}M_{32})\gamma_2^{-1}(x, t) \\
& + (-M_{12} + M_{22})M_{31}\gamma_2^*(x, t)]\gamma_3^{-1}(x, t) + \gamma_1^*(x, t)[M_{22}(-M_{11} + M_{31}) \\
& \times \gamma_2^{-1}(x, t) + (M_{12}M_{21} - M_{11}M_{22} - M_{12}M_{31} + M_{22}M_{31} + M_{11}M_{32} \\
& - M_{21}M_{32})\gamma_2^*(x, t) + (M_{11} - M_{21})M_{32}\gamma_3^{-1}(x, t)] + \gamma_1^{-1}(x, t)[M_{12}(M_{21} \\
& - M_{31})\gamma_1^*(x, t) + (M_{12}M_{21} - M_{11}M_{22})\gamma_2^{-1}(x, t) + M_{11}(-M_{22} + M_{32}) \\
& \times \gamma_2^*(x, t) + (-M_{12}M_{31} + M_{11}M_{32})\gamma_3^{-1}(x, t)]\} \\
& \div (M_{12}(-M_{23}M_{31} + M_{21}M_{33})\gamma_2^{-1}(x, t)\gamma_2^*(x, t)\gamma_3^{-1}(x, t) + \{(M_{13}M_{22} \\
& - M_{12}M_{23})M_{31}\gamma_2^*(x, t)\gamma_3^{-1}(x, t) + \gamma_2^{-1}(x, t)[M_{21}(-M_{13}M_{32} + M_{12}M_{33}) \\
& \times \gamma_2^*(x, t) + M_{13}(M_{22}M_{31} - M_{21}M_{32})\gamma_3^{-1}(x, t)]\}\gamma_3^*(x, t) + \gamma_1^{-1}(x, t) \\
& \times \{M_{22}(M_{13}M_{31} - M_{11}M_{33})\gamma_2^*(x, t)\gamma_3^{-1}(x, t) + [M_{11}(M_{23}M_{32} - M_{22}M_{33}) \\
& \times \gamma_2^*(x, t) + M_{23}(-M_{12}M_{31} + M_{11}M_{32})\gamma_3^{-1}(x, t)]\gamma_3^*(x, t) + \gamma_2^{-1}(x, t) \\
& \times [(-M_{13}M_{21} + M_{11}M_{23})M_{32}\gamma_2^*(x, t) + (M_{13}M_{22}M_{31} - M_{12}M_{23}M_{31} \\
& - M_{13}M_{21}M_{32} + M_{11}M_{23}M_{32} + M_{12}M_{21}M_{33} - M_{11}M_{22}M_{33})\gamma_3^{-1}(x, t) \\
& + (M_{12}M_{21} - M_{11}M_{22})M_{33}\gamma_3^*(x, t)] + \gamma_1^*(x, t)[(M_{13}M_{22} - M_{12}M_{23})M_{31} \\
& \times \gamma_2^{-1}(x, t) + M_{13}(M_{22}M_{31} - M_{21}M_{32})\gamma_2^*(x, t) + M_{21}(-M_{13}M_{32} + M_{12}M_{33}) \\
& \times \gamma_3^{-1}(x, t) + M_{12}(-M_{23}M_{31} + M_{21}M_{33})\gamma_3^*(x, t)]\} + \gamma_1^*(x, t)\{(-M_{13}M_{21} \\
& + M_{11}M_{23})M_{32}\gamma_3^{-1}(x, t)\gamma_3^*(x, t) + \gamma_2^{-1}(x, t)[M_{23}(-M_{12}M_{31} + M_{11}M_{32}) \\
& \times \gamma_2^*(x, t) + M_{11}(M_{23}M_{32} - M_{22}M_{33})\gamma_3^{-1}(x, t) + M_{22}(M_{13}M_{31} - M_{11}M_{33}) \\
& \times \gamma_3^*(x, t)] + \gamma_2^*(x, t)[(M_{12}M_{21} - M_{11}M_{22})M_{33}\gamma_3^{-1}(x, t) + (M_{13}M_{22}M_{31} \\
& - M_{12}M_{23}M_{31} - M_{13}M_{21}M_{32} + M_{11}M_{23}M_{32} + M_{12}M_{21}M_{33} \\
& - M_{11}M_{22}M_{33})\gamma_3^*(x, t)]\}),
\end{aligned}
$$

$a_1 > 0$,
$a_2 > 0$,
$M_{jk} = 1/(\lambda_j + \lambda_k^*)$,
$\gamma_j(x, t) = e^{\frac{\lambda_j}{\sqrt{2 a_1}}(x - x_{0j}) + i\,[\lambda_j^2\,(t - t_0)/2 + \phi_{0j}]}$,
$\lambda_j = \alpha_j + i\,\nu_j$,
α_j, ν_j, x_{0j}, t_0, and ϕ_{0j} are arbitrary real constants,
$N = 3$ in (2.35).

- *Reference*: [7].

***Solution* 37. *N*-Dark Solitons**

$$\psi(x,t) = \sqrt{\frac{-2}{a_2}}\left\{1 - 2i\sum_n \mu_n(t)\, e^{\frac{-2v_n}{\sqrt{a_1}}(x-x_0)}\left[-Q_{12}{}^n + (\lambda_n + i\,v_n)\left(Q_{11}{}^n - 1\right)\right]\right\} \times e^{-2i[t-t_0+\phi_0]}, \tag{2.39}$$

where

$v_j = \sqrt{1-\lambda_j^2}$,
$\mu_j(t) = e^{4v_j\lambda_j(t-t_0)}$,
Q_{11}^n and Q_{12}^n are obtained by solving the linear algebraic equations:

$$Q_{12}^j + \sum_n \frac{\mu_n(t)}{v_n+v_j} Q_{12}^n\, e^{\frac{-2v_n(x-x_0)}{\sqrt{a_1}}} - \sum_n Q_{11}^n \frac{\mu_n(t)(\lambda_n + i\,v_n)}{v_n+v_j} e^{\frac{-2v_n(x-x_0)}{\sqrt{a_1}}} = \sum_n \frac{\mu_n(t)(\lambda_n + i\,v_n)}{v_n+v_j} e^{\frac{-2v_n(x-x_0)}{\sqrt{a_1}}},$$

$$Q_{11}^j + \sum_n \frac{\mu_n(t)}{v_n+v_j} Q_{11}^n\, e^{\frac{-2v_n(x-x_0)}{\sqrt{a_1}}} + \sum_n Q_{12}^n \frac{\mu_n(t)(-\lambda_n + i\,v_n)}{v_n+v_j} e^{\frac{-2v_n(x-x_0)}{\sqrt{a_1}}} = \sum_n \frac{\mu_n(t)}{v_n+v_j} e^{\frac{-2v_n(x-x_0)}{\sqrt{a_1}}},$$

$a_1 > 0$,
$a_2 < 0$,
$-1 < \lambda_j < 1$,
x_0, t_0, and ϕ_0 are arbitrary real constants.

- *Reference*: [8], *we corrected the exponential prefactor.*

***Solution* 38. Two Dark Solitons**
(Figure 2.12)

$$\psi(x,t) = \left\{1 - \frac{2i}{p(x,t)}\left[\frac{2}{v_1+v_2}\left(\frac{1}{\lambda_1 + i\,v_1} + \frac{1}{\lambda_2 + i\,v_2}\right) - (\lambda_1 - i\,v_1)\,q_1(x,t) - (\lambda_2 - i\,v_2)\,q_2(x,t)\right]\right\}\sqrt{\frac{-2}{a_2}}\, e^{-2i[t-t_0+\phi_0]}, \tag{2.40}$$

where

$v_1 = \sqrt{1-\lambda_1^2}$,
$v_2 = \sqrt{1-\lambda_2^2}$,
$q_1(x,t) = \frac{1}{v_1} + e^{\frac{2v_1}{\sqrt{a_1}}(x-x_0) - 4v_1\lambda_1(t-t_0)}$,
$q_2(x,t) = \frac{1}{v_2} + e^{\frac{2v_2}{\sqrt{a_1}}(x-x_0) - 4v_2\lambda_2(t-t_0)}$,

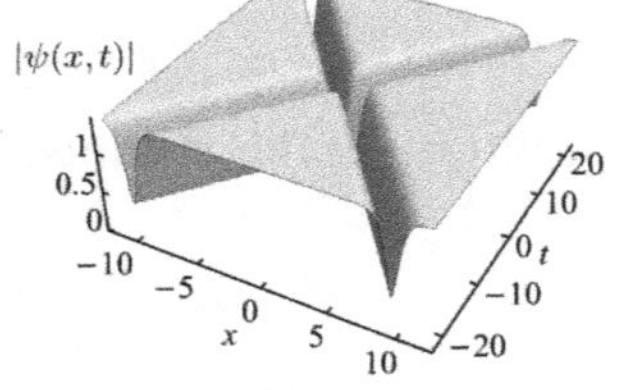

Figure 2.12. Two dark solitons, (2.40), with $a_1 = 1/2$, $a_2 = -1$, $\lambda_2 = 3/10$, and $x_0 = t_0 = \phi_0 = 0$.

$p(x, t) = (\lambda_1 - i\,v_1)(\lambda_2 - i\,v_2)\,q_1(x, t)\,q_2(x, t) - \frac{1}{(v_1 + v_2)^2}\left(\frac{1}{\lambda_1 + i\,v_1} + \frac{1}{\lambda_2 + i\,v_2}\right)^2,$
$a_1 > 0,$
$a_2 < 0,$
$\lambda_1 = -\lambda_2,$
$-1 < \lambda_2 < 1,$
x_0, t_0, and ϕ_0 are arbitrary real constants.

- *Reference*: [8], *we corrected the exponential prefactor.*

***Solution* 39. Generalized First-Order Breather (form I*)**

$$\psi(x, t) = \frac{1}{\sqrt{a_2}}\left\{\frac{\kappa^2\cosh[\delta\,(t - t_0)] + 2\,i\,\kappa\,\nu\,\sinh[\delta\,(t - t_0)]}{2\cosh[\delta\,(t - t_0)] - 2\,\nu\,\cos\left[\frac{\kappa}{\sqrt{2\,a_1}}\,(x - x_0)\right]} - 1\right\} e^{i\,[t - t_0 + \phi_0]}, \quad (2.41)$$

where
$a_1 > 0,$
$a_2 > 0,$
$\kappa = 2\sqrt{1 - \nu^2},$
$\delta = \kappa\,\nu,$
ν, x_0, t_0, and ϕ_0 are arbitrary real constants.

- *Reference*: [9].

***Solution* 40. Generalized First-Order Breather (form II*)**

$$\psi(x, t) = \frac{\cos(A_0)\cos[q_1(x, t) + 2\,i\,A_1] - \cosh(A_1)\cosh[q_2(x, t) + 2\,i\,A_0]}{\cos(A_0)\cos[q_1(x, t)] - \cosh(A_1)\cosh[q_2(x, t)]} \times e^{i\,[a_2\,(t - t_0) + \phi_0]}, \quad (2.42)$$

where
$q_1(x, t) = v_3\left[\sqrt{\frac{2\,a_2}{a_1}}\,(x - x_0) - a_2\,v_0\,(t - t_0)\right],$
$q_2(x, t) = v_2\left[\sqrt{\frac{2\,a_2}{a_1}}\,(x - x_0) - a_2\,v_1\,(t - t_0)\right],$
$v_0 = \frac{\sinh(2\,A_1)\cos(2\,A_0)}{\cosh(A_1)\sin(A_0)},$
$v_1 = -\frac{\cosh(2\,A_1)\sin(2\,A_0)}{\sinh(A_1)\cos(A_0)},$
$v_2 = -\sinh(A_1)\cos(A_0),$
$v_3 = \cosh(A_1)\sin(A_0),$
$a_1\,a_2 > 0,$
x_0, t_0, A_0, A_1, and ϕ_0 are arbitrary real constants.

- *Reference*: [10].

***Solution* 41. Generalized First-Order Breather (form III*)**

$$\psi(x,\,t) = \frac{A_0}{\sqrt{a_2}}\left(1 - \frac{\sqrt{8}\,\lambda_r}{A_0}\,p(x,\,t)\right)e^{i\left[A_0^2\,(t-t_0)+\phi_0\right]}, \tag{2.43}$$

where

$$p(x,\,t) = \frac{(A_0^2+\Gamma^2)\cos[q_1(x,t)] + i\,(A_0^2-\Gamma^2)\sin[q_1(x,t)] + 2\,A_0\,\{\Gamma_r\cosh[q_2(x,t)] - i\,\Gamma_i\sinh[q_2(x,t)]\}}{2\,A_0\,\Gamma_r\cos[q_1(x,t)] + (\Gamma^2 + A_0^2)\cosh[q_2(x,t)]},$$

$$q_1(x,\,t) = \delta_i + \sqrt{2}\left[\frac{x-x_0}{\sqrt{a_1}}\,\Delta_i - 2\,(t-t_0)\,(\Delta_i\,\lambda_i + \Delta_r\,\lambda_r)\right],$$

$$q_2(x,\,t) = \delta_r + \sqrt{2}\left[\frac{x-x_0}{\sqrt{a_1}}\,\Delta_r - 2\,(t-t_0)\,\Delta_r\,\lambda_i + 2\,(t-t_0)\,\Delta_i\,\lambda_r\right],$$

$$\Delta_r = \mathrm{Re}\left[\sqrt{2\,(\lambda_r - i\,\lambda_i)^2 - A_0^2}\right],$$

$$\Delta_i = \mathrm{Im}\left[\sqrt{2\,(\lambda_r - i\,\lambda_i)^2 - A_0^2}\right],$$

$$\Gamma_r = \Delta_r + \sqrt{2}\,\lambda_r,$$

$$\Gamma_i = \Delta_i - \sqrt{2}\,\lambda_i,$$

$$\Gamma = \sqrt{\Gamma_r^2 + \Gamma_i^2},$$

$a_1 > 0$,

$a_2 > 0$,

A_0, λ_r, λ_i, x_0, t_0, and ϕ_0 are arbitrary real constants.

- *Reference*: [11].

***Remark:** There are different forms of the breather in the literature, as given by Solutions 39, 40, 41. In appendix B.1.2 we derive a fourth expression valid for focusing and defocusing nonlinearities and the arbitrary constants are expressed in terms of physical parameters.

Solution* 42. Periodicity in *t* and Localization in *x* *Kuznetsov–Ma breather (Figure 2.13)

$$\psi(x,\,t) = \frac{1}{\sqrt{a_2}}\left\{\frac{-p^2\cos[\omega\,(t-t_0)] - 2\,i\,p\,\nu\,\sin[\omega\,(t-t_0)]}{2\cos[\omega\,(t-t_0)] - 2\,\nu\,\cosh\left[\frac{p}{\sqrt{2\,a_1}}\,(x-x_0)\right]} - 1\right\}e^{i\,[t-t_0+\phi_0]}, \tag{2.44}$$

where

$a_1 > 0$,

$a_2 > 0$,

$p = 2\sqrt{\nu^2 - 1}$,

$\omega = p\,\nu$,

$\nu > 1$,

x_0, t_0, and ϕ_0 are arbitrary real constants.

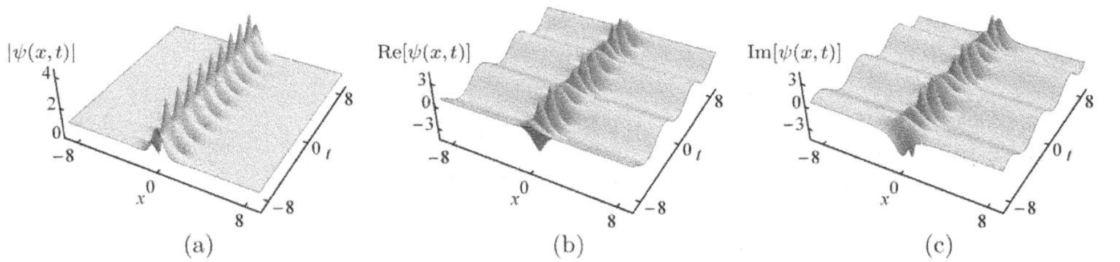

Figure 2.13. Kuznetsov–Ma breather (2.44), with $a_1 = a_2 = 1$, $\nu = 1.5$, and $x_0 = t_0 = \phi_0 = 0$. (a) Absolute value of (2.44), (b) real part of (2.44), and (c) imaginary part of (2.44). Animation available online at https://iopscience.iop.org/book/978-0-7503-2428-1.

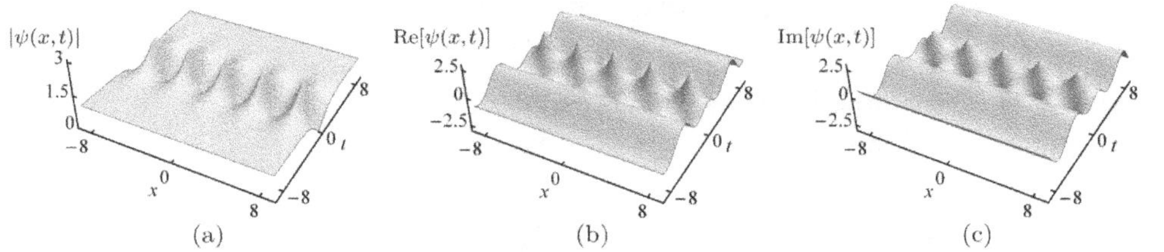

Figure 2.14. Akhmediev breather (2.45), with $a_1 = 1/2$, $a_2 = 1$, $\nu = 0.5$, and $x_0 = t_0 = \phi_0 = 0$. (a) Absolute value of (2.45), (b) real part of (2.45), and (c) imaginary part of (2.45). Animation available online at https://iopscience.iop.org/book/978-0-7503-2428-1.

- *Reference*: [9].
- *This solution can be generated from* (2.41) with $\nu > 1$.

Solution* 43. Periodicity in *x* and Localization in *t *Akhmediev breather* (Figure 2.14)

$$\psi(x,\, t) = \frac{1}{\sqrt{a_2}}\left\{\frac{\kappa^2 \cosh[\delta\,(t - t_0)] + 2\,i\,\kappa\,\nu\,\sinh[\delta\,(t - t_0)]}{2\cosh[\delta\,(t - t_0)] - 2\,\nu\,\cos\left[\frac{\kappa}{\sqrt{2\,a_1}}\,(x - x_0)\right]} - 1\right\} e^{i\,[t - t_0 + \phi_0]}, \quad (2.45)$$

where

$a_1 > 0,$
$a_2 > 0,$
$\kappa = 2\sqrt{1 - \nu^2},$
$\delta = \kappa\nu,$
$\nu < 1,$
x_0, t_0, and ϕ_0 are arbitrary real constants.

- *Reference*: [9].

Solution* 44. Localization in *x* and *t* *Peregrine soliton
(Figure 2.15)

$$\psi(x,\,t) = \frac{1}{\sqrt{a_2}}\left[\frac{4 + i\,8\,(t - t_0)}{1 + 4\,(t - t_0)^2 + \frac{2}{a_1}\,(x - x_0)^2} - 1\right]e^{i\,[t - t_0 + \phi_0]}, \tag{2.46}$$

where
$a_2 > 0$,
x_0, t_0, and ϕ_0 are arbitrary real constants.

- *Reference*: [9].
- *This solution can be generated from* (2.41) *in the limits* $\nu \to 1$ *and* $\kappa \to 0$.

***Solution* 45. Generalized Two-Breathers Solution**
(Figures 2.16–2.20)

$$\psi(x,\,t) = \frac{1}{\sqrt{a_2}}\left[1 + \frac{\alpha(x,\,t) + i\,\beta(x,\,t)}{\gamma(x,\,t)}\right]e^{i\,[t - t_0 + \phi_0]}, \tag{2.47}$$

where

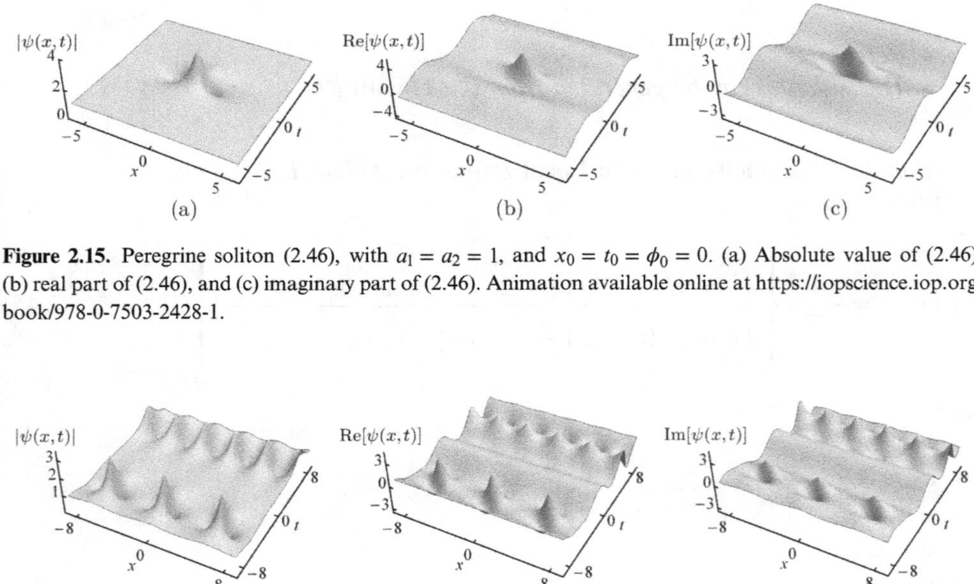

Figure 2.15. Peregrine soliton (2.46), with $a_1 = a_2 = 1$, and $x_0 = t_0 = \phi_0 = 0$. (a) Absolute value of (2.46), (b) real part of (2.46), and (c) imaginary part of (2.46). Animation available online at https://iopscience.iop.org/book/978-0-7503-2428-1.

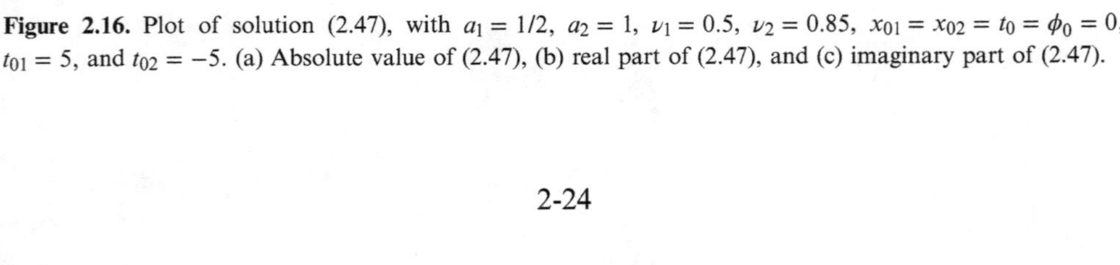

Figure 2.16. Plot of solution (2.47), with $a_1 = 1/2$, $a_2 = 1$, $\nu_1 = 0.5$, $\nu_2 = 0.85$, $x_{01} = x_{02} = t_0 = \phi_0 = 0$, $t_{01} = 5$, and $t_{02} = -5$. (a) Absolute value of (2.47), (b) real part of (2.47), and (c) imaginary part of (2.47).

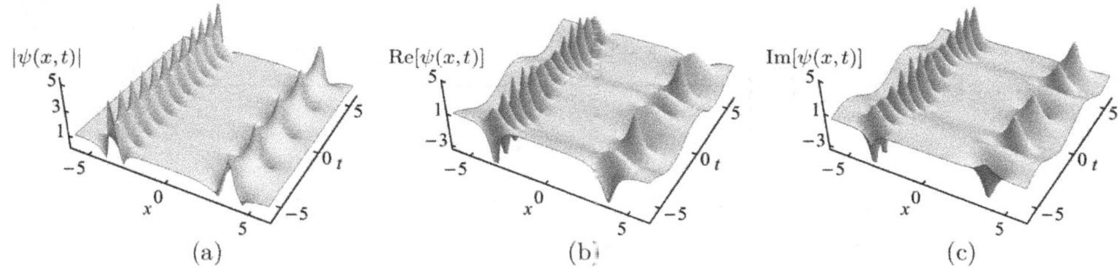

Figure 2.17. Plot of solution (2.47), with $a_1 = 1/2$, $a_2 = 1$, $\nu_1 = 1.3$, $\nu_2 = 1.85$, $x_{01} = 5$, $x_{02} = -5$, and $t_{01} = t_{02} = t_0 = \phi_0 = 0$. (a) Absolute value of (2.47), (b) real part of (2.47), and (c) imaginary part of (2.47).

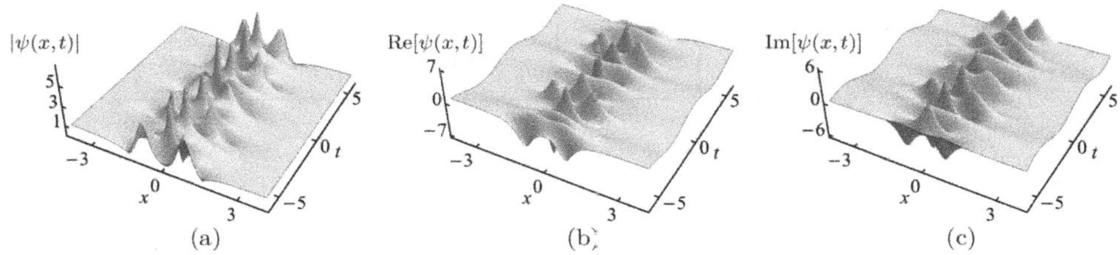

Figure 2.18. Plot of solution (2.47), with $a_1 = 1/2$, $a_2 = 1$, $\nu_1 = 1.3$, $\nu_2 = 1.85$, and $x_{01} = x_{02} = t_{01} = t_{02} = t_0 = \phi_0 = 0$. (a) Absolute value of (2.47), (b) real part of (2.47), and (c) imaginary part of (2.47).

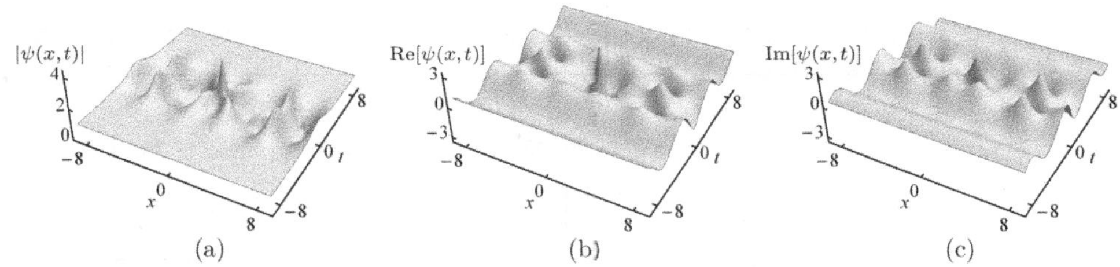

Figure 2.19. Plot of solution (2.47), with $a_1 = 1/2$, $a_2 = 1$, $\nu_1 = 0.5$, $\nu_2 = 0.85$, and $x_{01} = x_{02} = t_{01} = t_{02} = t_0 = \phi_0 = 0$. (a) Absolute value of (2.47), (b) real part of (2.47), and (c) imaginary part of (2.47).

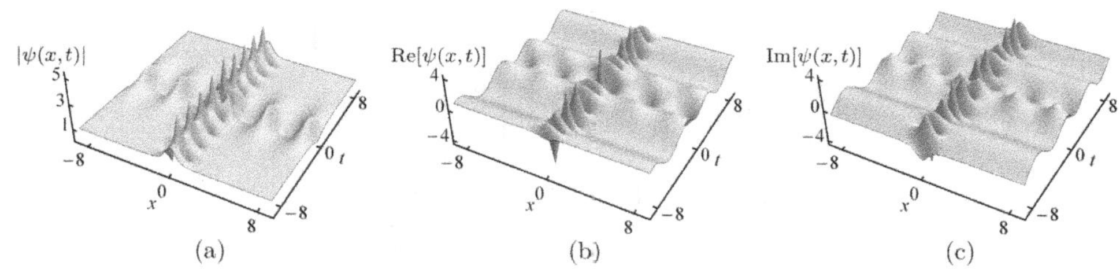

Figure 2.20. Plot of solution (2.47), with $a_1 = 1/2$, $a_2 = 1$, $\nu_1 = 0.5$, $\nu_2 = 1.5$, and $x_{01} = x_{02} = t_{01} = t_{02} = t_0 = \phi_0 = 0$. (a) Absolute value of (2.47), (b) real part of (2.47), and (c) imaginary part of (2.47).

$$\alpha(x,t)=\left(\kappa_2^2-\kappa_1^2\right)\left\{\frac{\delta_2\,\kappa_1^2\cos\left[\frac{\kappa_2}{\sqrt{2\,a_1}}(x-x_{02})\right]\cosh[\delta_1(t-t_{01})]}{\kappa_2}\right.$$

$$-\frac{\delta_1\,\kappa_2^2\cos\left[\frac{\kappa_1}{\sqrt{2\,a_1}}(x-x_{01})\right]\cosh[\delta_2(t-t_{02})]}{\kappa_1}$$

$$\left.-\left(\kappa_1^2-\kappa_2^2\right)\cosh[\delta_1(t-t_{01})]\cosh[\delta_2(t-t_{02})]\right\},$$

$$\beta(x,t)=-2\left(\kappa_1^2-\kappa_2^2\right)\left\{\frac{\delta_1\,\delta_2\cos\left[\frac{\kappa_2}{\sqrt{2\,a_1}}(x-x_{02})\right]\sinh[\delta_1(t-t_{01})]}{\kappa_2}\right.$$

$$-\delta_1\cosh[\delta_2(t-t_{02})]\sinh[\delta_1(t-t_{01})]$$
$$+\delta_2\cosh[\delta_1(t-t_{01})]\sinh[\delta_2(t-t_{02})]$$

$$\left.-\frac{\delta_1\,\delta_2\cos\left[\frac{\kappa_1}{\sqrt{2\,a_1}}(x-x_{01})\right]\sinh[\delta_2(t-t_{02})]}{\kappa_1}\right\},$$

$$\gamma(x,t)=\frac{2\,\delta_1\,\delta_2\left(\kappa_1^2+\kappa_2^2\right)\cos\left[\frac{\kappa_1}{\sqrt{2\,a_1}}(x-x_{01})\right]\cos\left[\frac{\kappa_2}{\sqrt{2\,a_1}}(x-x_{02})\right]}{\kappa_1\,\kappa_2}$$

$$-\left(2\,\kappa_1^2+2\,\kappa_2^2-\kappa_1^2\,\kappa_2^2\right)\cosh[\delta_1(t-t_{01})]\cosh[\delta_2(t-t_{02})]$$

$$-2\left(\kappa_1^2-\kappa_2^2\right)\left\{-\frac{\delta_2\cos\left[\frac{\kappa_2}{\sqrt{2\,a_1}}(x-x_{02})\right]\cosh[\delta_1(t-t_{01})]}{\kappa_2}\right.$$

$$\left.+\frac{\delta_1\cos\left[\frac{\kappa_1}{\sqrt{2\,a_1}}(x-x_{01})\right]\cosh[\delta_2(t-t_{02})]}{\kappa_1}\right\}$$

$$+4\,\delta_1\,\delta_2\left\{\sin\left[\frac{\kappa_1}{\sqrt{2\,a_1}}(x-x_{01})\right]\sin\left[\frac{\kappa_2}{\sqrt{2\,a_1}}(x-x_{02})\right]\right.$$

$$\left.+\sinh[\delta_1(t-t_{01})]\sinh[\delta_2(t-t_{02})]\right\},$$

$a_1 > 0,$
$a_2 > 0,$

$\kappa_j = 2\sqrt{1 - \nu_j^2}$,
$\delta_j = \frac{\kappa_j}{2}\sqrt{4 - \kappa_j^2}$,
ν_j, x_{0j}, t_{0j}, t_0, and ϕ_0 are arbitrary real constants, $j = 1, 2$.

- *Reference*: [9].

***Solution* 46. Specific Two-Breather Solution I** *Nonlinear superposition of Kuznetsov–Ma or Akhmediev breather with a Peregrine soliton*
(Figures 2.21–2.24)

$$\psi(x, t) = \frac{1}{\sqrt{a_2}}\left[1 + \frac{\alpha(x, t) + i\,\beta(x, t)}{\gamma(x, t)}\right] e^{i\,[t - t_0 + \phi_0]}, \qquad (2.48)$$

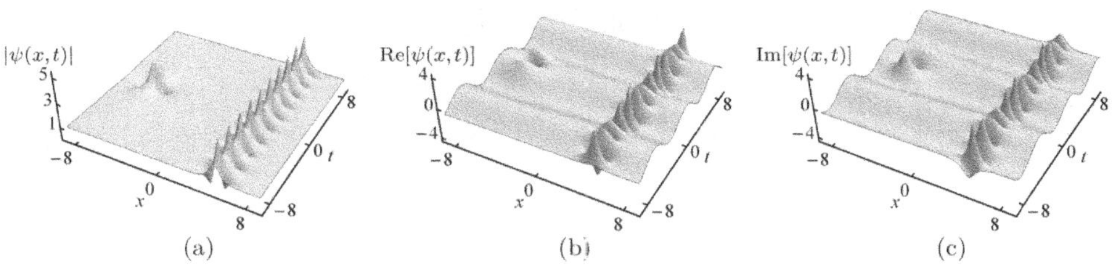

Figure 2.21. Plot of solution (2.48), with $a_1 = 1/2$, $a_2 = 1$, $\nu = 1.2$, $x_{01} = 5$, $x_{02} = -5$, and $t_{01} = t_{02} = t_0 = \phi_0 = 0$. (a) Absolute value of (2.48), (b) real part of (2.48), and (c) imaginary part of (2.48).

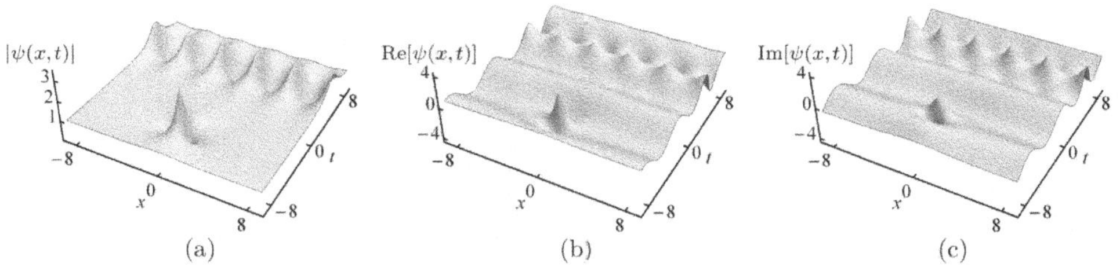

Figure 2.22. Plot of solution (2.48), with $a_1 = 1/2$, $a_2 = 1$, $\nu = 0.5$, $t_{01} = 5$, $t_{02} = -5$, and $x_{01} = x_{02} = t_0 = \phi_0 = 0$. (a) Absolute value of (2.48), (b) real part of (2.48), and (c) imaginary part of (2.48).

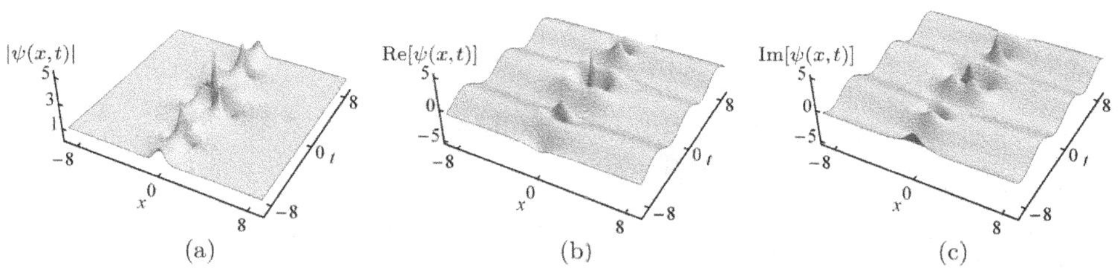

Figure 2.23. Plot of solution (2.48), with $a_1 = 1/2$, $a_2 = 1$, $\nu = 1.2$, and $x_{01} = x_{02} = t_{01} = t_{02} = t_0 = \phi_0 = 0$. (a) Absolute value of (2.48), (b) real part of (2.48), and (c) imaginary part of (2.48).

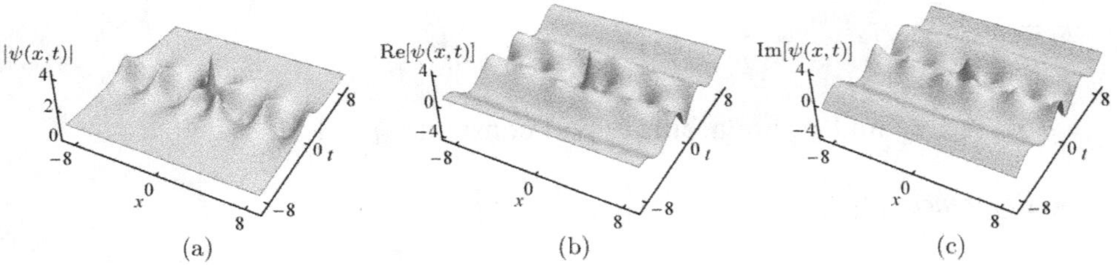

Figure 2.24. Plot of solution (2.48), with $a_1 = 1/2$, $a_2 = 1$, $\nu = 0.5$, $x_{01} = x_{02} = t_{01} = t_{02} = t_0 = \phi_0 = 0$. (a) Absolute value of (2.48), (b) real part of (2.48), and (c) imaginary part of (2.48).

where

$$\alpha(x, t) = \frac{\kappa}{8}\left(8\,\delta\cos\left[\frac{\kappa}{\sqrt{2\,a_1}}(x - x_{01})\right] + \kappa\left\{-8 + \left[1 + 4\,(t - t_{02})^2 + \frac{2}{a_1}(x - x_{02})^2\right]\kappa^2\right\}\cosh[\delta\,(t - t_{01})]\right),$$

$$\beta(x, t) = \frac{\kappa}{4}\left(8\,(t - t_{02})\left\{\delta\cos\left[\frac{\kappa}{\sqrt{2\,a_1}}(x - x_{01})\right] - \kappa\cosh[\delta\,(t - t_{01})]\right\} + \left[1 + 4\,(t - t_{02})^2 + \frac{2}{a_1}(x - x_{02})^2\right]\delta\,\kappa\sinh[\delta\,(t - t_{01})]\right),$$

$$\gamma(x, t) = -\frac{1}{4\,\kappa}\left[\delta\left\{-16 + \left[1 + 4\,(t - t_{02})^2 + \frac{2}{a_1}(x - x_{02})^2\right]\kappa^2\right\} \times\cos\left[\frac{\kappa}{\sqrt{2\,a_1}}(x - x_{01})\right] + \kappa\left(\left\{16 + [-3 + 4\,(t - t_{02})^2 + \frac{2}{a_1}(x - x_{02})^2]\,\kappa^2\right\}\cosh[\delta\,(t - t_{01})] - 16\,\delta\left\{\frac{1}{\sqrt{2a_1}}(x - x_{02}) \times\sin\left[\frac{\kappa}{\sqrt{2\,a_1}}(x - x_{01})\right] + (t - t_{02})\sinh[\delta\,(t - t_{01})]\right\}\right)\right],$$

$a_1 > 0$,
$a_2 > 0$,
$\kappa = 2\sqrt{1 - \nu^2}$,
$\delta = \frac{\kappa}{2}\sqrt{4 - \kappa^2}$,
ν, x_{0j}, t_{0j}, t_0, and ϕ_0 are arbitrary real constants, $j = 1, 2$.

- *Reference*: [9].
- *This solution can be generated from* (2.47) *in the limit* $\kappa_2 \to 0$ *with* $\kappa_1 \neq 0$.

***Solution* 47. Specific Two-Breather Solution II**
(Figures 2.25, 2.26)

$$\psi(x,\,t) = \frac{1}{\sqrt{a_2}}\left[1 + \frac{\alpha(x,\,t) + i\,\beta(x,\,t)}{\gamma(x,\,t)}\right]e^{i\,[t-t_0+\phi_0]}, \tag{2.49}$$

where

$$\begin{aligned}
\alpha(x,\,t) = &-\frac{2\,\kappa}{\delta}\left\{\cosh[\delta\,(t-t_0)]\left((\delta^2+\kappa^2)\cos\left[\kappa\,\frac{1}{\sqrt{2a_1}}\,(x-x_0)\right]\right.\right.\\
&-2\,\delta\,\kappa\cosh[\delta\,(t-t_0)] + \delta^2\,\kappa\,\frac{1}{\sqrt{2a_1}}\,(x-x_0)\sin\left[\kappa\,\frac{1}{\sqrt{2a_1}}\,(x-x_0)\right]\Bigg)\\
&\left.+\,\delta\,(2\,\delta^2-\kappa^2)\,(t-t_0)\cos\left[\kappa\,\frac{1}{\sqrt{2a_1}}\,(x-x_0)\right]\sinh[\delta\,(t-t_0)]\right\},
\end{aligned}$$

$$\begin{aligned}
\beta(x,\,t) = &-\frac{1}{2\,\delta\,\kappa}\left\{8\,\delta\,(2\,\delta^2-\kappa^2)\,(t-t_0)\left(-\kappa+\delta\cos\left[\kappa\,\frac{1}{\sqrt{2a_1}}\,(x-x_0)\right]\right.\right.\\
&\times\cosh[\delta\,(t-t_0)]\Big) + 8\,\delta^3\left(\cos\left[\kappa\,\frac{1}{\sqrt{2a_1}}\,(x-x_0)\right] + \kappa\,\frac{1}{\sqrt{2a_1}}\,(x-x_0)\right.\\
&\times\sin\left[\kappa\,\frac{1}{\sqrt{2a_1}}\,(x-x_0)\right]\Bigg)\sinh[\delta\,(t-t_0)]\\
&\left.+\,\kappa\,(\kappa^4-4\,\delta^2)\sinh[2\,\delta\,(t-t_0)]\right\},
\end{aligned}$$

$$\begin{aligned}
\gamma(x,\,t) = &-\frac{1}{4\,\delta^2\,\kappa^2}\left\{32\,\delta^4\,(\delta^2-\kappa^2)\,(t-t_0)^2 + \kappa^4\,(\delta^2+\kappa^2)\right.\\
&+8\,\delta^2\,\kappa^2\left(\delta^2\frac{1}{2a_1}\,(x-x_0)^2 + \kappa^2\,(t-t_0)^2\right)\\
&+4\left(\kappa^4\cosh[2\,\delta\,(t-t_0)] - \delta^4\cos\left[2\,\kappa\,\frac{1}{\sqrt{2a_1}}\,(x-x_0)\right]\right)\\
&-4\,\delta\,\kappa^2\cosh[\delta\,(t-t_0)]\left(\kappa^3\cos\left[\kappa\,\frac{1}{\sqrt{2a_1}}\,(x-x_0)\right]\right.\\
&+4\,\delta^2\,\frac{1}{\sqrt{2a_1}}\,(x-x_0)\sin\left[\kappa\,\frac{1}{\sqrt{2a_1}}\,(x-x_0)\right]\Bigg)\\
&\left.-\,16\,\delta^2\,\kappa\,(2\,\delta^2-\kappa^2)\,(t-t_0)\cos\left[\kappa\,\frac{1}{\sqrt{2a_1}}\,(x-x_0)\right]\sinh[\delta\,(t-t_0)]\right\},
\end{aligned}$$

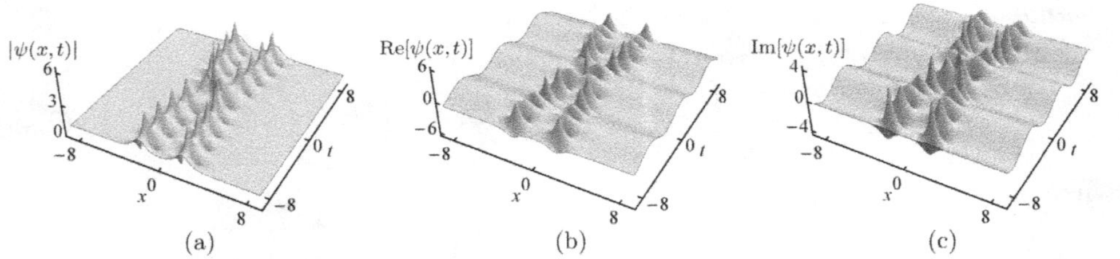

Figure 2.25. Plot of solution (2.49), with $a_1 = 1/2$, $a_2 = 1$, $\nu = 1.5$, and $x_0 = t_0 = \phi_0 = 0$. (a) Absolute value of (2.49), (b) real part of (2.49), and (c) imaginary part of (2.49).

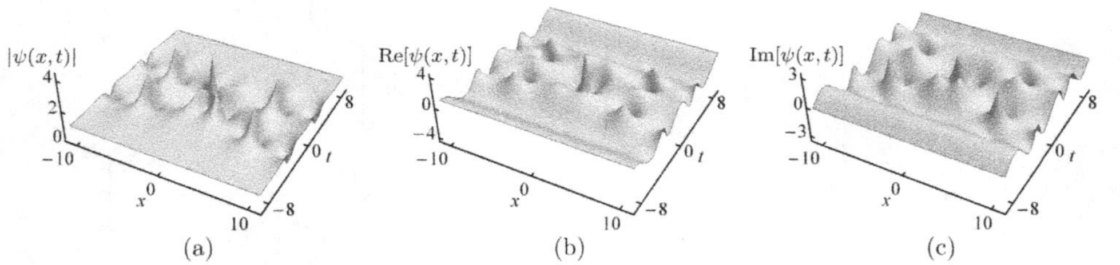

Figure 2.26. Plot of solution (2.49), with $a_1 = 1/2$, $a_2 = 1$, $\nu = 0.8$, and $x_0 = t_0 = 0$. (a) Absolute value of (2.49), (b) real part of (2.49), and (c) imaginary part of (2.49).

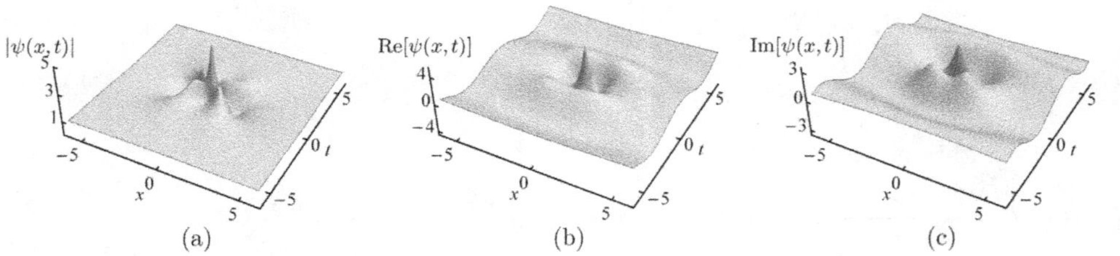

Figure 2.27. Second-order Peregrine soliton (2.50), with $a_1 = 1/2$, $a_2 = 1$, and $x_0 = t_0 = \phi_0 = 0$. (a) Absolute value of (2.50), (b) real part of (2.50), and (c) imaginary part of (2.50). Animation available online at https://iopscience.iop.org/book/978-0-7503-2428-1.

$a_1 > 0,$
$a_2 > 0,$
$\kappa = 2\sqrt{1 - \nu^2},$
$\delta = \frac{\kappa}{2}\sqrt{4 - \kappa^2},$
ν, x_0, t_0, and ϕ_0 are arbitrary real constants.

- *Reference*: [9].
- *This solution can be generated from* (2.47) *in the limit* $\kappa_2 \to \kappa_1$, $x_{01} = x_{02} = x_0$, *and* $t_{01} = t_{02} = t_0$.

***Solution* 48. Second-Order Peregrine Soliton**
(Figure 2.27)

$$\psi(x, t) = \frac{1}{\sqrt{a_2}}\left[1 + \frac{\alpha(x, t) + i\,\beta(x, t)}{\gamma(x, t)}\right] e^{i\,[t-t_0+\phi_0]}, \tag{2.50}$$

where

$$\alpha(x, t) = \frac{1}{96}\Bigg[-3 + 72\,(t-t_0)^2 + 80\,(t-t_0)^4 + \frac{12}{a_1}\,(x-x_0)^2 + \frac{48}{a_1}\,(x-x_0)^2\,(t-t_0)^2 + \frac{4}{a_1^2}\,(x-x_0)^4\Bigg],$$

$$\beta(x, t) = \frac{1}{48}\,(t-t_0)\Bigg[-15 + 8\,(t-t_0)^2 + 16\,(t-t_0)^4 - \frac{12}{a_1}\,(x-x_0)^2 + \frac{16}{a_1}\,(x-x_0)^2\,(t-t_0)^2 + \frac{4}{a_1^2}\,(x-x_0)^4\Bigg],$$

$$\gamma(x, t) = \frac{-1}{1152}\Bigg[9 + 396\,(t-t_0)^2 + 432\,(t-t_0)^4 + 64\,(t-t_0)^6 + \frac{54}{a_1}\,(x-x_0)^2 - \frac{144}{a_1}\,(x-x_0)^2\,(t-t_0)^2 + \frac{96}{a_1}\,(x-x_0)^2\,(t-t_0)^4 + \frac{12}{a_1^2}\,(x-x_0)^4 + \frac{48}{a_1^2}\,(x-x_0)^4\,(t-t_0)^2 + \frac{8}{a_1^3}\,(x-x_0)^6\Bigg],$$

$a_2 > 0$,
x_0, t_0, and ϕ_0 are arbitrary real constants.

- *Reference*: [9].
- *This solution can be generated from* (2.49) *in the limit* $\kappa \to 0$ *and* $\delta = 0$.

***Solution* 49. Rogue Wave Triplet**
(Figure 2.28)

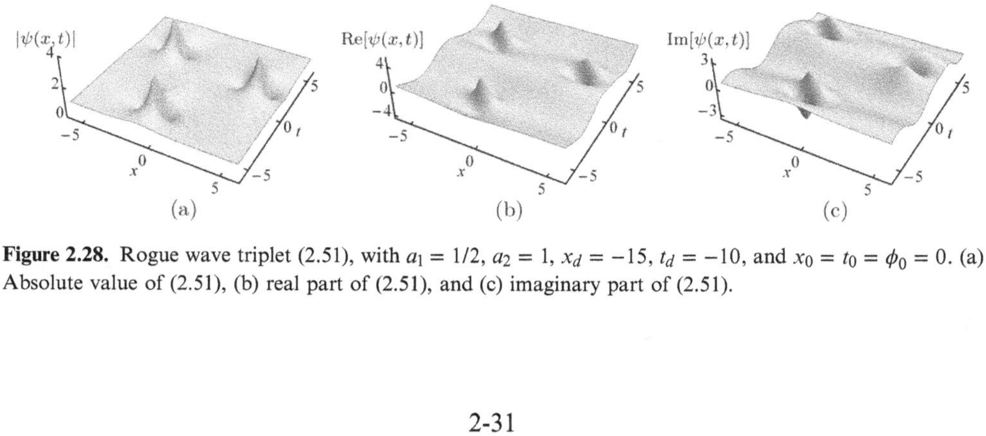

Figure 2.28. Rogue wave triplet (2.51), with $a_1 = 1/2$, $a_2 = 1$, $x_d = -15$, $t_d = -10$, and $x_0 = t_0 = \phi_0 = 0$. (a) Absolute value of (2.51), (b) real part of (2.51), and (c) imaginary part of (2.51).

$$\psi(x, t) = \frac{1}{\sqrt{a_2}} \left[1 + \frac{\alpha(x, t) + i\,\beta(x, t)}{\gamma(x, t)} \right] e^{i\,[t - t_0 + \phi_0]}, \tag{2.51}$$

where

$$\begin{aligned}
\alpha(x, t) = &- 12 \Bigg\{ -3 + \frac{4}{a_1^2} (x - x_0)^4 - 128\, t_d \frac{1}{\sqrt{2a_1}} (x - x_0) \\
&+ \frac{12}{a_1} (x - x_0)^2 [1 + 4\,(t - t_0)^2] \\
&+ 8\,[9 + 10\,(t - t_0)^2]\,(t - t_0)^2 - 128\, x_d\,(t - t_0) \Bigg\}, \\
\beta(x, t) = &- 24 \Bigg(-128 \frac{1}{\sqrt{2\,a_1}}\, t_d\,(x - x_0)\,(t - t_0) + (t - t_0) \Bigg\{ -15 + \frac{4}{a_1^2} (x - x_0)^4 \\
&+ 8\,(t - t_0)^2\,[1 + 2\,(t - t_0)^2] + \frac{4}{a_1} (x - x_0)^2\,[-3 + 4\,(t - t_0)^2] \Bigg\} \\
&+ 16 \left[1 + \frac{2}{a_1} (x - x_0)^2 - 4\,(t - t_0)^2 \right] x_d \Bigg), \\
\gamma(x, t) = &\, 9 + \frac{8}{a_1^3} (x - x_0)^6 + 396\,(t - t_0)^2 + 432\,(t - t_0)^4 + 64\,(t - t_0)^6 \\
&+ \frac{12}{a_1^2} (x - x_0)^4\,[1 + 4\,(t - t_0)^2] + \frac{6}{a_1} (x - x_0)^2 \\
&\times \{9 + 8\,(t - t_0)^2\,[-3 + 2\,(t - t_0)^2]\} \\
&+ 128 \Bigg(8\,t_d^2 + t_d \Bigg\{ 4 \left(\frac{1}{2\,a_1} \right)^{3/2} (x - x_0)^3 \\
&+ 3 \frac{1}{\sqrt{2\,a_1}} (x - x_0)\,[-1 - 4\,(t - t_0)^2] \Bigg\} \\
&+ x_d \left\{ (t - t_0) \left[-9 + \frac{6}{a_1} (x - x_0)^2 - 4\,(t - t_0)^2 \right] + 8\,x_d \right\} \Bigg),
\end{aligned}$$

$a_1 > 0$,
$a_2 > 0$,
x_d, t_d, x_0, t_0, and ϕ_0 are arbitrary real constants.

- *Reference*: [12], *solution* (8) *with* $\delta = 0$.

2.2 Summary of Subsection 2.1.1

Note: For lengthy conditions, the reader is referred to the solutions in subsection 2.1.1.

Equation

$$i\,\psi_t + a_1\,\psi_{xx} + a_2\,|\psi|^2\,\psi = 0$$

#	Solution	Conditions	Name	Eq. #
1.	$\psi(x,t) = A_0\,e^{i\left[a_2 A_0^2 (t-t_0)+\phi_0\right]}$	A_0, t_0, and ϕ_0 are arbitrary real constants	continuous wave, t-dependent phase	(2.2)
2.	$\psi(x,t) = A_0\,e^{i\left[\pm A_0 \sqrt{\frac{a_2}{a_1}}\,(x-x_0)+\phi_0\right]}$	$a_1\,a_2 > 0$, A_0, x_0, and ϕ_0 are arbitrary real constants	continuous wave, x-dependent phase	(2.3)
3.	$\psi(x,t) = A_0\,e^{i\left[A_1 (x-x_0)+(A_0^2 a_2 - A_1^2 a_1)(t-t_0)+\phi_0\right]}$	A_0, A_1, x_0, t_0, and ϕ_0 are arbitrary real constants	continuous wave, t- and x-dependent phase	(2.4)
4.	$\psi(x,t) = \frac{A_0}{\sqrt{A_1+t-t_0}}\,e^{i\left\{\frac{[a_1 A_2+x-x_0]^2}{4\,a_1\,[A_1+t-t_0]}+a_2 A_0^2 \ln[A_1+t-t_0]+\phi_0\right\}}$	A_0, A_1, A_2, t_0, x_0, and ϕ_0 are arbitrary real constants	decaying wave	(2.5)
5.	$\psi(x,t) = \sqrt{\frac{-2\,a_1}{a_2}}\,\frac{1}{x-x_0}\,e^{i\,\phi_0}$	$a_1\,a_2 < 0$, x_0, and ϕ_0 are arbitrary real constants	—	(2.6)
6.	$\psi(x,t) = \frac{1}{\sqrt{-a_2}}\,\frac{q_1(x,t)}{q_2(x,t)}$	See text	higher order of Solution 5	(2.7)
7.	$\psi(x,t) = A_0\sqrt{\frac{-2\,a_1}{a_2}}\,\sec[A_0\,(x-x_0)]\,e^{-i\left[a_1 A_0^2 (t-t_0)+\phi_0\right]}$	$a_1\,a_2 < 0$, A_0, x_0, t_0, and ϕ_0 are arbitrary real constants	—	(2.8)

(Continued)

(Continued)

#	Solution	Conditions	Name	Eq. #
	Equation $i\,\psi_t + a_1\,\psi_{xx} + a_2\,\lvert\psi\rvert^2\,\psi = 0$			
8.	$\psi(x, t) = A_0\sqrt{\frac{-2\,a_1}{a_2}}\,\csc[A_0\,(x - x_0)]\,e^{-i\left[a_1\,A_0^2\,(t-t_0)+\phi_0\right]}$	$a_1\,a_2 < 0$, A_0, x_0, t_0, and ϕ_0 are arbitrary real constants	–	(2.9)
9.	$\psi(x, t) = A_0\sqrt{\frac{-2\,a_1}{a_2}}\,\tan[A_0\,(x - x_0)]\,e^{i\left[2\,a_1\,A_0^2\,(t-t_0)+\phi_0\right]}$	$a_1\,a_2 < 0$, A_0, x_0, t_0, and ϕ_0 are arbitrary real constants	–	(2.10)
10.	$\psi(x, t) = A_0\sqrt{\frac{-2\,a_1}{a_2}}\,\cot[A_0\,(x - x_0)]\,e^{i\left[2\,a_1\,A_0^2\,(t-t_0)+\phi_0\right]}$	$a_1\,a_2 < 0$, A_0, x_0, t_0, and ϕ_0 are arbitrary real constants	—	(2.11)
11.	$\psi(x, t) = A_0\sqrt{\frac{2\,a_1}{a_2}}\,\text{sech}[A_0\,(x - x_0)]\,e^{i\left[a_1\,A_0^2\,(t-t_0)+\phi_0\right]}$	$a_1\,a_2 > 0$, A_0, x_0, t_0, and ϕ_0 are arbitrary real constants	bright soliton	(2.12)
12.	$\psi(x, t) = A_0\sqrt{\frac{-2\,a_1}{a_2}}\,\text{csch}[A_0\,(x - x_0)]\,e^{i\left[a_1\,A_0^2\,(t-t_0)+\phi_0\right]}$	$a_1\,a_2 < 0$, A_0, x_0, t_0, and ϕ_0 are arbitrary real constants	–	(2.13)
13.	$\psi(x, t) = A_0\sqrt{\frac{-2\,a_1}{a_2}}\,\tanh[A_0\,(x - x_0)]\,e^{-i\left[2\,a_1\,A_0^2\,(t-t_0)+\phi_0\right]}$	$a_1\,a_2 < 0$, A_0, x_0, t_0, and ϕ_0 are arbitrary real constants	dark soliton	(2.14)
14.	$\psi(x, t) = A_0\sqrt{\frac{-2\,a_1}{a_2}}\,\coth[A_0\,(x - x_0)]\,e^{-i\left[2\,a_1\,A_0^2\,(t-t_0)+\phi_0\right]}$	$a_1\,a_2 < 0$, A_0, x_0, t_0, and ϕ_0 are arbitrary real constants	–	(2.15)
15.	$\psi(x, t) = \left(A_0 + i\,A_1\tan\left\{A_1\left[\sqrt{\frac{-a_2}{2\,a_1}}\,(x - x_0) + a_2\,A_0\,(t - t_0)\right]\right\}\right) \times e^{i\left[a_2\,(A_0^2-A_1^2)\,(t-t_0)+\phi_0\right]}$	$a_1a_2 < 0$, A_0, A_1, x_0, t_0, and ϕ_0 are arbitrary real constants	–	(2.16)
16.	$\psi(x, t) = A_0\sqrt{\frac{-a_1}{2\,a_2}}\,\{\sec[A_0\,(x - x_0)] + \tan[A_0\,(x - x_0)]\} \times e^{i\left[\frac{a_1\,A_0^2}{2}\,(t-t_0)+\phi_0\right]}$	$a_1a_2 < 0$, A_0, x_0, t_0, and ϕ_0 are arbitrary real constants	–	(2.17)

17. $\psi(x,t) = A_0 \sqrt{\frac{-2}{a_2}} \left(\cot\left\{ A_0 \left[\frac{x-x_0}{\sqrt{a_1}} - 2 A_1 (t-t_0) \right] \right\} - \tan\left\{ A_0 \left[\frac{x-x_0}{\sqrt{a_1}} - 2 A_1 (t-t_0) \right] \right\} \right) e^{i\left[\frac{A_1}{\sqrt{a_1}}(x-x_0)-(A_1^2-8A_0^2)(t-t_0)+\phi_0\right]}$	$a_1 > 0$, $a_2 < 0$, A_0, A_1, x_0, t_0, and ϕ_0 are arbitrary real constants	–	(2.18)
18. $\psi(x,t) = A_0 \sqrt{\frac{-a_1}{2a_2}} \left\{ \frac{\cos[A_0(x-x_0)]}{1-\sin[A_0(x-x_0)]} \right\} e^{i\left[\frac{a_1 A_0^2}{2}(t-t_0)+\phi_0\right]}$	$a_1 a_2 < 0$, A_0, x_0, t_0, and ϕ_0 are arbitrary real constants	–	(2.19)
19. $\psi(x,t) = A_0 \sqrt{\frac{-2a_1}{a_2}} \left\{ \frac{1+\tanh^2[A_0(x-x_0)]}{\tanh[A_0(x-x_0)]} \right\} e^{-i\left[8 a_1 A_0^2(t-t_0)+\phi_0\right]}$	$a_1 a_2 < 0$, A_0, x_0, t_0, and ϕ_0 are arbitrary real constants	–	(2.20)
20. $\psi(x,t) = A_0 \sqrt{\frac{-2}{a_2}} \left(\frac{b_0 \tan\left\{ A_0 \left[\frac{x-x_0}{\sqrt{a_1}} - 2A_1(t-t_0) \right] \right\} - b_1}{b_0 + b_1 \tan\left\{ A_0 \left[\frac{x-x_0}{\sqrt{a_1}} - 2A_1(t-t_0) \right] \right\}} \right) \times e^{i\left[\frac{A_1(x-x_0)}{\sqrt{a_1}}-(A_1^2-2A_0^2)(t-t_0)+\phi_0\right]}$	$a_1 > 0$, $a_2 < 0$, A_0, A_1, b_0, b_1, x_0, t_0, and ϕ_0 are arbitrary real constants	–	(2.21)
21. $\psi(x,t) = \frac{q_1(x,t)}{q_2(x,t)} e^{i\left[\frac{A_0}{\sqrt{a_1}}(x-x_0)-A_0^2(t-t_0)+\phi_0\right]}$	See text	–	(2.22)
22. $\psi(x,t) = A_0 \sqrt{\frac{2m}{-a_2(1+m)}}\, \mathrm{sn}\left[\frac{A_0}{\sqrt{a_1(1+m)}}(x-x_0), m \right] e^{-i\left[A_0^2(t-t_0)+\phi_0\right]}$	$a_1(1+m) > 0$, $a_2 m(1+m) < 0$, $m \neq -1$, A_0, t_0, x_0, and ϕ_0 are arbitrary real constants	solitary wave	(2.23)
23. $\psi(x,t) = A_0 \sqrt{\frac{2a_1}{a_2}}\, \mathrm{sn}[A_0(x-x_0), -1]\, e^{i\phi_0}$	$a_1 a_2 > 0$, A_0, x_0, and ϕ_0 are arbitrary real constants	solitary wave	(2.24)
24. $\psi(x,t) = A_0 \sqrt{\frac{2m}{a_2(2m-1)}}\, \mathrm{cn}\left[\frac{A_0}{\sqrt{a_1(2m-1)}}(x-x_0), m \right] e^{i\left[A_0^2(t-t_0)+\phi_0\right]}$	$a_1(2m-1) > 0$, $a_2 m(2m-1) > 0$, $m \neq 1/2$, A_0, t_0, x_0, and ϕ_0 are arbitrary real constants	solitary wave	(2.25)

(Continued)

(*Continued*)

Equation

$i\,\psi_t + a_1\,\psi_{xx} + a_2\,|\psi|^2\,\psi = 0$

#	Solution	Conditions	Name	Eq. #
25.	$\psi(x,t) = A_0\sqrt{\frac{a_1}{a_2}}\,\mathrm{cn}\left[A_0\,(x - x_0),\, \frac{1}{2}\right] e^{i\,\phi_0}$	$a_1\,a_2 > 0$, $m = 1/2$, A_0, x_0, and ϕ_0 are arbitrary real constants	solitary wave	(2.26)
26.	$\psi(x,t) = A_0\sqrt{\frac{2}{a_2\,(2-m)}}\,\mathrm{dn}\left[\frac{A_0}{\sqrt{a_1\,(2-m)}}\,(x - x_0),\, m\right] e^{i\left[A_0^2\,(t-t_0)+\phi_0\right]}$	$a_1\,(2-m) > 0$, $a_2\,(2-m) > 0$, $m \neq 2$, A_0, t_0, x_0, and ϕ_0 are arbitrary real constants	solitary wave	(2.27)
27.	$\psi(x,t) = A_0\sqrt{\frac{2\,a_1}{a_2}}\,\mathrm{dn}\,[A_0\,(x - x_0),\, 2]\,e^{i\,\phi_0}$	$a_1\,a_2 > 0$, A_0, x_0, and ϕ_0 are arbitrary real constants	solitary wave	(2.28)
28.	$\psi(x,t) = A_0\sqrt{1-m}\,\mathrm{nd}\left[\frac{A_1}{\sqrt{a_1}}\,(x - x_0),\, m\right] e^{-i\,[A_2\,(t-t_0)+\phi_0]}$	$A_2 = (m-2)\,A_1^2$, $a_1 > 0$, $a_2 = \frac{2\,A_1^2}{A_0^2}$, $0 < m \leqslant 1$, A_0, A_1, x_0, t_0, and ϕ_0 are arbitrary real constants	solitary wave	(2.29)
29.	$\psi(x,t) = A_0\sqrt{m\,(1-m)}\,\mathrm{sd}\left[\frac{A_1}{\sqrt{a_1}}\,(x - x_0),\, m\right] e^{-i\,[A_2\,(t-t_0)+\phi_0]}$	$A_2 = (1-2\,m)\,A_1^2$, $a_1 > 0$, $a_2 = \frac{2\,A_1^2}{A_0^2}$, $0 < m \leqslant 1$, A_0, A_1, x_0, t_0, and ϕ_0 are arbitrary real constants	solitary wave	(2.30)
30.	$\psi(x,t) = A_0\sqrt{m}\,\mathrm{cd}\left[\frac{A_1}{\sqrt{a_1}}\,(x - x_0),\, m\right] e^{-i\,[A_2\,(t-t_0)+\phi_0]}$	$A_2 = (m+1)\,A_1^2$, $a_1 > 0$, $a_2 = \frac{-2\,A_1^2}{A_0^2}$, $m > 0$, A_0, A_1, x_0, t_0, and ϕ_0 are arbitrary real constants	solitary wave	(2.31)

31. $\psi(x, t) = A(x)\, e^{i\,[\phi(x,\, t)+\phi_0]}$, $A(x) = \sqrt{R_3 + (R_2 - R_3)\,\mathrm{sn}^2[A_0\,(x - x_0),\, m]}$, $\phi(x, t) = \dfrac{\sqrt{2}\,\lambda_0\,\Pi\left\{\frac{R_3 - R_2}{R_3},\, \mathrm{am}[A_0\,(x - x_0), m], m\right\}\mathrm{dn}[A_0\,(x - x_0), m]}{\sqrt{2}\,R_3\,A_0\sqrt{1 - m\,\mathrm{sn}^2[A_0\,(x - x_0), m]}} + \lambda_2\,(t - t_0)$	See text	solitary wave on a finite background	(2.32)
32. $\psi(x, t) = \left\{\frac{A_0}{2}\,\mathrm{dn}\left[\frac{A_1}{\sqrt{a_1}}(x - x_0),\, m\right] + \frac{B_0\sqrt{m}}{2}\,\mathrm{cn}\left[\frac{A_1}{\sqrt{a_1}}\,(x - x_0),\, m\right]\right\}\, e^{-i\,[A_2\,(t-t_0)+\phi_0]}$	$A_2 = \frac{-(m+1)\,A_1^2}{2}$, $B_0 = \pm A_0$, $a_1 > 0$, $a_2 = \frac{2\,A_1^2}{A_0^2}$, $0 < m \leqslant 1$, A_0, A_1, x_0, t_0, and ϕ_0 are arbitrary real constants	solitary wave	(2.33)
33. $\psi(x, t) = A_0\, m\, \dfrac{\mathrm{cn}\left[\frac{A_1}{\sqrt{a_1}}\,(x - x_0), m\right]\mathrm{sn}\left[\frac{A_1}{\sqrt{a_1}}\,(x - x_0), m\right]}{\mathrm{dn}\left[\frac{A_1}{\sqrt{a_1}}\,(x - x_0), m\right]}\, e^{-i\,[A_2\,(t-t_0)+\phi_0]}$	$A_2 = 2\,(2 - m)\,A_1^2$, $a_1 > 0$, $a_2 = -\frac{2\,A_1^2}{A_0^2}$, $m > 0$, A_0, A_1, x_0, t_0, and ϕ_0 are arbitrary real constants	solitary wave	(2.34)
34. $\psi(x, t) = \frac{1}{\sqrt{a_2}}\sum_{j=1}^{N}\psi_j(x, t)$, $\times\sum_{k=1}^{N} M_{j\,k}\left[\gamma_j^{-1}(x, t) + \gamma_k^{*}(x, t)\right]\psi_k(x, t) = 1, \quad j = 0, 1, 2, \ldots, N$	See text	N-bright solitons	(2.35)
35. $\psi(x, t) = \frac{1}{\sqrt{a_2}}\,[\psi_1(x, t) + \psi_2(x, t)]$	See text	two bright solitons	(2.37)
36. $\psi(x, t) = \frac{1}{\sqrt{a_2}}\,[\psi_1(x, t) + \psi_2(x, t) + \psi_3(x, t)]$	See text	three bright solitons	(2.38)

(Continued)

(Continued)

Equation

$i\,\psi_t + a_1\,\psi_{xx} + a_2\,|\psi|^2\,\psi = 0$

#	Solution	Conditions	Name	Eq. #
37.	$\psi(x,t)=\sqrt{\frac{-2}{a_2}}\left\{1-2\,i\sum_n \mu_n(t)\,e^{\frac{-2\,v_n}{\sqrt{a_1}}(x-x_0)}\,[-Q_{12}{}^n+(\lambda_n+i\,v_n)\,(Q_{11}{}^n-1)]\right\}$ $\times\,e^{-2\,i\,[t-t_0+\phi_0]}$	See text	N-dark solitons	(2.39)
38.	$\psi(x,t)=\sqrt{\frac{-2}{a_2}}\left\{1-\frac{2\,i}{p(x,t)}\left[\frac{2}{v_1+v_2}\left(\frac{1}{\lambda_1+i\,v_1}+\frac{1}{\lambda_2+i\,v_2}\right)\right.\right.$ $\left.\left.-(\lambda_1-i\,v_1)\,q_1(x,t)-(\lambda_2-i\,v_2)\,q_2(x,t)\right]\right\}\,e^{-2\,i\,[t-t_0+\phi_0]}$	See text	two dark solitons	(2.40)
39.	$\psi(x,t)=\frac{1}{\sqrt{a_2}}\left\{\frac{\kappa^2\cosh[\delta\,(t-t_0)]+2\,i\,\kappa\,\nu\sinh[\delta\,(t-t_0)]}{2\cosh[\delta\,(t-t_0)]-2\,\nu\cos\left[\frac{\kappa}{\sqrt{2\,a_1}}\,(x-x_0)\right]}-1\right\}e^{i\,[t-t_0+\phi_0]}$	$a_1>0$, $a_2>0$, $\kappa=2\sqrt{1-\nu^2}$, $\delta=\kappa\,\nu$, ν, x_0, t_0, and ϕ_0 are arbitrary real constants	generalized first-order breather I	(2.41)
40.	$\psi(x,t)=\frac{\cos(A_0)\cos[q_1(x,t)+2\,i\,A_1]-\cosh(A_1)\cosh[q_2(x,t)+2\,i\,A_0]}{\cos(A_0)\cos[q_1(x,t)]-\cosh(A_1)\cosh[q_2(x,t)]}$ $\times\,e^{i\,[a_2\,(t-t_0)+\phi_0]}$	See text	generalized first-order breather II	(2.42)
41.	$\psi(x,t)=\frac{A_0}{\sqrt{a_2}}\left(1-\frac{\sqrt{8}\,\lambda_r}{A_0}\,p(x,t)\right)e^{i\left[A_0^2\,(t-t_0)+\phi_0\right]}$	See text	generalized first-order breather III	(2.43)
42.	$\psi(x,t)=\frac{1}{\sqrt{a_2}}\left\{\frac{-p^2\cos[\omega\,(t-t_0)]-2\,i\,p\,\nu\sin[\omega\,(t-t_0)]}{2\cos[\omega\,(t-t_0)]-2\,\nu\cosh\left[\frac{p}{\sqrt{2\,a_1}}\,(x-x_0)\right]}-1\right\}e^{i\,[t-t_0+\phi_0]}$	$a_1>0$, $a_2>0$, $p=2\sqrt{\nu^2-1}$, $\omega=p\,\nu$, $\nu>1$, x_0, t_0, and ϕ_0 are arbitrary real constants	Kuznetsov–Ma breather	(2.44)

43. $\psi(x, t) = \frac{1}{\sqrt{a_2}} \left\{ \frac{\kappa^2 \cosh[\delta (t - t_0)] + 2 i \kappa \nu \sinh[\delta (t - t_0)]}{2 \cosh[\delta (t - t_0)] - 2 \nu \cos\left[\frac{\kappa}{\sqrt{2 a_1}} (x - x_0)\right]} - 1 \right\} e^{i [t - t_0 + \phi_0]}$	$a_1 > 0$, $a_2 > 0$, $\kappa = 2 \sqrt{1 - \nu^2}$, $\delta = \kappa \nu$, $\nu < 1$, x_0, t_0, and ϕ_0 are arbitrary real constants	Akhmediev breather	(2.45)
44. $\psi(x, t) = \frac{1}{\sqrt{a_2}} \left[\frac{4 + i 8 (t - t_0)}{1 + 4 (t - t_0)^2 + \frac{2}{a_1} (x - x_0)^2} - 1 \right] e^{i [t - t_0 + \phi_0]}$	$a_2 > 0$, x_0, t_0, and ϕ_0 are arbitrary real constants	Peregrine soliton	(2.46)
45. $\psi(x, t) = \frac{1}{\sqrt{a_2}} \left[1 + \frac{\alpha(x, t) + i \beta(x, t)}{\gamma(x, t)} \right] e^{i [t - t_0 + \phi_0]}$	See text	generalized two-breathers	(2.47)
46. $\psi(x, t) = \frac{1}{\sqrt{a_2}} \left[1 + \frac{\alpha(x, t) + i \beta(x, t)}{\gamma(x, t)} \right] e^{i [t - t_0 + \phi_0]}$	See text	specific two-breathers I	(2.48)
47. $\psi(x, t) = \frac{1}{\sqrt{a_2}} \left[1 + \frac{\alpha(x, t) + i \beta(x, t)}{\gamma(x, t)} \right] e^{i [t - t_0 + \phi_0]}$	See text	specific two-breathers II	(2.49)
48. $\psi(x, t) = \frac{1}{\sqrt{a_2}} \left[1 + \frac{\alpha(x, t) + i \beta(x, t)}{\gamma(x, t)} \right] e^{i [t - t_0 + \phi_0]}$	See text	second-order Peregrine soliton	(2.50)
49. $\psi(x, t) = \frac{1}{\sqrt{a_2}} \left[1 + \frac{\alpha(x, t) + i \beta(x, t)}{\gamma(x, t)} \right] e^{i [t - t_0 + \phi_0]}$	See text	rogue wave triplet	(2.51)

2.2.1 Complex Dispersion and Nonlinearity Coefficients

Here,

$a_1 = a_{1r} + i\, a_{1i}$,

$a_2 = a_{2r} + i\, a_{2i}$,

where a_{1r}, a_{1i}, a_{2r}, and a_{2i} are real constants.

***Solution* 1. Constant Amplitude I** *CW, t-dependent phase*

$$\psi(x,\,t) = A_0\, e^{i\left[\,a_{2r}\, A_0^2\,(t-t_0)+\phi_0\right]}, \tag{2.52}$$

where

$a_{2i} = 0$,

A_0, t_0, and ϕ_0 are arbitrary real constants.

- *Derived in appendix* A.

***Solution* 2. Constant Amplitude II** *CW, x-dependent phase*

$$\psi(x,\,t) = A_0\, e^{i\left[\pm A_0\,\sqrt{\frac{a_{2r}}{a_{1r}}}\,(x-x_0)+\phi_0\right]}, \tag{2.53}$$

where

$a_{1r}\, a_{2r} > 0$,

$a_{1i} = \frac{a_{1r}\, a_{2i}}{a_{2r}}$,

A_0, x_0, and ϕ_0 are arbitrary real constants.

- *Derived in appendix* A.

***Solution* 3. Rational Solution I** *DW*

$$\psi(x,\,t) = \frac{1}{\sqrt{2\, a_{2i}\,(t-t_0)}}\, e^{i\,\phi_0}, \tag{2.54}$$

where

$a_{2i} > 0$,

$a_{2r} = 0$,

t_0 and ϕ_0 are arbitrary real constants.

- *Derived in appendix* A.

***Solution* 4. Rational Solution II** *DW*

$$\psi(x,\,t) = \frac{1}{\sqrt{2\, a_{2i}\,(t-t_0)}}\, e^{i\left\{A_0\,(x-x_0)-a_{1r}\, A_0^2\,(t-t_0)+\frac{a_{2r}}{2\, a_{2i}}\,\ln[2\, a_{2i}\,(t-t_0)]+\phi_0\right\}}, \tag{2.55}$$

where
$a_{2i} > 0$,
$a_{1i} = 0$,
A_0, t_0, x_0, and ϕ_0 are arbitrary real constants.

- *Derived in appendix* A.

***Solution* 5.**

$$\psi(x, t) = A_0 \sqrt{\frac{a_{1r}}{e^{2\, a_{1r}\, A_0^2\, (t-t_0)} - a_{2i}}} \times e^{i \left\{ A_0\, (x-x_0) - \frac{a_{1r}\, (a_{2i}+a_{2r})\, A_0^2}{a_{2i}}\, (t-t_0) + \frac{a_{2r}}{2\, a_{2i}} \ln\left[-a_{2i} + e^{2\, a_{1r}\, A_0^2\, (t-t_0)} \right] + \phi_0 \right\}}, \tag{2.56}$$

where
$a_{1i} = -a_{1r}$,
A_0, t_0, x_0, and ϕ_0 are arbitrary real constants.

- *Derived in appendix* A.

***Solution* 6.**

$$\psi(x, t) = A_0 \frac{\sqrt{a_{1i}}\; e^{a_{1i}\, A_0^2\, (t-t_0)}}{\sqrt{a_{2i}\, e^{2\, a_{1i}\, A_0^2\, (t-t_0)} - 1}} \times e^{i \left(A_0\, (x-x_0) - a_{1r}\, A_0^2\, (t-t_0) + \frac{a_{2r}}{2\, a_{2i}} \ln \left\{ e^{2\, a_{1i}\, A_0^2\, A_1} \left[a_{2i}\, e^{2\, a_{1i}\, A_0^2\, (t-t_0)} - 1 \right] \right\} + \phi_0 \right)}, \tag{2.57}$$

where A_0, A_1, t_0, x_0, and ϕ_0 are arbitrary real constants.

- *Derived in appendix* A.

***Solution* 7.**

$$\psi(x, t) = A_0\, e^{-a_{1r}\, A_1^2\, (t-t_0)}\, e^{i \left[A_1\, (x-x_0) - a_{1r}\, A_1^2\, (t-t_0) - \frac{a_{2r}\, A_0^2}{2\, a_{1r}\, A_1^2}\, e^{-2\, a_{1r}\, A_1^2\, (t-t_0)} + \phi_0 \right]}, \tag{2.58}$$

where
$a_{1i} = -a_{1r}$,
$a_{2i} = 0$,
A_0, A_1, t_0, x_0, and ϕ_0 are arbitrary real constants.

- *Derived in appendix* A.

***Solution* 8. sn(x, −1)** *SW*

$$\psi(x, t) = A_0 \sqrt{\frac{2\, a_{1r}}{a_{2r}}}\, \mathrm{sn}\left[A_0\, (x - x_0),\, -1\right] e^{i\, \phi_0}, \tag{2.59}$$

where

$a_{1r}\, a_{2r} > 0$,
$a_{1i} = \frac{a_{1r}\, a_{2i}}{a_{2r}}$,
A_0, x_0, and ϕ_0 are arbitrary real constants.

- *Derived in appendix* A.

***Solution* 9. cn(*x*, 1/2)** *SW*

$$\psi(x,\, t) = A_0 \sqrt{\frac{a_{1r}}{a_{2r}}}\, \mathrm{cn}\left[A_0\,(x - x_0),\, \frac{1}{2}\right] e^{i\,\phi_0}, \tag{2.60}$$

where

$a_{1r}\, a_{2r} > 0$,
$a_{1i} = \frac{a_{1r}\, a_{2i}}{a_{2r}}$,
A_0, x_0, and ϕ_0 are arbitrary real constants.

- *Derived in appendix* A.

***Solution* 10. dn(*x*, 2)** *SW*

$$\psi(x,\, t) = A_0 \sqrt{\frac{2\, a_{1r}}{a_{2r}}}\, \mathrm{dn}\left[A_0\,(x - x_0),\, 2\right] e^{i\,\phi_0}, \tag{2.61}$$

where

$a_{1r}\, a_{2r} > 0$,
$a_{1i} = \frac{a_{1r}\, a_{2i}}{a_{2r}}$,
A_0, x_0, and ϕ_0 are arbitrary real constants.

- *Derived in appendix* A.

2.3 Summary of Subsection 2.2.1

#	Solution	Conditions	Name	Eq. #
	Equation $i\,\psi_t + (a_{1r} + i\,a_{1i})\,\psi_{xx} + (a_{2r} + i\,a_{2i})\,\lvert\psi\rvert^2\,\psi = 0$			
1.	$\psi(x,t) = A_0\,e^{i\left[a_{2r}\,A_0^2\,(t-t_0)+\phi_0\right]}$	$a_{2i}=0$, A_0, t_0, and ϕ_0 are arbitrary real constants	continuous wave, t-dependent phase	(2.52)
2.	$\psi(x,t) = A_0\,e^{i\left[\pm A_0\sqrt{\frac{a_{2r}}{a_{1r}}}\,(x-x_0)+\phi_0\right]}$	$a_{1r}\,a_{2r}>0$, $a_{1i}=\frac{a_{1r}\,a_{2i}}{a_{2r}}$, A_0, x_0, and ϕ_0 are arbitrary real constants	continuous wave, x-dependent phase	(2.53)
3.	$\psi(x,t) = \frac{1}{\sqrt{2\,a_{2i}\,(t-t_0)}}\,e^{i\,\phi_0}$	$a_{2i}>0$, $a_{2r}=0$, t_0 and ϕ_0 are arbitrary real constants	decaying wave	(2.54)
4.	$\psi(x,t) = \frac{1}{\sqrt{2\,a_{2i}\,(t-t_0)}}\,e^{i\left\{A_0\,(x-x_0)-a_{1r}\,A_0^2\,(t-t_0)+\frac{a_{2r}}{2\,a_{2i}}\ln[2\,a_{2i}\,(t-t_0)]+\phi_0\right\}}$	$a_{2i}>0$, $a_{1i}=0$, A_0, t_0, x_0, and ϕ_0 are arbitrary real constants	decaying wave	(2.55)
5.	$\psi(x,t) = A_0\sqrt{\frac{a_{1r}}{e^{2\,a_{1r}\,A_0^2\,(t-t_0)}-a_{2i}}}$ $\times\,e^{i\left\{A_0\,(x-x_0)-\frac{a_{1r}\,(a_{2i}+a_{2r})\,A_0^2}{a_{2i}}\,(t-t_0)+\frac{a_{2r}}{2\,a_{2i}}\ln\left[-a_{2i}+e^{2\,a_{1r}\,A_0^2\,(t-t_0)}\right]+\phi_0\right\}}$	$a_{1i}=-a_{1r}$, A_0, t_0, x_0, and ϕ_0 are arbitrary real constants	–	(2.56)
6.	$\psi(x,t) = A_0\,\frac{\sqrt{a_{1i}}\,e^{a_{1i}\,A_0^2\,(t-t_0)}}{\sqrt{a_{2i}\,e^{2\,a_{1i}\,A_0^2\,(t-t_0)}-1}}$ $\times\,e^{i\left(A_0\,(x-x_0)-a_{1r}\,A_0^2\,(t-t_0)+\frac{a_{2r}}{2\,a_{2i}}\ln\left\{e^{2\,a_{1i}\,A_0^2\,A_1}\left[a_{2i}\,e^{2\,a_{1i}\,A_0^2\,(t-t_0)}-1\right]\right\}+\phi_0\right)}$	A_0, A_1, t_0, x_0, and ϕ_0 are arbitrary real constants	–	(2.57)
7.	$\psi(x,t) = A_0\,e^{-a_{1r}\,A_1^2\,(t-t_0)}\,e^{i\left[A_1\,(x-x_0)-a_{1r}\,A_1^2\,(t-t_0)-\frac{a_{2r}\,A_0^2}{2\,a_{1r}\,A_1^2}\,e^{-2\,a_{1r}\,A_1^2\,(t-t_0)}+\phi_0\right]}$	$a_{1i}=-a_{1r}$, $a_{2i}=0$, A_0, A_1, t_0, x_0, and ϕ_0 are arbitrary real constants	–	(2.58)

(Continued)

(*Continued*)

Equation

$$i\,\psi_t + (a_{1r} + i\,a_{1i})\,\psi_{xx} + (a_{2r} + i\,a_{2i})\,|\psi|^2\,\psi = 0$$

#	Solution	Conditions	Name	Eq. #
8.	$\psi(x, t) = A_0\sqrt{\frac{2\,a_{1r}}{a_{2r}}}\ \mathrm{sn}\,[A_0\,(x - x_0), -1]\,e^{i\,\phi_0}$	$a_{1r}\,a_{2r} > 0$, $a_{1i} = \frac{a_{1r}\,a_{2i}}{a_{2r}}$, A_0, x_0, and ϕ_0 are arbitrary real constants	solitary wave	(2.59)
9.	$\psi(x, t) = A_0\sqrt{\frac{a_{1r}}{a_{2r}}}\ \mathrm{cn}\left[A_0\,(x - x_0), \frac{1}{2}\right]e^{i\,\phi_0}$	$a_{1r}\,a_{2r} > 0$, $a_{1i} = \frac{a_{1r}\,a_{2i}}{a_{2r}}$, A_0, x_0, and ϕ_0 are arbitrary real constants	solitary wave	(2.60)
10.	$\psi(x, t) = A_0\sqrt{\frac{2\,a_{1r}}{a_{2r}}}\ \mathrm{dn}\,[A_0\,(x - x_0), 2]\,e^{i\,\phi_0}$	$a_{1r}\,a_{2r} > 0$, $a_{1i} = \frac{a_{1r}\,a_{2i}}{a_{2r}}$, A_0, x_0, and ϕ_0 are arbitrary real constants	solitary wave	(2.61)

References

[1] Zaitsev V F and Polyanin A D 2003 *Handbook of Nonlinear Partial Differential Equations* (New York: Chapman and Hall)

[2] He B and Meng Q 2016 Qualitative analysis and explicit exact solitary, kink and anti-kink wave solutions of the generalized nonlinear Schrödinger equation with parabolic law nonlinearity *Commun. Theor. Phys.* **65** 1–10

[3] Zayed E M and Al-Nowehy A G 2017 Exact solutions for the perturbed nonlinear Schrödinger equation with power law nonlinearity and Hamiltonian perturbed terms *Optik* **139** 123–44

[4] Ma W X and Chen M 2009 Direct search for exact solutions to the nonlinear Schrödinger equation *Appl. Math. Comput.* **215** 2835–42

[5] Bronski J C, Carr L D, Deconinck B and Kutz J N 2001 Bose–Einstein condensates in standing waves: The cubic nonlinear Schrödinger equation with a periodic potential *Phys. Rev. Lett.* **86** 1402–5

[6] Khare A and Saxena A 2015 Periodic and hyperbolic soliton solutions of a number of nonlocal nonlinear equations *J. Math. Phys.* **56** 032104–27

[7] Gordon J P 1983 Interaction forces among solitons in optical fibers *Opt. Lett.* **8** 596–8

[8] Blow K J and Doran N J 1985 Multiple dark soliton solutions of the nonlinear Schrödinger equation *Phys. Lett.* A **107** 55–8

[9] Kedziora D J, Ankiewicz A and Akhmediev N 2012 Second-order nonlinear Schrödinger equation breather solutions in the degenerate and rogue wave limits *Phys. Rev.* E **85** 066601–9

[10] Kharif C, Pelinovsky E and Slunyaev A 2010 Rogue waves in the ocean *Advances in Geophysical and Environmental Mechanics and Mathematics* (Berlin: Springer)

[11] Al Khawaja U and Taki M 2013 Rogue waves management by external potentials *Phys. Lett.* A **37** 2944–9

[12] Chowdury A, Kedziora D J, Ankiewicz A and Akhmediev N 2015 Breather solutions of the integrable quintic nonlinear Schrödinger equation and their interactions *Phys. Rev.* E **91** 022919–11

IOP Publishing

Handbook of Exact Solutions to the Nonlinear Schrödinger Equations

Usama Al Khawaja and Laila Al Sakkaf

Chapter 3

Nonlinear Schrödinger Equation with Power Law and Dual Power Law Nonlinearities

A Glance at Chapter 3

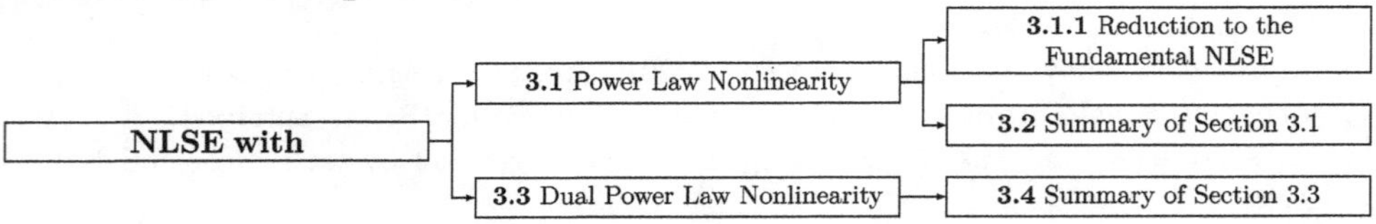

A Statistical View of Chapter 3

	Equation	Solutions
1	$i\,\psi_t + a_1\,\psi_{xx} + a_2\,\lvert\psi\rvert^n\,\psi = 0$	13
2	$i\,\psi_t + a_1\,\psi_{xx} + a_2\,\lvert\psi\rvert^n\,\psi + a_3\,\lvert\psi\rvert^m\,\psi = 0$	14
Total	2	27

3.1 NLSE with Power Law Nonlinearity

Equation:

$$i\,\psi_t + a_1\,\psi_{xx} + a_2\,\lvert\psi\rvert^n\,\psi = 0, \qquad (3.1)$$

where

$\psi = \psi(x,\,t)$ is the complex function profile,
x and t are its two independent variables,
n, a_1, and a_2 are arbitrary real constants.

3.1.1 Reduction to the Fundamental NLSE

Case I: CW profile

If

$$\psi(x,\, t) = A_0\, e^{i\left[A_1(x-x_0)+\left(A_0^2\, a_2 - A_1^2\, a_1\right)(t-t_0)+\phi_0\right]} \tag{3.2}$$

is a stationary solution of the fundamental NLSE with cubic nonlinearity, (2.1), then

$$\psi(x,\, t) = A_0^{2/n}\, e^{i\left[A_1(x-x_0)+\left(A_0^2\, a_2 - A_1^2\, a_1\right)(t-t_0)+\phi_0\right]} \tag{3.3}$$

is a stationary solution to the NLSE with power law nonlinearity, (3.1).

Case II: x-dependent profile

If

$$\phi(x,\, t) = u(x)\, e^{i\lambda t} \tag{3.4}$$

is a stationary solution of the fundamental NLSE with cubic nonlinearity, (2.1), then

$$\psi(x,\, t) = \left(\frac{n+2}{n^2}\right)^{1/n} u(x)^{2/n}\, e^{i\,\frac{4\lambda}{n^2}\,t} \tag{3.5}$$

is a stationary solution to the NLSE with power law nonlinearity, (3.1).

Solutions:

***Solution* 1. Constant Amplitude I** *continuous wave (CW), t-dependent phase*

$$\psi(x,\, t) = A_0^{2/n}\, e^{i\left[A_0^2\, a_2\,(t-t_0)+\phi_0\right]}, \tag{3.6}$$

where A_0, t_0, and ϕ_0 are arbitrary real constants.

- *Reference*: [1].

***Solution* 2. Constant Amplitude II** *CW, x-dependent phase*

$$\psi(x,\, t) = A_0^{2/n}\, e^{i\left[A_0\sqrt{\frac{a_2}{a_1}}\,(x-x_0)+\phi_0\right]}, \tag{3.7}$$

where

$a_1\, a_2 > 0$,

A_0, x_0, and ϕ_0 are arbitrary real constants.

- *Derived in appendix* A.

***Solution* 3. Constant Amplitude III** *CW, t- and x-dependent phase*

$$\psi(x,\, t) = A_0^{2/n}\, e^{i\left[A_1\,(x-x_0)+\left(A_0^2\, a_2 - A_1^2\, a_1\right)(t-t_0)+\phi_0\right]}, \tag{3.8}$$

where A_0, A_1, x_0, t_0, and ϕ_0 are arbitrary real constants.

- *Reference*: [1].

***Solution* 4. Rational Solution I** *decaying wave (DW)*

$$\psi(x, t) = \frac{A_0}{\sqrt{t - t_0}}\, e^{i\left[\frac{2\, a_2\, |A_0|^n\, (t-t_0)^{\frac{2-n}{2}}}{2-n} + \frac{(x-x_0)^2}{4\, a_1\, (t-t_0)} + \phi_0\right]}, \tag{3.9}$$

where A_0, x_0, t_0, and ϕ_0 are arbitrary real constants.

- *Reference*: [1].

***Solution* 5. Rational Solution II** *DW*

$$\psi(x, t) = \left[\frac{\pm 1}{n\sqrt{\frac{-a_2}{2\, a_1\, (2+n)}}\,(x - x_0)}\right]^{\frac{2}{n}} e^{i\,\phi}, \tag{3.10}$$

where
$a_1\, a_2\, (2 + n) < 0$,
A_0, x_0, and ϕ_0 are arbitrary real constants.

- *Reference*: [2].

***Solution* 6. sec(*x*)**

$$\psi(x, t) = \left\{\frac{-2\, A_0^2\, a_1\, (n + 2)}{a_2 n^2}\, \sec^2[A_0\, (x - x_0)]\right\}^{\frac{1}{n}} e^{-i\left[\frac{4\, a_1\, A_0^2}{n^2}\,(t-t_0)+\phi_0\right]}, \tag{3.11}$$

where
$a_1\, a_2\, (n + 2) < 0$,
A_0, x_0, t_0, and ϕ_0 are arbitrary real constants.

- *Reference*: [3] *with* $\alpha = \gamma = \lambda = \nu = 0$.

***Solution* 7. csc(*x*)**

$$\psi(x, t) = \left\{\frac{-2\, A_0^2\, a_1\, (n + 2)}{a_2 n^2}\, \csc^2[A_0\, (x - x_0)]\right\}^{\frac{1}{n}} e^{-i\left[\frac{4\, a_1\, A_0^2}{n^2}\,(t-t_0)+\phi_0\right]}, \tag{3.12}$$

where
$a_1\, a_2\, (n + 2) < 0$,
A_0, x_0, t_0, and ϕ_0 are arbitrary real constants.

- *Reference*: [3] *with* $\alpha = \gamma = \lambda = \nu = 0$.

***Solution* 8. sech(*x*)** *bright soliton*

$$\psi(x, t) = \left\{\frac{2\, A_0^2\, a_1\, (n + 2)}{a_2 n^2}\, \operatorname{sech}^2[A_0\, (x - x_0)]\right\}^{\frac{1}{n}} e^{i\left[\frac{4\, a_1\, A_0^2}{n^2}\,(t-t_0)+\phi_0\right]}, \tag{3.13}$$

where

$a_1\, a_2\, (n + 2) > 0$,
A_0, x_0, t_0, and ϕ_0 are arbitrary real constants.

- *Reference*: [3] *with* $\alpha = \gamma = \lambda = \nu = 0$.

***Solution* 9. csch(*x*)**

$$\psi(x,\, t) = \left\{\frac{-2\, A_0^2\, a_1\, (n + 2)}{a_2 n^2}\, \mathrm{csch}^2[A_0\, (x - x_0)]\right\}^{\frac{1}{n}} e^{i\left[\frac{4\, a_1\, A_0^2}{n^2}\, (t - t_0) + \phi_0\right]}, \tag{3.14}$$

where
$a_1\, a_2\, (n + 2) < 0$,
A_0, x_0, t_0, and ϕ_0 are arbitrary real constants.

- *Reference*: [3] *with* $\alpha = \gamma = \lambda = \nu = 0$.

***Solution* 10. Generalized Oscillatory Solution**

$$\psi(x,\, t) = A(x)\, e^{i\, \phi_0}, \tag{3.15}$$

where
$A(x) = Y^{-1}(x - x_0)$,
$Y[A(x)] = \frac{A(x)}{\sqrt{A_0}}\; {}_2F_1\!\left[\frac{1}{2},\, \frac{1}{n+2},\, \frac{n+3}{n+2},\, \frac{2\, a_2\, A^{n+2}(x)}{a_1\, A_0\, (n - 2)}\right]$,
${}_2F_1$ is the hypergeometric function and Y^{-1} is the inverse operator of the function $Y[A(x)]$,
$A_0 > 0$,
x_0 and ϕ_0 are arbitrary real constants.

- *Derived in appendix* A.

***Solution* 11. sn(*x*, *m*) for *n* = 1** *solitary wave (SW)*

$$\psi(x,\, t) = \left\{R_3 + (R_2 - R_3)\, \mathrm{sn}^2[A_0\, (x - x_0),\, m]\right\} e^{i\, [\lambda_1\, (t - t_0) + \phi_0]}, \tag{3.16}$$

where
$m = \frac{R_2 - R_3}{R_1 - R_3}$, R_j, $j = 1, 2, 3$, are the three roots of $Y(x) = 3\, \lambda_0 + \frac{3\, \lambda_1}{a_1}\, x^2 - \frac{2 a_2}{a_1}\, x^3$,
$A_0 = \sqrt{\frac{a_2\, (R_3 - R_1)}{6\, a_1}}$,
$a_1\, a_2\, (R_3 - R_1) > 0$,
$n = 1$,
λ_0, λ_1, x_0, t_0, and ϕ_0 are arbitrary real constants.

- *Derived in appendix* A.

Solution* 12. sn(*x*, *m*) for *n* = 4 *SW

$$\psi(x,\,t) = \frac{\sqrt{R_1}\,\mathrm{sn}[A_0\,(x - x_0),\,m]}{\sqrt{\frac{R_1 - R_3}{R_3} + \mathrm{sn}^2[A_0\,(x - x_0),\,m]}}\,e^{i\,[\lambda_1\,(t - t_0) + \phi_0]}, \tag{3.17}$$

where

$m = \frac{R_3\,(R_1 - R_2)}{R_2\,(R_1 - R_3)}$,

R_j, $j = 1, 2, 3$, are the three roots of $Y(x) = 3\,\lambda_0 + \frac{3\,\lambda_1}{a_1}\,x - \frac{a_2}{a_1}\,x^3$,

$A_0 = \sqrt{\frac{a_2\,R_2\,(R_1 - R_3)}{3\,a_1}}$,

$R_1 > 0$,

$a_1\,a_2\,R_2\,(R_1 - R_3) > 0$,

$n = 4$,

λ_0, λ_1, x_0, t_0, and ϕ_0 are arbitrary real constants.

- *Derived in appendix* A.

Solution* 13. sn²(*x*, *m*) for *n* = 4 *SW on a finite background

$$\psi(x,\,t) = \frac{\sqrt{R_1\,(R_2 - R_4) + R_2\,(R_4 - R_1)\,\mathrm{sn}^2[A_0\,(x - x_0),\,m]}}{\sqrt{-R_2 + R_4 + (R_1 - R_4)\,\mathrm{sn}^2[A_0\,(x - x_0),\,m]}}\,e^{i\,\phi(x,t)}, \tag{3.18}$$

where

$$\phi(x,\,t) = \frac{\lambda_0\left(\frac{(R_2 - R_1)}{A_0}\,\Pi\left\{\frac{R_2\,(R_1 - R_4)}{R_1\,(R_2 - R_4)},\,\mathrm{am}[A_0\,(x - x_0),\,m],\,m\right\}\mathrm{dn}[A_0\,(x - x_0),\,m]\right)}{R_1\,R_2\,\sqrt{1 - \frac{(R_2 - R_3)\,(R_1 - R_4)\,\mathrm{sn}^2[A_0\,(x - x_0),\,m]}{(R_1 - R_3)\,(R_2 - R_4)}}} + \frac{\lambda_0}{R_2}\,(x - x_0) + \lambda_2\,(t - t_0) + \phi_0,$$

$m = \frac{(R_2 - R_3)\,(R_1 - R_4)}{(R_1 - R_3)\,(R_2 - R_4)}$,

R_j, $j = 1, 2, 3, 4$, are the four roots of

$Y(x) = 3\,a_1\,\lambda_0^2 - 3\,a_1\,\lambda_1\,x - 3\,\lambda_2\,x^2 + a_2\,x^4$,

$A_0 = \sqrt{\frac{-a_2\,(R_1 - R_3)\,(R_2 - R_4)}{3\,a_1}}$,

Π is the incomplete elliptic integral,

am is the amplitude for Jacobi elliptic functions,

$a_1\,a_2\,(R_1 - R_3)\,(R_2 - R_4) < 0$,

$n = 4$,

λ_0, λ_1, λ_2, x_0, t_0, and ϕ_0 are arbitrary real constants.

- *Derived in appendix* A.

3.2 Summary of Section 3.1

Equation: $i\,\psi_t + a_1\,\psi_{xx} + a_2\,|\psi|^n\,\psi = 0$

#	Solution	Conditions	Name	Eq. #
1.	$\psi(x,t) = A_0^{2/n}\,e^{i\left[A_0^2\,a_2\,(t-t_0)+\phi_0\right]}$	A_0, t_0, and ϕ_0 are arbitrary real constants	continuous wave, t-dependent phase	(3.6)
2.	$\psi(x,t) = A_0^{2/n}\,e^{i\left[A_0\sqrt{\frac{a_2}{a_1}}\,(x-x_0)+\phi_0\right]}$	$a_1\,a_2 > 0$, A_0, x_0, and ϕ_0 are arbitrary real constants	continuous wave, x-dependent phase	(3.7)
3.	$\psi(x,t) = A_0^{2/n}\,e^{i\left[A_1\,(x-x_0)+(A_0^2\,a_2-A_1^2\,a_1)\,(t-t_0)+\phi_0\right]}$	A_0, A_1, x_0, t_0, and ϕ_0 are arbitrary real constants	continuous wave, t- and x-dependent phase	(3.8)
4.	$\psi(x,t) = \frac{A_0}{\sqrt{t-t_0}}\,e^{i\left[\frac{2\,a_2\,\lvert A_0\rvert^n\,(t-t_0)^{\frac{2-n}{2}}}{2-n}+\frac{(x-x_0)^2}{4\,a_1\,(t-t_0)}+\phi_0\right]}$	A_0, x_0, t_0, and ϕ_0 are arbitrary real constants	decaying wave	(3.9)
5.	$\psi(x,t) = \left[\frac{\pm 1}{n\sqrt{\frac{-a_2}{2\,a_1\,(2+n)}}\,(x-x_0)}\right]^{\frac{2}{n}} e^{i\phi}$	$a_1\,a_2\,(2+n) < 0$, A_0, x_0, and ϕ_0 are arbitrary real constants	–	(3.10)
6.	$\psi(x,t) = \left\{\frac{-2\,A_0^2\,a_1\,(n+2)}{a_2\,n^2}\,\sec^2[A_0\,(x-x_0)]\right\}^{\frac{1}{n}} e^{-i\left[\frac{4\,a_1\,A_0^2}{n^2}\,(t-t_0)+\phi_0\right]}$	$a_1\,a_2\,(n+2) < 0$, A_0, x_0, t_0, and ϕ_0 are arbitrary real constants	–	(3.11)
7.	$\psi(x,t) = \left\{\frac{-2\,A_0^2\,a_1\,(n+2)}{a_2\,n^2}\,\csc^2[A_0\,(x-x_0)]\right\}^{\frac{1}{n}} e^{-i\left[\frac{4\,a_1\,A_0^2}{n^2}\,(t-t_0)+\phi_0\right]}$	$a_1\,a_2\,(n+2) < 0$, A_0, x_0, t_0, and ϕ_0 are arbitrary real constants	–	(3.12)
8.	$\psi(x,t) = \left\{\frac{2\,A_0^2\,a_1\,(n+2)}{a_2\,n^2}\,\mathrm{sech}^2[A_0\,(x-x_0)]\right\}^{\frac{1}{n}} e^{i\left[\frac{4\,a_1\,A_0^2}{n^2}\,(t-t_0)+\phi_0\right]}$	$a_1\,a_2\,(n+2) > 0$, A_0, x_0, t_0, and ϕ_0 are arbitrary real constants	bright soliton	(3.13)

9. $\psi(x, t) = \left\{\frac{-2 A_0^2 a_1 (n+2)}{a_2 n^2} \operatorname{csch}^2[A_0 (x - x_0)]\right\}^{\frac{1}{n}} e^{i\left[\frac{4 a_1 A_0^2}{n^2}(t-t_0)+\phi_0\right]}$	$a_1 a_2 (n + 2) < 0$, A_0, x_0, t_0, and ϕ_0 are arbitrary real constants	–	(3.14)
10. $\psi(x, t) = A(x)\, e^{i\,\phi_0}$, $A(x) = Y^{-1}(x - x_0)$	$Y[A(x)] = \frac{A(x)}{\sqrt{A_0}}\, {}_2F_1\left[\frac{1}{2}, \frac{1}{n+2}, \frac{n+3}{n+2}, \frac{2 a_2 A^{n+2}(x)}{a_1 A_0 (n+2)}\right]$, ${}_2F_1$ is the hypergeometric function and Y^{-1} is the inverse operator of the function $Y[A(x)]$, $A_0 > 0$, x_0 and ϕ_0 are arbitrary real constants	generalized oscillatory solution	(3.15)
11. $\psi(x, t) = \{R_3 + (R_2 - R_3)\operatorname{sn}^2[A_0 (x - x_0), m]\}\, e^{i\,[\lambda_1 (t-t_0)+\phi_0]}$	$m = \frac{R_2 - R_3}{R_1 - R_3}$, R_j, $j = 1, 2, 3$ are the three roots of $Y(x) = 3\lambda_0 + \frac{3\lambda_1}{a_1}x^2 - \frac{2a_2}{a_1}x^3$, $A_0 = \sqrt{\frac{a_2 (R_3 - R_1)}{6 a_1}}$, $a_1 a_2 (R_3 - R_1) > 0$, $n = 1$, λ_0, λ_1, x_0, t_0, and ϕ_0 are arbitrary real constants	solitary wave	(3.16)
12. $\psi(x, t) = \frac{\sqrt{R_1}\,\operatorname{sn}[A_0 (x - x_0), m]}{\sqrt{\frac{R_1 - R_3}{R_3} + \operatorname{sn}^2[A_0 (x - x_0), m]}}\, e^{i\,[\lambda_1 (t-t_0)+\phi_0]}$	$m = \frac{R_3 (R_1 - R_2)}{R_2 (R_1 - R_3)}$, R_j, $j = 1, 2, 3$ are the three roots of $Y(x) = 3\lambda_0 + \frac{3\lambda_1}{a_1}x - \frac{a_2}{a_1}x^3$, $A_0 = \sqrt{\frac{a_2 R_2 (R_1 - R_3)}{3 a_1}}$, $n = 4$, $R_1 > 0$, $a_1 a_2 R_2 (R_1 - R_3) > 0$, λ_0, λ_1, x_0, t_0, and ϕ_0 are arbitrary real constants	solitary wave	(3.17)
13. $\psi(x, t) = \frac{\sqrt{R_1 (R_2 - R_4) + R_2 (R_4 - R_1)\operatorname{sn}^2[A_0 (x - x_0), m]}}{\sqrt{-R_2 + R_4 + (R_1 - R_4)\operatorname{sn}^2[A_0 (x - x_0), m]}}\, e^{i\phi(x,t)}$, $\phi(x, t) = \frac{\lambda_0\left(\frac{(R_2 - R_1)}{A_0}\Pi\left\{\frac{R_2 (R_1 - R_4)}{R_1 (R_2 - R_4)}, \operatorname{am}[A_0 (x - x_0), m], m\right\}\operatorname{dn}[A_0 (x - x_0), m]\right)}{R_1 R_2 \sqrt{1 - \frac{(R_2 - R_3)(R_1 - R_4)\operatorname{sn}^2[A_0 (x - x_0), m]}{(R_1 - R_3)(R_2 - R_4)}}} + \frac{\lambda_0}{R_2}(x - x_0) + \lambda_2 (t - t_0) + \phi_0$	$m = \frac{(R_2 - R_3)(R_1 - R_4)}{(R_1 - R_3)(R_2 - R_4)}$, R_j, $j = 1, 2, 3, 4$ are the four roots of $Y(x) = 3 a_1 \lambda_0^2 - 3 a_1 \lambda_1 x - 3\lambda_2 x^2 + a_2 x^4$, $A_0 = \sqrt{\frac{-a_2 (R_1 - R_3)(R_2 - R_4)}{3 a_1}}$, Π is the incomplete elliptic integral, am is the amplitude for Jacobi elliptic functions, $a_1 a_2 (R_1 - R_3)(R_2 - R_4) < 0$, $n = 4$, λ_0, λ_1, λ_2, x_0, t_0, and ϕ_0 are arbitrary real constants	solitary wave	(3.18)

3.3 NLSE with Dual Power Law Nonlinearity

Equation:

$$i\,\psi_t + a_1\,\psi_{xx} + a_2\,|\psi|^n\,\psi + a_3\,|\psi|^m\,\psi = 0, \tag{3.19}$$

where

$\psi = \psi(x, t)$ is the complex function profile,
x and t are its two independent variables,
n, m, a_1, a_2, and a_3 are arbitrary real constants.

Solutions:

***Solution* 1. Constant Amplitude I** *CW, t-dependent phase*

$$\psi(x, t) = A_0\,e^{i\,[(|A_0|^n\,a_2+|A_0|^m\,a_3)\,(t-t_0)+\phi_0]}, \tag{3.20}$$

where

A_0, t_0, and ϕ_0 are arbitrary real constants.

- *Derived in appendix* A.

***Solution* 2. Constant Amplitude II** *CW, x-dependent phase*

$$\psi(x, t) = A_0\,e^{i\left[\sqrt{\frac{|A_0|^n\,a_2+|A_0|^m\,a_3}{a_1}}\,(x-x_0)+\phi_0\right]}, \tag{3.21}$$

where

A_0, x_0, and ϕ_0 are arbitrary real constants.

- *Derived in appendix* A.

***Solution* 3. Constant Amplitude III** *CW, t- and x-dependent phase*

$$\psi(x, t) = A_0\,e^{i\left[A_1\,(x-x_0)+(|A_0|^n\,a_2+|A_0|^m\,a_3-a_1\,A_1^2)\,(t-t_0)+\phi_0\right]}, \tag{3.22}$$

where

A_0, A_1, x_0, t_0, and ϕ_0 are arbitrary real constants.

- *Derived in appendix* A.

***Solution* 4. Rational Solution I** *DW*

$$\psi(x, t) = \frac{A_0}{\sqrt{t-t_0}}\,e^{i\left\{-\frac{2|A_0|^m\,a_3\,(t-t_0)^{\frac{2-m}{2}}}{m-2}-\frac{2|A_0|^n\,a_2(t-t_0)^{\frac{2-n}{2}}}{n-2}+\frac{[2\,a_1\,A_1+(x-x_0)]^2}{4\,a_1(t-t_0)}+\phi_0\right\}}, \tag{3.23}$$

where

$n \neq 2$,
$m \neq 2$,
A_0, A_1, x_0, t_0, and ϕ_0 are arbitrary real constants.

- *Derived in appendix* A.

***Solution* 5. Rational Solution II**

$$\psi(x,\, t) = \left[\frac{-2\, a_1\, a_2\, (n+1)\,(n+2)}{a_1\, a_3\,(n+2)^2 + a_2^2\, n^2\,(1+n)\,(x-x_0)^2}\right]^{\frac{1}{n}} e^{i\,\phi_0}, \tag{3.24}$$

where
$a_1\, a_2 < 0$,
$a_1\, a_3 > 0$,
$m = 2\, n$,
x_0 and ϕ_0 are arbitrary real constants.

- *Reference*: [2].

***Solution* 6.**

$$\psi(x,\, t) = \sqrt{\frac{-a_2^3}{2\, a_3}}\, \frac{x - x_0}{\sqrt{12\, a_1\, a_3 + a_2^2\,(x-x_0)^2}}\, e^{-i\left[\frac{a_2^2}{4\, a_3}\,(t-t_0)+\phi_0\right]}, \tag{3.25}$$

where
$n = 2$,
$m = 4$,
$a_2\, a_3 < 0$,
$a_1\, a_3 > 0$,
x_0, t_0, and ϕ_0 are arbitrary real constants.

- *Reference*: [2] *with* $m = 1$.

***Solution* 7.**

$$\psi(x,\, t) = \left\{\frac{n+1}{(n+2)\,[1 + 2\, e^{\sqrt{A_0}\,(x-x_0)}]}\right\}^{\frac{1}{n}} e^{i\,[A_1\,(t-t_0)+\phi_0]}, \tag{3.26}$$

where
$A_0 = \frac{a_2\, n^2\,(n+1)}{a_1\,(n+2)^2}$,
$A_1 = \frac{a_1\, A_0}{n^2}$,
$m = 2\, n$,
$a_3 = -a_2$,

$a_1\, a_2\, (n + 1) > 0$,
x_0, t_0, and ϕ_0 are arbitrary real constants.

- *Reference*: [2].

***Solution* 8. sech(*x*)** *flat-top soliton*
(Figure 3.1)

$$\psi(x,\, t) = \left(\frac{A_0\, A_1\, (n + 2)}{a_2 \left\{A_0 + 2\cosh\left[n\sqrt{\frac{A_1}{a_1}}\,(x - x_0)\right]\right\}}\right)^{\frac{1}{n}} e^{i\,[A_1\,(t - t_0) + \phi_0]}, \tag{3.27}$$

where

$A_0 = \sqrt{\frac{4\, a_2^2\, (n+1)}{A_1\, \delta\, (n+2)^2}}$,
$\delta = a_3 + \frac{a_2^2\, (n+1)}{A_1\, (n+2)^2}$,
$m = 2\, n$,
$a_1\, A_1 > 0$,
$A_1\, \delta\, (n + 1) > 0$,
x_0, t_0, A_1, and ϕ_0 are arbitrary real constants.

- *Reference*: [4].

***Solution* 9. tanh(*x*)** *dark soliton*
(Figure 3.2)

$$\psi(x,\, t) = \left(\frac{2\, a_1\, A_0^2\, (n + 2)}{a_2\, n^2}\, \{1 - \tanh[A_0\, (x - x_0)]\}\right)^{\frac{1}{n}} e^{i\left[\frac{4\, a_1\, A_0^2}{n^2}\,(t - t_0) + \phi_0\right]}, \tag{3.28}$$

where

$A_0 = \sqrt{\frac{-a_2^2\, n^2\, (1+n)}{4\, a_1\, a_3\, (2+n)^2}}$,

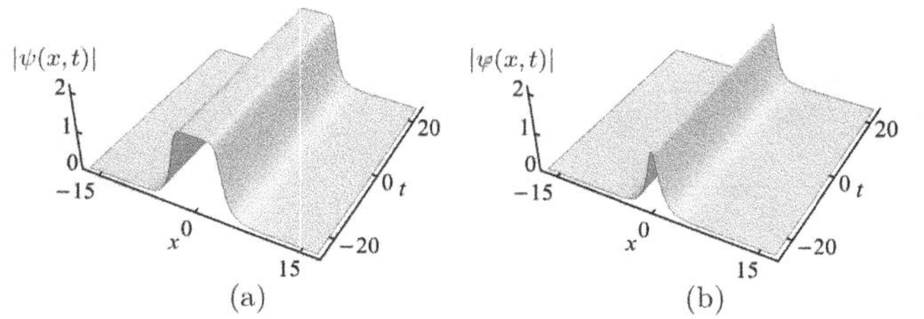

Figure 3.1. Plot of solution (3.27). (a) Flat-top soliton with $a_3 = -0.069444444444$, (b) bright soliton with $a_3 = 0.03055555555$. For the other parameters: $a_1 = A_1 = 2$, $a_2 = 1$, $n = 4$, and $x_0 = t_0 = \phi_0 = 0$.

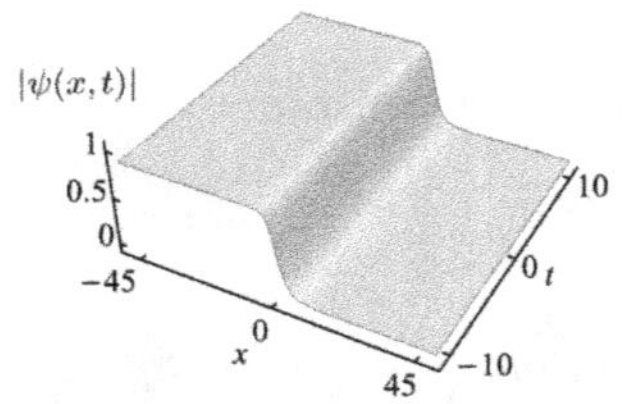

Figure 3.2. Dark soliton (3.28) with $a_1 = a_2 = 1$, $a_3 = -1$, $n = 2$, and $x_0 = t_0 = \phi_0 = 0$.

$m = 2\,n$,
$a_1\,a_2\,(n + 2) > 0$,
$a_1\,a_3\,(n + 1) < 0$,
x_0, t_0, and ϕ_0 are arbitrary real constants.

- *Reference*: [5].

***Solution* 10. coth(*x*)**

$$\psi(x,\,t) = \left(\frac{2\,a_1\,A_0^2\,(n+2)}{a_2\,n^2}\,\{1 - \coth[A_0\,(x - x_0)]\}\right)^{\frac{1}{n}} e^{i\left[\frac{4\,a_1\,A_0^2}{n^2}(t-t_0)+\phi_0\right]}, \qquad (3.29)$$

where

$A_0 = \sqrt{\dfrac{-a_2^2\,n^2\,(1+n)}{4\,a_1\,a_3\,(2+n)^2}}$,
$m = 2\,n$,
$a_1\,a_2\,(n + 2) > 0$,
$a_1\,a_3\,(n + 1) < 0$,
x_0, t_0, and ϕ_0 are arbitrary real constants.

- *Reference*: [6].

***Solution* 11. sinh(*x*)**

$$\psi(x,\,t) = \pm\frac{\mu_1\,\mu_2}{\sqrt{\mu_1^2 + \left(\mu_1^2 + 4\,\mu_2^2\right)\sinh^2[A_0\,(x - x_0)]}}\, e^{i\left[\frac{A_0^2\,A_1\,a_1}{a_3\,\mu_1^2\,\mu_2^2}(t-t_0)+\phi_0\right]}, \qquad (3.30)$$

where

$\mu_1 = \sqrt{\dfrac{3\,a_2}{a_3} + \sqrt{\dfrac{9\,a_2^2}{a_3^2} + \dfrac{4\,A_1}{a_3}}}$,

$\mu_2 = \dfrac{1}{2}\sqrt{\dfrac{-3\,a_2}{a_3} + \sqrt{\dfrac{9\,a_2^2}{a_3^2} + \dfrac{4\,A_1}{a_3}}}$,

$A_0 = \sqrt{\dfrac{a_3\,\mu_1^2\,\mu_2^2}{12\,a_1}}$,

$\frac{3 a_2}{a_3} < \sqrt{\frac{9 a_2^2}{a_3^2} + \frac{4 A_1}{a_3}}$,
$a_1 a_3 > 0$,
$n = 2$,
$m = 4$,
x_0, t_0, A_1, and ϕ_0 are arbitrary real constants.

- *Reference*: [2] *with* $m = 1$.

***Solution* 12. cosh(x)**

$$\psi(x, t) = \left(\frac{n+1}{2 a_2^n a_3 n^2}\right)^{\frac{1}{2n}} \times \left\{\frac{A_0^2}{\sqrt{\frac{2 a_2 n^2 (n+1)}{a_3 (n+2)^2} + A_0^2} \cosh\left[\frac{A_0}{\sqrt{2 a_1}} (x - x_0)\right] + \sqrt{\frac{2 a_2 n^2 (n+1)}{a_3 (n+2)^2}}}\right\}^{\frac{1}{n}} \times e^{i\left[\frac{A_0^2}{2 n^2} (t-t_0)+\phi_0\right]}, \tag{3.31}$$

where
$a_1 > 0$,
$a_2 = 1$,
$a_3 (n + 1) > 0$,
$m = 2 n$,
x_0, t_0, A_0, and ϕ_0 are arbitrary real constants.

- *Reference*: [7] *with* $A_2 = B_2 = 0$.

***Solution* 13. sin(x)**

$$\psi(x, t) = \left(\frac{-1}{2}\right)^{\frac{1}{n}} \left[\frac{2 (n+1)}{a_2^n a_3 n^2}\right]^{\frac{1}{2n}} \times \left\{\frac{A_0^2}{\sqrt{\frac{2 a_2 n^2 (n+1)}{a_3 (n+2)^2} - A_0^2} \sin\left[\frac{A_0}{\sqrt{2 a_1}} (x - x_0)\right] + \sqrt{\frac{2 a_2 n^2 (n+1)}{a_3 (n+2)^2}}}\right\}^{\frac{1}{n}} \times e^{i\left[\frac{-A_0^2}{2 n^2} (t-t_0)+\phi_0\right]}, \tag{3.32}$$

where
$a_1 > 0$,
$a_2 = 1$,
$a_3\,(n + 1) > 0$,
$m = 2\,n$,
x_0, t_0, A_0, and ϕ_0 are arbitrary real constants.

- *Reference*: [7] *with* $A_2 = B_2 = 0$.

***Solution* 14. cos(*x*)**

$$\psi(x,\,t) = \left(\frac{-1}{2}\right)^{\frac{1}{n}} \left[\frac{2\,(n+1)}{a_2^n\, a_3\, n^2}\right]^{\frac{1}{2\,n}} \times \left\{\frac{A_0^2}{\sqrt{\frac{2\,a_2\,n^2\,(n+1)}{a_3\,(n+2)^2} - A_0^2}\,\cos\left[\frac{A_0}{\sqrt{2\,a_1}}\,(x - x_0)\right] + \sqrt{\frac{2\,a_2\,n^2\,(n+1)}{a_3\,(n+2)^2}}}\right\}^{\frac{1}{n}} \times e^{i\left[\frac{-A_0^2}{2\,n^2}\,(t-t_0)+\phi_0\right]}, \tag{3.33}$$

where
$a_1 > 0$,
$a_2 = 1$,
$a_3\,(n + 1) > 0$,
$m = 2\,n$,
x_0, t_0, A_0, and ϕ_0 are arbitrary real constants.

- *Reference*: [7] *with* $A_2 = B_2 = 0$.

3.4 Summary of Section 3.3

Equation: $i\,\psi_t + a_1\,\psi_{xx} + a_2\,|\psi|^n\,\psi + a_3\,|\psi|^m\,\psi = 0$

#	Solution	Conditions	Name	Eq. #
1.	$\psi(x,t) = A_0\,e^{i\,[(|A_0|^n\,a_2+|A_0|^m\,a_3)\,(t-t_0)+\phi_0]}$	A_0, t_0, and ϕ_0 are arbitrary real constants	continuous wave, t-dependent phase	(3.20)
2.	$\psi(x,t) = A_0\,e^{i\left[\sqrt{\frac{|A_0|^n\,a_2+|A_0|^m\,a_3}{a_1}}\,(x-x_0)+\phi_0\right]}$	A_0, x_0, and ϕ_0 are arbitrary real constants	continuous wave, x-dependent phase	(3.21)
3.	$\psi(x,t) = A_0\,e^{i\left[A_1\,(x-x_0)+(|A_0|^n\,a_2+|A_0|^m\,a_3-a_1\,A_1^2)\,(t-t_0)+\phi_0\right]}$	A_0, A_1, x_0, t_0, and ϕ_0 are arbitrary real constants	continuous wave, t- and x-dependent phase	(3.22)
4.	$\psi(x,t) = \frac{A_0}{\sqrt{t-t_0}} \times e^{i\left\{\frac{-2\,A_0^m\,a_3\,(t-t_0)^{\frac{2-m}{2}}}{m-2}-\frac{2\,A_0^n\,a_2\,(t-t_0)^{\frac{2-n}{2}}}{n-2}+\frac{[2\,a_1\,A_1+(x-x_0)]^2}{4\,a_1\,(t-t_0)}+\phi_0\right\}}$	$n \neq 2$, $m \neq 2$, A_0, A_1, x_0, t_0, and ϕ_0 are arbitrary real constants	decaying wave	(3.23)
5.	$\psi(x,t) = \left[\frac{-2\,a_1\,a_2\,(n+1)\,(n+2)}{a_1\,a_3\,(n+2)^2+a_2^2\,n^2\,(1+n)\,(x-x_0)^2}\right]^{\frac{1}{n}} e^{i\,\phi_0}$	$a_1\,a_2 < 0$, $a_1\,a_3 > 0$, $m = 2\,n$, x_0 and ϕ_0 are arbitrary real constants	–	(3.24)
6.	$\psi(x,t) = \sqrt{\frac{-a_2^3}{2\,a_3}}\,\frac{x-x_0}{\sqrt{12\,a_1\,a_3+a_2^2\,(x-x_0)^2}}\,e^{-i\left[\frac{a_2^2}{4\,a_3}\,(t-t_0)+\phi_0\right]}$	$n = 2$, $m = 4$, $a_2\,a_3 < 0$, $a_1\,a_3 > 0$, x_0, t_0, and ϕ_0 are arbitrary real constants	–	(3.25)

7. $\psi(x,t)=\left\{\frac{n+1}{(n+2)\left[1+2e^{\sqrt{A_0}(x-x_0)}\right]}\right\}^{\frac{1}{n}} e^{i[A_1(t-t_0)+\phi_0]}$	$A_0=\frac{a_2 n^2(n+1)}{a_1(n+2)^2}$, $A_1=\frac{a_1 A_0}{n^2}$, $m=2n$, $a_3=-a_2$, $a_1 a_2(n+1)>0$, x_0, t_0, and ϕ_0 are arbitrary real constants	–	(3.26)
8. $\psi(x,t)=\left(\frac{A_0 A_1(n+2)}{a_2\left\{A_0+2\cosh\left[n\sqrt{\frac{A_1}{a_1}}(x-x_0)\right]\right\}}\right)^{\frac{1}{n}} e^{i[A_1(t-t_0)+\phi_0]}$	$A_0=\sqrt{\frac{4a_2^2(n+1)}{A_1\delta(n+2)^2}}$, $\delta=a_3+\frac{a_2^2(n+1)}{A_1(n+2)^2}$, $m=2n$, $a_1 A_1>0$, $A_1\delta(n+1)>0$, x_0, t_0, A_1, and ϕ_0 are arbitrary real constants	flat-top soliton	(3.27)
9. $\psi(x,t)=\left(\frac{2a_1A_0^2(n+2)}{a_2n^2}\{1-\tanh[A_0(x-x_0)]\}\right)^{\frac{1}{n}} \times e^{i\left[\frac{4a_1A_0^2}{n^2}(t-t_0)+\phi_0\right]}$	$A_0=\sqrt{\frac{-a_2^2n^2(1+n)}{4a_1a_3(2+n)^2}}$, $m=2n$, $a_1a_2(n+2)>0$, $a_1a_3(n+1)<0$, x_0, t_0, and ϕ_0 are arbitrary real constants	dark soliton	(3.28)
10. $\psi(x,t)=\left(\frac{2a_1A_0^2(n+2)}{a_2n^2}\{1-\coth[A_0(x-x_0)]\}\right)^{\frac{1}{n}} \times e^{i\left[\frac{4a_1A_0^2}{n^2}(t-t_0)+\phi_0\right]}$	$A_0=\sqrt{\frac{-a_2^2n^2(1+n)}{4a_1a_3(2+n)^2}}$, $m=2n$, $a_1a_2(n+2)>0$, $a_1a_3(n+1)<0$, x_0, t_0, and ϕ_0 are arbitrary real constants	–	(3.29)
11. $\psi(x,t)=\pm\frac{\mu_1\mu_2}{\sqrt{\mu_1^2+(\mu_1^2+4\mu_2^2)\sinh^2[A_0(x-x_0)]}} \times e^{i\left[\frac{A_0^2A_1a_1}{a_3\mu_1^2\mu_2^2}(t-t_0)+\phi_0\right]}$	$\mu_1=\sqrt{\frac{3a_2}{a_3}+\sqrt{\frac{9a_2^2}{a_3^2}+\frac{4A_1}{a_3}}}$, $\mu_2=\frac{1}{2}\sqrt{\frac{-3a_2}{a_3}+\sqrt{\frac{9a_2^2}{a_3^2}+\frac{4A_1}{a_3}}}$, $A_0=\sqrt{\frac{a_3\mu_1^2\mu_2^2}{12a_1}}$, $\frac{3a_2}{a_3}<\sqrt{\frac{9a_2^2}{a_3^2}+\frac{4A_1}{a_3}}$, $a_1a_3>0$, $n=2$, $m=4$, x_0, t_0, A_1, and ϕ_0 are arbitrary real constants	–	(3.30)
12. $\psi(x,t)=\left(\frac{n+1}{2a_2^n a_3 n^2}\right)^{\frac{1}{2n}} \times\left\{\frac{A_0^2}{\sqrt{\frac{2a_2n^2(n+1)}{a_3(n+2)^2}+A_0^2}\cosh\left[\frac{A_0}{\sqrt{2a_1}}(x-x_0)\right]+\sqrt{\frac{2a_2n^2(n+1)}{a_3(n+2)^2}}}\right\}^{\frac{1}{n}} \times e^{i\left[\frac{A_0^2}{2n^2}(t-t_0)+\phi_0\right]}$	$a_1>0$, $a_2=1$, $a_3(n+1)>0$, $m=2n$, x_0, t_0, A_0, and ϕ_0 are arbitrary real constants	–	(3.31)

(*Continued*)

(Continued)

Equation: $i\,\psi_t + a_1\,\psi_{xx} + a_2\,|\psi|^n\,\psi + a_3\,|\psi|^m\,\psi = 0$

#	Solution	Conditions	Name	Eq. #
13.	$\psi(x,t) = \left(\frac{-1}{2}\right)^{\frac{1}{n}} \left[\frac{2\,(n+1)}{a_2^n\,a_3\,n^2}\right]^{\frac{1}{2n}} \times \left\{\frac{A_0^2}{\sqrt{\frac{2\,a_2\,n^2\,(n+1)}{a_3\,(n+2)^2} - A_0^2}\,\sin\left[\frac{A_0}{\sqrt{2\,a_1}}\,(x-x_0)\right] + \sqrt{\frac{2\,a_2\,n^2\,(n+1)}{a_3\,(n+2)^2}}}\right\}^{\frac{1}{n}} \times e^{i\left[\frac{-A_0^2}{2\,n^2}\,(t-t_0)+\phi_0\right]}$	$a_1 > 0$, $a_2 = 1$, $a_3\,(n+1) > 0$, $m = 2\,n$, x_0, t_0, A_0, and ϕ_0 are arbitrary real constants	–	(3.32)
14.	$\psi(x,t) = \left(\frac{-1}{2}\right)^{\frac{1}{n}} \left[\frac{2\,(n+1)}{a_2^n\,a_3\,n^2}\right]^{\frac{1}{2n}} \times \left\{\frac{A_0^2}{\sqrt{\frac{2\,a_2\,n^2\,(n+1)}{a_3\,(n+2)^2} - A_0^2}\,\cos\left[\frac{A_0}{\sqrt{2\,a_1}}\,(x-x_0)\right] + \sqrt{\frac{2\,a_2\,n^2\,(n+1)}{a_3\,(n+2)^2}}}\right\}^{\frac{1}{n}} \times e^{i\left[\frac{-A_0^2}{2\,n^2}\,(t-t_0)+\phi_0\right]}$	$a_1 > 0$, $a_2 = 1$, $a_3\,(n+1) > 0$, $m = 2\,n$, x_0, t_0, A_0, and ϕ_0 are arbitrary real constants	–	(3.33)

References

[1] Zaitsev V F and Polyanin A D 2003 *Handbook of Nonlinear Partial Differential Equations* (New York: Chapman and Hall)

[2] He B and Meng Q 2016 Qualitative analysis and explicit exact solitary, kink and anti-kink wave solutions of the generalized nonlinear Schrödinger equation with parabolic law nonlinearity *Commun. Theor. Phys.* **65** 1–10

[3] Zayed E M and Al-Nowehy A G 2017 Exact solutions for the perturbed nonlinear Schrödinger equation with power law nonlinearity and Hamiltonian perturbed terms *Optik* **139** 123–44

[4] Al Khawaja U and Bahlouli H 2019 Integrability conditions and solitonic solutions of the nonlinear Schrödinger equation with generalized dual-power nonlinearities, PT-symmetric potentials, and space-and time-dependent coefficients *Commun. Nonlinear Sci. Numer. Simul.* **69** 248–60

[5] Triki H and Biswas A 2011 Dark solitons for a generalized nonlinear Schrödinger equation with parabolic law and dual-power law nonlinearities *Math. Methods Appl. Sci.* **34** 958–62

[6] Mirzazadeh M, Eslami M, Milovic D and Biswas A 2014 Topological solitons of resonant nonlinear Schödinger'sequation with dual-power law nonlinearity by GG-expansion technique *Optik* **125** 5480–9

[7] Zhang L H and Si J G 2010 New soliton and periodic solutions of (1 + 2)-dimensional nonlinear Schrödinger equation with dual-power law nonlinearity *Commun. Nonlinear Sci. Numer. Simul.* **15** 2747–54

IOP Publishing

Handbook of Exact Solutions to the Nonlinear Schrödinger Equations

Usama Al Khawaja and Laila Al Sakkaf

Chapter 4

Nonlinear Schrödinger Equation with Higher Order Terms

A Glance at Chapter 4

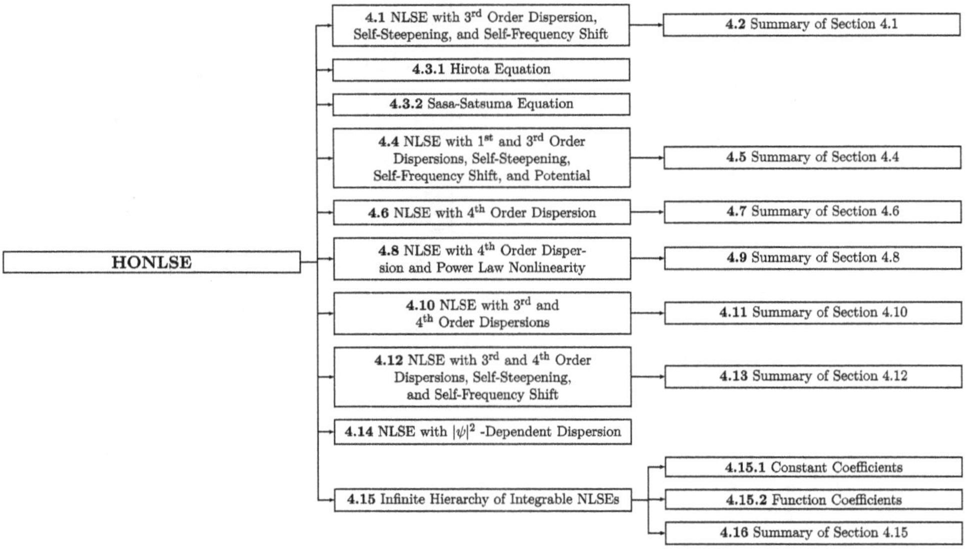

A Statistical View of Chapter 4

	Equation	Solutions
1	$i\,\psi_t + a_1\,\psi_{xx} + a_2\,\|\psi\|^2\,\psi + i\,a_3\,\psi_{xxx} + i\,a_4\,(\|\psi\|^2\,\psi)_x + i\,a_5\,(\|\psi\|^2)_x\,\psi = 0$	13
2	$i\,\psi_t + a_1\,\psi_{xx} + a_2\,\|\psi\|^2\,\psi + i\,a_3\,\psi_{xxx} + i\,a_4\,\|\psi\|^2\,\psi_x = 0$	0
3	$i\,\psi_t + a_1\,\psi_{xx} + a_2\,\|\psi\|^2\,\psi + i\,a_4\,[\psi_{xxx} + (\|\psi\|^2)_x\,\psi + \|\psi\|^2\,\psi_x] = 0$	0
4	$i\,\psi_t + i\,a_1\,\psi_x + a_2\,\psi_{xx} - i\,a_3\,\psi_{xxx} + a_4\,\|\psi\|^2\,\psi - i\,a_5\,\|\psi\|^2\,\psi_x - i\,a_6\,\psi^2\,\psi_x^* - a_7\,\psi = 0$	4
5	$i\,\psi_t + a_1\,\psi_{xx} + a_2\,\|\psi\|^2\,\psi + a_3\,\psi_{xxxx} = 0$	5
6	$i\,\psi_t + a_1\,\psi_{xx} + a_2\,\|\psi\|^{2\,n}\,\psi + a_3\,\psi_{xxxx} = 0$	5
7	$i\,\psi_t + a_1\,\psi_{xx} + a_2\,\|\psi\|^2\,\psi + a_3\,\|\psi\|^4\,\psi + i\,a_4\,\psi_{xxx} + a_5\,\psi_{xxxx} = 0$	9
8	$i\,\psi_t + a_1\,\psi_{xx} + a_2\,\|\psi\|^2\,\psi + a_3\,\|\psi\|^4\,\psi + i\,a_4\,\psi_{xxx} + a_5\,\psi_{xxxx} + i\,a_6\,(\|\psi\|^2\,\psi)_x + i\,a_7\,(\|\psi\|^2)_x\,\psi = 0$	9
9	$i\,\psi_t + (1 - \|\mu\|\,\|\psi\|^2)\,\psi_{xx} + 2\,(1 - \|\mu\|)\,\|\psi\|^2\,\psi = 0$	2
10	$i\,\psi_t + a_2\,k_2 - i\,a_3\,k_3 + a_4\,k_4 - i\,a_5\,k_5 + \ldots\, m = 0$	8
11	$i\,\psi_t + a_2(t)\,k_2 - i\,a_3(t)\,k_3 + a_4(t)\,k_4 - i\,a_5(t)\,k_5 + \ldots\, m = 0$	2
Total	11	57

4.1 NLSE with Third Order Dispersion, Self-Steepening, and Self-Frequency Shift

Equation:

$$i\,\psi_t + a_1\,\psi_{xx} + a_2\,|\psi|^2\,\psi + i\,a_3\,\psi_{xxx} + i\,a_4\,(|\psi|^2\,\psi)_x + i\,a_5\,(|\psi|^2)_x\,\psi = 0, \quad (4.1)$$

where

$\psi = \psi(x, t)$ is the complex function profile,
x and t are its two independent variables,
a_j are arbitrary real constants, $j = 1, 2, 3, 4, 5$.

Solutions:

***Solution* 1. Constant Amplitude** *continuous wave (CW), t- and x-dependent phase*

$$\psi(x, t) = c\,e^{i\,[c\,(x-x_0)+c^2\,(-a_1+a_2+c\,a_3-c\,a_4)\,(t-t_0)+\phi_0]}, \quad (4.2)$$

where x_0, t_0, c, and ϕ_0 are arbitrary real constants.

***Solution* 2. sec(*x*,*t*)**

$$\psi(x, t) = \pm\sqrt{-\mu_1}\,\sec\left\{\sqrt{\mu_2}\,[x - x_0 + c_1\,(t - t_0)]\right\}e^{i\,[c_2\,(x-x_0)+c_3\,(t-t_0)+\phi_0]}, \quad (4.3)$$

where

$$\mu_1 = \frac{6\,(c_1 + 2\,a_1\,c_2 - 3\,a_3\,c_2^2)}{3\,a_4 + 2\,a_5} < 0,$$
$$\mu_2 = \frac{c_1 + 2\,a_1\,c_2 - 3\,a_3\,c_2^2}{a_3} > 0,$$
$$c_2 = \frac{-3\,a_2\,a_3 + a_1\,(3\,a_4 + 2\,a_5)}{6\,a_3\,(a_4 + a_5)},$$
$$c_3 = 8\,a_1\,c_2^2 - 8\,a_3\,c_2^3 - \frac{a_1\,c_1 + 2\,a_1^2\,c_2}{a_3} + 3\,c_1\,c_2,$$

x_0, t_0, c_1, and ϕ_0 are arbitrary real constants.

- *Reference*: [1].

***Solution* 3. csc(*x*,*t*)**

$$\psi(x, t) = \pm\sqrt{-\mu_1}\,\csc\left\{\sqrt{\mu_2}\,[x - x_0 + c_1\,(t - t_0)]\right\}e^{i\,[c_2\,(x-x_0)+c_3\,(t-t_0)+\phi_0]}, \quad (4.4)$$

where

$$\mu_1 = \frac{6\,(c_1 + 2\,a_1\,c_2 - 3\,a_3\,c_2^2)}{(3\,a_4 + 2\,a_5)} < 0,$$
$$\mu_2 = \frac{c_1 + 2\,a_1\,c_2 - 3\,a_3\,c_2^2}{a_3} > 0,$$
$$c_2 = \frac{-3\,a_2\,a_3 + a_1\,(3\,a_4 + 2\,a_5)}{6\,a_3\,(a_4 + a_5)},$$

$c_3 = 8\,a_1\,c_2^2 - 8\,a_3\,c_2^3 - \frac{(a_1\,c_1 + 2\,a_1^2\,c_2)}{a_3} + 3\,c_1\,c_2,$
x_0, t_0, c_1, and ϕ_0 are arbitrary real constants.

- *Reference*: [1].

***Solution* 4. tan(*x*,*t*)**

$$\psi(x,\,t) = \pm\sqrt{\mu_1}\,\tan\left\{\sqrt{-\mu_2}\,[x - x_0 + c_1\,(t - t_0)]\right\}e^{i\,[c_2\,(x-x_0)+c_3\,(t-t_0)+\phi_0]}, \tag{4.5}$$

where

$\mu_1 = \frac{3\,(c_1 + 2\,a_1\,c_2 - 3\,a_3\,c_2^2)}{(3\,a_4 + 2\,a_5)} > 0,$
$\mu_2 = \frac{c_1 + 2\,a_1\,c_2 - 3\,a_3\,c_2^2}{2\,a_3} < 0,$
$c_2 = \frac{-3\,a_2\,a_3 + a_1\,(3\,a_4 + 2\,a_5)}{6\,a_3\,(a_4 + a_5)},$
$c_3 = 8\,a_1\,c_2^2 - 8\,a_3\,c_2^3 - \frac{(a_1\,c_1 + 2\,a_1^2\,c_2)}{a_3} + 3\,c_1\,c_2,$
x_0, t_0, c_1, and ϕ_0 are arbitrary real constants.

- *Reference*: [1].

***Solution* 5. cot(*x*,*t*)**

$$\psi(x,\,t) = \pm\sqrt{\mu_1}\,\cot\left\{\sqrt{-\mu_2}\,[x - x_0 + c_1\,(t - t_0)]\right\}e^{i\,[c_2\,(x-x_0)+c_3\,(t-t_0)+\phi_0]}, \tag{4.6}$$

where

$\mu_1 = \frac{3\,(c_1 + 2\,a_1\,c_2 - 3\,a_3\,c_2^2)}{(3\,a_4 + 2\,a_5)} > 0,$
$\mu_2 = \frac{c_1 + 2\,a_1\,c_2 - 3\,a_3\,c_2^2}{2\,a_3} < 0,$
$c_2 = \frac{-3\,a_2\,a_3 + a_1\,(3\,a_4 + 2\,a_5)}{6\,a_3\,(a_4 + a_5)},$
$c_3 = 8\,a_1\,c_2^2 - 8\,a_3\,c_2^3 - \frac{(a_1\,c_1 + 2\,a_1^2\,c_2)}{a_3} + 3\,c_1\,c_2,$
x_0, t_0, c_1, and ϕ_0 are arbitrary real constants.

- *Reference*: [1].

***Solution* 6. sech(*x*,*t*)** *bright soliton*
(Figure 4.1)

$$\psi(x,\,t) = \pm\sqrt{-\mu_1}\,\text{sech}\left\{\sqrt{-\mu_2}\,[x - x_0 + c_1\,(t - t_0)]\right\}e^{i\,[c_2\,(x-x_0)+c_3\,(t-t_0)+\phi_0]}, \tag{4.7}$$

where

$\mu_1 = \frac{6\,(c_1 + 2\,a_1\,c_2 - 3\,a_3\,c_2^2)}{(3\,a_4 + 2\,a_5)} < 0,$
$\mu_2 = \frac{c_1 + 2\,a_1\,c_2 - 3\,a_3\,c_2^2}{a_3} < 0,$

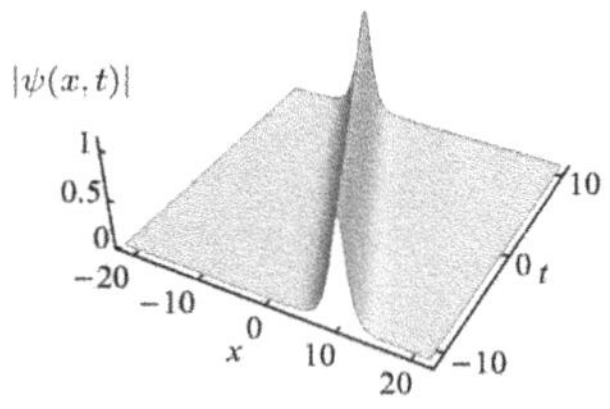

Figure 4.1. Bright soliton (4.7) with $a_1 = 1/2$, $a_2 = 1$, $a_3 = -1$, $a_4 = -3$, $a_5 = 2$, $c_1 = 9/10$, and $x_0 = t_0 = \phi_0 = 0$.

$$c_2 = \frac{-3\,a_2\,a_3 + a_1\,(3\,a_4 + 2\,a_5)}{6\,a_3\,(a_4 + a_5)},$$
$$c_3 = 8\,a_1\,c_2^2 - 8\,a_3\,c_2^3 - \frac{(a_1\,c_1 + 2\,a_1^2\,c_2)}{a_3} + 3\,c_1\,c_2,$$
x_0, t_0, c_1, and ϕ_0 are arbitrary real constants.

- *Reference*: [1].

***Solution* 7. csch(*x*,*t*)**

$$\psi(x,\,t) = \pm\sqrt{\mu_1}\,\mathrm{csch}\left\{\sqrt{-\mu_2}\,[x - x_0 + c_1\,(t - t_0)]\right\}e^{i\,[c_2\,(x-x_0)+c_3\,(t-t_0)+\phi_0]}, \qquad (4.8)$$

where

$$\mu_1 = \frac{6\,(c_1 + 2\,a_1\,c_2 - 3\,a_3\,c_2^2)}{(3\,a_4 + 2\,a_5)} > 0,$$
$$\mu_2 = \frac{c_1 + 2\,a_1\,c_2 - 3\,a_3\,c_2^2}{a_3} < 0,$$
$$c_2 = \frac{-3\,a_2\,a_3 + a_1\,(3\,a_4 + 2\,a_5)}{6\,a_3\,(a_4 + a_5)},$$
$$c_3 = 8\,a_1\,c_2^2 - 8\,a_3\,c_2^3 - \frac{(a_1\,c_1 + 2\,a_1^2\,c_2)}{a_3} + 3\,c_1\,c_2,$$
x_0, t_0, c_1, and ϕ_0 are arbitrary real constants.

- *Reference*: [1].

***Solution* 8. tanh(*x*,*t*)** *dark soliton*
(Figure 4.2)

$$\psi(x,\,t) = \pm\sqrt{-\mu_1}\,\tanh\left\{\sqrt{\mu_2}\,[x - x_0 + c_1\,(t - t_0)]\right\}e^{i\,[c_2\,(x-x_0)+c_3\,(t-t_0)+\phi_0]}, \qquad (4.9)$$

where

$$\mu_1 = \frac{3\,(c_1 + 2\,a_1\,c_2 - 3\,a_3\,c_2^2)}{(3\,a_4 + 2\,a_5)} < 0,$$
$$\mu_2 = \frac{c_1 + 2\,a_1\,c_2 - 3\,a_3\,c_2^2}{2\,a_3} > 0,$$
$$c_2 = \frac{-3\,a_2\,a_3 + a_1\,(3\,a_4 + 2\,a_5)}{6\,a_3\,(a_4 + a_5)},$$

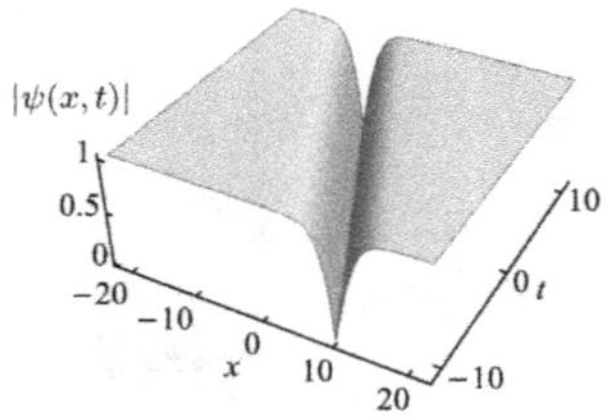

Figure 4.2. Dark soliton (4.9) with $a_1 = -4, a_2 = 1, a_3 = -1, a_4 = -1, a_5 = 2, c_1 = 9/10$, and $x_0 = t_0 = \phi_0 = 0$.

$c_3 = 8\,a_1\,c_2^2 - 8\,a_3\,c_2^3 - \frac{(a_1\,c_1 + 2\,a_1^2\,c_2)}{a_3} + 3\,c_1\,c_2$,
x_0, t_0, c_1, and ϕ_0 are arbitrary real constants.

- *Reference*: [1].

***Solution* 9. coth(*x*,*t*)**

$$\psi(x,\,t) = \pm\sqrt{-\mu_1}\,\coth\left\{\sqrt{\mu_2}\,[x - x_0 + c_1\,(t - t_0)]\right\}e^{i\,[c_2\,(x-x_0)+c_3\,(t-t_0)+\phi_0]}, \tag{4.10}$$

where

$\mu_1 = \frac{3\,(c_1 + 2\,a_1\,c_2 - 3\,a_3\,c_2^2)}{(3\,a_4 + 2\,a_5)} < 0$,

$\mu_2 = \frac{c_1 + 2\,a_1\,c_2 - 3\,a_3\,c_2^2}{2\,a_3} > 0$,

$c_2 = \frac{-3\,a_2\,a_3 + a_1\,(3\,a_4 + 2\,a_5)}{6\,a_3\,(a_4 + a_5)}$,

$c_3 = 8\,a_1\,c_2^2 - 8\,a_3\,c_2^3 - \frac{(a_1\,c_1 + 2\,a_1^2\,c_2)}{a_3} + 3\,c_1\,c_2$,
x_0, t_0, c_1, and ϕ_0 are arbitrary real constants.

- *Reference*: [1].

***Solution* 10. Rational Solution** *decaying wave (DW)*

$$\psi(x,\,t) = \pm\left\{\frac{\sqrt{-6\,a_3}}{\sqrt{3\,a_4 + 2\,a_5}\left[x - x_0 + \left(3\,a_3\,c_2^2 - 2\,a_1\,c_2\right)(t - t_0) + c_1\right]}\right\} \times e^{i\,[c_2\,(x-x_0)+c_3\,(t-t_0)+\phi_0]}, \tag{4.11}$$

where

$c_2 = \frac{a_1(3\,a_4 + 2\,a_5) - 3\,a_2\,a_3}{6\,a_3\,(a_4 + a_5)}$,

$c_3 = a_3\,c_2^3 - a_1\,c_2^2$,

$a_3 < 0$,
x_0, t_0, c_1, and ϕ_0 are arbitrary real constants.

- *Reference*: [1].

***Solution* 11.**

$$\psi(x, t) = \pm \sqrt{\frac{\mu_1}{\mu_2}} \left(\frac{\cot\left\{\sqrt{-2\,\mu_1}\,[x - x_0 - c_1\,(t - t_0)]\right\}}{\csc\left\{\sqrt{-2\,\mu_1}\,[x - x_0 - c_1\,(t - t_0)]\right\} + 1} \right) \times e^{i\,[c_2\,(x - x_0) + c_3\,(t - t_0) + \phi_0]}, \tag{4.12}$$

where

$\mu_1 = \frac{-c_1 + 2\,a_1\,c_2 - 3\,a_3\,c_2^2}{a_3} < 0,$

$\mu_2 = \frac{3\,a_4 + 2\,a_5}{3\,a_3} < 0,$

$c_2 = \frac{-3\,a_2\,a_3 + a_1\,(3\,a_4 + 2\,a_5)}{6\,a_3\,(a_4 + a_5)},$

$c_3 = 8\,a_1\,c_2^2 - 8\,a_3\,c_2^3 + \frac{(a_1\,c_1 - 2\,a_1^2\,c_2)}{a_3} - 3\,c_1\,c_2,$

x_0, t_0, c_1, and ϕ_0 are arbitrary real constants.

- *Reference*: [1].

***Solution* 12.**

$$\psi(x, t) = \pm \left(\frac{\sqrt{-3}\,\sec\left\{\sqrt{2\,\mu_1}\,[x - x_0 - c_1\,(t - t_0)]\right\} + \tan\left\{\sqrt{2\,\mu_1}\,[x - x_0 - c_1\,(t - t_0)]\right\}}{2\,\sec\left\{\sqrt{2\,\mu_1}\,[x - x_0 - c_1\,(t - t_0)]\right\} + 1} \right) \times \sqrt{\frac{\mu_1}{\mu_2}}\, e^{i\,[c_2\,(x - x_0) + c_3\,(t - t_0) + \phi_0]}, \tag{4.13}$$

where

$\mu_1 = \frac{c_1 - 2\,a_1\,c_2 + 3\,a_3\,c_2^2}{a_3} > 0,$

$\mu_2 = \frac{-3\,a_4 - 2\,a_5}{3\,a_3} > 0,$

$c_2 = -\frac{3\,a_2\,a_3 - a_1\,(3\,a_4 + 2\,a_5)}{6\,a_3\,(a_4 + a_5)},$

$c_3 = 8\,a_1\,c_2^2 - 8\,a_3\,c_2^3 + \frac{(a_1\,c_1 - 2\,a_1^2\,c_2)}{a_3} - 3\,c_1\,c_2,$

x_0, t_0, c_1, and ϕ_0 are arbitrary real constants.

- *Reference*: [1].

Solution **13.**

$$\psi(x, t) = \pm \left(\frac{\sqrt{5}\,\mathrm{csch}\left\{\sqrt{2\,\mu_1}\,[x - x_0 - c_1\,(t - t_0)]\right\} + \coth\left\{\sqrt{2\,\mu_1}\,[x - x_0 - c_1\,(t - t_0)]\right\}}{2\,\mathrm{csch}\left\{\sqrt{2\,\mu_1}\,[x - x_0 - c_1\,(t - t_0)]\right\} + 1} \right) \times \sqrt{-\frac{\mu_1}{\mu_2}}\, e^{i\,[c_2\,(x - x_0) + c_3\,(t - t_0) + \phi_0]}, \tag{4.14}$$

where

$\mu_1 = \frac{-c_1 + 2\,a_1\,c_2 - 3\,a_3\,c_2^2}{a_3} > 0,$

$\mu_2 = \frac{3\,a_4 + 2\,a_5}{3\,a_3} < 0,$

$c_2 = \frac{a_1\,(3\,a_4 + 2\,a_5) - 3\,a_2\,a_3}{6\,a_3\,(a_4 + a_5)},$

$c_3 = 8\,a_1\,c_2^2 - 8\,a_3\,c_2^3 + \frac{(a_1\,c_1 - 2\,a_1^2\,c_2)}{a_3} - 3\,c_1\,c_2,$

x_0, t_0, c_1, and ϕ_0 are arbitrary real constants.

- *Reference*: [1].

4.2 Summary of Section 4.1

Equation

$$i\,\psi_t + a_1\,\psi_{xx} + a_2\,|\psi|^2\,\psi + i\,a_3\,\psi_{xxx} + i\,a_4\,(|\psi|^2\,\psi)_x + i\,a_5\,(|\psi|^2)_x\,\psi = 0$$

#	Solution	Conditions	Name	Eq. #
1.	$\psi(x,\,t) = c\,e^{i\,[c\,(x-x_0)+c^2\,(-a_1+c\,a_3+a_2-c\,a_4)\,(t-t_0)+\phi_0]}$	x_0, t_0, c, and ϕ_0 are arbitrary real constants	continuous wave, t- and x- dependent phase	(4.2)
2.	$\psi(x,\,t) = \pm\,\sqrt{-\mu_1}\,\sec\{\sqrt{\mu_2}\,[x - x_0 + c_1\,(t - t_0)]\}$ $\times\,e^{i\,[c_2\,(x-x_0)+c_3\,(t-t_0)+\phi_0]}$	$\mu_1 = \frac{6\,(c_1+2\,a_1\,c_2-3\,a_3\,c_2^2)}{(3\,a_4+2\,a_5)} < 0$, $\mu_2 = \frac{c_1+2\,a_1\,c_2-3\,a_3\,c_2^2}{a_3} > 0$, $c_2 = \frac{-3\,a_2\,a_3+a_1\,(3\,a_4+2\,a_5)}{6\,a_3\,(a_4+a_5)}$, $c_3 = 8\,a_1\,c_2^2 - 8\,a_3\,c_2^3 - \frac{(a_1\,c_1+2\,a_1^2\,c_2)}{a_3} + 3\,c_1\,c_2$, x_0, t_0, c_1, and ϕ_0 are arbitrary real constants	—	(4.3)
3.	$\psi(x,\,t) = \pm\,\sqrt{-\mu_1}\,\csc\{\sqrt{\mu_2}\,[x - x_0 + c_1\,(t - t_0)]\}$ $\times\,e^{i\,[c_2\,(x-x_0)+c_3\,(t-t_0)+\phi_0]}$	$\mu_1 = \frac{6\,(c_1+2\,a_1\,c_2-3\,a_3\,c_2^2)}{(3\,a_4+2\,a_5)} < 0$, $\mu_2 = \frac{c_1+2\,a_1\,c_2-3\,a_3\,c_2^2}{a_3} > 0$, $c_2 = \frac{-3\,a_2\,a_3+a_1\,(3\,a_4+2\,a_5)}{6\,a_3\,(a_4+a_5)}$, $c_3 = 8\,a_1\,c_2^2 - 8\,a_3\,c_2^3 - \frac{(a_1\,c_1+2\,a_1^2\,c_2)}{a_3} + 3\,c_1\,c_2$, x_0, t_0, c_1, and ϕ_0 are arbitrary real constants	—	(4.4)
4.		$\mu_1 = \frac{3\,(c_1+2\,a_1\,c_2-3\,a_3\,c_2^2)}{(3\,a_4+2\,a_5)} > 0$,	—	(4.5)

(Continued)

$\psi(x,t) = \pm\sqrt{\mu_1}\,\tan\{\sqrt{-\mu_2}\,[x - x_0 + c_1(t - t_0)]\}$ $\times e^{i\,[c_2(x - x_0) + c_3(t - t_0) + \phi_0]}$	$\mu_2 = \frac{c_1 + 2a_1c_2 - 3a_3c_2^2}{2a_3} < 0,$ $c_2 = \frac{-3a_2a_3 + a_1(3a_4 + 2a_5)}{6a_3(a_4 + a_5)},$ $c_3 = 8a_1c_2^2 - 8a_3c_2^3 - \frac{(a_1c_1 + 2a_1^2c_2)}{a_3} + 3c_1c_2,$ x_0, t_0, c_1, and ϕ_0 are arbitrary real constants		
5. $\psi(x,t) = \pm\sqrt{\mu_1}\,\cot\{\sqrt{-\mu_2}\,[x - x_0 + c_1(t - t_0)]\}$ $\times e^{i\,[c_2(x - x_0) + c_3(t - t_0) + \phi_0]}$	$\mu_1 = \frac{3(c_1 + 2a_1c_2 - 3a_3c_2^2)}{(3a_4 + 2a_5)} > 0,$ $\mu_2 = \frac{c_1 + 2a_1c_2 - 3a_3c_2^2}{2a_3} < 0,$ $c_2 = \frac{-3a_2a_3 + a_1(3a_4 + 2a_5)}{6a_3(a_4 + a_5)},$ $c_3 = 8a_1c_2^2 - 8a_3c_2^3 - \frac{(a_1c_1 + 2a_1^2c_2)}{a_3} + 3c_1c_2,$ x_0, t_0, c_1, and ϕ_0 are arbitrary real constants	–	(4.6)
6. $\psi(x,t) = \pm\sqrt{-\mu_1}\,\mathrm{sech}\{\sqrt{-\mu_2}\,[x - x_0 + c_1(t - t_0)]\}$ $\times e^{i\,[c_2(x - x_0) + c_3(t - t_0) + \phi_0]}$	$\mu_1 = \frac{6(c_1 + 2a_1c_2 - 3a_3c_2^2)}{(3a_4 + 2a_5)} < 0,$ $\mu_2 = \frac{c_1 + 2a_1c_2 - 3a_3c_2^2}{a_3} < 0,$ $c_2 = \frac{-3a_2a_3 + a_1(3a_4 + 2a_5)}{6a_3(a_4 + a_5)},$ $c_3 = 8a_1c_2^2 - 8a_3c_2^3 - \frac{(a_1c_1 + 2a_1^2c_2)}{a_3} + 3c_1c_2,$ x_0, t_0, c_1, and ϕ_0 are arbitrary real constants	bright soliton	(4.7)
7. $\psi(x,t) = \pm\sqrt{\mu_1}\,\mathrm{csch}\{\sqrt{-\mu_2}\,[x - x_0 + c_1(t - t_0)]\}$ $\times e^{i\,[c_2(x - x_0) + c_3(t - t_0) + \phi_0]}$	$\mu_1 = \frac{6(c_1 + 2a_1c_2 - 3a_3c_2^2)}{(3a_4 + 2a_5)} > 0,$ $\mu_2 = \frac{c_1 + 2a_1c_2 - 3a_3c_2^2}{a_3} < 0,$ $c_2 = \frac{-3a_2a_3 + a_1(3a_4 + 2a_5)}{6a_3(a_4 + a_5)},$	–	(4.8)

	$c_3 = 8\,a_1\,c_2^2 - 8\,a_3\,c_2^3 - \frac{(a_1\,c_1 + 2\,a_1^2\,c_2)}{a_3} + 3\,c_1\,c_2$, x_0, t_0, c_1, and ϕ_0 are arbitrary real constants		
8. $\psi(x,\,t) = \pm\,\sqrt{-\mu_1}\,\tanh\{\sqrt{\mu_2}\,[x - x_0 + c_1\,(t - t_0)]\} \times e^{i\,[c_2\,(x-x_0)+c_3\,(t-t_0)+\phi_0]}$	$\mu_1 = \frac{3\,(c_1 + 2\,a_1\,c_2 - 3\,a_3\,c_2^2)}{(3\,a_4 + 2\,a_5)} < 0$, $\mu_2 = \frac{c_1 + 2\,a_1\,c_2 - 3\,a_3\,c_2^2}{2\,a_3} > 0$, $c_2 = \frac{-3\,a_2\,a_3 + a_1\,(3\,a_4 + 2\,a_5)}{6\,a_3\,(a_4 + a_5)}$, $c_3 = 8\,a_1\,c_2^2 - 8\,a_3\,c_2^3 - \frac{(a_1\,c_1 + 2\,a_1^2\,c_2)}{a_3} + 3\,c_1\,c_2$, x_0, t_0, c_1, and ϕ_0 are arbitrary real constants	dark soliton	(4.9)
9. $\psi(x,\,t) = \pm\,\sqrt{-\mu_1}\,\coth\{\sqrt{\mu_2}\,[x - x_0 + c_1\,(t - t_0)]\} \times e^{i\,[c_2\,(x-x_0)+c_3\,(t-t_0)+\phi_0]}$	$\mu_1 = \frac{3\,(c_1 + 2\,a_1\,c_2 - 3\,a_3\,c_2^2)}{(3\,a_4 + 2\,a_5)} < 0$, $\mu_2 = \frac{c_1 + 2\,a_1\,c_2 - 3\,a_3\,c_2^2}{2\,a_3} > 0$, $c_2 = \frac{-3\,a_2\,a_3 + a_1\,(3\,a_4 + 2\,a_5)}{6\,a_3\,(a_4 + a_5)}$, $c_3 = 8\,a_1\,c_2^2 - 8\,a_3\,c_2^3 - \frac{(a_1\,c_1 + 2\,a_1^2\,c_2)}{a_3} + 3\,c_1\,c_2$, x_0, t_0, c_1, and ϕ_0 are arbitrary real constants	—	(4.10)
10. $\psi(x,\,t) = \pm\left\{\frac{\sqrt{-6\,a_3}}{\sqrt{3\,a_4 + 2\,a_5}\,[x - x_0 + (3\,a_3\,c_2^2 - 2\,a_1\,c_2)\,(t - t_0) + c_1]}\right\} \times e^{i\,[c_2\,(x-x_0)+c_3\,(t-t_0)+\phi_0]}$	$c_2 = \frac{a_1(3\,a_4 + 2\,a_5) - 3\,a_2\,a_3}{6\,a_3\,(a_4 + a_5)}$, $c_3 = a_3\,c_2^3 - a_1\,c_2^2$, $a_3 < 0$, x_0, t_0, c_1, and ϕ_0 are arbitrary real constants	decaying wave	(4.11)
11. $\psi(x,\,t) = \pm\,\sqrt{\frac{\mu_1}{\mu_2}}\left(\frac{\cot\{\sqrt{-2\,\mu_1}\,[x - x_0 - c_1\,(t - t_0)]\}}{\csc\{\sqrt{-2\,\mu_1}\,[x - x_0 - c_1\,(t - t_0)]\} + 1}\right) \times e^{i\,[c_2\,(x-x_0)+c_3\,(t-t_0)+\phi_0]}$	$\mu_1 = \frac{-c_1 + 2\,a_1\,c_2 - 3\,a_3\,c_2^2}{a_3} < 0$, $\mu_2 = \frac{3\,a_4 + 2\,a_5}{3\,a_3} < 0$,	—	(4.12)

(Continued)

		$c_2 = \frac{-3\,a_2\,a_3 + a_1\,(3\,a_4 + 2\,a_5)}{6\,a_3\,(a_4 + a_5)}$, $c_3 = 8\,a_1\,c_2^2 - 8\,a_3\,c_2^3 + \frac{(a_1\,c_1 - 2\,a_1^2\,c_2)}{a_3} - 3\,c_1\,c_2$, x_0, t_0, c_1, and ϕ_0 are arbitrary real constants	
12.	$\psi(x,t) = \pm\left(\frac{\sqrt{-3}\,\sec\{\sqrt{2\,\mu_1}\,[x - x_0 - c_1\,(t - t_0)]\} + \tan\{\sqrt{2\,\mu_1}\,[x - x_0 - c_1\,(t - t_0)]\}}{2\,\sec\{\sqrt{2\,\mu_1}\,[x - x_0 - c_1\,(t - t_0)]\} + 1}\right) \times \sqrt{\frac{\mu_1}{\mu_2}}\,e^{i\,[c_2\,(x - x_0) + c_3\,(t - t_0) + \phi_0]}$	$\mu_1 = \frac{c_1 - 2\,a_1\,c_2 + 3\,a_3\,c_2^2}{a_3} > 0$, $\mu_2 = \frac{-3\,a_4 - 2\,a_5}{3\,a_3} > 0$, $c_2 = -\frac{3\,a_2\,a_3 - a_1\,(3\,a_4 + 2\,a_5)}{6\,a_3\,(a_4 + a_5)}$, $c_3 = 8\,a_1\,c_2^2 - 8\,a_3\,c_2^3 + \frac{(a_1\,c_1 - 2\,a_1^2\,c_2)}{a_3} - 3\,c_1\,c_2$, x_0, t_0, c_1, and ϕ_0 are arbitrary real constants	(4.13)
13.	$\psi(x,t) = \pm\sqrt{-\frac{\mu_1}{\mu_2}}\,e^{i\,[c_2\,(x - x_0) + c_3\,(t - t_0) + \phi_0]} \times \left(\frac{\sqrt{5}\,\mathrm{csch}\{\sqrt{2\,\mu_1}\,[x - x_0 - c_1\,(t - t_0)]\} + \coth\{\sqrt{2\,\mu_1}\,[x - x_0 - c_1\,(t - t_0)]\}}{2\,\mathrm{csch}\{\sqrt{2\,\mu_1}\,[x - x_0 - c_1\,(t - t_0)]\} + 1}\right)$	$\mu_1 = \frac{-c_1 + 2\,a_1\,c_2 - 3\,a_3\,c_2^2}{a_3} > 0$, $\mu_2 = \frac{3\,a_4 + 2\,a_5}{3\,a_3} < 0$, $c_2 = \frac{a_1\,(3\,a_4 + 2\,a_5) - 3\,a_2\,a_3}{6\,a_3\,(a_4 + a_5)}$, $c_3 = 8\,a_1\,c_2^2 - 8\,a_3\,c_2^3 + \frac{(a_1\,c_1 - 2\,a_1^2\,c_2)}{a_3} - 3\,c_1\,c_2$, x_0, t_0, c_1, and ϕ_0 are arbitrary real constants	(4.14)

4.3 Special Cases of Equation (4.1)

4.3.1 Case I: Hirota Equation (HE)

$$i\,\psi_t + a_1\,\psi_{xx} + a_2\,|\psi|^2\,\psi + i\,a_3\,\psi_{xxx} + i\,a_4\,|\psi|^2\,\psi_x = 0. \tag{4.15}$$

Solutions of (4.15) can be obtained from solutions of 4.1 for $a_5 = -a_4$.

4.3.2 Case II: Sasa–Satsuma Equation (SSE)

$$i\,\psi_t + a_1\,\psi_{xx} + a_2\,|\psi|^2\,\psi + i\,a_4\,[\psi_{xxx} + (|\psi|^2)_x\,\psi + |\psi|^2\,\psi_x] = 0. \tag{4.16}$$

Solutions of (4.16) can be obtained from solutions of 4.1 for $a_4 = a_3$ and $a_5 = 0$.

4.4 NLSE with First and Third Order Dispersions, Self-Steepening, Self-Frequency Shift, and Potential

Equation:

$$\begin{aligned} &i\,\psi_t + i\,a_1\,\psi_x + a_2\,\psi_{xx} - i\,a_3\,\psi_{xxx} + a_4\,|\psi|^2\,\psi - i\,a_5\,|\psi|^2\,\psi_x \\ &-i\,a_6\,\psi^2\,\psi_x^* - a_7\,\psi = 0, \end{aligned} \tag{4.17}$$

where

$\psi = \psi(x, t)$ is the complex function profile,
x and t are its two independent variables,
a_j are real constants, $j = 1, 2, \ldots, 7$.

Solutions:

***Solution* 1. Constant Amplitude** *CW, t- and x-dependent phase*

$$\psi(x, t) = A_0\, e^{i\,[A_1\,(x-x_0)+A_2\,(t-t_0)+\phi_0]}, \tag{4.18}$$

where

$A_2 = A_0^2\,[a_4 + A_1\,(a_5 - a_6)] - A_1\,[a_1 + A_1\,(a_2 + a_3\,A_1)] - a_7$,
x_0, t_0, A_0, A_1, ϕ_0, and a_j are arbitrary real constants, $j = 1, 2, \ldots, 7$.

***Solution* 2.**

$$\psi(x, t) = \lambda\,\tanh\{\eta\,[x - x_0 - \chi\,(t - t_0)]\} + i\,\rho\,\operatorname{sech}\{\eta\,[x - x_0 - \chi\,(t - t_0)]\}, \tag{4.19}$$

where

$a_1 = -2\,\alpha_1\,\Omega + 3\,a_3\,\Omega^2$,
$a_2 = \alpha_1 - 3\,a_3\,\Omega$,
$a_3 = \frac{\alpha_1\,\alpha_3}{3\,\alpha_2}$,

$a_4 = \alpha_2 - \alpha_3\,\Omega,$
$a_5 = 2\,\alpha_3 + \alpha_4,$
$a_6 = \alpha_3 + \alpha_4,$
$a_7 = \kappa + \alpha_1\,\Omega^2 - a_3\,\Omega^3,$
$\alpha_4 = -\frac{\alpha_3}{2},$
$\kappa = -\frac{2\,\alpha_1\,\alpha_2^2}{3\,\alpha_3^2},$
$\Omega = \frac{\alpha_2}{\alpha_3},$
$\chi = -(\alpha_1\,\Omega + \alpha_3\,\lambda^2) - a_3\,\eta^2,$
$\eta = \sqrt{\frac{\alpha_3}{3\,a_3}\,(\rho^2 - \lambda^2)},$
$\frac{\alpha_2}{\alpha_1}\,(\rho^2 - \lambda^2) > 0,$
x_0, t_0, α_1, α_2, α_3, ρ, and λ are arbitrary real constants.

- *Reference*: [2].

***Solution* 3.**

$$\psi(x,\,t) = i\,\beta + \lambda\,\tanh\{\eta\,[x - x_0 - \chi\,(t - t_0)]\} + i\,\lambda\,\mathrm{sech}\{\eta\,[x - x_0 - \chi\,(t - t_0)]\}, \tag{4.20}$$

where
$a_1 = -2\,\alpha_1\,\Omega + 3\,a_3\,\Omega^2,$
$a_2 = \alpha_1 - 3\,a_3\,\Omega,$
$a_3 = 0,$
$a_4 = \alpha_2 - \alpha_3\,\Omega,$
$a_5 = 2\,\alpha_3 + \alpha_4,$
$a_6 = \alpha_3 + \alpha_4,$
$a_7 = \kappa + \alpha_1\,\Omega^2 - a_3\,\Omega^3,$
$\alpha_4 = -\alpha_3,$
$\kappa = (\alpha_2 - \alpha_3\,\Omega)\,(\lambda^2 + \beta^2) - \alpha_1\,\Omega^2,$
$\chi = -(2\,\alpha_1\,\Omega + \alpha_3\,\beta^2),$
$\eta = -\frac{\alpha_3\,\beta\,\lambda}{\alpha_1},$
$\lambda = \frac{\sqrt{2\,\alpha_1\,(\alpha_3\,\Omega - \alpha_2)}}{\alpha_3},$
$\alpha_1\,(\alpha_3\,\Omega - \alpha_2) > 0,$
x_0, t_0, α_1, α_2, α_3, β, ρ, and Ω are arbitrary real constants.

- *Reference*: [2].

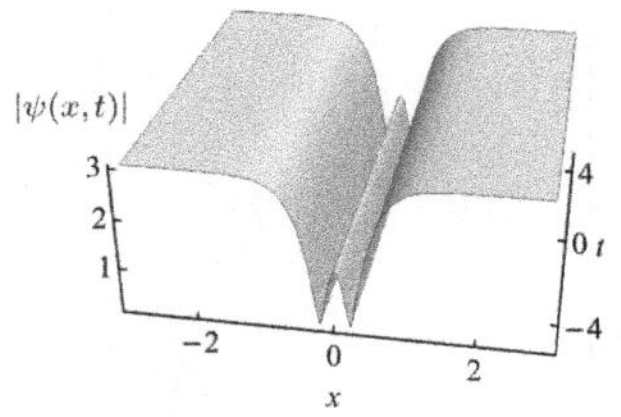

Figure 4.3. Plot of solution (4.21) with $\alpha_2 = 1$, $\alpha_4 = 3$, $\Omega = \lambda = 1/5$, $\beta = -3$, and $x_0 = t_0 = 0$.

***Solution* 4.**

(Figure 4.3)

$$\psi(x,\,t) = i\,\beta + \lambda\,\tanh[\eta\,(x - x_0)] + i\,\rho\,\text{sech}[\eta\,(x - x_0)], \tag{4.21}$$

where

$a_1 = -2\,\alpha_1\,\Omega + 3\,a_3\,\Omega^2$,

$a_2 = \alpha_1 - 3\,a_3\,\Omega$,

$a_3 = 0$,

$a_4 = \alpha_2 - \alpha_3\,\Omega$,

$a_5 = 2\,\alpha_3 + \alpha_4$,

$a_6 = \alpha_3 + \alpha_4$,

$a_7 = \kappa + \alpha_1\,\Omega^2 - a_3\,\Omega^3$,

$\alpha_1 = 0$,

$\alpha_4 = -\frac{3\,\alpha_3}{2}$,

$\kappa = (\alpha_2 - \alpha_3\,\Omega)\,(\lambda^2 + \beta^2)$,

$\eta = -\frac{\beta\,(\alpha_2 - \alpha_3\,\Omega)}{\lambda\,(\alpha_3 + \alpha_4)}$,

$\rho = \sqrt{\lambda^2 + 2\,\beta^2}$,

x_0, t_0, α_2, α_3, β, λ, and Ω are arbitrary real constants.

- *Reference*: [2].

4.5 Summary of Section 4.4

Equation

$$i\,\psi_t + i\,a_1\,\psi_x + a_2\,\psi_{xx} - i\,a_3\,\psi_{xxx} + a_4\,|\psi|^2\,\psi - i\,a_5\,|\psi|^2\,\psi_x - i\,a_6\,\psi^2\,\psi_x^* - a_7\,\psi = 0$$

# Solution	Conditions	Name	Eq. #
1. $\psi(x,t) = A_0\,e^{i\,[A_1\,(x-x_0)+A_2\,(t-t_0)+\phi_0]}$	$A_2 = A_0^2\,[a_4 + A_1\,(a_5 - a_6)] - A_1\,[a_1 + A_1\,(a_2 + a_3\,A_1)] - a_7$, x_0, t_0, A_0, A_1, ϕ_0, and a_j are arbitrary real constants, $j = 1, 2, \ldots, 7$	continuous wave t- and x-dependent phase	(4.18)
2. $\psi(x,t) = \lambda\,\tanh\{\eta\,[x - x_0 - \chi\,(t - t_0)]\}$ $+\,i\,\rho\,\text{sech}\{\eta\,[x - x_0 - \chi\,(t - t_0)]\}$	$a_1 = -2\,\alpha_1\,\Omega + 3\,a_3\,\Omega^2$, $a_2 = \alpha_1 - 3\,a_3\,\Omega$, $a_3 = \frac{\alpha_1\,\alpha_3}{3\,\alpha_2}$, $a_4 = \alpha_2 - \alpha_3\,\Omega$, $a_5 = 2\,\alpha_3 + \alpha_4$, $a_6 = \alpha_3 + \alpha_4$, $\frac{\alpha_2}{\alpha_1}\,(\rho^2 - \lambda^2) > 0$ $a_7 = \kappa + \alpha_1\,\Omega^2 - a_3\,\Omega^3$, $\alpha_4 = -\frac{\alpha_3}{2}$, $\kappa = -\frac{2\,\alpha_1\,\alpha_2^2}{3\,\alpha_3^2}$, $\Omega = \frac{\alpha_2}{\alpha_3}$, $\chi = -(\alpha_1\,\Omega + \alpha_3\,\lambda^2) - a_3\,\eta^2$, $\eta = \sqrt{\frac{\alpha_3}{3\,a_3}\,(\rho^2 - \lambda^2)}$, x_0, t_0, α_1, α_2, α_3, ρ, and λ are arbitrary real constants	–	(4.19)
3. $\psi(x,t) = i\,\beta + \lambda\,\tanh\{\eta\,[x - x_0 - \chi\,(t - t_0)]\}$ $+\,i\,\lambda\,\text{sech}\{\eta\,[x - x_0 - \chi\,(t - t_0)]\}$	$a_1 = -2\,\alpha_1\,\Omega + 3\,a_3\,\Omega^2$, $a_2 = \alpha_1 - 3\,a_3\,\Omega$, $a_3 = 0$, $a_4 = \alpha_2 - \alpha_3\,\Omega$, $a_5 = 2\,\alpha_3 + \alpha_4$, $a_6 = \alpha_3 + \alpha_4$, $\alpha_4 = -\alpha_3$, $a_7 = \kappa + \alpha_1\,\Omega^2 - a_3\,\Omega^3$, $\kappa = (\alpha_2 - \alpha_3\,\Omega)\,(\lambda^2 + \beta^2) - \alpha_1\,\Omega^2$, $\chi = -(2\,\alpha_1\,\Omega + \alpha_3\,\beta^2)$, $\eta = -\frac{\alpha_3\,\beta\,\lambda}{\alpha_1}$, $\lambda = \frac{\sqrt{2\,\alpha_1\,(\alpha_3\,\Omega - \alpha_2)}}{\alpha_3}$, $\alpha_1\,(\alpha_3\,\Omega - \alpha_2) > 0$, x_0, t_0, α_1, α_2, α_3, β, ρ, and Ω are arbitrary real constants	–	(4.20)
4. $\psi(x,t) = i\,\beta + \lambda\,\tanh[\eta\,(x - x_0)]$ $+\,i\,\rho\,\text{sech}[\eta\,(x - x_0)]$	$a_1 = -2\,\alpha_1\,\Omega + 3\,a_3\,\Omega^2$, $a_2 = \alpha_1 - 3\,a_3\,\Omega$, $a_3 = 0$, $a_4 = \alpha_2 - \alpha_3\,\Omega$, $a_5 = 2\,\alpha_3 + \alpha_4$, $a_6 = \alpha_3 + \alpha_4$, $a_7 = \kappa + \alpha_1\,\Omega^2 - a_3\,\Omega^3$, $\alpha_1 = 0$, $\alpha_4 = -\frac{3\,\alpha_3}{2}$, $\kappa = (\alpha_2 - \alpha_3\,\Omega)\,(\lambda^2 + \beta^2)$, $\eta = -\frac{\beta\,(\alpha_2 - \alpha_3\,\Omega)}{\lambda\,(\alpha_3 + \alpha_4)}$, $\rho = \sqrt{\lambda^2 + 2\,\beta^2}$, x_0, t_0, α_2, α_3, β, λ, and Ω are arbitrary real constants	–	(4.21)

4.6 NLSE with Fourth Order Dispersion

Equation:

$$i\,\psi_t + a_1\,\psi_{xx} + a_2\,|\psi|^2\,\psi + a_3\,\psi_{xxxx} = 0, \tag{4.22}$$

where

$\psi = \psi(x, t)$ is the complex function profile,
x and t are its two independent variables,
a_j are arbitrary real constants, $j = 1, 2, 3$.

Solutions:

***Solution* 1. Constant Amplitude** *CW, t- and x-dependent phase*

$$\psi(x, t) = c\,e^{i\left[\sqrt{\frac{a_1}{a_3}}\,(x-x_0)+c^2\,a_2\,(t-t_0)+\phi_0\right]}, \tag{4.23}$$

where

$a_1\,a_3 > 0$,
x_0, t_0, c and ϕ_0 are arbitrary real constants.

***Solution* 2. sec(*x*)**

$$\psi(x, t) = a_1\sqrt{\frac{-3}{10\,a_2\,a_3}}\,\sec^2\left[\sqrt{\frac{a_1}{20\,a_3}}\,(x - x_0)\right]e^{i\left[\frac{-4\,a_1^2}{25\,a_3}\,(t-t_0)+\phi_0\right]}, \tag{4.24}$$

where

$a_1\,a_3 > 0$,
$a_2\,a_3 < 0$,
x_0, t_0, and ϕ_0 are arbitrary real constants.

- *Reference*: [3].

***Solution* 3. csc(*x*)**

$$\psi(x, t) = a_1\sqrt{\frac{-3}{10\,a_2\,a_3}}\,\csc^2\left[\sqrt{\frac{a_1}{20\,a_3}}\,(x - x_0)\right]e^{i\left[\frac{-4\,a_1^2}{25\,a_3}\,(t-t_0)+\phi_0\right]}, \tag{4.25}$$

where

$a_1\,a_3 > 0$,
$a_2\,a_3 < 0$,
x_0, t_0, and ϕ_0 are arbitrary real constants.

- *Reference*: [3].

***Solution* 4.** **sech(x)** *bright soliton*

$$\psi(x, t) = a_1 \sqrt{\frac{-3}{10\, a_2\, a_3}}\ \text{sech}^2\left[\sqrt{\frac{-a_1}{20\, a_3}}\ (x - x_0)\right] e^{i\left[\frac{-4\, a_1^2}{25\, a_3}(t-t_0)+\phi_0\right]}, \tag{4.26}$$

where

$a_1\, a_3 < 0$,
$a_2\, a_3 < 0$,
x_0, t_0, and ϕ_0 are arbitrary real constants.

- *Reference*: [3].

***Solution* 5.** **csch(x)**

$$\psi(x, t) = a_1 \sqrt{\frac{-3}{10\, a_2\, a_3}}\ \text{csch}^2\left[\sqrt{\frac{-a_1}{20\, a_3}}\ (x - x_0)\right] e^{i\left[\frac{-4\, a_1^2}{25\, a_3}(t-t_0)+\phi_0\right]}, \tag{4.27}$$

where

$a_1\, a_3 < 0$,
$a_2\, a_3 < 0$,
x_0, t_0, and ϕ_0 are arbitrary real constants.

- *Reference*: [3].

4.7 Summary of Section 4.6

#	Solution	Conditions	Name	Eq. #
	Equation: $i\,\psi_t + a_1\,\psi_{xx} + a_2\,\lvert\psi\rvert^2\,\psi + a_3\,\psi_{xxxx} = 0$			
1.	$\psi(x,t) = c\,e^{i\left[\sqrt{\frac{a_1}{a_3}}\,(x-x_0)+c^2\,a_2\,(t-t_0)+\phi_0\right]}$	$a_1\,a_3 > 0$, x_0, t_0, c and ϕ_0 are arbitrary real constants	continuous wave, t- and x-dependent phase	(4.23)
2.	$\psi(x,t) = a_1\sqrt{\frac{-3}{10\,a_2\,a_3}}\,\sec^2\left[\sqrt{\frac{a_1}{20\,a_3}}\,(x-x_0)\right]e^{i\left[\frac{-4\,a_1^2}{25\,a_3}\,(t-t_0)+\phi_0\right]}$	$a_1\,a_3 > 0$, $a_2\,a_3 < 0$, x_0, t_0, and ϕ_0 are arbitrary real constants	–	(4.24)
3.	$\psi(x,t) = a_1\sqrt{\frac{-3}{10\,a_2\,a_3}}\,\csc^2\left[\sqrt{\frac{a_1}{20\,a_3}}\,(x-x_0)\right]e^{i\left[\frac{-4\,a_1^2}{25\,a_3}\,(t-t_0)+\phi_0\right]}$	$a_1\,a_3 > 0$, $a_2\,a_3 < 0$, x_0, t_0, and ϕ_0 are arbitrary real constants	–	(4.25)
4.	$\psi(x,t) = a_1\sqrt{\frac{-3}{10\,a_2\,a_3}}\,\mathrm{sech}^2\left[\sqrt{\frac{-a_1}{20\,a_3}}\,(x-x_0)\right]e^{i\left[\frac{-4\,a_1^2}{25\,a_3}\,(t-t_0)+\phi_0\right]}$	$a_1\,a_3 < 0$, $a_2\,a_3 < 0$, x_0, t_0, and ϕ_0 are arbitrary real constants	bright soliton	(4.26)
5.	$\psi(x,t) = a_1\sqrt{\frac{-3}{10\,a_2\,a_3}}\,\mathrm{csch}^2\left[\sqrt{\frac{-a_1}{20\,a_3}}\,(x-x_0)\right]e^{i\left[\frac{-4\,a_1^2}{25\,a_3}\,(t-t_0)+\phi_0\right]}$	$a_1\,a_3 < 0$, $a_2\,a_3 < 0$, x_0, t_0, and ϕ_0 are arbitrary real constants	–	(4.27)

4.8 NLSE with Fourth Order Dispersion and Power Law Nonlinearity

Equation:

$$i\,\psi_t + a_1\,\psi_{xx} + a_2\,|\psi|^{2\,n}\,\psi + a_3\,\psi_{xxxx} = 0, \tag{4.28}$$

where
$\psi = \psi(x, t)$ is the complex function profile,
x and t are its two independent variables,
n and a_j are arbitrary real constants, $j = 1, 2, 3$.

Solutions:

***Solution* 1. Constant Amplitude** *CW, t- and x-dependent phase*

$$\psi(x, t) = c\, e^{i\left[\sqrt{\frac{a_1}{a_3}}\,(x-x_0)+c^{2\,n}\,a_2\,(t-t_0)+\phi_0\right]}, \tag{4.29}$$

where
$a_1\,a_3 > 0$,
x_0, t_0, c, and ϕ_0 are arbitrary real constants.

***Solution* 2. sec(x)**

$$\psi(x, t) = \left\{\sqrt{-\mu_1}\,\sec^2\left[\sqrt{\mu_2}\,(x - x_0)\right]\right\}^{\frac{1}{n}}\,e^{-i\,[\mu_3\,(t-t_0)+\phi_0]}, \tag{4.30}$$

where
$\mu_1 = \frac{a_1^2\,(n+1)\,(n+2)\,(3\,n+2)}{4\,a_3\,a_2\,(n^2+2\,n+2)^2} < 0,$
$\mu_2 = \frac{a_1\,n^2}{4\,a_3\,(n^2+2\,n+2)} > 0,$
$\mu_3 = \frac{a_1^2\,(n+1)^2}{a_3\,(n^2+2\,n+2)^2},$
x_0, t_0, and ϕ_0 are arbitrary real constants.

- *Reference*: [3].

***Solution* 3. csc(x)**

$$\psi(x, t) = \left\{\sqrt{-\mu_1}\,\csc^2\left[\sqrt{\mu_2}\,(x - x_0)\right]\right\}^{\frac{1}{n}}\,e^{-i\,[\mu_3\,(t-t_0)+\phi_0]}, \tag{4.31}$$

where
$\mu_1 = \frac{a_1^2\,(n+1)\,(n+2)\,(3\,n+2)}{4\,a_3\,a_2\,(n^2+2\,n+2)^2} < 0,$
$\mu_2 = \frac{a_1\,n^2}{4\,a_3\,(n^2+2\,n+2)} > 0,$

$\mu_3 = \frac{a_1^2\,(n+1)^2}{a_3\,(n^2+2\,n+2)^2}$,
x_0, t_0, and ϕ_0 are arbitrary real constants.

- *Reference*: [3].

***Solution* 4. sech(x)** *bright soliton*

$$\psi(x,\,t) = \left\{\sqrt{-\mu_1}\;\mathrm{sech}^2\!\left[\sqrt{-\mu_2}\;(x - x_0)\right]\right\}^{\frac{1}{n}}\, e^{-i\,[\mu_3\,(t-t_0)+\phi_0]}, \tag{4.32}$$

where
$\mu_1 = \frac{a_1^2\,(n+1)\,(n+2)\,(3\,n+2)}{4\,a_3\,a_2\,(n^2+2\,n+2)^2} < 0$,
$\mu_2 = \frac{a_1\,n^2}{4\,a_3\,(n^2+2\,n+2)} < 0$,
$\mu_3 = \frac{a_1^2\,(n+1)^2}{a_3\,(n^2+2\,n+2)^2}$,
x_0, t_0, and ϕ_0 are arbitrary real constants.

- *Reference*: [3].

***Solution* 5. csch(x)**

$$\psi(x,\,t) = \left\{\sqrt{-\mu_1}\;\mathrm{csch}^2\!\left[\sqrt{-\mu_2}\;(x - x_0)\right]\right\}^{\frac{1}{n}}\, e^{-i\,[\mu_3\,(t-t_0)+\phi_0]}, \tag{4.33}$$

where
$\mu_1 = \frac{a_1^2\,(n+1)\,(n+2)\,(3\,n+2)}{4\,a_3\,a_2\,(n^2+2\,n+2)^2} < 0$,
$\mu_2 = \frac{a_1\,n^2}{4\,a_3\,(n^2+2\,n+2)} < 0$,
$\mu_3 = \frac{a_1^2\,(n+1)^2}{a_3\,(n^2+2\,n+2)^2}$,
x_0, t_0, and ϕ_0 are arbitrary real constants.

- *Reference*: [3].

4.9 Summary of Section 4.8

Equation

$$i\,\psi_t + a_1\,\psi_{xx} + a_2\,|\psi|^{2n}\,\psi + a_3\,\psi_{xxxx} = 0$$

#	Solution	Conditions	Name	Eq. #
1.	$\psi(x,t) = c\,e^{i\left[\sqrt{\frac{a_1}{a_3}}\,(x-x_0)+c^{2n}\,a_2\,(t-t_0)+\phi_0\right]}$	$a_1\,a_3 > 0$, x_0, t_0, c, and ϕ_0 are arbitrary real constants	continuous wave, t- and x-dependent phase	(4.29)
2.	$\psi(x,t) = \left\{\sqrt{-\mu_1}\,\sec^2\left[\sqrt{\mu_2}\,(x-x_0)\right]\right\}^{\frac{1}{n}}$ $\times\,e^{-i\,[\mu_3\,(t-t_0)+\phi_0]}$	$\mu_1 = \frac{a_1^2\,(n+1)\,(n+2)\,(3n+2)}{4\,a_3\,a_2\,(n^2+2n+2)^2} < 0$, $\mu_2 = \frac{a_1\,n^2}{4\,a_3\,(n^2+2n+2)} > 0$, $\mu_3 = \frac{a_1^2\,(n+1)^2}{a_3\,(n^2+2n+2)^2}$, x_0, t_0, and ϕ_0 are arbitrary real constants	–	(4.30)
3.	$\psi(x,t) = \left\{\sqrt{-\mu_1}\,\csc^2\left[\sqrt{\mu_2}\,(x-x_0)\right]\right\}^{\frac{1}{n}}$ $\times\,e^{-i\,[\mu_3\,(t-t_0)+\phi_0]}$	$\mu_1 = \frac{a_1^2\,(n+1)\,(n+2)\,(3n+2)}{4\,a_3\,a_2\,(n^2+2n+2)^2} < 0$, $\mu_2 = \frac{a_1\,n^2}{4\,a_3\,(n^2+2n+2)} > 0$, $\mu_3 = \frac{a_1^2\,(n+1)^2}{a_3\,(n^2+2n+2)^2}$, x_0, t_0, and ϕ_0 are arbitrary real constants	–	(4.31)
4.	$\psi(x,t) = \left\{\sqrt{-\mu_1}\,\mathrm{sech}^2\left[\sqrt{-\mu_2}\,(x-x_0)\right]\right\}^{\frac{1}{n}}$ $\times\,e^{-i\,[\mu_3\,(t-t_0)+\phi_0]}$	$\mu_1 = \frac{a_1^2\,(n+1)\,(n+2)\,(3n+2)}{4\,a_3\,a_2\,(n^2+2n+2)^2} < 0$, $\mu_2 = \frac{a_1\,n^2}{4\,a_3\,(n^2+2n+2)} < 0$, $\mu_3 = \frac{a_1^2\,(n+1)^2}{a_3\,(n^2+2n+2)^2}$, x_0, t_0, and ϕ_0 are arbitrary real constants	bright soliton	(4.32)

5.	$\psi(x, t) = \left\{\sqrt{-\mu_1}\ \mathrm{csch}^2\left[\sqrt{-\mu_2}\ (x - x_0)\right]\right\}^{\frac{1}{n}} \times e^{-i\,[\mu_3\,(t-t_0)+\phi_0]}$	$\mu_1 = \frac{a_1^2\,(n+1)\,(n+2)\,(3\,n+2)}{4\,a_3\,a_2\,(n^2+2\,n+2)^2} < 0$, $\mu_2 = \frac{a_1\,n^2}{4\,a_3\,(n^2+2\,n+2)} < 0$, $\mu_3 = \frac{a_1^2\,(n+1)^2}{a_3\,(n^2+2\,n+2)^2}$, x_0, t_0, and ϕ_0 are arbitrary real constants	(4.33)

4.10 NLSE with Third and Fourth Order Dispersions and Cubic and Quintic Nonlinearities

Equation:

$$i\,\psi_t + a_1\,\psi_{xx} + a_2\,|\psi|^2\,\psi + a_3\,|\psi|^4\,\psi + i\,a_4\,\psi_{xxx} + a_5\,\psi_{xxxx} = 0, \tag{4.34}$$

where
$\psi = \psi(x, t)$ is the complex function profile,
x and t are its two independent variables,
a_j are arbitrary real constants, $j = 1, 2, 3, 4, 5$.

Solutions:

***Solution* 1. Constant Amplitude** *CW, t- and x-dependent phase*

$$\psi(x, t) = c_1\,e^{i\,\left\{c_2\,(x-x_0)+\left[c_1^2\,\left(a_2+c_1^2\,a_3\right)+c_2^2\,\left(-a_1+c_2\,a_4+c_2^2\,a_5\right)\right]\,(t-t_0)+\phi_0\right\}}, \tag{4.35}$$

where x_0, t_0, c_1, c_2, and ϕ_0 are arbitrary real constants.

***Solution* 2. sec(*x*,*t*)**

$$\psi(x, t) = \pm 2\,c_1\,\sqrt{\frac{-6\,a_5}{a_3}}\,\sec[c_1\,(x - x_0) + c_2\,(t - t_0)]\,e^{i\,[c_3\,(x-x_0)+c_4\,(t-t_0)+\phi_0]}, \tag{4.36}$$

where

$$c_1 = -\sqrt{\frac{3\,a_4^2 + 8\,a_5\left(a_1 + a_2\,\sqrt{\frac{-6\,a_5}{a_3}}\right)}{80\,a_5^2}},$$

$$c_2 = -c_1\left(\frac{a_4^3 + 4\,a_1\,a_4\,a_5}{192\,a_5^3}\right),$$

$$c_3 = \frac{-a_4}{4\,a_5},$$

$$c_4 = a_5\,c_1^4 - \frac{(8\,a_5\,a_1 + 3\,a_4^2)}{8\,a_5}\,c_1^2 - \frac{3\,a_4^4 + 16\,a_1\,a_4^2\,a_5}{256\,a_5^3},$$

$$a_3\,a_5 < 0,$$

$$3\,a_4^2 + 8\,a_5\left(a_1 + a_2\,\sqrt{\frac{-6\,a_5}{a_3}}\right) > 0,$$

x_0, t_0, and ϕ_0 are arbitrary real constants.

- *Reference*: [4].

***Solution* 3. csc(*x*,*t*)**

$$\psi(x, t) = \pm 2\,c_1\,\sqrt{\frac{-6\,a_5}{a_3}}\,\csc[c_1\,(x - x_0) + c_2\,(t - t_0)]\,e^{i\,[c_3\,(x-x_0)+c_4\,(t-t_0)+\phi_0]}, \tag{4.37}$$

where

$$c_1 = -\sqrt{\frac{3\,a_4^2 + 8\,a_5\left(a_1 + a_2\sqrt{\frac{-6\,a_5}{a_3}}\right)}{80\,a_5^2}},$$

$$c_2 = -c_1\left(\frac{a_4^3 + 4\,a_1\,a_4\,a_5}{192\,a_5^3}\right),$$

$$c_3 = \frac{-a_4}{4\,a_5},$$

$$c_4 = a_5\,c_1^4 - \frac{(8\,a_5\,a_1 + 3\,a_4^2)}{8\,a_5}\,c_1^2 - \frac{3\,a_4^4 + 16\,a_1\,a_4^2\,a_5}{256\,a_5^3},$$

$$a_3\,a_5 < 0,$$

$$3\,a_4^2 + 8\,a_5\left(a_1 + a_2\sqrt{\frac{-6\,a_5}{a_3}}\right) > 0,$$

x_0, t_0, and ϕ_0 are arbitrary real constants.

- *Reference*: [4].

***Solution* 4. tan(*x*,*t*)**

$$\psi(x,\,t) = \pm 2\,c_1\sqrt{\frac{-6\,a_5}{a_3}}\,\tan[c_1\,(x - x_0) + c_2\,(t - t_0)]\,e^{i\,[c_3\,(x-x_0)+c_4\,(t-t_0)+\phi_0]}, \quad (4.38)$$

where

$$c_1 = -\sqrt{\frac{-3\,a_4^2 - 8\,a_5\left(a_1 + a_2\sqrt{\frac{-6\,a_5}{a_3}}\right)}{160\,a_5^2}},$$

$$c_2 = -c_1\left(\frac{a_4^3 + 4\,a_1\,a_4\,a_5}{192\,a_5^3}\right),$$

$$c_3 = \frac{-a_4}{4\,a_5},$$

$$c_4 = 16\,a_5\,c_1^4 + \frac{(8\,a_5\,a_1 + 3\,a_4^2)}{4\,a_5}\,c_1^2 - \frac{3\,a_4^4 + 16\,a_1\,a_4^2\,a_5}{256\,a_5^3},$$

$$a_3\,a_5 < 0,$$

$$3\,a_4^2 + 8\,a_5\left(a_1 + a_2\sqrt{\frac{-6\,a_5}{a_3}}\right) < 0,$$

x_0, t_0, and ϕ_0 are arbitrary real constants.

- *Reference*: [4].

***Solution* 5. cot(*x*,*t*)**

$$\psi(x,\,t) = \pm 2\,c_1\sqrt{\frac{-6\,a_5}{a_3}}\,\cot[c_1\,(x - x_0) + c_2\,(t - t_0)]\,e^{i\,[c_3\,(x-x_0)+c_4\,(t-t_0)+\phi_0]}, \quad (4.39)$$

where

$$c_1 = -\sqrt{\frac{-3\,a_4^2 - 8\,a_5\left(a_1 + a_2\sqrt{\frac{-6\,a_5}{a_3}}\right)}{160\,a_5^2}},$$

$$c_2 = -c_1\left(\frac{a_4^3 + 4\,a_1\,a_4\,a_5}{192\,a_5^3}\right),$$

$c_3 = \frac{-a_4}{4\,a_5}$,

$c_4 = 16\,a_5\,c_1^4 + \frac{(8\,a_5\,a_1 + 3\,a_4^2)}{4\,a_5}\,c_1^2 - \frac{3\,a_4^4 + 16\,a_1\,a_4^2\,a_5}{256\,a_5^3}$,

$a_3\,a_5 < 0$,

$3\,a_4^2 + 8\,a_5\left(a_1 + a_2\,\sqrt{\frac{-6\,a_5}{a_3}}\right) < 0$,

x_0, t_0, and ϕ_0 are arbitrary real constants.

- *Reference*: [4].

***Solution* 6. csch(*x*,*t*)**

$$\psi(x,\,t) = \pm 2\,c_1\sqrt{\frac{-6\,a_5}{a_3}}\,\text{csch}[c_1\,(x - x_0) + c_2\,(t - t_0)]\,e^{i\,[c_3\,(x-x_0)+c_4\,(t-t_0)+\phi_0]}, \quad (4.40)$$

where

$c_1 = -\sqrt{\frac{-3\,a_4^2 - 4\,a_5\left(2\,a_1 + a_2\,\sqrt{\frac{-24\,a_5}{a_3}}\right)}{80\,a_5^2}}$,

$c_2 = -c_1\left(\frac{a_4^3 + 4\,a_1\,a_4\,a_5}{192\,a_5^3}\right)$,

$c_3 = \frac{-a_4}{4\,a_5}$,

$c_4 = a_5\,c_1^4 + \frac{(8\,a_5\,a_1 + 3\,a_4^2)}{8\,a_5}\,c_1^2 - \frac{3\,a_4^4 + 16\,a_1\,a_4^2\,a_5}{256\,a_5^3}$,

$a_3\,a_5 < 0$,

$-3\,a_4^2 - 4\,a_5\left(2\,a_1 + a_2\,\sqrt{\frac{-24\,a_5}{a_3}}\right) > 0$,

x_0, t_0, and ϕ_0 are arbitrary real constants.

- *Reference*: [4], *we corrected the constant prefactor.*

***Solution* 7. sech(*x*,*t*)** *bright soliton*

$$\psi(x,\,t) = \pm 2\,c_1\sqrt{\frac{-6\,a_5}{a_3}}\,\text{sech}[c_1\,(x - x_0) + c_2\,(t - t_0)]\,e^{i\,[c_3\,(x-x_0)+c_4\,(t-t_0)+\phi_0]}, \quad (4.41)$$

where

$c_1 = -\sqrt{\frac{-3\,a_4^2 - 8\,a_5\left(a_1 - a_2\,\sqrt{\frac{-6\,a_5}{a_3}}\right)}{80\,a_5^2}}$,

$c_2 = -c_1\left(\frac{a_4^3 + 4\,a_1\,a_4\,a_5}{192\,a_5^3}\right)$,

$c_3 = \frac{-a_4}{4\,a_5}$,

$c_4 = a_5\, c_1^4 + \frac{(8\, a_5\, a_1 + 3\, a_4^2)}{8\, a_5}\, c_1^2 - \frac{3\, a_4^4 + 16\, a_1\, a_4^2\, a_5}{256\, a_5^3}$,
$a_3\, a_5 < 0$,
$-3\, a_4^2 - 8\, a_5 \left(a_1 - a_2 \sqrt{\frac{-6\, a_5}{a_3}}\right) > 0$,
x_0, t_0, and ϕ_0 are arbitrary real constants.

- *Reference*: [4].

***Solution* 8. tanh(*x*,*t*)** *dark soliton*

$$\psi(x,\, t) = \pm 2\, c_1 \sqrt{\frac{-6\, a_5}{a_3}}\, \tanh[c_1\, (x - x_0) + c_2\, (t - t_0)]\, e^{i\, [c_3\, (x - x_0) + c_4\, (t - t_0) + \phi_0]}, \tag{4.42}$$

where

$c_1 = -\sqrt{\frac{3\, a_4^2 + 8\, a_5 \left(a_1 + a_2 \sqrt{\frac{-6\, a_5}{a_3}}\right)}{160\, a_5^2}}$,
$c_2 = -c_1 \left(\frac{a_4^3 + 4\, a_1\, a_4\, a_5}{192\, a_5^3}\right)$,
$c_3 = \frac{-a_4}{4\, a_5}$,
$c_4 = 16\, a_5\, c_1^4 - \frac{(8\, a_5\, a_1 + 3\, a_4^2)}{4\, a_5}\, c_1^2 - \frac{3\, a_4^4 + 16\, a_1\, a_4^2\, a_5}{256\, a_5^3}$,
$a_3\, a_5 < 0$,
$3\, a_4^2 + 8\, a_5 \left(a_1 + a_2 \sqrt{\frac{-6\, a_5}{a_3}}\right) > 0$,
x_0, t_0, and ϕ_0 are arbitrary real constants.

- *Reference*: [4].

***Solution* 9. coth(*x*,*t*)**

$$\psi(x,\, t) = \pm 2\, c_1 \sqrt{\frac{-6\, a_5}{a_3}}\, \coth[c_1\, (x - x_0) + c_2\, (t - t_0)]\, e^{i\, [c_3\, (x - x_0) + c_4\, (t - t_0) + \phi_0]}, \tag{4.43}$$

where

$c_1 = -\sqrt{\frac{3\, a_4^2 + 8\, a_5 \left(a_1 + a_2 \sqrt{\frac{-6\, a_5}{a_3}}\right)}{160\, a_5^2}}$,
$c_2 = -c_1 \left(\frac{a_4^3 + 4\, a_1\, a_4\, a_5}{192\, a_5^3}\right)$,

$c_3 = \frac{-a_4}{4\,a_5}$,

$c_4 = 16\,a_5\,c_1^4 - \frac{(8\,a_5\,a_1 + 3\,a_4^2)}{4\,a_5}\,c_1^2 - \frac{3\,a_4^4 + 16\,a_1\,a_4^2\,a_5}{256\,a_5^3}$,

$a_3\,a_5 < 0$,

$3\,a_4^2 + 8\,a_5\left(a_1 + a_2\sqrt{\frac{-6\,a_5}{a_3}}\right) > 0$,

x_0, t_0, and ϕ_0 are arbitrary real constants.

- *Reference*: [4].

4.11 Summary of Section 4.10

Equation: $i\,\psi_t + a_1\,\psi_{xx} + a_2\,|\psi|^2\,\psi + a_3\,|\psi|^4\,\psi + i\,a_4\,\psi_{xxx} + a_5\,\psi_{xxxx} = 0$

# Solution	Conditions	Name	Eq. #
1. $\psi(x,t) = c_1\,e^{i\,\{c_2\,(x-x_0)+[c_1^2\,(a_2+c_1^2\,a_3)+c_2^2\,(-a_1+c_2\,a_4+c_2^2\,a_5)]\,(t-t_0)+\phi_0\}}$	x_0, t_0, c_1, c_2, and ϕ_0 are arbitrary real constants	continuous wave, t- and x-dependent phase	(4.35)
2. $\psi(x,t) = \pm 2\,c_1\sqrt{\frac{-6\,a_5}{a_3}}\,\sec[c_1\,(x-x_0)+c_2\,(t-t_0)]$ $\times\,e^{i\,[c_3\,(x-x_0)+c_4\,(t-t_0)+\phi_0]}$	$c_1 = -\sqrt{\frac{3\,a_4^2+8\,a_5\left(a_1+a_2\sqrt{\frac{-6\,a_5}{a_3}}\right)}{80\,a_5^2}}$, $c_2 = -c_1\left(\frac{a_4^3+4\,a_1\,a_4\,a_5}{192\,a_5^3}\right)$, $c_3 = \frac{-a_4}{4\,a_5}$, $c_4 = a_5\,c_1^4 - \frac{(8\,a_5\,a_1+3\,a_4^2)}{8\,a_5}\,c_1^2 - \frac{3\,a_4^4+16\,a_1\,a_4^2\,a_5}{256\,a_5^3}$, $a_3\,a_5<0$, $3\,a_4^2+8\,a_5\,(a_1+a_2\sqrt{\frac{-6\,a_5}{a_3}})>0$, x_0, t_0, and ϕ_0 are arbitrary real constants	—	(4.36)
3. $\psi(x,t) = \pm 2\,c_1\sqrt{\frac{-6\,a_5}{a_3}}\,\csc[c_1\,(x-x_0)+c_2\,(t-t_0)]$ $\times\,e^{i\,[c_3\,(x-x_0)+c_4\,(t-t_0)+\phi_0]}$	$c_1 = -\sqrt{\frac{3\,a_4^2+8\,a_5\left(a_1+a_2\sqrt{\frac{-6\,a_5}{a_3}}\right)}{80\,a_5^2}}$, $c_2 = -c_1\left(\frac{a_4^3+4\,a_1\,a_4\,a_5}{192\,a_5^3}\right)$, $c_3 = \frac{-a_4}{4\,a_5}$, $c_4 = a_5\,c_1^4 - \frac{(8\,a_5\,a_1+3\,a_4^2)}{8\,a_5}\,c_1^2 - \frac{3\,a_4^4+16\,a_1\,a_4^2\,a_5}{256\,a_5^3}$, $a_3\,a_5<0$, $3\,a_4^2+8\,a_5\,(a_1+a_2\sqrt{\frac{-6\,a_5}{a_3}})>0$,	—	(4.37)

(Continued)

x_0, t_0, and ϕ_0 are arbitrary real constants

4. $\psi(x,t) = \pm 2\, c_1 \sqrt{\frac{-6\,a_5}{a_3}}\, \tan[c_1\,(x - x_0) + c_2\,(t - t_0)] \times e^{i\,[c_3\,(x-x_0)+c_4\,(t-t_0)+\phi_0]}$ — (4.38)

$$c_1 = -\sqrt{\frac{-3\,a_4^2 - 8\,a_5\left(a_1 + a_2\sqrt{\frac{-6\,a_5}{a_3}}\right)}{160\,a_5^2}},$$

$$c_2 = -c_1\left(\frac{a_4^3 + 4\,a_1\,a_4\,a_5}{192\,a_5^3}\right),\ c_3 = \frac{-a_4}{4\,a_5},$$

$$c_4 = 16\,a_5\,c_1^4 + \frac{(8\,a_5\,a_1 + 3\,a_4^2)}{4\,a_5}\,c_1^2 - \frac{3\,a_4^4 + 16\,a_1\,a_4^2\,a_5}{256\,a_5^3},$$

$$a_3\,a_5 < 0,\ 3\,a_4^2 + 8\,a_5\,(a_1 + a_2\sqrt{\frac{-6\,a_5}{a_3}}) < 0,$$

x_0, t_0, and ϕ_0 are arbitrary real constants

---

5. $\psi(x,t) = \pm 2\, c_1 \sqrt{\frac{-6\,a_5}{a_3}}\, \cot[c_1\,(x - x_0) + c_2\,(t - t_0)] \times e^{i\,[c_3\,(x-x_0)+c_4\,(t-t_0)+\phi_0]}$ — (4.39)

$$c_1 = -\sqrt{\frac{-3\,a_4^2 - 8\,a_5\left(a_1 + a_2\sqrt{\frac{-6\,a_5}{a_3}}\right)}{160\,a_5^2}},$$

$$c_2 = -c_1\,(\frac{a_4^3 + 4\,a_1\,a_4\,a_5}{192\,a_5^3}),\ c_3 = \frac{-a_4}{4\,a_5},$$

$$c_4 = 16\,a_5\,c_1^4 + \frac{(8\,a_5\,a_1 + 3\,a_4^2)}{4\,a_5}\,c_1^2 - \frac{3\,a_4^4 + 16\,a_1\,a_4^2\,a_5}{256\,a_5^3},$$

$$a_3\,a_5 < 0,\ 3\,a_4^2 + 8\,a_5\,(a_1 + a_2\sqrt{\frac{-6\,a_5}{a_3}}) < 0,$$

x_0, t_0, and ϕ_0 are arbitrary real constants

---

6. $\psi(x,t) = \pm 2\, c_1 \sqrt{\frac{-6\,a_5}{a_3}}\, \mathrm{csch}[c_1\,(x - x_0) + c_2\,(t - t_0)] \times e^{i\,[c_3\,(x-x_0)+c_4\,(t-t_0)+\phi_0]}$ — (4.40)

$$c_1 = -\sqrt{\frac{-3\,a_4^2 - 4\,a_5\left(2\,a_1 + a_2\sqrt{\frac{-24\,a_5}{a_3}}\right)}{80\,a_5^2}},$$

$$c_2 = -c_1\left(\frac{a_4^3 + 4\,a_1\,a_4\,a_5}{192\,a_5^3}\right),\ c_3 = \frac{-a_4}{4\,a_5},$$

$$c_4 = a_5\,c_1^4 + \frac{(8\,a_5\,a_1 + 3\,a_4^2)}{8\,a_5}\,c_1^2 - \frac{3\,a_4^4 + 16\,a_1\,a_4^2\,a_5}{256\,a_5^3},$$

$$a_3\,a_5 < 0,\ -3\,a_4^2 - 4\,a_5\,(2\,a_1 + a_2\sqrt{\frac{-24\,a_5}{a_3}}) > 0,$$

x_0, t_0, and ϕ_0 are arbitrary real constants

---

7. $\psi(x, t) = \pm 2\, c_1 \sqrt{\frac{-6\, a_5}{a_3}}\, \text{sech}[c_1\,(x - x_0) + c_2\,(t - t_0)]$ $\times\, e^{i\,[c_3\,(x-x_0)+c_4\,(t-t_0)+\phi_0]}$	$c_1 = -\sqrt{\frac{-3\, a_4^2 - 8\, a_5 \left(a_1 - a_2 \sqrt{\frac{-6\, a_5}{a_3}}\right)}{80\, a_5^2}}$, $c_2 = -c_1 \left(\frac{a_4^3 + 4\, a_1\, a_4\, a_5}{192\, a_5^3}\right)$, $c_3 = \frac{-a_4}{4\, a_5}$, $c_4 = a_5\, c_1^4 + \frac{(8\, a_5\, a_1 + 3\, a_4^2)}{8\, a_5}\, c_1^2 - \frac{3\, a_4^4 + 16\, a_1\, a_4^2\, a_5}{256\, a_5^3}$, $a_3\, a_5 < 0, -3\, a_4^2 - 8\, a_5\, (a_1 - a_2 \sqrt{\frac{-6\, a_5}{a_3}}) > 0$, x_0, t_0, and ϕ_0 are arbitrary real constants	bright soliton	(4.41)
8. $\psi(x, t) = \pm 2\, c_1 \sqrt{\frac{-6\, a_5}{a_3}}\, \tanh[c_1\,(x - x_0) + c_2\,(t - t_0)]$ $\times\, e^{i\,[c_3\,(x-x_0)+c_4\,(t-t_0)+\phi_0]}$	$c_1 = -\sqrt{\frac{3\, a_4^2 + 8\, a_5 \left(a_1 + a_2 \sqrt{\frac{-6\, a_5}{a_3}}\right)}{160\, a_5^2}}$, $c_2 = -c_1 \left(\frac{a_4^3 + 4\, a_1\, a_4\, a_5}{192\, a_5^3}\right)$, $c_3 = \frac{-a_4}{4\, a_5}$, $c_4 = 16\, a_5\, c_1^4 - \frac{(8\, a_5\, a_1 + 3\, a_4^2)}{4\, a_5}\, c_1^2 - \frac{3\, a_4^4 + 16\, a_1\, a_4^2\, a_5}{256\, a_5^3}$, $a_3\, a_5 < 0$, $3\, a_4^2 + 8\, a_5 \left(a_1 + a_2 \sqrt{\frac{-6\, a_5}{a_3}}\right) > 0$, x_0, t_0, and ϕ_0 are arbitrary real constants	dark soliton	(4.42)
9. $\psi(x, t) = \pm 2\, c_1 \sqrt{\frac{-6\, a_5}{a_3}}\, \coth[c_1\,(x - x_0) + c_2\,(t - t_0)]$ $\times\, e^{i\,[c_3\,(x-x_0)+c_4\,(t-t_0)+\phi_0]}$	$c_1 = -\sqrt{\frac{3\, a_4^2 + 8\, a_5 \left(a_1 + a_2 \sqrt{\frac{-6\, a_5}{a_3}}\right)}{160\, a_5^2}}$, $c_2 = -c_1 \left(\frac{a_4^3 + 4\, a_1\, a_4\, a_5}{192\, a_5^3}\right)$, $c_3 = \frac{-a_4}{4\, a_5}$, $c_4 = 16\, a_5\, c_1^4 - \frac{(8\, a_5\, a_1 + 3\, a_4^2)}{4\, a_5}\, c_1^2 - \frac{3\, a_4^4 + 16\, a_1\, a_4^2\, a_5}{256\, a_5^3}$, $a_3\, a_5 < 0, 3\, a_4^2 + 8\, a_5\, (a_1 + a_2 \sqrt{\frac{-6\, a_5}{a_3}}) > 0$, x_0, t_0, and ϕ_0 are arbitrary real constants	—	(4.43)

4.12 NLSE with Third and Fourth Order Dispersions, Self-Steepening, Self-Frequency Shift, and Cubic and Quintic Nonlinearities

Equation:

$$i\,\psi_t + a_1\,\psi_{xx} + a_2\,|\psi|^2\,\psi + a_3\,|\psi|^4\,\psi + i\,a_4\,\psi_{xxx} + a_5\,\psi_{xxxx} \\ +i\,a_6\,(|\psi|^2\,\psi)_x + i\,a_7\,(|\psi|^2)_x\,\psi = 0, \tag{4.44}$$

where
$\psi = \psi(x, t)$ is the complex function profile,
x and t are its two independent variables,
a_j are arbitrary real constants, $j = 1, 2, 3, 4, 5, 6, 7$.

Solutions:

***Solution* 1. Constant Amplitude** *CW, t- and x-dependent phase*

$$\psi(x, t) = c_1\,e^{i\left[c_2\,(x-x_0)+\left(c_1^2\,a_2+c_1^4\,a_3\right)\,(t-t_0)+\phi_0\right]}, \tag{4.45}$$

where
$a_4 = \frac{a_6\,c_1^2 + a_1\,c_2 - a_5\,c_2^3}{c_2^2}$,
x_0, t_0, c_1, c_2, and ϕ_0 are arbitrary real constants.

***Solution* 2. sec(*x*,*t*)**

$$\psi(x, t) = \sqrt{\frac{-6\,(4\,a_5\,c_3 + a_4)}{3\,a_6 + 2\,a_7}}\,\sec[x - x_0 + c_1\,(t - t_0)]\,e^{i\,[c_3\,(x-x_0)+c_2\,(t-t_0)+\phi_0]}, \tag{4.46}$$

where
$c_1 = -2\,a_1\,c_3 + a_4\,(3\,c_3^2 + 1) + 4\,a_5\,(c_3^3 + c_3)$,
$c_2 = -a_1\,(c_3^2 + 1) + a_4\,(c_3^3 + 3\,c_3) + a_5\,(c_3^4 + 6\,c_3^2 + 1)$,
$a_1 = \frac{(4\,a_5\,c_3 + a_4)\,(3\,a_2 + 2\,c_3\,a_7)}{3\,a_6 + 2\,a_7} + 2\,a_4\,c_3 + 2\,a_5\,(c_3^2 + 5)$,
$a_3 = \frac{-2\,a_5\,(3\,a_6 + 2\,a_7)^2}{3\,(a_4 + 4\,a_5\,c_3)^2}$,
$(4\,a_5\,c_3 + a_4)\,(3\,a_6 + 2\,a_7) < 0$,
x_0, t_0, c_3, and ϕ_0 are arbitrary real constants.

- *Reference*: [5].

***Solution* 3. csc(*x*,*t*)**

$$\psi(x, t) = \sqrt{\frac{-6\,(4\,a_5\,c_3 + a_4)}{3\,a_6 + 2\,a_7}}\,\csc[x - x_0 + c_1\,(t - t_0)]\,e^{i\,[c_3\,(x-x_0)+c_2\,(t-t_0)+\phi_0]}, \tag{4.47}$$

where

$c_1 = -2\, a_1\, c_3 + a_4\, (3\, c_3^2 + 1) + 4\, a_5\, (c_3^3 + c_3),$
$c_2 = -a_1\, (c_3^2 + 1) + a_4\, (c_3^3 + 3\, c_3) + a_5\, (c_3^4 + 6\, c_3^2 + 1),$
$a_1 = \frac{(4\, a_5\, c_3 + a_4)\, (3\, a_2 + 2\, c_3\, a_7)}{3\, a_6 + 2\, a_7} + 2\, a_4\, c_3 + 2\, a_5\, (c_3^2 + 5),$
$a_3 = \frac{-2\, a_5\, (3\, a_6 + 2\, a_7)^2}{3\, (a_4 + 4\, a_5\, c_3)^2},$
$(4\, a_5\, c_3 + a_4)\, (3\, a_6 + 2\, a_7) < 0,$
x_0, t_0, c_3, and ϕ_0 are arbitrary real constants.

- *Reference*: [5].

***Solution* 4. tan(*x*,*t*)**

$$\psi(x,\, t) = \sqrt{\frac{-6\, (4\, a_5\, c_3 + a_4)}{3\, a_6 + 2\, a_7}}\, \tan[x - x_0 + c_1\, (t - t_0)]\, e^{i\, [c_3\, (x - x_0) + c_2\, (t - t_0) + \phi_0]}, \quad (4.48)$$

where

$c_1 = -2\, a_1\, c_3 + a_4\, (3\, c_3^2 - 2) + 4\, a_5\, (c_3^3 - 2\, c_3),$
$c_2 = a_1\, (2 - c_3^2) + a_4\, (c_3^3 - 6\, c_3) + a_5\, (c_3^4 - 12\, c_3^2 + 16),$
$a_1 = \frac{(4\, a_5\, c_3 + a_4)\, (3\, a_2 + 2\, c_3\, a_7)}{3\, a_6 + 2\, a_7} + 2\, a_4\, c_3 + 2\, a_5\, (c_3^2 - 10),$
$a_3 = \frac{-2\, a_5\, (3\, a_6 + 2\, a_7)^2}{3\, (a_4 + 4\, a_5\, c_3)^2},$
$(4\, a_5\, c_3 + a_4)\, (3\, a_6 + 2\, a_7) < 0,$
x_0, t_0, c_3, and ϕ_0 are arbitrary real constants.

- *Reference*: [5].

***Solution* 5. cot(*x*,*t*)**

$$\psi(x,\, t) = \sqrt{\frac{-6\, (4\, a_5\, c_3 + a_4)}{3\, a_6 + 2\, a_7}}\, \cot[x - x_0 + c_1\, (t - t_0)]\, e^{i\, [c_3\, (x - x_0) + c_2\, (t - t_0) + \phi_0]}, \quad (4.49)$$

where

$c_1 = -2\, a_1\, c_3 + a_4\, (3\, c_3^2 - 2) + 4\, a_5\, (c_3^3 - 2\, c_3),$
$c_2 = a_1\, (2 - c_3^2) + a_4\, (c_3^3 - 6\, c_3) + a_5\, (c_3^4 - 12\, c_3^2 + 16),$
$a_1 = \frac{(4\, a_5\, c_3 + a_4)\, (3\, a_2 + 2\, c_3\, a_7)}{3\, a_6 + 2\, a_7} + 2\, a_4\, c_3 + 2\, a_5\, (c_3^2 - 10),$
$a_3 = \frac{-2\, a_5\, (3\, a_6 + 2\, a_7)^2}{3\, (a_4 + 4\, a_5\, c_3)^2},$
$(4\, a_5\, c_3 + a_4)\, (3\, a_6 + 2\, a_7) < 0,$
x_0, t_0, c_3, and ϕ_0 are arbitrary real constants.

- *Reference*: [5].

***Solution* 6. sech(x,t)** *bright soliton*

$$\psi(x, t) = \sqrt{\frac{6\,(4\,a_5\,c_3 + a_4)}{3\,a_6 + 2\,a_7}}\ \text{sech}[x - x_0 + c_1\,(t - t_0)]\, e^{i\,[c_3\,(x - x_0) + c_2\,(t - t_0) + \phi_0]}, \quad (4.50)$$

where

$c_1 = -2\,a_1\,c_3 + a_4\,(3\,c_3^2 - 1) + 4\,a_5\,(c_3^3 - c_3)$,
$c_2 = a_1\,(1 - c_3^2) + a_4\,(c_3^3 - 3\,c_3) + a_5\,(c_3^4 - 6\,c_3^2 + 1)$,
$a_1 = \frac{(4\,a_5\,c_3 + a_4)\,(3\,a_2 + 2\,c_3\,a_7)}{3\,a_6 + 2\,a_7} + 2\,a_4\,c_3 + 2\,a_5\,(c_3^2 - 5)$,
$a_3 = \frac{-2\,a_5\,(3\,a_6 + 2\,a_7)^2}{3\,(a_4 + 4\,a_5\,c_3)^2}$,
$(4\,a_5\,c_3 + a_4)\,(3\,a_6 + 2\,a_7) > 0$,
x_0, t_0, c_3, and ϕ_0 are arbitrary real constants.

- *Reference*: [5].

***Solution* 7. csch(x,t)**

$$\psi(x, t) = \sqrt{\frac{-6\,(4\,a_5\,c_3 + a_4)}{3\,a_6 + 2\,a_7}}\ \text{csch}[x - x_0 + c_1\,(t - t_0)]\, e^{i\,[c_3\,(x - x_0) + c_2\,(t - t_0) + \phi_0]}, \quad (4.51)$$

where

$c_1 = -2\,a_1\,c_3 + a_4\,(3\,c_3^2 - 1) + 4\,a_5\,(c_3^3 - c_3)$,
$c_2 = a_1\,(1 - c_3^2) + a_4\,(c_3^3 - 3\,c_3) + a_5\,(c_3^4 - 6\,c_3^2 + 1)$,
$a_1 = \frac{(4\,a_5\,c_3 + a_4)\,(3\,a_2 + 2\,c_3\,a_7)}{3\,a_6 + 2\,a_7} + 2\,a_4\,c_3 + 2\,a_5\,(c_3^2 - 5)$,
$a_3 = \frac{-2\,a_5\,(3\,a_6 + 2\,a_7)^2}{3\,(a_4 + 4\,a_5\,c_3)^2}$,
$(4\,a_5\,c_3 + a_4)\,(3\,a_6 + 2\,a_7) < 0$,
x_0, t_0, c_3, and ϕ_0 are arbitrary real constants.

- *Reference*: [5].

***Solution* 8. tanh(x,t)** *dark soliton*

$$\psi(x, t) = \sqrt{\frac{-6\,(4\,a_5\,c_3 + a_4)}{3\,a_6 + 2\,a_7}}\ \tanh[x - x_0 + c_1\,(t - t_0)]\, e^{i\,[c_3\,(x - x_0) + c_2\,(t - t_0) + \phi_0]}, \quad (4.52)$$

where

$c_1 = -2\,a_1\,c_3 + a_4\,(3\,c_3^2 + 2) + 4\,a_5\,(c_3^3 + 2\,c_3)$,
$c_2 = -a_1\,(c_3^2 + 2) + a_4\,(6\,c_3 + c_3^3) + a_5\,(c_3^4 + 12\,c_3^2 + 16)$,
$a_1 = \frac{(4\,a_5\,c_3 + a_4)\,(3\,a_2 + 2\,c_3\,a_7)}{3\,a_6 + 2\,a_7} + 2\,a_4\,c_3 + 2\,a_5\,(c_3^2 + 10)$,
$a_3 = \frac{-2\,a_5\,(3\,a_6 + 2\,a_7)^2}{3\,(a_4 + 4\,a_5\,c_3)^2}$,

$(4\,a_5\,c_3 + a_4)\,(3\,a_6 + 2\,a_7) < 0$,
x_0, t_0, c_3, and ϕ_0 are arbitrary real constants.

- *Reference*: [5].

***Solution* 9. coth(*x*,*t*)**

$$\psi(x,\,t) = \sqrt{\frac{-6\,(4\,a_5\,c_3 + a_4)}{3\,a_6 + 2\,a_7}}\,\coth[x - x_0 + c_1\,(t - t_0)]\,e^{i\,[c_3\,(x-x_0)+c_2\,(t-t_0)+\phi_0]}, \quad (4.53)$$

where

$c_1 = -2\,a_1\,c_3 + a_4\,(3\,c_3^2 + 2) + 4\,a_5\,(c_3^3 + 2\,c_3)$,
$c_2 = -a_1\,(c_3^2 + 2) + a_4\,(6\,c_3 + c_3^3) + a_5\,(c_3^4 + 12\,c_3^2 + 16)$,
$a_1 = \frac{(4\,a_5\,c_3 + a_4)\,(3\,a_2 + 2\,c_3\,a_7)}{3\,a_6 + 2\,a_7} + 2\,a_4\,c_3 + 2\,a_5\,(c_3^2 + 10)$,
$a_3 = \frac{-2\,a_5\,(3\,a_6 + 2\,a_7)^2}{3\,(a_4 + 4\,a_5\,c_3)^2}$,
$(4\,a_5\,c_3 + a_4)\,(3\,a_6 + 2\,a_7) < 0$,
x_0, t_0, c_3, and ϕ_0 are arbitrary real constants.

- *Reference*: [5].

4.13 Summary of Section 4.12

Equation

$$i\,\psi_t + a_1\,\psi_{xx} + a_2\,|\psi|^2\,\psi + a_3\,|\psi|^4\,\psi + i\,a_4\,\psi_{xxx} + a_5\,\psi_{xxxx} + i\,a_6\,(|\psi|^2\,\psi)_x + i\,a_7\,(|\psi|^2)_x\,\psi = 0$$

#	Solution	Conditions	Name	Eq.#
1.	$\psi(x,t) = c_1\,e^{i\,[c_2\,(x-x_0)+(c_1^2\,a_2+c_1^4\,a_3)\,(t-t_0)+\phi_0]}$	$a_4 = \frac{a_6\,c_1^2 + a_1\,c_2 - a_5\,c_2^3}{c_2^2}$, x_0, t_0, c_1, c_2, and ϕ_0 are arbitrary real constants	continuous wave, t- and x-dependent phase	(4.45)
2.	$\psi(x,t) = \sqrt{\frac{-6\,(4\,a_5\,c_3+a_4)}{3\,a_6+2\,a_7}}\,\sec[x - x_0 + c_1\,(t-t_0)] \times e^{i\,[c_3\,(x-x_0)+c_2\,(t-t_0)+\phi_0]}$	$c_1 = -2\,a_1\,c_3 + a_4\,(3\,c_3^2+1) + 4\,a_5\,(c_3^3+c_3)$, $c_2 = -a_1\,(c_3^2+1) + a_4\,(c_3^3+3\,c_3) + a_5\,(c_3^4+6\,c_3^2+1)$, $a_1 = \frac{(4\,a_5\,c_3+a_4)\,(3\,a_2+2\,c_3\,a_7)}{3\,a_6+2\,a_7} + 2\,a_4\,c_3 + 2\,a_5\,(c_3^2+5)$, $a_3 = \frac{-2\,a_5\,(3\,a_6+2\,a_7)^2}{3\,(a_4+4\,a_5\,c_3)^2}$, $(4\,a_5\,c_3+a_4)\,(3\,a_6+2\,a_7) < 0$, x_0, t_0, c_3, and ϕ_0 are arbitrary real constants	–	(4.46)
3.	$\psi(x,t) = \sqrt{\frac{-6\,(4\,a_5\,c_3+a_4)}{3\,a_6+2\,a_7}}\,\csc[x - x_0 + c_1\,(t-t_0)] \times e^{i\,[c_3\,(x-x_0)+c_2\,(t-t_0)+\phi_0]}$	$c_1 = -2\,a_1\,c_3 + a_4\,(3\,c_3^2+1) + 4\,a_5\,(c_3^3+c_3)$, $c_2 = -a_1\,(c_3^2+1) + a_4\,(c_3^3+3\,c_3) + a_5\,(c_3^4+6\,c_3^2+1)$, $a_1 = \frac{(4\,a_5\,c_3+a_4)\,(3\,a_2+2\,c_3\,a_7)}{3\,a_6+2\,a_7} + 2\,a_4\,c_3 + 2\,a_5\,(c_3^2+5)$, $a_3 = \frac{-2\,a_5\,(3\,a_6+2\,a_7)^2}{3\,(a_4+4\,a_5\,c_3)^2}$, $(4\,a_5\,c_3+a_4)\,(3\,a_6+2\,a_7) < 0$, x_0, t_0, c_3, and ϕ_0 are arbitrary real constants	–	(4.47)
4.	$\psi(x,t) = \sqrt{\frac{-6\,(4\,a_5\,c_3+a_4)}{3\,a_6+2\,a_7}}\,\tan[x - x_0 + c_1\,(t-t_0)] \times e^{i\,[c_3\,(x-x_0)+c_2\,(t-t_0)+\phi_0]}$	$c_1 = -2\,a_1\,c_3 + a_4\,(3\,c_3^2-2) + 4\,a_5\,(c_3^3-2\,c_3)$, $c_2 = a_1\,(2-c_3^2) + a_4\,(c_3^3-6\,c_3) + a_5\,(c_3^4-12\,c_3^2+16)$, $a_1 = \frac{(4\,a_5\,c_3+a_4)\,(3\,a_2+2\,c_3\,a_7)}{3\,a_6+2\,a_7} + 2\,a_4\,c_3 + 2\,a_5\,(c_3^2-10)$, $a_3 = \frac{-2\,a_5\,(3\,a_6+2\,a_7)^2}{3\,(a_4+4\,a_5\,c_3)^2}$, $(4\,a_5\,c_3+a_4)\,(3\,a_6+2\,a_7) < 0$,	–	(4.48)

		x_0, t_0, c_3, and ϕ_0 are arbitrary real constants		
5.	$\psi(x,t)=\sqrt{\frac{-6(4a_5c_3+a_4)}{3a_6+2a_7}}\cot[x-x_0+c_1(t-t_0)]$ $\times e^{i[c_3(x-x_0)+c_2(t-t_0)+\phi_0]}$	$c_1=-2a_1c_3+a_4(3c_3^2-2)+4a_5(c_3^3-2c_3)$, $c_2=a_1(2-c_3^2)+a_4(c_3^3-6c_3)+a_5(c_3^4-12c_3^2+16)$, $a_1=\frac{(4a_5c_3+a_4)(3a_2+2c_3a_7)}{3a_6+2a_7}+2a_4c_3+2a_5(c_3^2-10)$, $a_3=\frac{-2a_5(3a_6+2a_7)^2}{3(a_4+4a_5c_3)^2}$, $(4a_5c_3+a_4)(3a_6+2a_7)<0$, x_0, t_0, c_3, and ϕ_0 are arbitrary real constants	–	(4.49)
6.	$\psi(x,t)=\sqrt{\frac{6(4a_5c_3+a_4)}{3a_6+2a_7}}\operatorname{sech}[x-x_0+c_1(t-t_0)]$ $\times e^{i[c_3(x-x_0)+c_2(t-t_0)+\phi_0]}$	$c_1=-2a_1c_3+a_4(3c_3^2-1)+4a_5(c_3^3-c_3)$, $c_2=a_1(1-c_3^2)+a_4(c_3^3-3c_3)+a_5(c_3^4-6c_3^2+1)$, $a_1=\frac{(4a_5c_3+a_4)(3a_2+2c_3a_7)}{3a_6+2a_7}+2a_4c_3+2a_5(c_3^2-5)$, $a_3=\frac{-2a_5(3a_6+2a_7)^2}{3(a_4+4a_5c_3)^2}$, $(4a_5c_3+a_4)(3a_6+2a_7)>0$, x_0, t_0, c_3, and ϕ_0 are arbitrary real constants	bright soliton	(4.50)
7.	$\psi(x,t)=\sqrt{\frac{-6(4a_5c_3+a_4)}{3a_6+2a_7}}\operatorname{csch}[x-x_0+c_1(t-t_0)]$ $\times e^{i[c_3(x-x_0)+c_2(t-t_0)+\phi_0]}$	$c_1=-2a_1c_3+a_4(3c_3^2-1)+4a_5(c_3^3-c_3)$, $c_2=a_1(1-c_3^2)+a_4(c_3^3-3c_3)+a_5(c_3^4-6c_3^2+1)$, $a_1=\frac{(4a_5c_3+a_4)(3a_2+2c_3a_7)}{3a_6+2a_7}+2a_4c_3+2a_5(c_3^2-5)$, $a_3=\frac{-2a_5(3a_6+2a_7)^2}{3(a_4+4a_5c_3)^2}$, $(4a_5c_3+a_4)(3a_6+2a_7)<0$, x_0, t_0, c_3, and ϕ_0 are arbitrary real constants	–	(4.51)
8.	$\psi(x,t)=\sqrt{\frac{-6(4a_5c_3+a_4)}{3a_6+2a_7}}\tanh[x-x_0+c_1(t-t_0)]$ $\times e^{i[c_3(x-x_0)+c_2(t-t_0)+\phi_0]}$	$c_1=-2a_1c_3+a_4(3c_3^2+2)+4a_5(c_3^3+2c_3)$, $c_2=-a_1(c_3^2+2)+a_4(6c_3+c_3^3)+a_5(c_3^4+12c_3^2+16)$, $a_1=\frac{(4a_5c_3+a_4)(3a_2+2c_3a_7)}{3a_6+2a_7}+2a_4c_3+2a_5(c_3^2+10)$, $a_3=\frac{-2a_5(3a_6+2a_7)^2}{3(a_4+4a_5c_3)^2}$, $(4a_5c_3+a_4)(3a_6+2a_7)<0$, x_0, t_0, c_3, and ϕ_0 are arbitrary real constants	dark soliton	(4.52)

(Continued)

(Continued)

Equation

$$i\,\psi_t + a_1\,\psi_{xx} + a_2\,|\psi|^2\,\psi + a_3\,|\psi|^4\,\psi + i\,a_4\,\psi_{xxx} + a_5\,\psi_{xxxx} + i\,a_6\,(|\psi|^2\,\psi)_x + i\,a_7\,(|\psi|^2)_x\,\psi = 0$$

#	Solution	Conditions	Name	Eq.#
9.	$\psi(x, t) = \sqrt{\frac{-6\,(4\,a_5\,c_3 + a_4)}{3\,a_6 + 2\,a_7}}\,\coth[x - x_0 + c_1\,(t - t_0)]$ $\times\, e^{i\,[c_3\,(x - x_0) + c_2\,(t - t_0) + \phi_0]}$	$c_1 = -2\,a_1\,c_3 + a_4\,(3\,c_3^2 + 2) + 4\,a_5\,(c_3^3 + 2\,c_3)$, $c_2 = -a_1\,(c_3^2 + 2) + a_4\,(6\,c_3 + c_3^3) + a_5\,(c_3^4 + 12\,c_3^2 + 16)$, $a_1 = \frac{(4\,a_5\,c_3 + a_4)\,(3\,a_2 + 2\,c_3\,a_7)}{3\,a_6 + 2\,a_7} + 2\,a_4\,c_3 + 2\,a_5\,(c_3^2 + 10)$, $a_3 = \frac{-2\,a_5\,(3\,a_6 + 2\,a_7)^2}{3\,(a_4 + 4\,a_5\,c_3)^2}$, $(4\,a_5\,c_3 + a_4)\,(3\,a_6 + 2\,a_7) < 0$, x_0, t_0, c_3, and ϕ_0 are arbitrary real constants	–	(4.53)

4.14 NLSE with $|\psi|^2$-Dependent Dispersion

Equation:

$$i\,\psi_t + (1 - |\mu|\,|\psi|^2)\,\psi_{xx} + 2\,(1 - |\mu|)\,|\psi|^2\,\psi = 0, \tag{4.54}$$

where
$\psi = \psi(x, t)$ is the complex function profile,
x and t are its two independent variables,
μ is an arbitrary real constant.

Solutions:

***Solution* 1. Constant Amplitude** *CW, t- and x-dependent phase*

$$\psi(x, t) = A_0\,e^{i\,[A_1\,(x-x_0)+A_2\,(t-t_0)+\phi_0]}, \tag{4.55}$$

where
$A_2 = 2\,A_0^2 + A_0^2\,|\mu|\,(A_1^2 - 2) - A_1^2$,
x_0, t_0, A_0, A_1, μ, and ϕ_0 are arbitrary real constants.

***Solution* 2.** *Peakon*
(Figure 4.4)

$$\psi(x, t) = A_0\,e^{-\sqrt{\frac{2-2\,|\mu|}{|\mu|}}\,|x-x_0|}\,e^{-i\,[A_1\,(t-t_0)+\phi_0]}, \tag{4.56}$$

where
$2 - 2\,|\mu| > 0$,
$|\mu| < 1$, (for peakon solution)
$A_1 = 2 - \frac{2}{|\mu|}$,
x_0, t_0, A_0, and ϕ_0 are arbitrary real constants.

- *Reference*: [7], *we corrected the prefactors of x and t.*

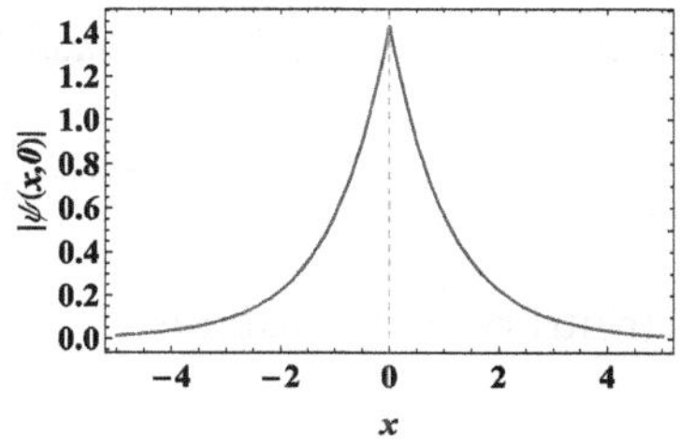

Figure 4.4. Peakon solution (4.56) at $t = 0$, with $A_0 = 1$, $a_2 = -1/2$, $\mu = 0.7$, and $\phi_0 = 0$.

4.15 Infinite Hierarchy of Integrable NLSEs with Higher Order Terms

4.15.1 Constant Coefficients

Equation:

$$i\,\psi_t + a_2\,k_2 - i\,a_3\,k_3 + a_4\,k_4 - i\,a_5\,k_5 + \ldots\, m = 0, \tag{4.57}$$

where

$\psi = \psi(x, t)$ is the complex function profile,
x and t are its two independent variables,
a_j are in general arbitrary real constants, $j = 2, 3, 4, 5, \ldots m$,
$k_j = (-1)^j \frac{\delta}{\delta\,\psi^*} \int p_{j+1}\,dx$,
$p_j = \psi\,\frac{\partial}{\partial\,x}\left(\frac{p_{j-1}}{\psi}\right) + \sum_{j_1+j_2=j-1} p_{j_1}\,p_{j_2}.$
Starting with $p_1 = |\psi|^2$, the next three p_j terms are then
$p_2 = \psi\,\psi_x^*$,
$p_3 = |\psi|^4 + \psi\,\psi_{x\,x}^*$,
$p_4 = \psi(\psi_x\,\psi^{*^2} + 4\,|\psi|^2\,\psi_x^* + \psi_{x\,x\,x}^*)$,
while the first four k_j terms are
$k_2 = \psi_{x\,x} + 2\,|\psi|^2\,\psi$,
$k_3 = \psi_{x\,x\,x} + 6\,|\psi|^2\,\psi_x$,
$k_4 = \psi_{x\,x\,x\,x} + 8\,|\psi|^2\,\psi_{x\,x} + 6\,|\psi|^4\,\psi + 4\,|\psi_x|^2\,\psi + 6\,\psi_x^2\,\psi^* + 2\,\psi^2\,\psi_{x\,x}^*$,
$k_5 = \psi_{x\,x\,x\,x\,x} + 10\,|\psi|^2\,\psi_{x\,x\,x} + 10(|\psi_x|^2\,\psi)_x + 20\,\psi^*\,\psi_x\,\psi_{x\,x} + 30\,|\psi|^4\,\psi_x.$

Hierarchy of Integrable NLSEs:

$j = 2$,

$$i\psi_t + a_2(\psi_{xx} + 2|\psi|^2\psi) = 0, \tag{4.58}$$

$j = 3$,

$$i\psi_t + a_2\,(\psi_{xx} + 2|\psi|^2\psi) - i\,a_3\,(\psi_{xxx} + 6|\psi|^2\psi_x) = 0, \tag{4.59}$$

…

Solutions to all Equations in the Hierarchy:

***Solution* 1. Constant Amplitude** *CW, t-dependent phase*

$$\psi(x,\,t) = c\,e^{i\left[c^2 \sum\limits_{n=1}^{\infty} \frac{(2\,n)!}{(n\,!)^2}\,a_{2\,n}\,c^{2\,n-2}\,(t-t_0)+\phi_0\right]}, \tag{4.60}$$

where x_0, t_0, c, and ϕ_0 are arbitrary real constants.

- *Reference*: [6].

***Solution* 2.** **sech(x,t)** *bright soliton*

$$\psi(x,\, t) = c\ \mathrm{sech}[c\,(x - x_0) + v\,(t - t_0)]\, e^{i\,[\phi_1\,(t-t_0)+\phi_0]}, \tag{4.61}$$

where

$$\phi_1 = \sum_{n=1}^{\infty} a_{2\,n}\, c^{2\,n},$$
$$v = \sum_{n=1}^{\infty} a_{2\,n+1}\, c^{2\,n+1},$$

x_0, t_0, c, and ϕ_0 are arbitrary real constants.

- *Reference*: [6].

***Solution* 3.** **Localization in x and t** *Peregrine soliton*

$$\psi(x,\, t) = c\left\{4\left[\frac{1 + 2\,i\,b\,(t - t_0)}{d(x,\, t)}\right] - 1\right\} e^{i\,[\phi_1\,(t-t_0)+\phi_0]}, \tag{4.62}$$

where

$$\phi_1 = c^2 \sum_{n=1}^{\infty} \frac{(2\,n)!}{(n\,!)^2}\, a_{2\,n}\, c^{2\,n-2},$$
$$d(x,\, t) = 1 + 4\,b^2\,(t - t_0)^2 + 4\,[c\,(x - x_0) + v\,(t - t_0)]^2,$$
$$b = \sum_{n=1}^{\infty} \frac{n(2\,n)!}{(n!)^2}\, a_{2\,n}\, c^{2\,n},$$
$$v = \sum_{n=1}^{\infty} \frac{(2\,n+1)!}{(n\,!)^2}\, a_{2\,n+1}\, c^{2\,n+1},$$

x_0, t_0, c, and ϕ_0 are arbitrary real constants.

- *Reference*: [6].

***Solution* 4.** **Rational Solution** *DW*

$$\psi(x,\, t) = c\left\{\frac{4}{1 + 4[c\,(x - x_0) + v\,(t - t_0)]^2} - 1\right\} e^{i\,\phi_0}, \tag{4.63}$$

where

$$v = \sum_{n=1}^{\infty} \frac{(2\,n+1)!}{(n\,!)^2}\, a_{2\,n+1}\, c^{2\,n+1},$$

x_0, t_0, c, and ϕ_0 are arbitrary real constants.

- *Reference*: [6].

Solution* 5. Periodicity in *x* and Localization in *t *Akhmediev breather*

$$\psi(x, t) = c\left\{1 + \frac{\kappa^2 \cosh\left[b\,\kappa\,\sqrt{1-\frac{\kappa^2}{4}}\,(t-t_0)\right] + i\,\kappa\,\sqrt{4-\kappa^2}\,\sinh\left[b\,\kappa\,\sqrt{1-\frac{\kappa^2}{4}}\,(t-t_0)\right]}{\sqrt{4-\kappa^2}\,\cos[\kappa\,[c\,(x-x_0)+v\,(t-t_0)]] - 2\cosh\left[b\,\kappa\,\sqrt{1-\frac{\kappa^2}{4}}\,(t-t_0)\right]}\right\} \times e^{i\,[\phi_1\,(t-t_0)+\phi_0]}, \tag{4.64}$$

where

$$v = \sum_{n=1}^{\infty} \frac{(2\,n+1)!}{n\,!}\,a_{2\,n+1}\,c^{2\,n+1}\left(\sum_{r=0}^{n} \frac{(-1)^r \kappa^{2\,r}\,r\,!}{(n-r)!\,(2\,r+1)!}\right),$$

$$\phi_1 = \sum_{n=1}^{\infty} \frac{(2\,n)!}{(n\,!)^2}\,a_{2\,n}\,c^{2\,n},$$

$$b = 2\sum_{n=0}^{\infty} \frac{(2\,n+1)!}{n\,!}\,a_{2\,n+2}\,c^{2\,n+2}\left(\sum_{r=0}^{n} \frac{(-1)^r \kappa^{2\,r}\,r\,!}{(n-r)!\,(2\,r+1)!}\right),$$

$0 < \kappa < 2$,

x_0, t_0, c, and ϕ_0 are arbitrary real constants.

- *Reference*: [6].

Solution* 6. Periodicity in *t* and Localization in *x *Kuznetsov–Ma breather*

$$\psi(x, t) = c\,\sqrt{2}\left[\frac{2(1-\kappa)\,d_1(t) - \sqrt{2\,\kappa}\,d_3(x, t) + 2\,i\,\sqrt{1-2\,\kappa}\,d_2(t)}{\sqrt{2}\,d_3(x, t) - 2\,\sqrt{\kappa}\,d_1(t)}\right] \times e^{i\,[\phi_1\,(t-t_0)+\phi_0]}, \tag{4.65}$$

where

$d_1(t) = \cos[2\,\sqrt{1-2\,\kappa}\,b\,(t-t_0)]$,

$d_2(t) = \sin[2\,\sqrt{1-2\,\kappa}\,b\,(t-t_0)]$,

$d_3(x, t) = \cosh[2\,\sqrt{1-2\,\kappa}\,[c\,(x-x_0)+v\,(t-t_0)]]$,

$$v = \sum_{n=1}^{\infty} 4^n\,a_{2\,n+1}\,c^{2\,n+1}\left(1 + \sum_{r=1}^{n} \frac{(2\,r-1)!!\,\kappa^r}{r\,!}\right),$$

$$\phi_1 = \sum_{n=1}^{\infty} \frac{(2\,n)!}{(n\,!)^2}\,a_{2\,n}\,c^{2\,n}\,(2\,\kappa)^n,$$

$$b = 2\sum_{n=0}^{\infty} 4^n\,a_{2\,n+2}\,c^{2\,n+2}\left(1 + \sum_{r=1}^{n} \frac{(2\,r-1)!!\,\kappa^r}{r\,!}\right),$$

$0 < \kappa < 1/2$,

x_0, t_0, c, and ϕ_0 are arbitrary real constants.

- *Reference*: [6].

***Solution* 7. dn(*x*,*t*)** *solitary wane (SW)*

$$\psi(x, t) = c\ \mathrm{dn}[c\ (x - x_0) + v\ (t - t_0), m]\ e^{i\,[\phi_1\,(t-t_0)+\phi_0]}, \tag{4.66}$$

where

$v = \sum_{n=1}^{\infty} a_{2\,n+1}\, c^{2\,n+1}\, m^n\, P_n\!\left(\frac{2}{m} - 1\right),$

$\phi_1 = \sum_{n=1}^{\infty} a_{2\,n}\, c^{2\,n}\, m^n\, P_n\!\left(\frac{2}{m} - 1\right),$

P_n is the set of orthogonal Legendre polynomials of the first kind,
$0 < m < 1$,
x_0, t_0, c, and ϕ_0 are arbitrary real constants.

- *Reference*: [6].

***Solution* 8. cn(*x*,*t*)** *SW*

$$\psi(x, t) = \frac{c\ \coth(\kappa)}{\sqrt{2}}\ \mathrm{cn}\!\left[\frac{c\ (x - x_0) + v\ (t - t_0)}{\sinh(\kappa)}, m\right] e^{i\,[\phi_1\,(t-t_0)+\phi_0]}, \tag{4.67}$$

where

$v = \sum_{n=1}^{\infty} a_{2\,n+1}\, c^{2\,n+1}\, \sinh^{-2\,n}(\kappa)\, P_n[\sinh^2(\kappa)],$

$\phi_1 = \sum_{n=1}^{\infty} a_{2\,n}\, c^{2\,n}\, \sinh^{-2\,n}(\kappa)\, P_n[\sinh^2(\kappa)],$

P_n is the set of orthogonal Legendre polynomials of the first kind,
$0 < [m = \frac{1}{2}\cosh^2(\kappa)] < 1$ with κ real, example: $\kappa = 1/2$,
x_0, t_0, c, and ϕ_0 are arbitrary real constants.

- *Reference*: [6].

4.15.2 Function Coefficients

If $\psi(x, t)$ is a solution of an infinite hierarchy with constant coefficients

$$i\,\psi_t + a_{02}\, k_2 - i\, a_{03}\, k_3 + a_{04}\, k_4 - i\, a_{05}\, k_5 + \ldots\ m = 0, \tag{4.68}$$

then $\psi(x, t)$ will be also a solution of the same infinite hierarchy with t-dependent coefficients

$$i\,\psi_t + a_2(t)\, k_2 - i\, a_3(t)\, k_3 + a_4(t)\, k_4 - i\, a_5(t)\, k_5 + \ldots\ m = 0, \tag{4.69}$$

after making the following replacements in ψ:

$a_{02j}\,(t-t_0) \rightarrow \int a_{2j}(t)\,dt,$

$a_{02j+1}\,(t-t_0) \rightarrow \int a_{2j+1}(t)\,dt,\ j = 1, 2, 3, 4, \ldots,$

where

a_{0n} are in general arbitrary real constants,

$a_n(t)$ are arbitrary real functions, $n = 2, 3, 4, 5, \ldots$.

Example 1. **sech(x,t)** *bright soliton*

Given

$$\psi(x, t) = c\ \mathrm{sech}[c\,(x-x_0) + v\,(t-t_0)]\, e^{i\,[\phi_1\,(t-t_0)+\phi_0]}$$

is a solution of (4.68), where

$$\phi_1 = \sum_{n=1}^{\infty} a_{02\,n}\, c^{2\,n},$$

$$v = \sum_{n=1}^{\infty} a_{02\,n+1}\, c^{2\,n+1},\ \text{then}$$

$$\psi(x, t) = c\ \mathrm{sech}\left\{c\,(x-x_0) + \sum_{n=1}^{\infty} \int a_{2\,n+1}(t)\,dt c^{2n+1}\right\} e^{i\left(\sum_{n=1}^{\infty} \int a_{2\,n}(t)\,dt c^{2n} + \phi_0\right)} \tag{4.70}$$

is a solution of (4.69), where x_0, t_0, c, and ϕ_0 are arbitrary real constants.

- *Reference*: [6].

Example 2. **Localization in (x,t)** *Peregrine soliton*

Given

$$\psi(x, t) = c\left\{4\left[\frac{1 + 2\,i\,b\,(t-t_0)}{d(x,t)}\right] - 1\right\} e^{i\,[\phi_1\,(t-t_0)+\phi_0]}$$

is a solution of (4.68), where

$$\phi_1 = c^2 \sum_{n=1}^{\infty} \frac{(2\,n)!}{(n\,!)^2}\, a_{02\,n}\, c^{2\,n-2},$$

$$d(x, t) = 1 + 4\,b^2\,(t-t_0)^2 + 4\,[c\,(x-x_0) + v\,(t-t_0)]^2,$$

$$b = \sum_{n=1}^{\infty} \frac{n(2\,n)!}{(n!)^2}\, a_{02\,n}\, c^{2\,n},$$

$$v = \sum_{n=1}^{\infty} \frac{(2\,n+1)!}{(n\,!)^2}\, a_{02\,n+1}\, c^{2\,n+1},\ \text{then}$$

$$\psi(x, t) = c\left\{4\left[\frac{1 + 2i\sum_{n=1}^{\infty}\frac{n(2n)!}{(n!)^2}\int a_{2n}(t)\,dtc^{2n}}{d(x, t)}\right] - 1\right\} \times e^{i\left(c^2\sum_{n=1}^{\infty}\frac{(2n)!}{(n!)^2}\int a_{2n}(t)\,dtc^{2n-2} + \phi_0\right)}, \tag{4.71}$$

is a solution of (4.69), where

$$d(x, t) = 1 + 4\left[\sum_{n=1}^{\infty}\frac{n(2n)!}{(n!)^2}\int a_{2n}(t)\,dtc^{2n}\right]^2 + 4\left[c(x - x_0) + \sum_{n=1}^{\infty}\frac{(2n+1)!}{(n!)^2}\int a_{2n+1}(t)\,dtc^{2n+1}\right]^2,$$

x_0, t_0, c, and ϕ_0 are arbitrary real constants.

- *Reference*: [6].

4.16 Summary of Section 4.15

Constant Coefficients

Equation: $i\,\psi_t + a_2\,k_2 - i\,a_3\,k_3 + a_4\,k_4 - i\,a_5\,k_5 + \ldots\, m = 0$

#	Solution	Conditions	Name	Eq. #
1.	$\psi(x,t) = c\,e^{i\left[c^2\sum_{n=1}^{\infty}\frac{(2n)!}{(n!)^2}a_{2n}\,c^{2n-2}\,(t-t_0)+\phi_0\right]}$	x_0, t_0, c, and ϕ_0 are arbitrary real constants	continuous wave, t-dependent phase	(4.60)
2.	$\psi(x,t) = c\,\text{sech}[c\,(x-x_0) + v\,(t-t_0)]\,e^{i\,[\phi_1\,(t-t_0)+\phi_0]}$	$\phi_1 = \sum_{n=1}^{\infty} a_{2n}\,c^{2n}$, $v = \sum_{n=1}^{\infty} a_{2n+1}\,c^{2n+1}$, x_0, t_0, c, and ϕ_0 are arbitrary real constants	bright soliton	(4.61)
3.	$\psi(x,t) = c\left\{4\left[\frac{1+2\,i\,b\,(t-t_0)}{d(x,t)}\right]-1\right\}e^{i\,[\phi_1\,(t-t_0)+\phi_0]}$	$\phi_1 = c^2\sum_{n=1}^{\infty}\frac{(2n)!}{(n!)^2}a_{2n}\,c^{2n-2}$, $d(x,t) = 1 + 4\,b^2\,(t-t_0)^2 + 4\,[c\,(x-x_0)+v\,(t-t_0)]^2$, $b = \sum_{n=1}^{\infty}\frac{n(2n)!}{(n!)^2}a_{2n}\,c^{2n}$, $v = \sum_{n=1}^{\infty}\frac{(2n+1)!}{(n!)^2}a_{2n+1}\,c^{2n+1}$, x_0, t_0, c, and ϕ_0 are arbitrary real constants	Peregrine soliton	(4.62)
4.	$\psi(x,t) = c\left\{\frac{4}{1+4[c\,(x-x_0)+v\,(t-t_0)]^2}-1\right\}e^{i\,\phi_0}$	$v = \sum_{n=1}^{\infty}\frac{(2n+1)!}{(n!)^2}a_{2n+1}\,c^{2n+1}$, x_0, t_0, c, and ϕ_0 are arbitrary real constants	decaying wave	(4.63)
5.	$\psi(x,t) = c\left\{1+\frac{q_1(t)}{q_2(x,t)}\right\}e^{i\,[\phi_1\,(t-t_0)+\phi_0]}$, $q_1(t) = \kappa^2\cosh[b\,\kappa\sqrt{1-\frac{\kappa^2}{4}}\,(t-t_0)]$ $+\,i\,\kappa\sqrt{4-\kappa^2}\,\sinh[b\,\kappa\sqrt{1-\frac{\kappa^2}{4}}\,(t-t_0)]$,	$v = \sum_{n=1}^{\infty}\frac{(2n+1)!}{n!}a_{2n+1}\,c^{2n+1}\left(\sum_{r=0}^{n}\frac{(-1)^r\kappa^{2r}\,r!}{(n-r)!\,(2r+1)!}\right)$ $\phi_1 = \sum_{n=1}^{\infty}\frac{(2n)!}{(n!)^2}a_{2n}\,c^{2n}$,	Akhmediev breather	(4.64)

	$q_2(x, t) = \sqrt{4 - \kappa^2} \cos[\kappa\, [c\,(x - x_0) + v\,(t - t_0)]]$ $- 2\cosh[b\,\kappa \sqrt{1 - \frac{\kappa^2}{4}}\,(t - t_0)]$,	$b = 2 \sum_{n=0}^{\infty} \frac{(2n+1)!}{n!} a_{2n+2}\, c^{2n+2} \left(\sum_{r=0}^{n} \frac{(-1)^r \kappa^{2r}\, r!}{(n-r)!\,(2r+1)!} \right)$, $0 < \kappa < 2$, x_0, t_0, c, and ϕ_0 are arbitrary real constants		
6.	$\psi(x, t) = c\,\sqrt{2} \left[\frac{2(1-\kappa)\, d_1(t) - \sqrt{2\kappa}\, d_3(x,t) + 2\,i\,\sqrt{1-2\kappa}\, d_2(t)}{\sqrt{2}\, d_3(x,t) - 2\sqrt{\kappa}\, d_1(t)} \right]$ $\times\, e^{i\,[\phi_1\,(t - t_0) + \phi_0]}$	$d_1(t) = \cos[2\sqrt{1 - 2\kappa}\, b\,(t - t_0)]$, $d_2(t) = \sin[2\sqrt{1 - 2\kappa}\, b\,(t - t_0)]$, $d_3(x, t) = \cosh[2\sqrt{1 - 2\kappa}\,[c\,(x - x_0) + v\,(t - t_0)]]$, $v = \sum_{n=1}^{\infty} 4^n a_{2n+1}\, c^{2n+1} \left(1 + \sum_{r=1}^{n} \frac{(2r-1)!!\, \kappa^r}{r!} \right)$, $\phi_1 = \sum_{n=1}^{\infty} \frac{(2n)!}{(n!)^2} a_{2n}\, c^{2n}\,(2\kappa)^n$, $0 < \kappa < 1/2$, $b = 2 \sum_{n=0}^{\infty} 4^n a_{2n+2}\, c^{2n+2} \left(1 + \sum_{r=1}^{n} \frac{(2r-1)!!\, \kappa^r}{r!} \right)$, x_0, t_0, c, and ϕ_0 are arbitrary real constants	Kuznetsov–Ma breather	(4.65)
7.	$\psi(x, t) = c\, \mathrm{dn}[c\,(x - x_0) + v\,(t - t_0), m]\, e^{i\,[\phi_1\,(t - t_0) + \phi_0]}$	$v = \sum_{n=1}^{\infty} a_{2n+1}\, c^{2n+1}\, m^n\, P_n(\frac{2}{m} - 1)$, $\phi_1 = \sum_{n=1}^{\infty} a_{2n}\, c^{2n}\, m^n\, P_n(\frac{2}{m} - 1)$, P_n is the set of orthogonal Legendre polynomials of the first kind, $0 < m < 1$, x_0, t_0, c, and ϕ_0 are arbitrary real constants	solitary wave solitary wave	(4.66)

(*Continued*)

(Continued)

Constant Coefficients

Equation: $i\,\psi_t + a_2\,k_2 - i\,a_3\,k_3 + a_4\,k_4 - i\,a_5\,k_5 + \ldots\ m = 0$

#	Solution	Conditions	Name	Eq. #
8.	$\psi(x,t) = \frac{c\coth(\kappa)}{\sqrt{2}}\,\mathrm{cn}\left[\frac{c\,(x-x_0)+v\,(t-t_0)}{\sinh(\kappa)},\,m\right]e^{i\,[\phi_1\,(t-t_0)+\phi_0]}$	$v = \sum_{n=1}^{\infty} a_{2n+1}\,c^{2n+1}\,\sinh^{-2n}(\kappa)\,P_n[\sinh^2(\kappa)]$, $\phi_1 = \sum_{n=1}^{\infty} a_{2n}\,c^{2n}\,\sinh^{-2n}(\kappa)\,P_n[\sinh^2(\kappa)]$, P_n is the set of orthogonal Legendre polynomials of the first kind, $0 < [m = \frac{1}{2}\cosh^2(\kappa)] < 1$ with κ real, example: $\kappa = 1/2$		(4.67)

Function Coefficients

Equation: $i\,\psi_t + a_2(t)\,k_2 - i\,a_3(t)\,k_3 + a_4(t)\,k_4 - i\,a_5(t)\,k_5 + \ldots\ m = 0$

#	Solution	Conditions	Name	Eq. #
1.	$\psi(x,t) = c\,\mathrm{sech}\left\{c\,(x-x_0) + \sum_{n=1}^{\infty}\int a_{2n+1}(t)\,dtc^{2n+1}\right\} \times e^{i\left(\sum_{n=1}^{\infty}\int a_{2n}(t)\,dtc^{2n}+\phi_0\right)}$	x_0, c, and ϕ_0 are arbitrary real constants	bright soliton	(4.70)
2.	$\psi(x,t) = c\left\{4\left[\frac{1+2\,i\sum_{n=1}^{\infty}\frac{n(2n)!}{(n!)^2}\int a_{2n}(t)\,dtc^{2n}}{d(x,t)}\right]-1\right\} \times e^{i\left(c^2\sum_{n=1}^{\infty}\frac{(2n)!}{(n!)^2}\int a_{2n}(t)\,dtc^{2n-2}+\phi_0\right)}$	$d(x,t) = 1 + 4\left[\sum_{n=1}^{\infty}\frac{n(2n)!}{(n!)^2}\int a_{2n}(t)\,dtc^{2n}\right]^2 + 4\left[c\,(x-x_0) + \sum_{n=1}^{\infty}\frac{(2n+1)!}{(n!)^2}\int a_{2n+1}(t)\,dtc^{2n+1}\right]^2$, x_0, c, and ϕ_0 are arbitrary real constants	Peregrine soliton	(4.71)

References

[1] Zayed E M and Al-Nowehy A G 2017 Exact solutions for the perturbed nonlinear Schrödinger equation with power law nonlinearity and Hamiltonian perturbed terms *Optik* **139** 123–44

[2] Li Z, Li L, Tian H and Zhou G 2000 New types of solitary wave solutions for the higher order nonlinear Schrödinger equation *Phys. Rev. Lett.* **84** 4096-9

[3] Wazwaz A M 2006 Exact solutions for the fourth order nonlinear Schrödinger equations with cubic and power law nonlinearities *Math. Comput. Modelling* **43** 802–8

[4] Huang Y and Liu P 2014 New exact solutions for a class of high-order dispersive cubic-quintic nonlinear Schrödinger equation *J. Math. Res.* **6** 104–8

[5] Zayed E M and Al-Nowehy A G 2017 Exact solutions and optical soliton solutions for the nonlinear Schrödinger equation with fourth-order dispersion and cubic-quintic nonlinearity *Ric. Mat.* **66** 531–52

[6] Ankiewicz A, Kedziora D J, Chowdury A, Bandelow U and Akhmediev N 2016 Infinite hierarchy of nonlinear Schrödinger equations and their solutions *Phys. Rev.* E **93** 012206

[7] Kevrekidis P G 2009 *The Discrete Nonlinear Schrödinger Equation: Mathematical Analysis Numerical Computations and Physical Perspectives* (Springer Tracts in Modern Physics vol 232) (Berlin: Springer)

Usama Al Khawaja and Laila Al Sakkaf

Chapter 5

Scaling Transformations

Known also as similarity transformations

A Glance at Chapter 5

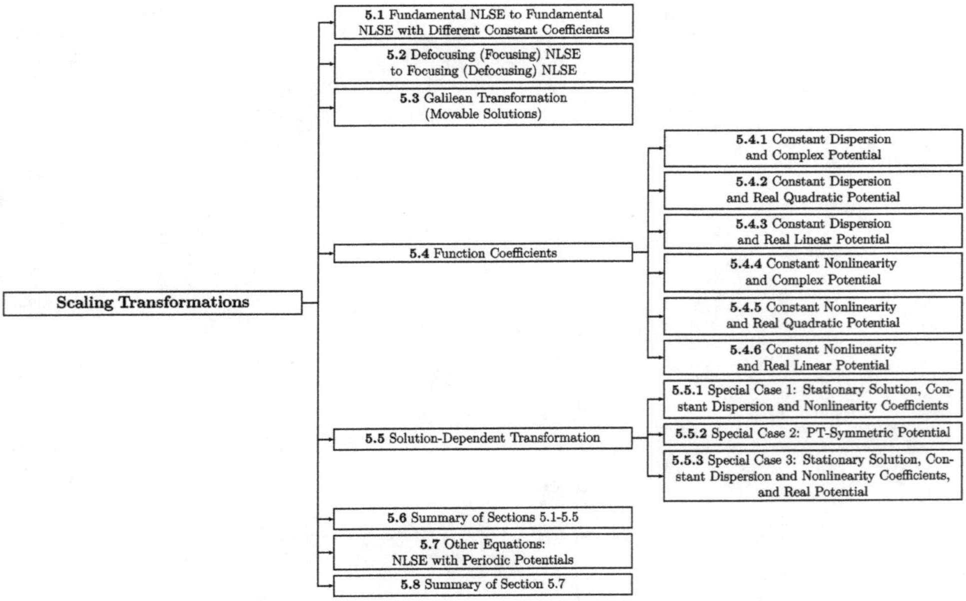

doi:10.1088/978-0-7503-2428-1ch5

A Statistical View of Chapter 5

	Equation	Solution
1	$i\,\Phi_t + a_{11}\,\Phi_{xx} + a_{22}\,\lvert\Phi\rvert^2\,\Phi = 0$	2
2	$i\,\Phi_t + a_1\,\Phi_{xx} - a_2\,\lvert\Phi\rvert^2\,\Phi = 0$	2
3	$i\,\Phi_t + a_1\,\Phi_{xx} + a_2\,\lvert\Phi\rvert^2\,\Phi = 0$	3
4	$i\,\Phi_t + a_1\,\Phi_{xx} + a_2\,\lvert\Phi\rvert^n\,\Phi = 0$	1
5	$i\,\Phi_t + a_1\,\Phi_{xx} + a_2\,\lvert\Phi\rvert^n\,\Phi + a_3\,\lvert\Phi\rvert^m\,\Phi = 0$	1
6	$i\,\Phi_t + b_1(x, t)\,\Phi_{xx} + b_2(x, t)\,\lvert\Phi\rvert^2\,\Phi + [b_{3r}(x, t) + i\,b_{3i}(x, t)]\,\Phi = 0$	0
7	$i\,\Phi_t + b_{10}\,\Phi_{xx} + b_2(x, t)\,\lvert\Phi\rvert^2\,\Phi + [b_{3r}(x, t) + i\,b_{3i}(x, t)]\,\Phi = 0$	0
8	$i\,\Phi_t + b_{10}\,\Phi_{xx} + \frac{a_2\,b_{10}\,g_5(t)}{a_1\,c_2^2}\,\lvert\Phi\rvert^2\,\Phi - \frac{g_5(t)\,g_5''(t) - 2\,g_5'^2(t)}{4\,b_{10}\,g_5^2(t)}\,x^2\,\Phi = 0$	3
9	$i\,\Phi_t + b_{10}\,\Phi_{xx} + \frac{a_2\,b_{10}\,[\alpha + \beta\,\sin(\gamma\,t)]}{a_1\,c_2^2}\,\lvert\Phi\rvert^2\,\Phi + \frac{\beta\,\gamma^2\,[3\,\beta + \beta\,\cos(2\,\gamma\,t) + 2\,\alpha\,\sin(\gamma\,t)]}{8\,b_{10}\,[\alpha + \beta\,\sin(\gamma\,t)]^2}\,x^2\,\Phi = 0$	3
10	$i\,\Phi_t + b_{10}\,\Phi_{xx} + \frac{a_2\,b_{10}\,e^{\gamma\,t}}{a_1\,c_2^2}\,\lvert\Phi\rvert^2\,\Phi + \frac{\gamma^2}{4\,b_{10}}\,x^2\,\Phi = 0$	3
11	$i\,\Phi_t + b_{10}\,\Phi_{xx} + \frac{a_2\,b_{10}\,c_5}{a_1\,c_2^2\,c_7}\,\lvert\Phi\rvert^2\,\Phi - \frac{c_7\,g_4''(t)}{2\,b_{10}\,c_5}\,x\,\Phi = 0$	4
12	$i\,\Phi_t + b_1(x, t)\,\Phi_{xx} + b_{20}\,\lvert\Phi\rvert^2\,\Phi + [b_{3r}(x, t) + i\,b_{3i}(x, t)]\,\Phi = 0$	0
13	$i\,\Phi_t + \frac{a_1\,b_{20}\,c_2^2}{a_2\,g_5(t)}\,\Phi_{xx} + b_{20}\,\lvert\Phi\rvert^2\,\Phi + \frac{a_2\,[g_5'^2(t) - g_5(t)\,g_5''(t)]}{4\,a_1\,b_{20}\,c_2^2\,g_5(t)}\,x^2\,\Phi = 0$	0

14	$i\,\Phi_t + \frac{a_1\, b_{20}\, c_2^2\, e^{-c_6\, t}}{a_2\, c_5}\,\Phi_{xx} + b_{20}\, \lvert\Phi\rvert^2\,\Phi + \frac{a_2\,[c_6\, g_4'(t) - g_4''(t)]}{2\, a_1\, b_{20}\, c_2^2}\, x\,\Phi = 0$	0
15	$i\,\Phi_t + b_{10}\,\Phi_{xx} + b_{20}\, \lvert\Phi\rvert^2\,\Phi + \left[B_t(x,\, t) + b_{10}\, B_x^2(x,\, t) - i\, b_{10}\left(\frac{\sqrt{\frac{4\, a_1}{b_{10}}}\, f'[\sqrt{\frac{a_1}{b_{10}}}\,(x - x_0)]\, B_x(x, t)}{f[\sqrt{\frac{a_1}{b_{10}}}\,(x - x_0)]} + B_{xx}(x,\, t)\right)\right]\Phi = 0$	0
16	$i\,\Phi_t + b_{10}\,\Phi_{xx} + b_{20}\,\lvert\Phi\rvert^2\,\Phi + [V_{\text{even}}(x) + i\; V_{\text{odd}}(x)]\Phi = 0$	0
17	$i\,\Phi_t + b_{10}\,\Phi_{xx} + b_{20}\,\lvert\Phi\rvert^2\,\Phi + b_{10}\,[\cos^2(x - x_0) + i\,\sin(x - x_0)]\,\Phi = 0$	0
18	$i\Phi_t + b_{10}\,\Phi_{xx} + b_{20}\,\lvert\Phi\rvert^2\,\Phi + \left\{\frac{b_{10}\, g_1^2(t)}{f^4\left[\sqrt{\frac{a_1}{b_{10}}}\,(x - x_0)\right]} + g_1'(t)\int\frac{dx}{f^2\left[\sqrt{\frac{a_1}{b_{10}}}\,(x - x_0)\right]} + g_2'(t)\right\}\Phi = 0$	0
19	$i\,\psi_t + \frac{1}{2}\,\psi_{xx} - \lvert\psi\rvert^2\,\psi + V_0\,\mathrm{sn}^2(x,\, m)\,\psi = 0$	3
20	$i\,\psi_t + \frac{1}{2}\,\psi_{xx} - \lvert\psi\rvert^2\,\psi + V_0\,\sin^2(x)\,\psi = 0$	2
Total	20	26

5.1 Fundamental NLSE to Fundamental NLSE with Different Constant Coefficients

If $\psi(x, t)$ is a solution of the fundamental NLSE, (2.1),

$$i\,\psi_t + a_1\,\psi_{xx} + a_2\,|\psi|^2\,\psi = 0,$$

then

$$\Phi(x, t) = \sqrt{\frac{a_2}{a_{22}}}\,\psi\left(\sqrt{\frac{a_1}{a_{11}}}\,x,\, t\right) \tag{5.1}$$

is a solution of

$$i\,\Phi_t + a_{11}\,\Phi_{xx} + a_{22}\,|\Phi|^2\,\Phi = 0, \tag{5.2}$$

with arbitrary real constants a_1, a_2, a_{11} and a_{22}.

***Example* 1. sech(x)** *bright soliton*

Given

$$\psi(x, t) = A_0\,\sqrt{\frac{2\,a_1}{a_2}}\,\text{sech}[A_0\,(x - x_0)]\,e^{i\left[a_1\,A_0^2\,(t-t_0)+\phi_0\right]}$$

is a solution of (2.1), then

$$\Phi(x, t) = A_0\,\sqrt{\frac{2\,a_1}{a_{22}}}\,\text{sech}\left[A_0\,\sqrt{\frac{a_1}{a_{11}}}\,(x - x_0)\right]e^{i\left[a_1\,A_0^2\,(t-t_0)+\phi_0\right]} \tag{5.3}$$

is a solution of (5.2), where

$a_1\,a_{22} > 0$,
$a_1\,a_{11} > 0$,
A_0, x_0, t_0, and ϕ_0 are arbitrary real constants.

***Example* 2. tanh(x)** *dark soliton*

Given

$$\psi(x, t) = A_0\,\sqrt{\frac{-2\,a_1}{a_2}}\,\tanh[A_0\,(x - x_0)]\,e^{-i\left[2\,a_1\,A_0^2\,(t-t_0)+\phi_0\right]}$$

is a solution of (2.1), then

$$\Phi(x, t) = A_0\,\sqrt{\frac{-2\,a_1}{a_{22}}}\,\tanh\left[A_0\,\sqrt{\frac{a_1}{a_{11}}}\,(x - x_0)\right]e^{-i\left[2\,a_1\,A_0^2\,(t-t_0)+\phi_0\right]} \tag{5.4}$$

is a solution of (5.2), where

$a_1\,a_{22} < 0$,
$a_1\,a_{11} > 0$,
A_0, x_0, t_0, and ϕ_0 are arbitrary real constants.

5.2 Defocusing (Focusing) NLSE to Focusing (Defocusing) NLSE

If $\psi(x, t)$ is a solution of the fundamental NLSE, (2.1),

$$i\,\psi_t + a_1\,\psi_{xx} + a_2\,|\psi|^2\,\psi = 0,$$

then

$$\Phi(x, t) = \psi(i\,x, -t) \tag{5.5}$$

is a solution of

$$i\,\Phi_t + a_1\,\Phi_{xx} - a_2\,|\Phi|^2\,\Phi = 0, \tag{5.6}$$

with arbitrary real constants a_1 and a_2. Here $\psi(x, t)$ should be an even function in x.

***Example* 1.**
(Figure 5.1)
Given

$$\psi(x, t) = A_0\,\sqrt{\frac{2\,a_1}{a_2}}\;\text{sech}[A_0\,(x - x_0)]\,e^{i\left[a_1\,A_0^2\,(t-t_0)+\phi_0\right]}$$

is a solution of (2.1), then

$$\Phi(x, t) = A_0\,\sqrt{\frac{2\,a_1}{a_2}}\;\text{sech}[i\,A_0\,(x - x_0)]\,e^{i\left[-a_1\,A_0^2\,(t-t_0)+\phi_0\right]} \tag{5.7}$$

is a solution of (5.6), where
$a_1\,a_2 > 0$,
A_0, x_0, t_0, and ϕ_0 are arbitrary real constants.

***Example* 2.**
(Figure 5.2)
Given

$$\psi(x, t) = \frac{1}{\sqrt{a_2}}\left[\frac{4 + i\,8\,(t - t_0)}{1 + 4\,(t - t_0)^2 + \frac{2}{a_1}\,(x - x_0)^2} - 1\right]e^{i\,[t-t_0+\phi_0]}$$

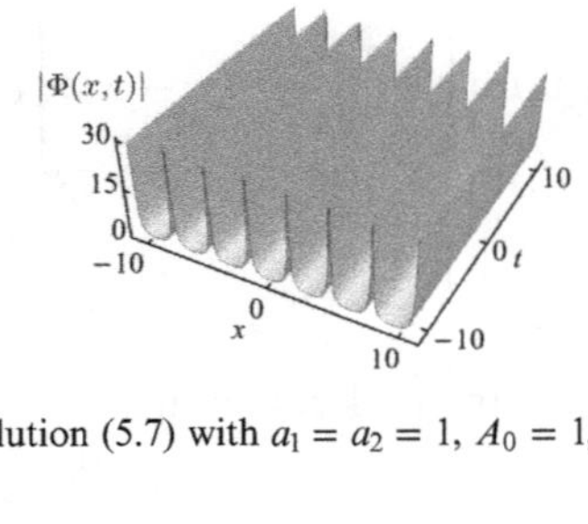

Figure 5.1. Plot of solution (5.7) with $a_1 = a_2 = 1$, $A_0 = 1$, and $x_0 = t_0 = \phi_0 = 0$.

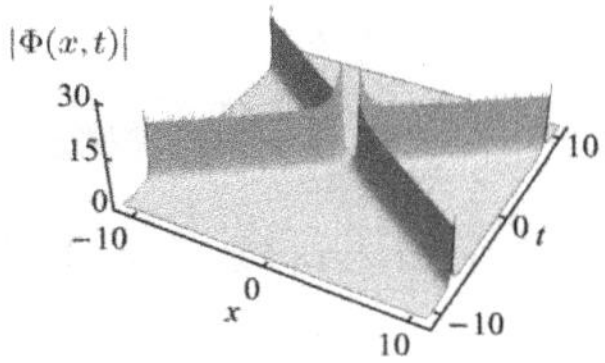

Figure 5.2. Plot of solution (5.8) with $a_1 = a_2 = 1$ and $x_0 = t_0 = \phi_0 = 0$. Animation available online at https://iopscience.iop.org/book/978-0-7503-2428-1.

is a solution of (2.1), then

$$\Phi(x, t) = \frac{1}{\sqrt{a_2}} \left[\frac{4 - i\,8\,(t - t_0)}{1 + 4\,(t - t_0)^2 - \dfrac{2}{a_1}\,(x - x_0)^2} - 1 \right] e^{i\,[-(t-t_0)+\phi_0]} \tag{5.8}$$

is a solution of (5.6), where

$a_2 > 0$,

x_0, t_0, and ϕ_0 are arbitrary real constants.

5.3 Galilean Transformation (Movable Solutions)

If $\psi(x, t)$ is a solution of one of the three equations, fundamental NLSE, (2.1), NLSE with power law nonlinearity, (3.1), and NLSE with dual power law nonlinearity, (3.19),

$$i\,\psi_t + a_1\,\psi_{xx} + a_2\,|\psi|^2\,\psi = 0,$$
$$i\,\psi_t + a_1\,\psi_{xx} + a_2\,|\psi|^n\,\psi = 0,$$
$$i\,\psi_t + a_1\,\psi_{xx} + a_2\,|\psi|^n\,\psi + a_3\,|\psi|^m\,\psi = 0,$$

then

$$\Phi(x, t) = \psi(x - v\,t,\, t)\, e^{i\left[\frac{v}{2\,a_1}(x-x_0) - \frac{v^2}{4\,a_1}(t-t_0)\right]} \tag{5.9}$$

is a movable solution of the same equation

$$i\,\Phi_t + a_1\,\Phi_{xx} + a_2\,|\Phi|^2\,\Phi = 0, \tag{5.10}$$

$$i\,\Phi_t + a_1\,\Phi_{xx} + a_2\,|\Phi|^n\,\Phi = 0, \tag{5.11}$$

$$i\,\Phi_t + a_1\,\Phi_{xx} + a_2\,|\Phi|^n\,\Phi + a_3\,|\Phi|^m\,\Phi = 0, \tag{5.12}$$

respectively, with real constants x_0, t_0, v, a_1, a_2, a_3, n, and m.

***Example* 1. sech(*x*,*t*)** *moving bright soliton*
(Figure 5.3)
Given

$$\psi(x,\, t) = A_0\,\sqrt{\frac{2\,a_1}{a_2}}\;\mathrm{sech}\{A_0\,(x - x_0)\}\; e^{i\left[a_1\,A_0^2\,(t-t_0)+\phi_0\right]}$$

is a static solution of (2.1), then

$$\Phi(x,\, t) = A_0\,\sqrt{\frac{2\,a_1}{a_2}}\;\mathrm{sech}\{A_0\,[x - (x_0 + v\,t)]\}\; e^{i\left[\frac{v}{2\,a_1}\,(x-x_0)+\frac{4a_1^2\,A_0^2-v^2}{4\,a_1}\,(t-t_0)+\phi_0\right]} \tag{5.13}$$

is a movable solution of (5.10), where
$a_1\,a_2 > 0$,
A_0, x_0, t_0, v, and ϕ_0 are arbitrary real constants.

***Example* 2. tanh(*x*,*t*)** *moving dark soliton*
(Figure 5.4)
Given

$$\psi(x,\, t) = A_0\,\sqrt{\frac{-2\,a_1}{a_2}}\;\tanh\{A_0\,(x - x_0)\}\; e^{-i\left[2\,a_1\,A_0^2\,(t-t_0)+\phi_0\right]}$$

is a static solution of (2.1), then

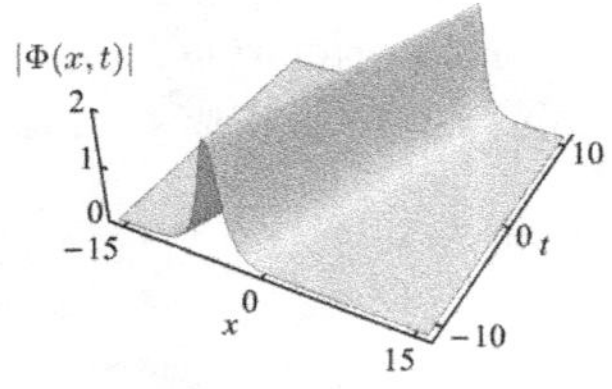

Figure 5.3. Moving bright soliton (5.13) with $a_1 = 1$, $a_2 = 1/2$, $A_0 = 1$, $v = 1/2$, and $x_0 = t_0 = \phi_0 = 0$. Animation available online at https://iopscience.iop.org/book/978-0-7503-2428-1.

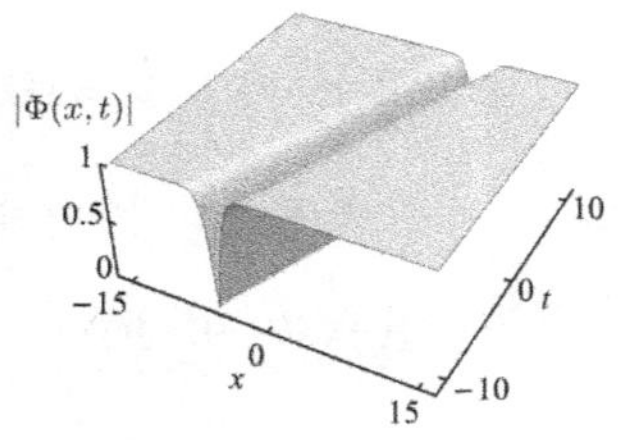

Figure 5.4. Moving dark soliton (5.14) with $a_1 = 1/2$, $a_2 = -1$, $A_0 = 1$, $v = 1/2$, and $x_0 = t_0 = \phi_0 = 0$. Animation available online at https://iopscience.iop.org/book/978-0-7503-2428-1.

$$\Phi(x,\,t) = A_0\,\sqrt{\frac{-2\,a_1}{a_2}}\,\tanh\{A_0\,[x-(x_0+v\,t)]\}$$
$$\times\, e^{-i\left[-\frac{v}{2\,a_1}(x-x_0)+\frac{8\,a_1^2\,A_0^2+v^2}{4\,a_1}(t-t_0)+\phi_0\right]} \tag{5.14}$$

is a movable solution of (5.10), where

$a_1\,a_2 < 0$,

A_0, x_0, t_0, v and ϕ_0 are arbitrary real constants.

Example* 3. Localization in *x* and *t *moving Peregrine soliton*
(Figure 5.5)
Given

$$\psi(x,\,t) = \frac{1}{\sqrt{a_2}}\left[\frac{4+i\,8\,(t-t_0)}{1+4\,(t-t_0)^2+\frac{2}{a_1}(x-x_0)^2}-1\right]e^{i\,[t-t_0+\phi_0]}$$

is a static solution of (2.1), then

$$\Phi(x,\,t) = \frac{1}{\sqrt{a_2}}\left[\frac{4+i\,8\,(t-t_0)}{1+4\,(t-t_0)^2+\dfrac{2}{a_1}(x-x_0-v\,t)^2}-1\right]$$
$$\times\, e^{i\left[\frac{v}{2\,a_1}(x-x_0)-\frac{v^2}{4\,a_1}(t-t_0)\right]}\,e^{i\,[t-t_0+\phi_0]} \tag{5.15}$$

is a moving solution of (5.10), where

$a_2 > 0$,

x_0, t_0, v, and ϕ_0 are arbitrary real constants.

***Example* 4. sech(*x*,*t*)** *moving bright soliton*
Given

$$\psi(x,\,t) = \left\{\frac{2\,A_0^2\,a_1\,(n+2)}{a_2\,n^2}\,\text{sech}^2[A_0\,(x-x_0)]\right\}^{\frac{1}{n}}\,e^{i\left[\frac{4\,a_1\,A_0^2}{n^2}(t-t_0)+\phi_0\right]}$$

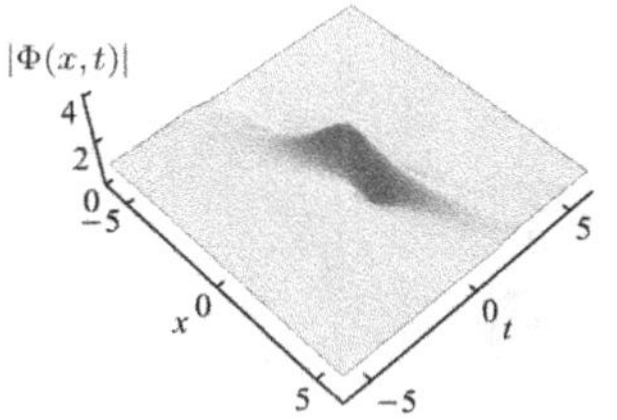

Figure 5.5. Moving Peregrine soliton (5.15) with $a_1 = a_2 = 1$, $A_0 = 1$, $v = 2$, and $x_0 = t_0 = \phi_0 = 0$. Animation available online at https://iopscience.iop.org/book/978-0-7503-2428-1.

is a static solution of (3.1), then

$$\Phi(x,\,t) = \left(\frac{2\,A_0^2\,a_1\,(n+2)}{a_2\,n^2}\,\mathrm{sech}^2\{A_0\,[x-(x_0+v\,t)]\}\right)^{\frac{1}{n}} \times e^{i\left[\frac{v}{2\,a_1}(x-x_0)-\frac{v^2}{4\,a_1}(t-t_0)\right]}\,e^{i\left[\frac{4\,a_1\,A_0^2}{n^2}(t-t_0)+\phi_0\right]} \tag{5.16}$$

is a movable solution of (3.19), where
$a_1\,a_2\,(n+2) > 0$,
A_0, x_0, t_0, v, and ϕ_0 are arbitrary real constants.

***Example* 5. sech(*x*,*t*)** *moving flat-top soliton*
(Figure 5.6)
Given

$$\psi(x,\,t) = \left(\frac{A_0\,A_1\,(n+2)}{a_2\left\{A_0 + 2\cosh\left[n\sqrt{\frac{A_1}{a_1}}\,(x-x_0)\right]\right\}}\right)^{\frac{1}{n}} e^{i\,[A_1\,(t-t_0)+\phi_0]}$$

is a static solution of (3.19), then

$$\Phi(x,\,t) = \left[\frac{A_0\,A_1\,(n+2)}{a_2\left(A_0 + 2\cosh\left\{n\sqrt{\frac{A_1}{a_1}}\,[x-(x_0+v\,t)]\right\}\right)}\right]^{\frac{1}{n}} \times e^{i\left[\frac{v}{2\,a_1}(x-x_0)-\frac{v^2}{4\,a_1}(t-t_0)\right]}\,e^{i\,[A_1\,(t-t_0)+\phi_0]} \tag{5.17}$$

is a movable solution of (5.12), where
$A_0 = \sqrt{\frac{4\,a_2^2\,(n+1)}{A_1\,\delta\,(n+2)^2}}$,
$\delta = a_3 + \frac{a_2^2\,(n+1)}{A_1\,(n+2)^2}$,
$m = 2\,n$,
$a_1\,A_1 > 0$,

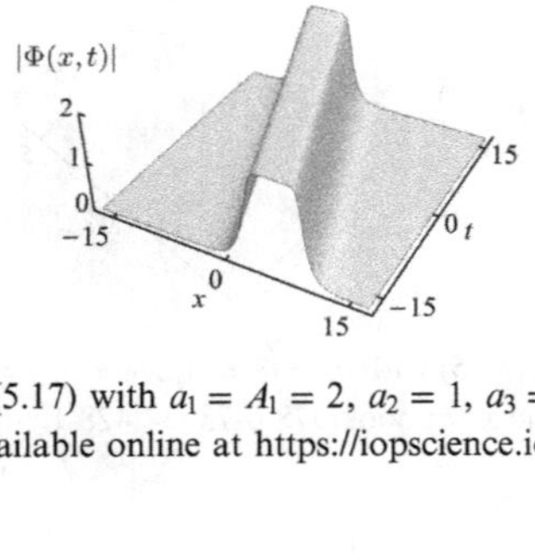

Figure 5.6. Moving flat-top soliton (5.17) with $a_1 = A_1 = 2$, $a_2 = 1$, $a_3 = -0.06944444444$, $n = 4$, $v = -1/3$, and $x_0 = t_0 = \phi_0 = 0$. Animation available online at https://iopscience.iop.org/book/978-0-7503-2428-1.

$A_1\, \delta\, (n + 1) > 0$,
x_0, t_0, A_1, v, and ϕ_0 are arbitrary real constants.

5.4 Function Coefficients

If $\psi(x, t)$ is a solution of the fundamental NLSE, (2.1),

$$i\, \psi_t + a_1\, \psi_{xx} + a_2\, |\psi|^2\, \psi = 0,$$

then

$$\Phi(x, t) = A(x, t)\, e^{i\, B(x,t)}\, \psi[X(x, t),\, T(x, t)] \tag{5.18}$$

is a solution of

$$i\, \Phi_t + b_1(x, t)\, \Phi_{xx} + b_2(x, t)\, |\Phi|^2\, \Phi + [b_{3r}(x, t) + i\, b_{3i}(x, t)]\, \Phi = 0, \tag{5.19}$$

where

$$T(x, t) = g_1(t), \tag{5.20}$$

$$A(x, t) = \frac{g_3(t)}{\sqrt{X_x(x, t)}}, \tag{5.21}$$

$$B(x, t) = g_2(t) - \int \frac{X_t(x, t)\, dx}{2\, b_1(x, t)\, X_x(x, t)}, \tag{5.22}$$

$$b_1(x, t) = \frac{a_1\, g_1'(t)}{X_x^2(x, t)}, \tag{5.23}$$

$$b_2(x, t) = \frac{a_2\, g_1'(t)}{A^2(x, t)}, \tag{5.24}$$

$$\begin{aligned} b_{3r}(x, t) = {} & \frac{\int \left\{g_1''(t)\, X_t(x, t)\, X_x(x, t) - g_1'(t)\, [X_{tt}(x, t)\, X_x(x, t) + X_t(x, t)\, X_{xt}(x, t)]\right\}\, dx}{2\, a_1\, g_1'^2(t)} \\ & + g_2'(t) + \frac{X_t^2(x, t)}{4\, a_1\, g_1'(t)} + \frac{a_1\, g_1'(t)\, [2\, X_x(x, t)\, X_{xxx}(x, t) - 3\, X_{xx}^2(x, t)]}{4\, X_x^4(x, t)}, \end{aligned} \tag{5.25}$$

$$b_{3i}(x, t) = \frac{X_{xt}(x, t)}{X_x(x, t)} - \frac{g_3'(t)}{g_3(t)}, \tag{5.26}$$

$g_1(t)$, $g_2(t)$, $g_3(t)$, and $X(x, t)$ are arbitrary real functions.

5.4.1 Constant Dispersion and Complex Potential

If $\psi(x, t)$ is a solution of the fundamental NLSE, (2.1),

$$i\,\psi_t + a_1\,\psi_{xx} + a_2\,|\psi|^2\,\psi = 0,$$

then

$$\Phi(x,\,t) = A(x,\,t)\,e^{i\,B(x,t)}\,\psi[X(x,\,t),\,T(x,\,t)] \tag{5.27}$$

is a solution of

$$i\,\Phi_t + b_{10}\,\Phi_{xx} + b_2(x,\,t)\,|\Phi|^2\,\Phi + [b_{3r}(x,\,t) + i\,b_{3i}(x,\,t)]\,\Phi = 0, \tag{5.28}$$

where

$$X(x,\,t) = g_4(t) + g_5(t)\,x, \tag{5.29}$$

$$T(x,\,t) = g_1(t) = c_0 + \frac{b_{10}}{a_1}\int g_5^2(t)\,dt, \tag{5.30}$$

$$A(x,\,t) = \frac{g_3(t)}{\sqrt{g_5(t)}}, \tag{5.31}$$

$$B(x,\,t) = g_2(t) - \frac{\left[2\,g_4'(t) + x\,g_5'(t)\right]x}{4\,b_{10}\,g_5(t)}, \tag{5.32}$$

$$b_2(x,\,t) = \frac{a_2\,b_{10}\,g_5^3(t)}{a_1\,g_3^2(t)}, \tag{5.33}$$

$$b_{3i}(x,\,t) = \frac{g_5'(t)}{g_5(t)} - \frac{g_3'(t)}{g_3(t)}, \tag{5.34}$$

$$\begin{aligned} b_{3r}(x,\,t) = g_2'(t) &+ \frac{g_4'^2(t)}{4\,b_{10}\,g_5^2(t)} + \frac{2\,g_4'(t)\,g_5'(t) - g_5(t)\,g_4''(t)}{2\,b_{10}\,g_5^2(t)}\,x \\ &+ \frac{2\,g_5'^2(t) - g_5(t)\,g_5''(t)}{4\,b_{10}\,g_5^2(t)}\,x^2, \end{aligned} \tag{5.35}$$

$g_2(t)$, $g_3(t)$, $g_4(t)$ and $g_5(t)$ are arbitrary real functions and b_{10}, a_1, a_2, and c_0 are arbitrary real constants.

5.4.2 Constant Dispersion and Real Quadratic Potential

If $\psi(x,\,t)$ is a solution of the fundamental NLSE, (2.1),

$$i\,\psi_t + a_1\,\psi_{xx} + a_2\,|\psi|^2\,\psi = 0,$$

then

$$\Phi(x,\,t) = c_2\,\sqrt{g_5(t)}\;e^{i\left[c_1 - \frac{c_4^2}{4\,b_{10}}\int g_5^2(t)\,d\,t - \frac{2\,c_4\,g_5^2(t) + g_5'(t)\,x}{4\,b_{10}\,g_5(t)}\,x\right]}$$
$$\times\,\psi\left[c_3 + g_5(t)\,x + c_4\int g_5^2(t)\,d\,t,\; c_0 + \frac{b_{10}}{a_1}\int g_5^2(t)\,dt\right] \tag{5.36}$$

is a solution of

$$i\,\Phi_t + b_{10}\,\Phi_{xx} + \frac{a_2\,b_{10}\,g_5(t)}{a_1\,c_2^2}\,|\Phi|^2\,\Phi - \frac{g_5(t)\,g_5''(t) - 2\,g_5'^2(t)}{4\,b_{10}\,g_5^2(t)}\,x^2\,\Phi = 0, \tag{5.37}$$

where

$$g_2(t) = c_1 - \frac{c_4^2}{4\,b_{10}}\int g_5^2(t)\,dt, \tag{5.38}$$

$$g_3(t) = c_2\,g_5(t), \tag{5.39}$$

$$g_4(t) = c_3 + c_4\int g_5^2(t)\,dt, \tag{5.40}$$

c_1, c_2, c_3, and c_4 are arbitrary real constants,
$g_5(t)$ is an arbitrary real function of t.

***Example* 1. sech(*x*)** *bright soliton*
Given

$$\psi(x,\,t) = A_0\,\sqrt{\frac{2\,a_1}{a_2}}\;\mathrm{sech}[A_0\,(x - x_0)]\;e^{i\left[a_1\,A_0^2\,(t - t_0) + \phi_0\right]}$$

is a solution of (2.1), then

$$\Phi(x,\,t) = \sqrt{\frac{2\,A_0^2\,a_1\,c_2^2\,g_5(t)}{a_2}}\;\mathrm{sech}\left\{A_0\,[c_3 - x_0 + g_5(t)\,x + c_4\int g_5^2(t)\,dt\right\}$$
$$\times\,e^{i\,\phi(x,t)} \tag{5.41}$$

is a solution of (5.37), where

$$\phi(x,\,t) = A_0^2\left[a_1\,(c_0 - t_0) + b_{10}\int g_5^2(t)\,dt\right] - \frac{c_4^2}{4\,b_{10}}\int g_5^2(t)\,dt - \frac{\left[2\,c_4\,g_5^2(t) + x\,g_5'(t)\right]x}{4\,b_{10}\,g_5(t)} + c_1 + \phi_0,$$

$a_1\, a_2 > 0$,
A_0, x_0, t_0, and ϕ_0 are arbitrary real constants.

Case I: $g_5(t) = \alpha + \beta \sin(\gamma\, t)$
(Figure 5.7)

$$\Phi(x,\, t) = \sqrt{\frac{2\, A_0^2\, a_1\, c_2^2\, [\alpha + \beta \sin(\gamma\, t)]}{a_2}}\, e^{i\,\phi(x,t)}\, \text{sech}\!\left(A_0 \left\{ c_3 - x_0 + [\alpha + \beta \sin(\gamma\, t)]\, x + \frac{c_4\, [2\,\gamma\, (2\,\alpha^2 + \beta^2)\, t - 8\,\alpha\,\beta \cos(\gamma\, t) - \beta^2 \sin(2\,\gamma\, t)]}{4\,\gamma} \right\} \right) \tag{5.42}$$

is a solution of

$$i\,\Phi_t + b_{10}\,\Phi_{xx} + \frac{a_2\, b_{10}\, [\alpha + \beta \sin(\gamma\, t)]}{a_1\, c_2^2}\, |\Phi|^2\, \Phi + \frac{\beta\,\gamma^2\, [3\,\beta + \beta \cos(2\,\gamma\, t) + 2\,\alpha \sin(\gamma\, t)]}{8\, b_{10}\, [\alpha + \beta \sin(\gamma\, t)]^2}\, x^2\, \Phi = 0, \tag{5.43}$$

where

$$\begin{aligned}\phi(x,\, t) = &-\frac{\beta\,\gamma \cos(\gamma\, t)\, x^2 + 2\, c_4\, [\alpha + \beta \sin(\gamma\, t)]^2\, x}{4\, b_{10}\, [\alpha + \beta \sin(\gamma\, t)]} \\ &- \frac{c_4^2\, [2\,\gamma\, (2\,\alpha^2 + \beta^2)\, t - 8\,\alpha\,\beta \cos(\gamma\, t) - \beta^2 \sin(2\,\gamma\, t)]}{16\, b_{10}\,\gamma} \\ &+ \frac{A_0^2}{4\,\gamma}\, [4\,\gamma\, a_1\, (c_0 - t_0) + b_{10}\, [2\,\gamma\, (2\,\alpha^2 + \beta^2)\, t \\ &-8\,\alpha\,\beta \cos(\gamma\, t) - \beta^2 \sin(2\,\gamma\, t)]] + c_1 + \phi_0,\end{aligned}$$

α, β, and γ are arbitrary real constants.

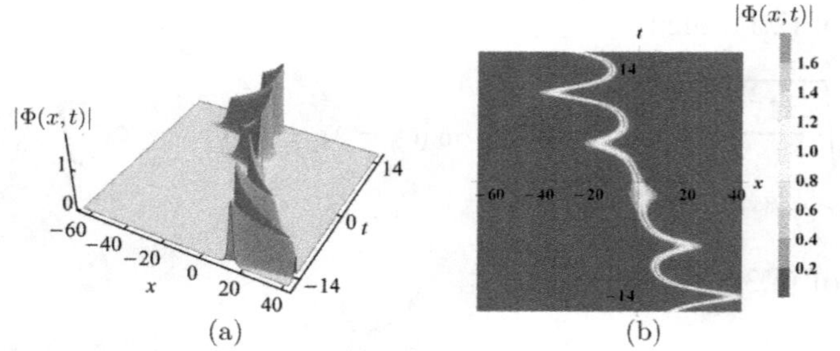

Figure 5.7. Plot of solution (5.42). (a) 3D plot, (b) contour plot, with $a_1 = a_2 = b_{10} = A_0 = \alpha = \gamma = c_0 = c_1 = c_2 = c_3 = c_4 = 1$, $\beta = 6/10$, and $x_0 = t_0 = \phi_0 = 0$. Animation available online at https://iopscience.iop.org/book/978-0-7503-2428-1.

Case II: $g_5(t) = e^{\gamma t}$
(Figure 5.8)

$$\Phi(x, t) = c_2 A_0 \sqrt{\frac{2 a_1}{a_2} e^{\gamma t}} \operatorname{sech}\left[A_0 \left(\frac{c_4 e^{2\gamma t}}{2\gamma} + e^{\gamma t} x - x_0 + c_3\right)\right] e^{i \phi(x,t)} \quad (5.44)$$

is a solution of

$$i \Phi_t + b_{10} \Phi_{xx} + \frac{a_2 b_{10} e^{\gamma t}}{a_1 c_2^2} |\Phi|^2 \Phi + \frac{\gamma^2}{4 b_{10}} x^2 \Phi = 0, \quad (5.45)$$

where

$$\phi(x, t) = \frac{1}{8 b_{10} \gamma} \{-c_4 e^{\gamma t} (c_4 e^{\gamma t} + 4 \gamma x) + 2\gamma [4 b_{10} (c_1 + \phi_0) - \gamma x^2] + 4 A_0^2 b_{10} \left[b_{10} e^{2\gamma t} + 2 a_1 \gamma (c_0 - t_0)\right]\} + \phi_0,$$

γ is an arbitrary real constant.

***Example* 2. tanh(*x*)** *dark soliton*
Given

$$\psi(x, t) = A_0 \sqrt{\frac{-2 a_1}{a_2}} \tanh[A_0 (x - x_0)] e^{-i\left[2 a_1 A_0^2 (t-t_0)+\phi_0\right]}$$

is a solution of (2.1), then

$$\Phi(x, t) = \sqrt{\frac{-2 A_0^2 a_1 c_2^2 g_5(t)}{a_2}} \tanh\left\{A_0 \left[c_3 - x_0 + g_5(t) x + c_4 \int g_5^2(t)\, dt\right]\right\} \times e^{-i \phi(x,t)} \quad (5.46)$$

is a solution of (5.37), where

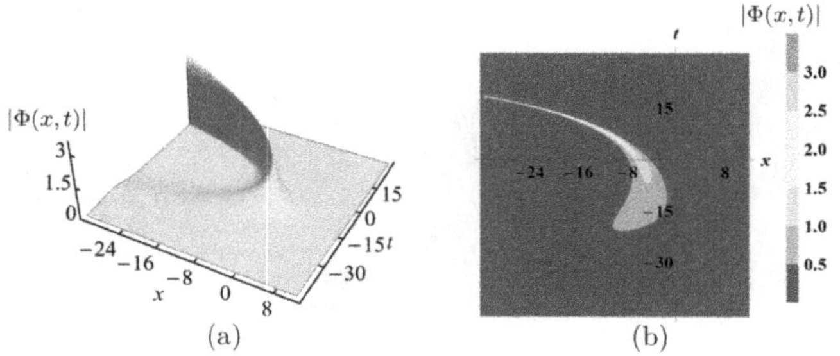

Figure 5.8. Plot of solution (5.44). (a) 3D plot, (b) contour plot, with $a_1 = a_2 = b_{10} = A_0 = c_0 = c_1 = c_2 = c_3 = c_4 = 1$, $\gamma = 1/10$, and $x_0 = t_0 = \phi_0 = 0$. Animation available online at https://iopscience.iop.org/book/978-0-7503-2428-1.

$$\phi(x, t) = 2\, A_0^2 \left[a_1\,(c_0 - t_0) + b_{10} \int g_5^2(t)\, dt\right] + \frac{c_4^2}{4\, b_{10}} \int g_5^2(t)\, dt + \frac{\left[2\, c_4\, g_5^2(t) + x\, g_5'(t)\right] x}{4\, b_{10}\, g_5(t)} - c_1 + \phi_0,$$

$a_1\, a_2 < 0$,
A_0, x_0, t_0, and ϕ_0 are arbitrary real constants.

Case I: $g_5(t) = \alpha + \beta \sin(\gamma\, t)$
(Figure 5.9)

$$\Phi(x, t) = \sqrt{\frac{-2\, A_0^2\, a_1\, c_2^2\, [\alpha + \beta \sin(\gamma\, t)]}{a_2}}\, e^{-i\,\phi(x,t)} \times \tanh\left[A_0\left(c_3 - x_0 + [\alpha + \beta \sin(\gamma\, t)]\, x + \frac{c_4\,[2\,\gamma\,(2\,\alpha^2 + \beta^2)\, t - 8\,\alpha\,\beta\cos(\gamma\, t) - \beta^2 \sin(2\,\gamma\, t)]}{4\,\gamma}\right)\right] \tag{5.47}$$

is a solution of (5.43), where

$$\phi(x, t) = \frac{\beta\,\gamma\cos(\gamma\, t)\, x^2 + 2\, c_4\,[\alpha + \beta \sin(\gamma\, t)]^2\, x}{4\, b_{10}\,[\alpha + \beta \sin(\gamma\, t)]} + \frac{c_4^2\,[2\,\gamma\,(2\,\alpha^2 + \beta^2)\, t - 8\,\alpha\,\beta\cos(\gamma\, t) - \beta^2 \sin(2\,\gamma\, t)]}{16\, b_{10}\,\gamma} + \frac{A_0^2}{2\,\gamma}[4\,\gamma\, a_1\,(c_0 - t_0) + b_{10}\,[2\,\gamma\,(2\,\alpha^2 + \beta^2)\, t - 8\,\alpha\,\beta\cos(\gamma\, t) - \beta^2 \sin(2\,\gamma\, t)]] - c_1 + \phi_0,$$

$a_1\, a_2 < 0$,
α, β, and γ are arbitrary real constants.

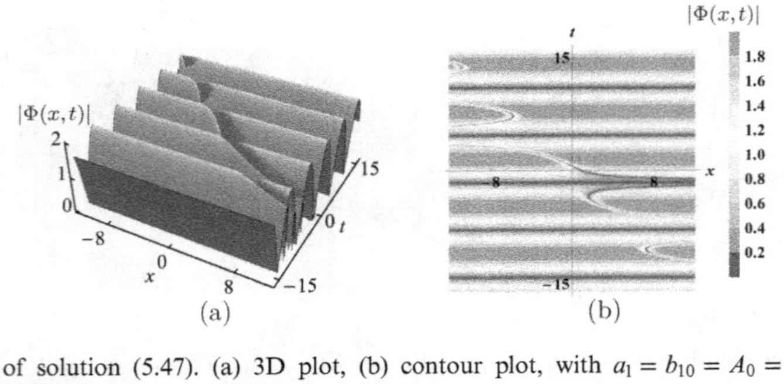

Figure 5.9. Plot of solution (5.47). (a) 3D plot, (b) contour plot, with $a_1 = b_{10} = A_0 = c_0 = c_1 = c_2 = c_3 = c_4 = \alpha = \beta = \gamma = 1$, $a_2 = -1$, and $x_0 = t_0 = \phi_0 = 0$. Animation available online at https://iopscience.iop.org/book/978-0-7503-2428-1.

Case II: $g_5(t) = e^{\gamma t}$
(Figure 5.10)

$$\Phi(x, t) = c_2 A_0 \sqrt{\frac{-2 a_1}{a_2} e^{\gamma t}} \tanh\left[A_0 \left(\frac{c_4 e^{2 \gamma t}}{2 \gamma} + e^{\gamma t} x - x_0 + c_3\right)\right] e^{i \phi(x,t)} \quad (5.48)$$

is a solution of (5.45), where

$$\phi(x, t) = \frac{-1}{8 b_{10} \gamma} \left\{c_4 e^{\gamma t} (c_4 e^{\gamma t} + 4 \gamma x) + 2 \gamma [4 b_{10} (\phi_0 - c_1) + \gamma x^2] + 8 A_0^2 b_{10} \left[b_{10} e^{2 \gamma t} + 2 a_1 \gamma (c_0 - t_0)\right]\right\} + \phi_0,$$

$a_1 a_2 < 0$,
γ is an arbitrary real constant.

***Example* 3. Two Bright Solitons**
Given

$$\psi(x, t) = \frac{1}{\sqrt{a_2}} [\psi_1(x, t) + \psi_2(x, t)]$$

is a solution of (2.1), then

$$\Phi(x, t) = c_2 \sqrt{\frac{g_5(t)}{a_2}} [\psi_1(x, t) + \psi_2(x, t)] e^{i\left[c_1 - \frac{2 c_4 g_5^2(t) x + c_4^2 g_5(t) \int g_5^2(t) d t + g_5'(t) x^2}{4 b_{10} g_5(t)}\right]}, \quad (5.49)$$

is a solution of (5.37), where

$$\psi_1(x, t) = \frac{M_{12}\left[\gamma_1^{-1}(x,t) + \gamma_2^*(x,t)\right] - M_{22}\left[\gamma_2^{-1}(x,t) + \gamma_2^*(x,t)\right]}{M_{12} M_{21}\left[\gamma_1^*(x,t) + \gamma_2^{-1}(x,t)\right]\left[\gamma_1^{-1}(x,t) + \gamma_2^*(x,t)\right] - M_{11} M_{22}\left[\gamma_1^{-1}(x,t) + \gamma_1^*(x,t)\right]\left[\gamma_2^{-1}(x,t) + \gamma_2^*(x,t)\right]},$$

$$\psi_2(x, t) = \frac{-M_{11}\left[\gamma_1^{-1}(x,t) + \gamma_1^*(x,t)\right] + M_{21}\left[\gamma_1^*(x,t) + \gamma_2^{-1}(x,t)\right]}{M_{12} M_{21}\left[\gamma_1^*(x,t) + \gamma_2^{-1}(x,t)\right]\left[\gamma_1^{-1}(x,t) + \gamma_2^*(x,t)\right] - M_{11} M_{22}\left[\gamma_1^{-1}(x,t) + \gamma_1^*(x,t)\right]\left[\gamma_2^{-1}(x,t) + \gamma_2^*(x,t)\right]},$$

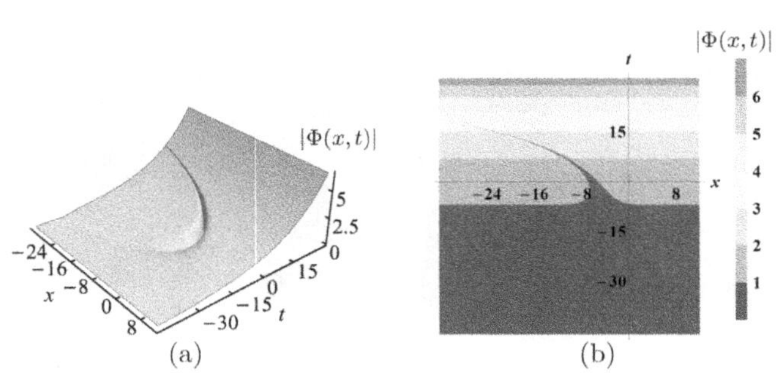

Figure 5.10. Plot of solution (5.48). (a) 3D plot, (b) contour plot, with $a_1 = b_{10} = A_0 = c_0 = c_1 = c_2 = c_3 = c_4 = 1$, $a_2 = -1$, $\gamma = 1/10$, and $x_0 = t_0 = \phi_0 = 0$. Animation available online at https://iopscience.iop.org/book/978-0-7503-2428-1.

$$M_{11} = 1/(\lambda_1 + \lambda_1^*),$$
$$M_{12} = 1/(\lambda_1 + \lambda_2^*),$$
$$M_{21} = 1/(\lambda_2 + \lambda_1^*),$$
$$M_{22} = 1/(\lambda_2 + \lambda_2^*),$$
$$\lambda_1 = \alpha_1 + i\,\nu_1,$$
$$\lambda_2 = \alpha_2 + i\,\nu_2,$$

$$\gamma_1(x,\, t) = e^{i\left\{\frac{\lambda_1^2}{2}\left[c_0 - t_0 + \frac{b_{10}}{a_1}\int g_5^2(t)\,dt\right] + \phi_{01}\right\} + \frac{\lambda_1\left[c_3 + g_5(t)\,x + c_4\int g_5^2(t)\,dt\right]}{\sqrt{2\,a_1}} - x_{01}\,\lambda_1},$$

$$\gamma_2(x,\, t) = e^{i\left\{\frac{\lambda_2^2}{2}\left[c_0 - t_0 + \frac{b_{10}}{a_1}\int g_5^2(t)\,dt\right] + \phi_{02}\right\} + \frac{\lambda_2\left[c_3 + g_5(t)\,x + c_4\int g_5^2(t)\,dt\right]}{\sqrt{2\,a_1}} - x_{02}\,\lambda_2}.$$

Case I: $g_5(t) = \alpha + \beta\,\sin(\gamma\,t)$
(Figure 5.11)

$$\Phi(x,\, t) = c_2\,\sqrt{\frac{\alpha + \beta\,\sin(\gamma\,t)}{a_2}}\,[\psi_1(x,\, t) + \psi_2(x,\, t)]\,e^{i\,\phi(x,t)} \tag{5.50}$$

is a solution of (5.43), where

$$\phi(x,\, t) = -\frac{\beta\,\gamma\,\cos(\gamma\,t)\,x^2 + 2\,c_4\,[\alpha + \beta\,\sin(\gamma\,t)]^2\,x}{4\,b_{10}\,[\alpha + \beta\,\sin(\gamma\,t)]}$$
$$- \frac{c_4^2\,[2\,\gamma\,(2\,\alpha^2 + \beta^2)\,t - 8\,\alpha\,\beta\,\cos(\gamma\,t) - \beta^2\,\sin(2\,\gamma\,t)]}{16\,b_{10}\,\gamma} + c_1,$$

$$\gamma_1(x,\, t) = e^{i\left\{\frac{\lambda_1^2}{2}\left[c_0 - t_0 + \frac{b_{10}}{4\,\gamma\,a_1}\,p(t)\right] + \phi_{01}\right\} + \frac{\lambda_1}{\sqrt{2\,a_1}}\left[c_3 + (\alpha + \beta\,\sin(\gamma\,t))\,x - \frac{c_4}{4\,\gamma}\,p(t)\right] - x_{01}\,\lambda_1}$$

$$\gamma_2(x,\, t) = e^{i\left\{\frac{\lambda_2^2}{2}\left[c_0 - t_0 + \frac{b_{10}}{4\,\gamma\,a_1}\,p(t)\right] + \phi_{02}\right\} + \frac{\lambda_2}{\sqrt{2\,a_1}}\left[c_3 + (\alpha + \beta\,\sin(\gamma\,t))\,x - \frac{c_4}{4\,\gamma}\,p(t)\right] - x_{02}\,\lambda_2},$$

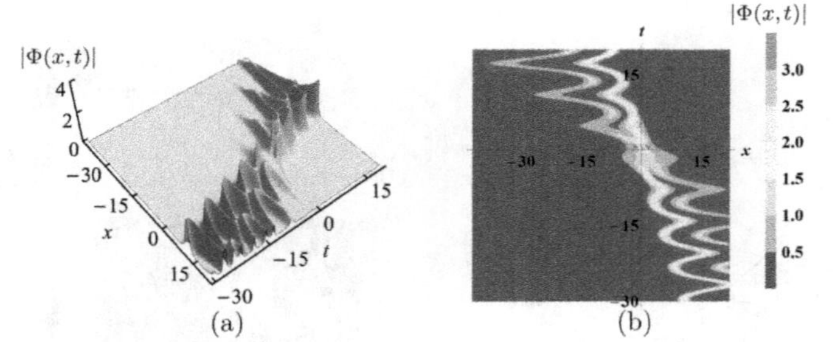

Figure 5.11. Plot of solution (5.50). (a) 3D plot, (b) contour plot, with $a_1 = A_0 = 2$, $a_2 = b_{10} = c_2 = c_4 = \alpha = \gamma = 1$, $\beta = 4/10$, $c_0 = c_1 = c_3 = 0$, $\alpha_1 = 1$, $\alpha_2 = 2$, $\nu_1 = 0$, $\nu_2 = 1/2$, and $x_{01} = x_{02} = \phi_{01} = \phi_{02} = 0$.

$$p(t) = 8\,\alpha\,\beta\cos(\gamma\,t) + \beta^2\sin(2\,\gamma\,t) - 2\,\gamma\,(2\,\alpha^2 + \beta^2)\,t,$$
$$2\,A_0^2\,a_1\,c_2^2\,[\alpha + \beta\sin(\gamma\,t)]/a_2 < 0,$$

α, β, and γ are arbitrary real constants.

Case II: $g_5(t) = e^{\gamma\,t}$
(Figure 5.12)

$$\Phi(x,\,t) = \sqrt{\frac{e^{\frac{t}{2}}}{a_2}}\,[\psi_1(x,\,t) + \psi_2(x,\,t)]\,e^{-i\,\frac{1}{8}\,(2\,e^t + 4\,e^{\frac{t}{2}}\,x + x^2)} \tag{5.51}$$

is a solution of (5.43), where

$$\gamma_1(x,\,t) = e^{i\left\{\frac{\lambda_1^2}{2}\left(c_0 - t_0 + \frac{b_{10}\,e^{2\,\gamma\,t}}{2\,a_1\,\gamma}\right) + \phi_{01}\right\} + \frac{\lambda_1}{\sqrt{2\,a_1}}\left[c_3 + e^{\gamma\,t}\,x + \frac{c_4\,e^{2\,\gamma\,t}}{2\,\gamma}\right] - x_{01}\,\lambda_1},$$

$$\gamma_2(x,\,t) = e^{i\left\{\frac{\lambda_2^2}{2}\left(c_0 - t_0 + \frac{b_{10}\,e^{2\,\gamma\,t}}{2\,a_1\,\gamma}\right) + \phi_{02}\right\} + \frac{\lambda_2}{\sqrt{2\,a_1}}\left[c_3 + e^{\gamma\,t}\,x + \frac{c_4\,e^{2\,\gamma\,t}}{2\,\gamma}\right] - x_{02}\,\lambda_2},$$

γ is an arbitrary real constant.

5.4.3 Constant Dispersion and Real Linear Potential

If $\psi(x,\,t)$ is a solution of the fundamental NLSE, (2.1),

$$i\,\psi_t + a_1\,\psi_{xx} + a_2\,|\psi|^2\,\psi = 0,$$

then

$$\Phi(x,\,t) = \sqrt{\frac{c_2^2\,c_5}{c_7}}\,\psi\left[\frac{c_5\,x}{c_7} + g_4(t),\,c_0 + \frac{b_{10}\,c_5^2\,t}{a_1\,c_7^2}\right]e^{i\left[c_1 - \frac{c_7^2}{4\,b_{10}\,c_5^2}\int g_4'^2(t)\,d\,t - \frac{c_7\,g_4'(t)\,x}{2\,b_{10}\,c_5}\right]} \tag{5.52}$$

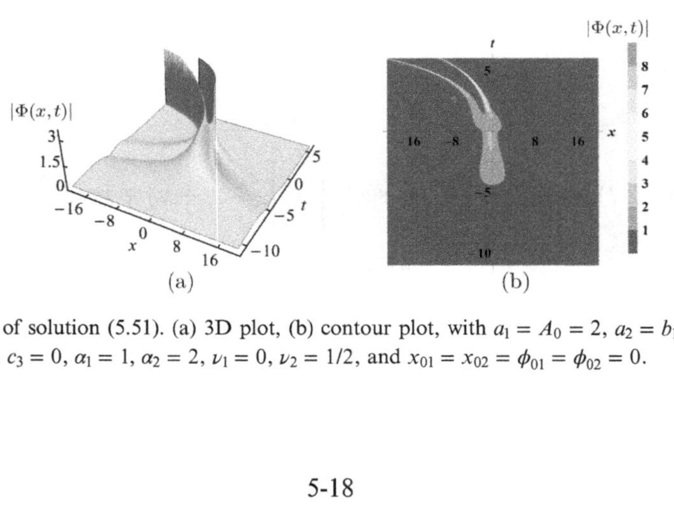

Figure 5.12. Plot of solution (5.51). (a) 3D plot, (b) contour plot, with $a_1 = A_0 = 2$, $a_2 = b_{10} = c_2 = c_4 = 1$, $\gamma = 1/2$, $c_0 = c_1 = c_3 = 0$, $\alpha_1 = 1$, $\alpha_2 = 2$, $\nu_1 = 0$, $\nu_2 = 1/2$, and $x_{01} = x_{02} = \phi_{01} = \phi_{02} = 0$.

is a solution of

$$i\,\Phi_t + b_{10}\,\Phi_{xx} + \frac{a_2\,b_{10}\,c_5}{a_1\,c_2^2\,c_7}\,|\Phi|^2\,\Phi - \frac{c_7\,g_4''(t)}{2\,b_{10}\,c_5}\,x\,\Phi = 0, \tag{5.53}$$

where

$$g_2(t) = c_1 - \frac{1}{4\,b_{10}\,c_5^2}\int\left[c_7^2\,g_4'^2(t) + 2\,c_6\,c_7\,t\,g_4'^2(t) + c_6^2\,t^2\,g_4'^2(t)\right]dt, \tag{5.54}$$

$$g_5(t) = \frac{c_5}{c_6\,t + c_7}, \tag{5.55}$$

$g_4(t)$ is an arbitrary real function of t,
c_1, c_2, c_5, c_6, and c_7 are arbitrary real constants. In (5.52) and (5.53), c_6 is taken to be zero to obtain a solution with a norm proportional to that of $\psi(x, t)$.

***Example* 1. sech(*x*)** *bright soliton*
Given

$$\psi(x, t) = A_0\sqrt{\frac{2\,a_1}{a_2}}\,\mathrm{sech}[A_0\,(x - x_0)]\,e^{i\left[a_1\,A_0^2\,(t-t_0)+\phi_0\right]}$$

is a solution of (2.1), then

$$\Phi(x, t) = c_2\,A_0\sqrt{\frac{2\,a_1\,c_5}{a_2\,c_7}}\,\mathrm{sech}\left\{A_0\left[\frac{c_5}{c_7}\,x - x_0 + g_4(t)\right]\right\} \times e^{i\left\{A_0^2\,a_1\left[c_0+\frac{b_{10}\,c_5^2}{a_1\,c_7^2}\,(t-t_0)\right]-\frac{c_7^2}{4\,b_{10}\,c_5^2}\int g_4'^2(t)\,d\,t-\frac{c_7\,g_4'(t)}{2\,b_{10}\,c_5}\,x+c_1+\phi_0\right\}} \tag{5.56}$$

is a solution of (5.53), where
$a_1\,a_2\,c_5\,c_7 > 0$,
A_0, x_0, t_0, and ϕ_0 are arbitrary real constants.

Case I: $g_4(t) = \alpha\,t^2$
(Figure 5.13)

$$\Phi(x, t) = c_2\,A_0\sqrt{\frac{2\,a_1\,c_5}{a_2\,c_7}}\,\mathrm{sech}\left\{A_0\left[\frac{c_5}{c_7}\,x - x_0 + \alpha\,t^2\right]\right\} \times e^{i\left\{A_0^2\,a_1\left[c_0+\frac{b_{10}\,c_5^2}{a_1\,c_7^2}\,(t-t_0)\right]-\frac{c_7^2\,\alpha^2\,t^3}{3\,b_{10}\,c_5^2}-\frac{c_7\,\alpha\,t}{b_{10}\,c_5}\,x+c_1+\phi_0\right\}} \tag{5.57}$$

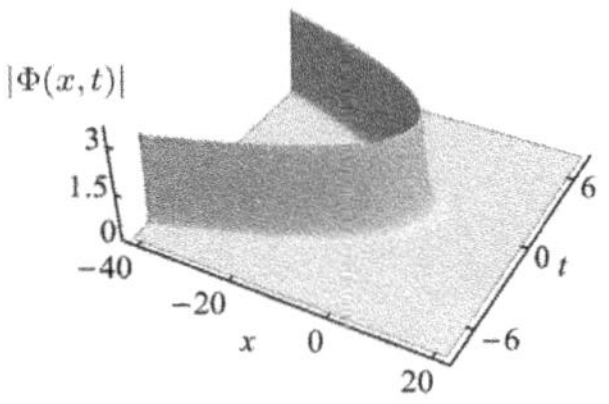

Figure 5.13. Plot of (5.57) with $a_1 = a_2 = A_0 = \alpha = c_5 = c_7 = 1$, $c_2 = b_{10} = 2$, and $c_0 = c_1 = x_0 = t_0 = \phi_0 = 0$. Animation available online at https://iopscience.iop.org/book/978-0-7503-2428-1.

is a solution of

$$i\,\Phi_t + b_{10}\,\Phi_{xx} + \frac{a_2\,b_{10}\,c_5}{a_1\,c_2^2\,c_7}\,|\Phi|^2\,\Phi - \frac{c_7\,\alpha}{b_{10}\,c_5}\,x\,\Phi = 0. \tag{5.58}$$

***Example* 2. tanh(*x*)** *dark soliton*

Given

$$\psi(x,\,t) = A_0\sqrt{\frac{-2\,a_1}{a_2}}\,\tanh[A_0\,(x - x_0)]\,e^{-i\left[2\,a_1\,A_0^2\,(t-t_0)+\phi_0\right]}$$

is a solution of (2.1)], then

$$\begin{aligned}\Phi(x,\,t) = {}& c_2\,A_0\sqrt{\frac{-2\,a_1\,c_5}{a_2\,c_7}}\,\tanh\left\{A_0\left[\frac{c_5}{c_7}\,x - x_0 + g_4(t)\right]\right\}\\ &\times e^{-i\left\{2\,A_0^2\,a_1\left[c_0+\frac{b_{10}\,c_5^2}{a_1\,c_7^2}\,(t-t_0)\right]+\frac{c_7^2}{4\,b_{10}\,c_5^2}\int g_4'^2(t)\,d\,t+\frac{c_7\,g_4'(t)}{2\,b_{10}\,c_5}\,x-c_1+\phi_0\right\}}\end{aligned} \tag{5.59}$$

is a solution of (5.53), where

$a_1\,a_2\,c_5\,c_7 < 0$,

A_0, x_0, t_0, and ϕ_0 are arbitrary real constants.

Case I: $g_4(t) = \alpha\,t^2$

(Figure 5.14)

$$\begin{aligned}\Phi(x,\,t) = {}& c_2\,A_0\sqrt{\frac{-2\,a_1\,c_5}{a_2\,c_7}}\,\tanh\left\{A_0\left[\frac{c_5}{c_7}\,x - x_0 + \alpha\,t^2\right]\right\}\\ &\times e^{-i\left\{2\,A_0^2\,a_1\left[c_0+\frac{b_{10}\,c_5^2}{a_1\,c_7^2}\,(t-t_0)\right]+\frac{c_7^2\,\alpha^2\,t^3}{3\,b_{10}\,c_5^2}+\frac{c_7\,\alpha\,t}{b_{10}\,c_5}\,x-c_1+\phi_0\right\}}\end{aligned} \tag{5.60}$$

is a solution of (5.58).

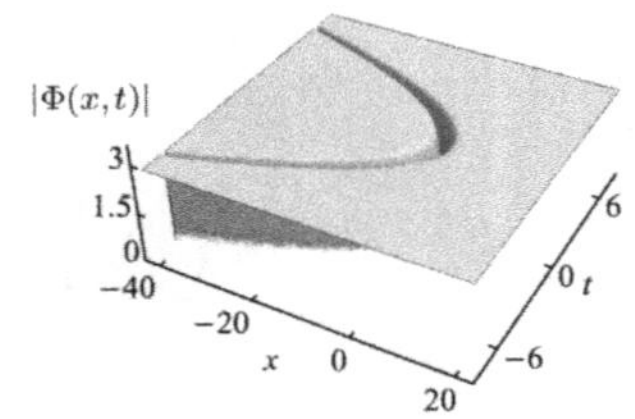

Figure 5.14. Plot of (5.60) with $a_1 = A_0 = \alpha = c_5 = c_7 = 1$, $a_2 = -1$, $c_2 = b_{10} = 2$, and $c_0 = c_1 = x_0 = t_0 = \phi_0 = 0$. Animation available online at https://iopscience.iop.org/book/978-0-7503-2428-1.

***Example* 3. Generalized First-Order Breather**

Given

$$\psi(x, t) = \frac{A_0}{\sqrt{a_2}}\left(1 - \frac{\sqrt{8}\,\lambda_r}{A_0}\,p(x, t)\right) e^{i\left[A_0^2\,(t-t_0)+\phi_0\right]}$$

is a solution of (2.1), then

$$\Phi(x, t) = A_0\sqrt{\frac{c_2^2\,c_5}{a_2\,c_7}}\left(1 - \frac{\sqrt{8}\,\lambda_r}{A_0}\,p(x, t)\right) \times e^{i\left[A_0^2\left(c_0+\frac{b_{10}\,c_5^2\,t}{a_1\,c_7^2}-t_0\right)-\frac{c_7^2}{4\,b_{10}\,c_5^2}\int g_4'^2(t)\,d\,t-\frac{c_7\,g_4'(t)\,x}{2\,b_{10}\,c_5}+c_1+\phi_0\right]} \tag{5.61}$$

is a solution of (5.53), where

$$p(x, t) = \frac{\left(A_0^2+\Gamma^2\right)\cos[q_1(x, t)] + i\left(A_0^2-\Gamma^2\right)\sin[q_1(x, t)] + 2\,A_0\,\{\Gamma_r\cosh[q_2(x, t)] - i\,\Gamma_i\sinh[q_2(x, t)]\}}{2\,A_0\,\Gamma_r\cos[q_1(x, t)] + \left(\Gamma^2+A_0^2\right)\cosh[q_2(x, t)]},$$

$$q_1(x, t) = \delta_i + \sqrt{2}\left[\frac{\frac{c_5\,x}{c_7}+g_4(t)-x_0}{\sqrt{a_1}}\,\Delta_i - 2\left(c_0 + \frac{b_{10}\,c_5^2\,t}{a_1\,c_7^2} - t_0\right)(\Delta_i\,\lambda_i + \Delta_r\,\lambda_r)\right],$$

$$q_2(x, t) = \delta_r + \sqrt{2}\left[\frac{\frac{c_5\,x}{c_7}+g_4(t)-x_0}{\sqrt{a_1}}\,\Delta_r - 2\left(c_0 + \frac{b_{10}\,c_5^2\,t}{a_1\,c_7^2} - t_0\right)\Delta_r\,\lambda_i + 2\left(c_0 + \frac{b_{10}\,c_5^2\,t}{a_1\,c_7^2} - t_0\right)\Delta_i\,\lambda_r\right],$$

$$\Delta_r = \mathrm{Re}\left[\sqrt{2\,(\lambda_r - i\,\lambda_i)^2 - A_0^2}\right],$$
$$\Delta_i = \mathrm{Im}\left[\sqrt{2\,(\lambda_r - i\,\lambda_i)^2 - A_0^2}\right],$$
$$\Gamma_r = \Delta_r + \sqrt{2}\,\lambda_r,$$
$$\Gamma_i = \Delta_i - \sqrt{2}\,\lambda_i,$$
$$\Gamma = \sqrt{\Gamma_r^2 + \Gamma_i^2},$$
$$a_1 > 0,$$

$a_2 > 0$,
A_0, λ_r, λ_i, x_0, t_0, and ϕ_0 are arbitrary real constants.

Case I: $g_4(t) = \alpha\, t^2$
(Figure 5.15)

$$\Phi(x, t) = A_0 \sqrt{\frac{c_2^2\, c_5}{a_2\, c_7}} \left(1 - \frac{\sqrt{8}\, \lambda_r}{A_0}\, p(x, t)\right) \times e^{i\left[A_0^2 \left(c_0 + \frac{b_{10}\, c_5^2\, t}{a_1\, c_7^2} - t_0\right) - \frac{c_7\, \alpha\, (3\, c_5\, x + c_7\, \alpha\, t^2)\, t}{3\, b_{10}\, c_5^2} + c_1 + \phi_0\right]} \tag{5.62}$$

is a solution of (5.58), where

$$p(x, t) = \frac{(A_0^2 + \Gamma^2) \cos[q_1(x, t)] + i\, (A_0^2 - \Gamma^2) \sin[q_1(x, t)] + 2\, A_0\, \{\Gamma_r \cosh[q_2(x, t)] - i\, \Gamma_i \sinh[q_2(x, t)]\}}{2\, A_0\, \Gamma_r \cos[q_1(x, t)] + (\Gamma^2 + A_0^2) \cosh[q_2(x, t)]},$$

$$q_1(x, t) = \delta_i + \sqrt{2} \left[\frac{\frac{c_5\, x}{c_7} + \alpha\, t^2 - x_0}{\sqrt{a_1}}\, \Delta_i - 2 \left(c_0 + \frac{b_{10}\, c_5^2\, t}{a_1\, c_7^2} - t_0\right) (\Delta_i\, \lambda_i + \Delta_r\, \lambda_r)\right],$$

$$q_2(x, t) = \delta_r + \sqrt{2} \left[\frac{\frac{c_5\, x}{c_7} + \alpha\, t^2 - x_0}{\sqrt{a_1}} \Delta_r - 2 \left(c_0 + \frac{b_{10}\, c_5^2\, t}{a_1\, c_7^2} - t_0\right) \Delta_r\, \lambda_i + 2 \left(c_0 + \frac{b_{10}\, c_5^2\, t}{a_1\, c_7^2} - t_0\right) \Delta_i\, \lambda_r\right],$$

$\Delta_r = \mathrm{Re}\left[\sqrt{2\, (\lambda_r - i\, \lambda_i)^2 - A_0^2}\right]$,
$\Delta_i = \mathrm{Im}\left[\sqrt{2\, (\lambda_r - i\, \lambda_i)^2 - A_0^2}\right]$,
$\Gamma_r = \Delta_r + \sqrt{2}\, \lambda_r$,
$\Gamma_i = \Delta_i - \sqrt{2}\, \lambda_i$,
$\Gamma = \sqrt{\Gamma_r^2 + \Gamma_i^2}$,
$a_1 > 0$,
$a_2 > 0$,
A_0, λ_r, λ_i, x_0, t_0, and ϕ_0 are arbitrary real constants.

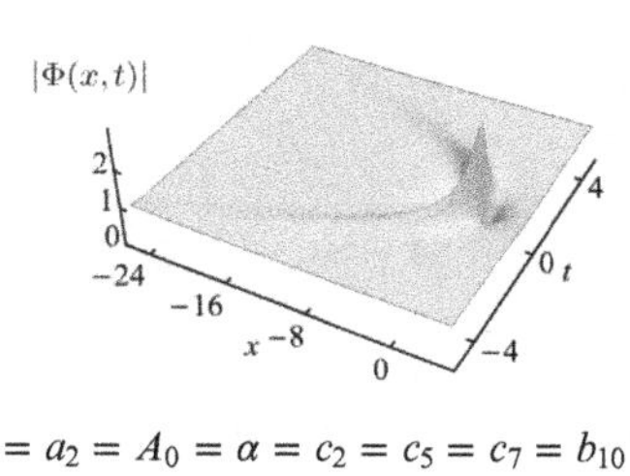

Figure 5.15. Plot of (5.62) with $a_1 = a_2 = A_0 = \alpha = c_2 = c_5 = c_7 = b_{10} = 1$, $\lambda_r = -1/\sqrt{2}$, $\lambda_i = 1/1000$, and $c_0 = c_1 = \delta_r = \delta_i = x_0 = t_0 = \phi_0 = 0$.

Case II. $g_4(t) = -\alpha \cos(\beta\, t)$
(Figure 5.16)

$$\Phi(x,\,t) \;=\; A_0 \sqrt{\frac{c_2^2\, c_5}{a_2\, c_7}}\; [1 - p(x,\,t)] \times e^{i\,\{A_0^2\,(c_0 + \frac{b_{10}\, c_5^2\, t}{a_1\, c_7^2} - t_0) - \frac{c_7\,\alpha\,\beta\,\sin(\beta\, t)\, x}{2\, b_{10}\, c_5} - \frac{c_7^2\,\alpha^2\,\beta\,[2\,\beta\, t - \sin(2\,\beta\, t)]}{16\, b_{10}\, c_5^2} + c_1 + \phi_0\}} \tag{5.63}$$

is a solution of (5.58), where

$$p(x,\,t) = 2\sqrt{2}\,\lambda_r\, \frac{(A_0^2 + \Gamma^2)\cos[q_1(x,t)] + i\,(A_0^2 - \Gamma^2)\sin[q_1(x,t)] + 2\,A_0\,\{\Gamma_r \cosh[q_2(x,t)] - i\,\Gamma_i \sinh[q_2(x,t)]\}}{2\,A_0^2\,\Gamma_r\cos[q_1(x,t)] + A_0\,(\Gamma^2 + A_0^2)\cosh[q_2(x,t)]},$$

$$q_1(x,\,t) = \delta_i + \sqrt{2}\left[\frac{\frac{c_5\, x}{c_7} - \alpha\cos(\beta\, t) - x_0}{\sqrt{a_1}}\,\Delta_i - 2\left(c_0 + \frac{b_{10}\, c_5^2\, t}{a_1\, c_7^2} - t_0\right)(\Delta_i\,\lambda_i + \Delta_r\,\lambda_r)\right],$$

$$q_2(x,\,t) = \delta_r + \sqrt{2}\left[\frac{\frac{c_5\, x}{c_7} - \alpha\cos(\beta\, t) - x_0}{\sqrt{a_1}}\,\Delta_r - 2\left(c_0 + \frac{b_{10}\, c_5^2\, t}{a_1\, c_7^2} - t_0\right)\Delta_r\,\lambda_i + 2\left(c_0 + \frac{b_{10}\, c_5^2\, t}{a_1\, c_7^2} - t_0\right)\Delta_i\,\lambda_r\right],$$

$\Delta_r = \mathrm{Re}[\sqrt{2\,(\lambda_r - i\,\lambda_i)^2 - \mathrm{A}_0^2}]$,
$\Delta_i = \mathrm{Im}[\sqrt{2\,(\lambda_r - i\,\lambda_i)^2 - \mathrm{A}_0^2}]$,
$\Gamma_r = \Delta_r + \sqrt{2}\,\lambda_r$,
$\Gamma_i = \Delta_i - \sqrt{2}\,\lambda_i$,
$\Gamma = \sqrt{\Gamma_r^2 + \Gamma_i^2}$,
$a_1 > 0$,
$a_2 > 0$,
A_0, λ_r, λ_i, x_0, t_0, and ϕ_0 are arbitrary real constants.

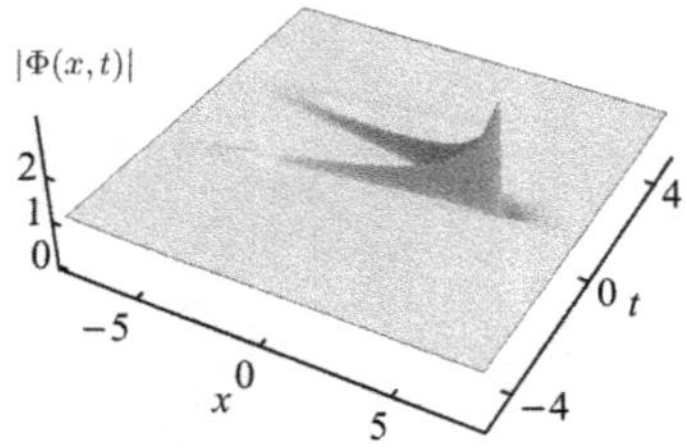

Figure 5.16. Plot of (5.63) with $a_1 = b_{10} = 1/2$, $a_2 = 2$, $A_0 = 3/2$, $\alpha = 18/4$, $c_2 = c_5 = c_7 = 1$, $\lambda_r = -3/2\sqrt{2}$, $\lambda_i = 1/100$, and $c_0 = c_1 = \delta_r = \delta_i = x_0 = t_0 = \phi_0 = 0$. (Used in the front cover page).

5.4.4 Constant Nonlinearity and Complex Potential

If $\psi(x, t)$ is a solution of the fundamental NLSE, (2.1),

$$i\,\psi_t + a_1\,\psi_{xx} + a_2\,|\psi|^2\,\psi = 0,$$

then

$$\Phi(x, t) = A(x, t)\,e^{i\,B(x,t)}\,\psi[X(x, t),\, T(x, t)] \tag{5.64}$$

is a solution of

$$i\,\Phi_t + b_1(x, t)\,\Phi_{xx} + b_{20}\,|\Phi|^2\,\Phi + [b_{3r}(x, t) + i\,b_{3i}(x, t)]\,\Phi = 0, \tag{5.65}$$

where

$$X(x, t) = g_4(t) + g_5(t)\,x, \tag{5.66}$$

$$T(x, t) = g_1(t) = c_0 + \frac{b_{20}}{a_2}\int \frac{g_3^2(t)}{g_5(t)}\,dt, \tag{5.67}$$

$$A(x, t) = \frac{g_3(t)}{\sqrt{g_5(t)}}, \tag{5.68}$$

$$B(x, t) = g_2(t) - \frac{a_2\left[2\,g_4'(t) + x\,g_5'(t)\right]x\,g_5^2(t)}{4\,a_1\,b_{20}\,g_3^2(t)}, \tag{5.69}$$

$$b_1(x, t) = \frac{a_1\,b_{20}\,g_3^2(t)}{a_2\,g_5^3(t)}, \tag{5.70}$$

$$b_{3i}(x, t) = \frac{g_5'(t)}{g_5(t)} - \frac{g_3'(t)}{g_3(t)}, \tag{5.71}$$

$$\begin{aligned} b_{3r}(x, t) = g_2'(t) &+ \frac{a_2\,g_4'^2(t)\,g_5(t)}{4\,a_1\,b_{20}\,g_3^2(t)} \\ &+ \frac{a_2\,g_5(t)\left[2\,g_5(t)\,g_3'(t)\,g_4'(t) - g_3(t)\,g_4'(t)\,g_5'(t) - g_3(t)\,g_4''(t)\,g_5(t)\right]}{2\,a_1\,b_{20}\,g_3^3(t)}\,x \\ &+ \frac{a_2\,g_5(t)\left[2\,g_3'(t)\,g_5(t)\,g_5'(t) - g_3(t)\,g_5'^2(t) - g_3(t)\,g_5(t)\,g_5''(t)\right]}{4\,a_1\,b_{20}\,g_3^3(t)}\,x^2, \end{aligned} \tag{5.72}$$

$g_4(t)$ and $g_5(t)$ are arbitrary real functions,
b_{20} and c_0 are arbitrary real constants.

5.4.5 Constant Nonlinearity and Real Quadratic Potential

If $\psi(x, t)$ is a solution of the fundamental NLSE, (2.1),

$$i\,\psi_t + a_1\,\psi_{xx} + a_2\,|\psi|^2\,\psi = 0,$$

then

$$\Phi(x, t) = c_2\,\sqrt{g_5(t)}\;e^{i\left\{c_1 - \frac{a_2\left[2\,c_4\,g_5(t)\,x + g_5'(t)\,x^2 + c_4^2\,\int g_5(t)\,dt\right]}{4\,a_1\,b_{20}\,c_2^2}\right\}} \times \psi\left[c_3 + g_5(t)\,x + c_4\int g_5(t)\,dt,\; c_0 + \frac{b_{20}\,c_2^2}{a_2}\int g_5(t)\,dt\right] \tag{5.73}$$

is a solution of

$$i\,\Phi_t + \frac{a_1\,b_{20}\,c_2^2}{a_2\,g_5(t)}\,\Phi_{xx} + b_{20}\,|\Phi|^2\,\Phi + \frac{a_2\left[g_5'^2(t) - g_5(t)\,g_5''(t)\right]}{4\,a_1\,b_{20}\,c_2^2\,g_5(t)}\,x^2\,\Phi = 0, \tag{5.74}$$

where

$g_5(t)$ is an arbitrary real function of t,
c_1, c_2, c_3, and c_4 are arbitrary real constants.

5.4.6 Constant Nonlinearity and Real Linear Potential

If $\psi(x, t)$ is a solution of the fundamental NLSE, (2.1),

$$i\,\psi_t + a_1\,\psi_{xx} + a_2\,|\psi|^2\,\psi = 0,$$

then

$$\Phi(x, t) = c_2\,\sqrt{c_5\,e^{c_6\,t}}\;e^{i\left\{c_1 - \frac{a_2\left[c_5^2\,c_6\,e^{c_6\,t}\,x^2 + 2\,c_5\,g_4'(t)\,x + \int g_4'^2(t)\,e^{-c_6\,t}\,dt\right]}{4\,a_1\,b_{20}\,c_2^2\,c_5}\right\}} \times \psi\left[c_5\,e^{c_6\,t}\,x + g_4(t),\; c_0 + \frac{b_{20}\,c_2^2\,c_5\,(e^{c_6\,t} - e^{c_6})}{a_2\,c_6}\right] \tag{5.75}$$

is a solution of

$$i\,\Phi_t + \frac{a_1\,b_{20}\,c_2^2\,e^{-c_6\,t}}{a_2\,c_5}\,\Phi_{xx} + b_{20}\,|\Phi|^2\,\Phi + \frac{a_2\left[c_6\,g_4'(t) - g_4''(t)\right]}{2\,a_1\,b_{20}\,c_2^2}\,x\,\Phi = 0, \tag{5.76}$$

where

$$g_2(t) = c_1 - \frac{a_2}{4\, a_1\, b_{20}\, c_2^2\, c_5} \int g_4'(t)^2\, e^{-c_6\, t}\, dt, \tag{5.77}$$

$$g_5(t) = c_5\, e^{c_6\, t}, \tag{5.78}$$

$g_4(t)$ is an arbitrary real function of t,
c_1, c_2, c_5, and c_6 are arbitrary real constants.

5.5 Solution-Dependent Transformation

If $\psi(x, t)$ is a solution of the fundamental NLSE, (2.1),

$$i\, \psi_t + a_1\, \psi_{xx} + a_2\, |\psi|^2\, \psi = 0,$$

then

$$\Phi(x, t) = A(x, t)\, e^{i\, B(x,t)}\, \psi[X(x, t),\, T(x, t)] \tag{5.79}$$

is a solution of

$$i\, \Phi_t + b_1(x, t)\, \Phi_{xx} + b_2(x, t)\, |\Phi|^2\, \Phi + [b_{3r}(x, t) + i\, b_{3i}(x, t)]\, \Phi = 0, \tag{5.80}$$

where

$$T(x, t) = g_1(t), \tag{5.81}$$

$$b_1(x, t) = \frac{a_1\, g_1'(t)}{X_x^2(x, t)}, \tag{5.82}$$

$$b_2(x, t) = \frac{a_2\, g_1'(t)}{A^2(x, t)}, \tag{5.83}$$

$$\begin{aligned} b_{3r}(x, t) = {} & \frac{a_1\, g_1'(t)[A(x, t)\, B_x^2(x, t) - A_{xx}(x, t)]}{A(x, t)\, X_x^2(x, t)} + B_t(x, t) \\ & - \frac{a_1\, g1'(t)[2\, A_x(x, t)\, X_x(x, t) + A(x, t)\, X_{xx}(x, t)]}{A(x, t) X_x^2(x, t)}\, \mathrm{Re}\left[\frac{\psi_x(x, t)}{\psi(x, t)}\right] \\ & + \left[\frac{2\, a_1\, g_1'(t) B_x(x, t)}{X_x(x, t)} + X_t(x, t)\right] \mathrm{Im}\left[\frac{\psi_x(x, t)}{\psi(x, t)}\right], \end{aligned} \tag{5.84}$$

$$
\begin{aligned}
b_{3i}(x,\ t) = &- \frac{a_1\, g_1'(t)[2\, A_x(x,\ t)\, B_x(x,\ t) + A(x,\ t)\, B_{xx}(x,\ t)]}{X_x^2(x,\ t)\, A(x,\ t)} - \frac{A_t(x,\ t)}{A(x,\ t)} \\
&- \left[\frac{2\, a_1\, g_1'(t)\, B_x(x,\ t)}{X_x(x,\ t)} + X_t(x,\ t)\right] \mathrm{Re}\left[\frac{\psi_x(x,\ t)}{\psi(x,\ t)}\right] \\
&- \frac{a_1\, g_1'(t)\, [2\, A_x(x,\ t)\, X_x(x,\ t) + A(x,\ t)\, X_{xx}(x,\ t)]}{X_x^2(x,\ t)\, A(x,\ t)}\, \mathrm{Im}\left[\frac{\psi_x(x,\ t)}{\psi(x,\ t)}\right],
\end{aligned} \tag{5.85}
$$

$g_1(t)$, $X(x,\ t)$, and $B(x,\ t)$ are arbitrary real functions.

5.5.1 Special Case I: Stationary Solution, Constant Dispersion and Nonlinearity Coefficients

If $\psi(x,\ t) = A_0\, f(x)\, e^{i\, A_0^2\, (t-t_0)}$ is a stationary solution of the fundamental NLSE, (2.1),

$$
i\, \psi_t + a_1\, \psi_{xx} + a_2\, |\psi|^2\, \psi = 0,
$$

then

$$
\Phi(x,\ t) = \sqrt{\frac{a_2}{b_{20}}}\, e^{i\, B(x,t)}\, \psi\left[\sqrt{\frac{a_1}{b_{10}}}\, (x - x_0),\ (t - t_0)\right] \tag{5.86}
$$

$$
= A_0\, \sqrt{\frac{a_2}{b_{20}}}\, f\left[\sqrt{\frac{a_1}{b_{10}}}\, (x - x_0)\right] e^{i\, [B(x,t) + A_0^2\, (t-t_0)]} \tag{5.87}
$$

is a solution of

$$
\begin{aligned}
&i\, \Phi_t + b_{10}\, \Phi_{xx} + b_{20}\, |\Phi|^2\, \Phi \\
&+ \left[B_t(x,\ t) + b_{10}\, B_x^2(x,\ t) - i\, b_{10} \left(\frac{\sqrt{\frac{4\, a_1}{b_{10}}}\, f'\left[\sqrt{\frac{a_1}{b_{10}}}\, (x - x_0)\right] B_x(x,\ t)}{f\left[\sqrt{\frac{a_1}{b_{10}}}\, (x - x_0)\right]} + B_{xx}(x,\ t) \right) \right] \Phi = 0,
\end{aligned} \tag{5.88}
$$

where

$B(x,\ t)$ is an arbitrary real function,

b_{10}, b_{20}, A_0 and t_0 are arbitrary real constants.

5.5.2 Special Case II: PT-Symmetric Potential

If $\psi(x, t) = A_0 f(x)\, e^{i A_0^2 (t-t_0)}$ is a stationary solution of the fundamental NLSE, (2.1),

$$i\,\psi_t + a_2\,|\psi|^2\,\psi = 0,$$

then

$$\Phi(x, t) = \sqrt{\frac{a_2}{b_{20}}}\, e^{i\,B(x)}\,\psi\left[\sqrt{\frac{a_1}{b_{10}}}\,(x - x_0),\,(t - t_0)\right] \tag{5.89}$$

$$= A_0\sqrt{\frac{a_2}{b_{20}}}\, f\left[\sqrt{\frac{a_1}{b_{10}}}\,(x - x_0)\right] e^{i\,[B(x)+A_0^2\,(t-t_0)]} \tag{5.90}$$

is a solution of

$$i\,\Phi_t + b_{10}\,\Phi_{xx} + b_{20}\,|\Phi|^2\,\Phi + [V_{\text{even}}(x) + i\,V_{\text{odd}}(x)]\Phi = 0, \tag{5.91}$$

where

$$V_{\text{even}} = b_{10}\,B'^2(x), \tag{5.92}$$

and

$$V_{\text{odd}} = -b_{10}\left\{\frac{\sqrt{\frac{4\,a_1}{b_{10}}}\, f'\left[\sqrt{\frac{a_1}{b_{10}}}\,(x - x_0)\right] B'(x)}{f\left[\sqrt{\frac{a_1}{b_{10}}}\,(x - x_0)\right]} + B''(x)\right\}, \tag{5.93}$$

form a PT-symmetric potential for some choices of $B(x)$.

Case I: $B(x, t) = \sin(x - x_0)$

Given $\psi(x, t) = A_0\, e^{i\,[a_2\,A_0^2\,(t-t_0)+\phi_0]}$ is a stationary solution of the fundamental NLSE, (2.1),

$$i\,\psi_t + a_1\,\psi_{xx} + a_2\,|\psi|^2\,\psi = 0,$$

then

$$\Phi(x, t) = \sqrt{\frac{a_2}{b_{20}}}\, e^{i\,\sin(x-x_0)}\,\psi\left[\sqrt{\frac{a_1}{b_{10}}}\,(x - x_0),\,(t - t_0)\right] \tag{5.94}$$

is a solution of

$$i\,\Phi_t + b_{10}\,\Phi_{xx} + b_{20}\,|\Phi|^2\,\Phi + b_{10}\,[\cos^2(x - x_0) + i\,\sin(x - x_0)]\,\Phi = 0, \tag{5.95}$$

where

$B(x)$ is an arbitrary real function,
b_{10}, b_{20}, A_0, x_0, and t_0 are real constants.

5.5.3 Special Case III: Stationary Solution, Constant Dispersion and Nonlinearity Coefficients, and Real Potential

If $\psi(x, t) = A_0 f(x)\, e^{i\,[a_2 A_0^2 (t-t_0)+\phi_0]}$ is a stationary solution of

$$i\,\psi_t + a_1\,\psi_{xx} + a_2\,|\psi|^2\,\psi = 0,$$

then

$$\Phi(x, t) = \sqrt{\frac{a_2}{b_{20}}}\, e^{i\,B(x,t)}\, \psi\left[\sqrt{\frac{a_1}{b_{10}}}\,(x - x_0),\, (t - t_0)\right] \tag{5.96}$$

$$= A_0 \sqrt{\frac{a_2}{b_{20}}}\, f\left[\sqrt{\frac{a_1}{b_{10}}}\,(x - x_0)\right] e^{i\,[B(x,t)+A_0^2\,(t-t_0)]} \tag{5.97}$$

is a solution of

$$i\Phi_t + b_{10}\,\Phi_{xx} + b_{20}\,|\Phi|^2\,\Phi + \left\{ \frac{b_{10}\, g_1^2(t)}{f^4\left[\sqrt{\frac{a_1}{b_{10}}}\,(x - x_0)\right]} + g_1'(t) \int \frac{dx}{f^2\left[\sqrt{\frac{a_1}{b_{10}}}\,(x - x_0)\right]} + g_2'(t) \right\} \Phi = 0, \tag{5.98}$$

where

$$B(x, t) = g_1(t) \int \frac{dx}{f^2\left[\sqrt{\frac{a_1}{b_{10}}}\,(x - x_0)\right]} + g_2(t), \tag{5.99}$$

$g_1(t)$ and $g_2(t)$ are arbitrary real functions of t,
b_{10}, b_{20}, A_0, x_0, and t_0 are real constants.

5.6 Summary of Sections 5.1–5.5

Note: For lengthy conditions, the reader is referred to the solutions in sections 5.1–5.5.

Fundamental NLSE to Fundamental NLSE with Different Constant Coefficients

Transformation: $\Phi(x, t) = \sqrt{\frac{a_2}{a_{22}}}\, \psi\left(\sqrt{\frac{a_1}{a_{11}}}\, x, t\right)$, ψ is a solution of the fundamental NLSE (2.1)

Equation: $i\, \Phi_t + a_{11}\, \Phi_{xx} + a_{22}\, |\Phi|^2\, \Phi = 0$

#	Example	Conditions	Name	Eq. #
1.	$\Phi(x, t) = A_0 \sqrt{\frac{2\, a_1}{a_{22}}}\, \text{sech}\left[A_0 \sqrt{\frac{a_1}{a_{11}}}\, (x - x_0)\right] e^{i\left[a_1 A_0^2 (t-t_0)+\phi_0\right]}$	$a_1\, a_{22} > 0$, $a_1\, a_{11} > 0$, A_0, x_0, t_0, and ϕ_0 are arbitrary real constants	bright soliton	(5.3)
2.	$\Phi(x, t) = A_0 \sqrt{\frac{-2\, a_1}{a_{22}}}\, \tanh\left[A_0 \sqrt{\frac{a_1}{a_{11}}}\, (x - x_0)\right] \times e^{-i\left[2\, a_1 A_0^2 (t-t_0)+\phi_0\right]}$	$a_1\, a_{22} < 0$, $a_1\, a_{11} > 0$, A_0, x_0, t_0, and ϕ_0 are arbitrary real constants	dark soliton	(5.4)

Defocusing (Focusing) NLSE to Focusing (Defocusing) NLSE

Transformation: $\Phi(x, t) = \psi(i\, x, -t)$

Equation: $i\, \Phi_t + a_1\, \Phi_{xx} - a_2\, |\Phi|^2\, \Phi = 0$

#	Example	Conditions	Name	Eq. #
1.	$\Phi(x, t) = A_0 \sqrt{\frac{2\, a_1}{a_2}}\, \text{sech}[i\, A_0\, (x - x_0)]\, e^{i\left[-a_1 A_0^2 (t-t_0)+\phi_0\right]}$	$a_1\, a_2 > 0$, A_0, x_0, t_0, and ϕ_0 are arbitrary real constants	–	(5.7)
2.	$\Phi(x, t) = \frac{1}{\sqrt{a_2}}\left[\frac{4 - i\, 8\, (t - t_0)}{1 + 4\, (t - t_0)^2 - \frac{2}{a_1}\, (x - x_0)^2} - 1\right] e^{i\,[-(t-t_0)+\phi_0]}$	$a_2 > 0$, x_0, t_0, and ϕ_0 are arbitrary real constants	Peregrine soliton, two solitons	(5.8)

Galilean Transformation (Movable Solutions)

Transformation: $\Phi(x, t) = \psi(x - v\, t, t)\, e^{i\left[\frac{v}{2a_1}(x-x_0) - \frac{v^2}{4a_1}(t-t_0)\right]}$

Equation: $i\,\Phi_t + a_1\,\Phi_{xx} + a_2\,|\Phi|^2\,\Phi = 0$

#	Example	Conditions	Name	Eq. #
1.	$\Phi(x, t) = A_0\sqrt{\frac{2a_1}{a_2}}\,\text{sech}\{A_0[x - (x_0 + v\,t)]\} \times e^{i\left[\frac{v}{2a_1}(x-x_0) + \frac{4a_1^2 A_0^2 - v^2}{4a_1}(t-t_0) + \phi_0\right]}$	$a_1 a_2 > 0$, A_0, x_0, t_0, v, and ϕ_0 are arbitrary real constants	moving bright soliton	(5.13)
2.	$\Phi(x, t) = A_0\sqrt{\frac{-2a_1}{a_2}}\,\tanh\{A_0[x - (x_0 + v\,t)]\} \times e^{-i\left[-\frac{v}{2a_1}(x-x_0) + \frac{8a_1^2 A_0^2 + v^2}{4a_1}(t-t_0) + \phi_0\right]}$	$a_1 a_2 < 0$, A_0, x_0, t_0, v and ϕ_0 are arbitrary real constants	moving dark soliton	(5.14)
3.	$\Phi(x, t) = \frac{1}{\sqrt{a_2}}\left[\frac{4 + i\,8(t-t_0)}{1 + 4(t-t_0)^2 + \frac{2}{a_1}(x - x_0 - v\,t)^2} - 1\right] \times e^{i\left[\frac{v}{2a_1}(x-x_0) - \frac{v^2}{4a_1}(t-t_0)\right]}\, e^{i[t - t_0 + \phi_0]}$	$a_2 > 0$, x_0, t_0, v, and ϕ_0 are arbitrary real constants	moving Peregrine soliton	(5.15)

Equation: $i\,\Phi_t + a_1\,\Phi_{xx} + a_2\,|\Phi|^n\,\Phi = 0$

#	Example	Conditions	Name	Eq. #
1.	$\Phi(x, t) = \left(\frac{2A_0^2 a_1 (n+2)}{a_2 n^2}\,\text{sech}^2\{A_0[x - (x_0 + v\,t)]\}\right)^{\frac{1}{n}} \times e^{i\left[\frac{v}{2a_1}(x-x_0) - \frac{v^2}{4a_1}(t-t_0)\right]}\, e^{i\left[\frac{4a_1 A_0^2}{n^2}(t-t_0) + \phi_0\right]}$	$a_1 a_2 (n+2) > 0$, A_0, x_0, t_0, v, and ϕ_0 are arbitrary real constants	moving bright soliton	(5.16)

(*Continued*)

Equation: $i\,\Phi_t + a_1\,\Phi_{xx} + a_2\,|\Phi|^n\,\Phi + a_3\,|\Phi|^m\,\Phi = 0$

#	Example	Conditions	Name	Eq. #
1.	$\Phi(x,t) = \left[\dfrac{A_0\,A_1\,(n+2)}{a_2\left(A_0 + 2\cosh\left\{n\sqrt{\frac{A_1}{a_1}}\,[x-(x_0+v\,t)]\right\}\right)}\right]^{\frac{1}{n}} \times e^{i\left[\frac{v}{2a_1}(x-x_0)-\frac{v^2}{4a_1}(t-t_0)\right]}\,e^{i\,[A_1\,(t-t_0)+\phi_0]}$	$A_0 = \sqrt{\frac{4\,a_2^2\,(n+1)}{A_1\,\delta\,(n+2)^2}}$, $\delta = a_3 + \frac{a_2^2\,(n+1)}{A_1\,(n+2)^2}$, $m = 2\,n$, $a_1\,A_1 > 0$, $A_1\,\delta\,(n+1) > 0$, x_0, t_0, A_1, v, and ϕ_0 are arbitrary real constants	moving flat-top soliton	(5.17)

Function Coefficients

Transformation: $\Phi(x,t) = A(x,t)\,e^{i\,B(x,t)}\,\psi[X(x,t),\,T(x,t)]$

Equation: $i\,\Phi_t + b_1(x,t)\,\Phi_{xx} + b_2(x,t)\,|\Phi|^2\,\Phi + [b_{3r}(x,t) + i\,b_{3i}(x,t)]\,\Phi = 0$

Constant Dispersion and Complex Potential

Equation: $i\,\Phi_t + b_{10}\,\Phi_{xx} + b_2(x,t)\,|\Phi|^2\,\Phi + [b_{3r}(x,t) + i\,b_{3i}(x,t)]\,\Phi = 0$

Constant Dispersion and Real Quadratic Potential

Transformation:

$$\Phi(x,t) = c_2\sqrt{g_5(t)}\,e^{i\left[c_1 - \frac{c_4^2}{4b_{10}}\int g_5^2(t)dt - \frac{2c_4 g_5^2(t) + g_5'(t)x}{4b_{10}g_5(t)}x\right]} \times \psi\left[c_3 + g_5(t)x + c_4\int g_5^2(t)dt,\ c_0 + \frac{b_{10}}{a_1}\int g_5^2(t)dt\right]$$

Equation 1: $i\,\Phi_t + b_{10}\,\Phi_{xx} + \frac{a_2\,b_{10}\,g_5(t)}{a_1\,c_2^2}\,|\Phi|^2\,\Phi - \frac{g_5(t)\,g_5''(t) - 2\,g_5'^2(t)}{4\,b_{10}\,g_5^2(t)}\,x^2\,\Phi = 0$

#	Example	Conditions	Name	Eq. #
1.	$\Phi(x,t)=\sqrt{\frac{2A_0^2\,a_1\,c_2^2\,g_5(t)}{a_2}}\;e^{i\,\phi(x,t)}$ $\times\,\text{sech}\left\{A_0\,[c_3-x_0+g_5(t)\,x+c_4\int g_5^2(t)\,dt]\right\}$	$\phi(x,t)=A_0^2\left[a_1\,(c_0-t_0)+b_{10}\int g_5^2(t)\,dt\right]-\frac{c_4^2}{4\,b_{10}}\int g_5^2(t)\,dt-\frac{\left[2\,c_4\,g_5^2(t)+x\,g_5'(t)\right]x}{4\,b_{10}\,g_5(t)}+c_1+\phi_0$ $a_1\,a_2>0$, A_0, x_0, t_0, and ϕ_0 are arbitrary real constants	bright soliton	(5.41)
2.	$\Phi(x,t)=\sqrt{\frac{-2A_0^2\,a_1\,c_2^2\,g_5(t)}{a_2}}\;e^{-i\,\phi(x,t)}$ $\times\,\tanh\{A_0\,[c_3-x_0+g_5(t)\,x+c_4\int g_5^2(t)\,dt]\}$	$\phi(x,t)=2\,A_0^2\,[a_1\,(c_0-t_0)+b_{10}\int g_5^2(t)\,dt]+\frac{c_4^2}{4\,b_{10}}\int g_5^2(t)\,dt+\frac{[2\,c_4\,g_5^2(t)+x\,g_5'(t)]\,x}{4\,b_{10}\,g_5(t)}-c_1+\phi_0$ $a_1\,a_2<0$, A_0, x_0, t_0, and ϕ_0 are arbitrary real constants	dark soliton	(5.46)
3.	$\Phi(x,t)=c_2\sqrt{\frac{g_5(t)}{a_2}}\;[\psi_1(x,t)+\psi_2(x,t)]$ $\times\,e^{i\,[c_1-\frac{2\,c_4\,g_5^2(t)\,x+c_4^2\,g_5(t)\int g_5^2(t)\,d\,t+g_5'(t)x^2}{4\,b_{10}\,g_5(t)}]}$	See text.	two bright solitons	(5.49)

Equation 2: $i\,\Phi_t+b_{10}\,\Phi_{xx}+\frac{a_2\,b_{10}\,[\alpha+\beta\,\sin(\gamma\,t)]}{a_1\,c_2^2}\,|\Phi|^2\,\Phi+\frac{\beta\,\gamma^2\,[3\,\beta+\beta\,\cos(2\,\gamma\,t)+2\,\alpha\,\sin(\gamma\,t)]}{8\,b_{10}\,[\alpha+\beta\,\sin(\gamma\,t)]^2}\,x^2\,\Phi=0$

#	Example	Conditions	Name	Eq. #
1.	$\Phi(x,t)=\sqrt{\frac{2A_0^2\,a_1\,c_2^2\,[\alpha+\beta\,\sin(\gamma\,t)]}{a_2}}$ $\times\,\text{sech}(A_0\,\{c_3-x_0+[\alpha+\beta\,\sin(\gamma\,t)]\,x+\frac{c_4\,[2\,\gamma\,(2\,\alpha^2+\beta^2)\,t-8\,\alpha\,\beta\,\cos(\gamma\,t)-\beta^2\,\sin(2\,\gamma\,t)]}{4\,\gamma}\})$ $\times\,e^{i\,\phi(x,t)}$	$\phi(x,t)=-\frac{\beta\,\gamma\,\cos(\gamma\,t)\,x^2+2\,c_4\,[\alpha+\beta\,\sin(\gamma\,t)]^2\,x}{4\,b_{10}\,[\alpha+\beta\,\sin(\gamma\,t)]}-\frac{c_4^2\,[2\,\gamma\,(2\,\alpha^2+\beta^2)\,t-8\,\alpha\,\beta\,\cos(\gamma\,t)-\beta^2\,\sin(2\,\gamma\,t)]}{16\,b_{10}\,\gamma}+\frac{A_0^2}{4\,\gamma}\,[4\,\gamma\,a_1\,(c_0-t_0)+b_{10}\,[2\,\gamma\,(2\,\alpha^2+\beta^2)\,t-8\,\alpha\,\beta\,\cos(\gamma\,t)-\beta^2\,\sin(2\,\gamma\,t)]]+c_1+\phi_0$	bright soliton	(5.42)

(Continued)

2. $\Phi(x,t)=\sqrt{\frac{-2A_0^2\,a_1\,c_2^2\,[\alpha+\beta\sin(\gamma t)]}{a_2}}\times\tanh[A_0\,(c_3-x_0+[\alpha+\beta\sin(\gamma t)]\,x+\frac{c_4\,[2\gamma\,(2\alpha^2+\beta^2)\,t-8\,\alpha\,\beta\cos(\gamma t)-\beta^2\sin(2\gamma t)]}{4\gamma})]\times e^{-i\,\phi(x,t)}$	$\phi(x,t)=\frac{\beta\,\gamma\cos(\gamma t)\,x^2+2\,c_4\,[\alpha+\beta\sin(\gamma t)]^2\,x}{4\,b_{10}\,[\alpha+\beta\sin(\gamma t)]}+\frac{c_4^2\,[2\gamma\,(2\alpha^2+\beta^2)\,t-8\,\alpha\,\beta\cos(\gamma t)-\beta^2\sin(2\gamma t)]}{16\,b_{10}\,\gamma}+\frac{A_0^2}{2\gamma}[4\,\gamma\,a_1\,(c_0-t_0)+b_{10}\,[2\,\gamma\,(2\,\alpha^2+\beta^2)\,t-8\,\alpha\,\beta\cos(\gamma t)-\beta^2\sin(2\,\gamma\,t)]]-c_1+\phi_0$	dark soliton	(5.47)
3. $\Phi(x,t)=c_2\sqrt{\frac{\alpha+\beta\sin(\gamma t)}{a_2}}\,[\psi_1(x,t)+\psi_2(x,t)]\,e^{i\,\phi(x,t)}$	See text.	two bright solitons	(5.50)

Equation 3: $i\,\Phi_t+b_{10}\,\Phi_{xx}+\frac{a_2\,b_{10}\,e^{\gamma t}}{a_1\,c_2^2}\,|\Phi|^2\,\Phi+\frac{\gamma^2}{4\,b_{10}}\,x^2\,\Phi=0$

# Example	Conditions	Name	Eq. #
1. $\Phi(x,t)=c_2\,A_0\sqrt{\frac{2\,a_1}{a_2}\,e^{\gamma t}}\,\mathrm{sech}[A_0\,(\frac{c_4\,e^{2\gamma t}}{2\gamma}+e^{\gamma t}\,x-x_0+c_3)]\times e^{i\,\phi(x,t)}$	$\phi(x,t)=\frac{1}{8\,b_{10}\,\gamma}\{-c_4\,e^{\gamma t}\,(c_4\,e^{\gamma t}+4\,\gamma\,x)+2\,\gamma\,[4\,b_{10}\,(c_1+\phi_0)-\gamma\,x^2]+4\,A_0^2\,b_{10}\,[b_{10}\,e^{2\gamma t}+2\,a_1\,\gamma\,(c_0-t_0)]\}+\phi_0$, γ is an arbitrary real constant	bright soliton	(5.44)
2. $\Phi(x,t)=c_2\,A_0\sqrt{\frac{-2\,a_1}{a_2}\,e^{\gamma t}}\,\tanh[A_0\,(\frac{c_4\,e^{2\gamma t}}{2\gamma}+e^{\gamma t}\,x-x_0+c_3)]\times e^{i\,\phi(x,t)}$	$\phi(x,t)=\frac{-1}{8\,b_{10}\,\gamma}\{c_4\,e^{\gamma t}\,(c_4\,e^{\gamma t}+4\,\gamma\,x)+2\,\gamma\,[4\,b_{10}\,(\phi_0-c_1)+\gamma\,x^2]+8\,A_0^2\,b_{10}\,[b_{10}\,e^{2\gamma t}+2\,a_1\,\gamma\,(c_0-t_0)]\}+\phi_0$, γ is an arbitrary real constant	dark soliton	(5.48)
3. $\Phi(x,t)=\sqrt{\frac{e^{\frac{t}{2}}}{a_2}}\,[\psi_1(x,t)+\psi_2(x,t)]\,e^{-i\,\frac{1}{8}\,(2\,e^t+4\,e^{\frac{t}{2}}\,x+x^2)}$	See text.	two bright solitons	(5.51)

Constant Dispersion and Real Linear Potential

Transformation: $\Phi(x,t)=\sqrt{\frac{c_2^2\,c_5}{c_7}}\,\psi[\frac{c_5\,x}{c_7}+g_4(t),\,c_0+\frac{b_{10}\,c_5^2\,t}{a_1\,c_7^2}]\,e^{i\,[c_1-\frac{c_7^2}{4\,b_{10}\,c_5^2}\int g_4'^{2}(t)\,dt-\frac{c_7\,g_4'(t)\,x}{2\,b_{10}\,c_5}]}$

Equation 1: $i\,\Phi_t + b_{10}\,\Phi_{xx} + \frac{a_2\,b_{10}\,c_5}{a_1\,c_2^2\,c_7}\,\lvert\Phi\rvert^2\,\Phi - \frac{c_7\,g_4''(t)}{2\,b_{10}\,c_5}\,x\,\Phi = 0$

1. $\Phi(x,t) = c_2\,A_0\sqrt{\frac{2\,a_1\,c_5}{a_2\,c_7}}\,\mathrm{sech}\{A_0\,[\frac{c_5}{c_7}\,x - x_0 + g_4(t)]\}\,e^{i\,\phi(x,t)}$	$\phi(x,t) = A_0^2\,a_1\,[c_0 + \frac{b_{10}\,c_5^2}{a_1\,c_7^2}\,(t - t_0)] - \frac{c_7^2}{4\,b_{10}\,c_5^2}\int g_4'^2(t)\,dt - \frac{c_7\,g_4'(t)}{2\,b_{10}\,c_5}\,x + c_1 + \phi_0,$	bright soliton	(5.56)
	$a_1\,a_2\,c_5\,c_7 > 0$, A_0, x_0, t_0, and ϕ_0 are arbitrary real constants		
2. $\Phi(x,t) = c_2\,A_0\sqrt{\frac{-2\,a_1\,c_5}{a_2\,c_7}}\,\tanh\{A_0\,[\frac{c_5}{c_7}\,x - x_0 + g_4(t)]\}\,e^{-i\,\phi(x,t)}$	$\phi(x,t) = 2\,A_0^2\,a_1\,[c_0 + \frac{b_{10}\,c_5^2}{a_1\,c_7^2}\,(t - t_0)] + \frac{c_7^2}{4\,b_{10}\,c_5^2}\int g_4'^2(t)\,dt + \frac{c_7\,g_4'(t)}{2\,b_{10}\,c_5}\,x - c_1 + \phi_0$	dark soliton	(5.59)
3. $\Phi(x,t) = A_0\sqrt{\frac{c_2^2\,c5}{a_2\,c_7}}\,(1 - \frac{\sqrt{8}\,\lambda_r}{A_0}\,p(x,t)) \times e^{i\,[A_0^2\,(c_0+\frac{b_{10}c_5^2 t}{a_1\,c_7^2}-t_0)-\frac{c_7^2}{4\,b_{10}\,c_5^2}\int g_4'^2(t)d\,t-\frac{c_7\,g_4'(t)\,x}{2\,b_{10}\,c_5}+c_1+\phi_0]}$	See text.	generalized first-order breather	(5.61)

Equation 2: $i\,\Phi_t + b_{10}\,\Phi_{xx} + \frac{a_2\,b_{10}\,c_5}{a_1\,c_2^2\,c_7}\,\lvert\Phi\rvert^2\,\Phi - \frac{c_7\,\alpha}{b_{10}\,c_5}\,x\,\Phi = 0$

1. $\Phi(x,t) = c_2\,A_0\sqrt{\frac{2\,a_1\,c_5}{a_2\,c_7}}\,\mathrm{sech}\{A_0\,[\frac{c_5}{c_7}\,x - x_0 + \alpha\,t^2]\} \times e^{i\,\{A_0^2\,a_1\,[c_0+\frac{b_{10}\,c_5^2}{a_1\,c_7^2}\,(t-t_0)]-\frac{c_7^2\,\alpha^2\,t^3}{3\,b_{10}\,c_5^2}-\frac{c_7\,\alpha\,t}{b_{10}\,c_5}\,x+c_1+\phi_0\}}$	—	bright soliton	(5.57)
2. $\Phi(x,t) = c_2\,A_0\sqrt{\frac{-2\,a_1\,c_5}{a_2\,c_7}}\,\tanh\{A_0\,[\frac{c_5}{c_7}\,x - x_0 + \alpha\,t^2]\} \times e^{-i\,\{2\,A_0^2\,a_1\,[c_0+\frac{b_{10}\,c_5^2}{a_1\,c_7^2}\,(t-t_0)]+\frac{c_7^2\,\alpha^2\,t^3}{3\,b_{10}\,c_5^2}+\frac{c_7\,\alpha\,t}{b_{10}\,c_5}\,x-c_1+\phi_0\}}$	—	dark soliton	(5.60)
3. $\Phi(x,t) = A_0\sqrt{\frac{c_2^2\,c5}{a_2\,c_7}}\,(1 - \frac{\sqrt{8}\,\lambda_r}{A_0}\,p(x,t)) \times e^{i\,[A_0^2\,(c_0+\frac{b_{10}\,c_5^2\,t}{a_1\,c_7^2}-t_0)-\frac{c_7\,\alpha\,(3\,c_5\,x+c_7\,\alpha\,t^2)\,t}{3\,b_{10}\,c_5^2}+c_1+\phi_0]}$	See text.	generalized first-order breather	(5.62)

(Continued)

Constant Nonlinearity and Complex Potential

Transformation: $\Phi(x, t) = A(x, t)\, e^{i\, B(x,t)}\, \psi[X(x, t),\, T(x, t)]$

Equation: $i\, \Phi_t + b_1(x, t)\, \Phi_{xx} + b_{20}\, |\Phi|^2\, \Phi + [b_{3r}(x, t) + i\, b_{3i}(x, t)]\, \Phi = 0$

Constant Nonlinearity and Real Quadratic Potential

Transformation: $\Phi(x, t) = c_2 \sqrt{g_5(t)}\, e^{i\{c_1 - \frac{a_2[2 c_4 g_5(t) x + g_5'(t) x^2 + c_4^2 \int g_5(t) dt]}{4 a_1 b_{20} c_2^2}\}}\, \psi[c_3 + g_5(t) x + c_4 \int g_5(t) dt,\, c_0 + \frac{b_{20} c_2^2}{a_2} \int g_5(t) dt]$

Equation: $i\, \Phi_t + \frac{a_1\, b_{20}\, c_2^2}{a_2\, g_5(t)}\, \Phi_{xx} + b_{20}\, |\Phi|^2\, \Phi + \frac{a_2\, [g_5'^2(t) - g_5(t)\, g_5''(t)]}{4\, a_1\, b_{20}\, c_2^2\, g_5(t)}\, x^2\, \Phi = 0$

Constant Nonlinearity and Real Linear Potential

Transformation: $\Phi(x, t) = c_2 \sqrt{c_5 e^{c_6\, t}}\, e^{i\, \{c_1 - \frac{a_2[c_5^2 c_6 e^{c_6 t} x^2 + 2 c_5 g_4'(t) x + \int g_4'^2(t) e^{-c_6 t} dt]}{4 a_1 b_{20} c_2^2 c_5}\}}\, \psi[c_5 e^{c_6\, t} x + g_4(t),\, c_0 + \frac{b_{20}\, c_2^2\, c_5\, (e^{c_6 t} - e^{c_6})}{a_2\, c_6}]$

Solution-Dependent Transformation

Special Case 1: Stationary Solution, Constant Dispersion and Nonlinearity Coefficients

Transformation: $\Phi(x, t) = \sqrt{\frac{a_2}{b_{20}}}\, e^{i\, B(x,t)}\, \psi[\sqrt{\frac{a_1}{b_{10}}}\, (x - x_0),\, (t - t_0)] = A_0\, \sqrt{\frac{a_2}{b_{20}}}\, f[\sqrt{\frac{a_1}{b_{10}}}\, (x - x_0)]$

$e^{i\, [B(x,t) + A_0^2\, (t - t_0)]}$

Equation: $i\, \Phi_t + b_{10}\, \Phi_{xx} + b_{20}\, |\Phi|^2$

$$\Phi + [B_t(x, t) + b_{10}\, B_x^2(x, t) - i\, b_{10}\, (\frac{\sqrt{\frac{4 a_1}{b_{10}}}\, f'[\sqrt{\frac{a_1}{b_{10}}}\, (x - x_0)]\, B_x(x, t)}{f[\sqrt{\frac{a_1}{b_{10}}}\, (x - x_0)]} + B_{xx}(x, t))]\, \Phi = 0$$

Special Case 2: PT-symmetric Potential

Transformation 1: $\Phi(x, t) = \sqrt{\frac{a_2}{b_{20}}}\, e^{i\, B(x)}\, \psi[\sqrt{\frac{a_1}{b_{10}}}\, (x - x_0),\, (t - t_0)] = A_0\, \sqrt{\frac{a_2}{b_{20}}}\, f[\sqrt{\frac{a_1}{b_{10}}}\, (x - x_0)]\, e^{i\, [B(x) + A_0^2\, (t - t_0)]}$

Equation: $i\, \Phi_t + b_{10}\, \Phi_{xx} + b_{20}\, |\Phi|^2\, \Phi + [V_{\text{even}}(x) + i\, V_{\text{odd}}(x)]\Phi = 0$

Transformation 2: $\Phi(x, t) = \sqrt{\frac{a_2}{b_{20}}}\, e^{i \sin(x-x_0)}\, \psi[\sqrt{\frac{a_1}{b_{10}}}\,(x - x_0), (t - t_0)]$

Equation: $i\, \Phi_t + b_{10}\, \Phi_{xx} + b_{20}\, |\Phi|^2\, \Phi + b_{10}\, [\cos^2(x - x_0) + i\, \sin(x - x_0)]\, \Phi = 0$

Special Case 3: Stationary Solution, Constant Dispersion and Nonlinearity Coefficients, and Real Potential

Transformation: $\Phi(x, t) = \sqrt{\frac{a_2}{b_{20}}}\, e^{i\, B(x,t)}\, \psi[\sqrt{\frac{a_1}{b_{10}}}\,(x - x_0), (t - t_0)] = A_0 \sqrt{\frac{a_2}{b_{20}}}\, f[\sqrt{\frac{a_1}{b_{10}}}\,(x - x_0)]$
$e^{i\,[B(x,t)+A_0^2\,(t-t_0)]}$

Equation: $i\Phi_t + b_{10}\, \Phi_{xx} + b_{20}\, |\Phi|^2\, \Phi + \{\frac{b_{10}\, g_1^2(t)}{f^4[\sqrt{\frac{a_1}{b_{10}}}\,(x-x_0)]} + g_1'(t) \int \frac{dx}{f^2[\sqrt{\frac{a_1}{b_{10}}}\,(x-x_0)]} + g_2'(t)\}\Phi = 0$

5.7 Other Equations: NLSE with Periodic Potentials

5.7.1 General Case: $\mathrm{sn}^2(x, m)$ Potential

Equation:

$$i\,\psi_t + \frac{1}{2}\,\psi_{xx} - |\psi|^2\,\psi + V_0\,\mathrm{sn}^2(x, m)\,\psi = 0, \tag{5.100}$$

where

$\psi = \psi(x, t)$ is the complex function profile,
x and t are its two independent variables,
V_0 is a real constant.

Solutions:

***Solution* 1. sn(*x*,*m*)** *solitary wave (SW)*

$$\psi(x, t) = \sqrt{V_0 + m}\;\mathrm{sn}(x - x_0, m)\,e^{-i\left[\frac{(1+m)\,(t-t_0)}{2} + \phi_0\right]}, \tag{5.101}$$

where

$V_0 + m \geqslant 0$,
x_0, t_0, and ϕ_0 are arbitrary real constants.

- *Reference*: [1], *we corrected the constant prefactor and the exponential term.*

***Solution* 2. cn(*x*,*m*)** *SW*

$$\psi(x, t) = \sqrt{-(V_0 + m)}\;\mathrm{cn}(x - x_0, m)\,e^{i\left[\left(V_0 + m - \frac{1}{2}\right)(t-t_0) + \phi_0\right]}, \tag{5.102}$$

where

$V_0 + m \leqslant 0$,
x_0, t_0, and ϕ_0 are arbitrary real constants.

- *Reference*: [1], *we corrected the constant prefactor and the exponential term.*

***Solution* 3. dn(*x*,*m*)** *SW*

$$\psi(x, t) = \sqrt{-\left(1 + \frac{V_0}{m}\right)}\;\mathrm{dn}(x - x_0, m)\,e^{i\left[\left(1 + \frac{V_0}{m} - \frac{m}{2}\right)(t-t_0) + \phi_0\right]}, \tag{5.103}$$

where

$1 + \frac{V_0}{m} \leqslant 0$,
x_0, t_0, and ϕ_0 are arbitrary real constants.

- *Reference*: [1], *we corrected the constant prefactor and the exponential term.*

5.7.2 Specific Case: $\sin^2(x)$ Potential

Equation:

$$i\,\psi_t + \frac{1}{2}\,\psi_{xx} - |\psi|^2\,\psi + V_0 \sin^2(x)\,\psi = 0, \tag{5.104}$$

where V_0 is a real constant.

Solutions:

***Solution* 1. sin(*x*)**

$$\psi(x,\,t) = \sqrt{V_0}\,\sin(x - x_0)\,e^{-i\left(\frac{t-t_0}{2}+\phi_0\right)}, \tag{5.105}$$

where
$V_0 > 0$,
x_0, t_0, and ϕ_0 are arbitrary real constants.

- *Reference*: [1].

***Solution* 2. cos(*x*)**

$$\psi(x,\,t) = \sqrt{-V_0}\,\cos(x - x_0)\,e^{-i\left[\left(V_0-\frac{1}{2}\right)(t-t_0)+\phi_0\right]}, \tag{5.106}$$

where
$V_0 < 0$,
x_0, t_0, and ϕ_0 are arbitrary real constants.

- *Reference*: [1].

5.8 Summary of Section 5.7

Equation			
$i\,\psi_t + \frac{1}{2}\,\psi_{xx} - \lvert\psi\rvert^2\,\psi + V_0\,\mathrm{sn}^2(x,\,m)\,\psi = 0$			
# Solution	**Conditions**	**Name**	**Eq. #**
1. $\psi(x,\,t) = \sqrt{V_0 + m}\,\mathrm{sn}(x - x_0,\,m) \times e^{-i\,[\frac{(1+m)\,(t-t_0)}{2} + \phi_0]}$	$V_0 + m \geqslant 0$, x_0, t_0, and ϕ_0 are arbitrary real constants	solitary wave	(5.101)
2. $\psi(x,\,t) = \sqrt{-(V_0 + m)}\,\mathrm{cn}(x - x_0,\,m) \times e^{i\,[(V_0+m-\frac{1}{2})\,(t-t_0)+\phi_0]}$	$V_0 + m \leqslant 0$,	solitary wave	(5.102)
	x_0, t_0, and ϕ_0 are arbitrary real constants		
3. $\psi(x,\,t) = \sqrt{-(1 + \frac{V_0}{m})}\,\mathrm{dn}(x - x_0,\,m) \times e^{i\,[(1+\frac{V_0}{m}-\frac{m}{2})\,(t-t_0)+\phi_0]}$,	solitary wave		(5.103)
	x_0, t_0, and ϕ_0 are arbitrary real constants		

Equation			
$i\,\psi_t + \frac{1}{2}\,\psi_{xx} - \lvert\psi\rvert^2\,\psi + V_0\,\sin^2(x)\,\psi = 0$			
# Solution	**Conditions**	**Name**	**Eq. #**
1. $\psi(x,\,t) = \sqrt{V_0}\,\sin(x - x_0)\,e^{-i\,[\frac{(t-t_0)}{2} + \phi_0]}$	$V_0 > 0$,	–	(5.105)
	x_0, t_0, and ϕ_0 are arbitrary real constants		
2. $\psi(x,\,t) = \sqrt{-V_0}\,\cos(x - x_0)\,e^{-i\,[(V_0-\frac{1}{2})\,(t-t_0)+\phi_0]}$	$V_0 < 0$,	–	(5.106)
	x_0, t_0, and ϕ_0 are arbitrary real constants		

Reference

[1] Bronski J C, Carr L D, Deconinck B and Kutz J N 2001 Bose–Einstein condensates in standing waves: The cubic nonlinear Schrödinger equation with a periodic potential *Phys. Rev. Lett.* **86** 1402–5

IOP Publishing

Handbook of Exact Solutions to the Nonlinear Schrödinger Equations

Usama Al Khawaja and Laila Al Sakkaf

Chapter 6

Nonlinear Schrödinger Equation in $(N + 1)$-Dimensions

A Glance at Chapter 6

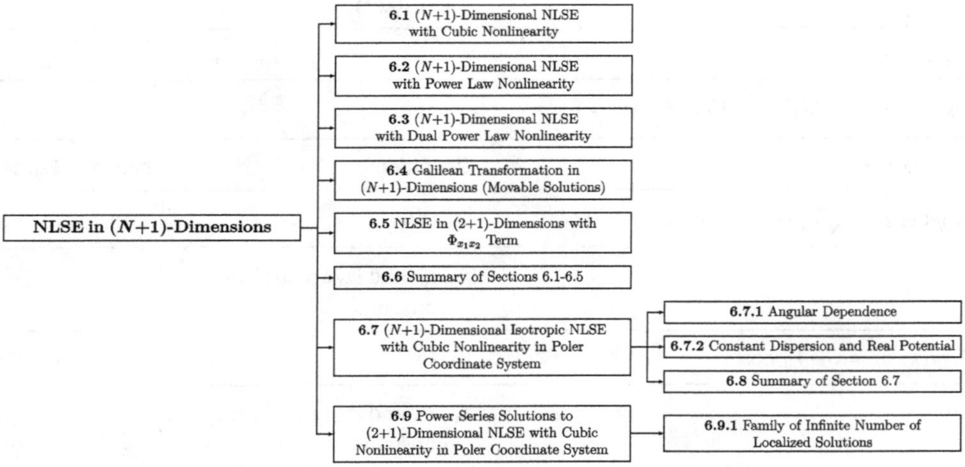

doi:10.1088/978-0-7503-2428-1ch6

A Statistical View of Chapter 6

	Equation	Solutions
1	$i\,\Phi_t + \sum_{k=1}^{N} \alpha_k\, \Phi_{x_k x_k} + a_2\, \lvert\Phi\rvert^2\, \Phi = 0$	14
2	$i\,\Phi_t + \sum_{k=1}^{N} \alpha_k\, \Phi_{x_k x_k} + a_2\, \lvert\Phi\rvert^n\, \Phi = 0$	4
3	$i\,\Phi_t + \sum_{k=1}^{N} \alpha_k\, \Phi_{x_k x_k} + a_2\, \lvert\Phi\rvert^n\, \Phi + a_3\, \lvert\Phi\rvert^m\, \Phi = 0$	8
4	$i\,\Phi_t + b_1\, \Phi_{x_1x_1} + b_2\, \Phi_{x_2x_2} + b_3\, \Phi_{x_1x_2} + b_4\, \lvert\Phi\rvert^2\, \Phi = 0$	2
5	$i\,\Phi_t + b_1(r,\, t)\left(\Phi_{rr} + \frac{N-1}{r}\, \Phi_r\right) + b_2(r,\, t)\, \lvert\Phi\rvert^2\, \Phi + [b_{3r}(r,\, t) + i\, b_{3i}(r,\, t)]\Phi = 0$	0
6	$i\,\Phi_t + a_1\left(\Phi_{rr} + \frac{N-1}{r}\, \Phi_r + \frac{1}{r^2}\, \Phi_{\theta\theta}\right) + a_2\, r^{N-1}\, \lvert\Phi\rvert^2\, \Phi = 0$, for $N = 1$: $\Phi = \Phi(r)$	1
7	$i\,\Phi_t + a_1\left(\Phi_{rr} + \frac{1}{r}\, \Phi_r - \frac{1}{4r^2}\, \Phi\right) + a_2\, r\, \lvert\Phi\rvert^2\, \Phi = 0$	1
8	$i\,\Phi_t + a_1\left(\Phi_{rr} + \frac{2}{r}\, \Phi_r\right) + a_2\, r^2\, \lvert\Phi\rvert^2\, \Phi = 0$	1

9	$i\,\Phi_t + b_{10}\,\Phi_{rr} + \frac{b_{10}\,(N-1)}{r}\,\Phi_r + \frac{a_2\,\sqrt{g_1'(t)}}{c_0^2\,r^{1-N}}\,\vert\Phi\vert^2\,\Phi + \left\{\frac{b_{10}\,(N-1)\,(N-3)}{4\,r^2} + \frac{g_2'^2(t)}{4\,a_1\,g_1'(t)} + g_4'(t) + \frac{\left[g_2'(t)\,g_1''(t) - g_1'(t)\,g_2''(t)\right]r}{2\,\sqrt{a_1\,b_{10}}\,g_1'^{3/2}(t)} + \frac{\left[3\,g_1''^2(t) - 2\,g_1'(t)\,g_1'''(t)\right]r^2}{16\,b_{10}\,g_1'^2(t)}\right\}\Phi = 0$	2
10	$-\lambda\,Z(r) + a_1\left[Z''(r) + \frac{1}{r}\,Z'(r) - \frac{\alpha^2}{r^2}\,Z(r)\right] + a_2\,Z^3(r) = 0$	1
Total	10	34

6.1 (N + 1)-Dimensional NLSE with Cubic Nonlinearity

If $\psi(x, t; a_1, a_2)$ is a solution to the fundamental NLSE, (2.1),

$$i\,\psi_t + a_1\,\psi_{xx} + a_2\,|\psi|^2\,\psi = 0,$$

then

$$\begin{aligned}&\Phi(x_1, x_2, \dots, x_N, t; \alpha_1, \alpha_2, \dots, \alpha_N, a_2)\\ &= \psi\big[\,c_1\,(x_1 - x_{01}) + c_2\,(x_2 - x_{02}) + \cdots + c_N\,(x_N - x_{0N}),\ t;\\ &\quad c_1^2\,\alpha_1 + c_1^2\,\alpha_2 + \cdots + c_N^2\,\alpha_N,\ a_2\big]\end{aligned} \tag{6.1}$$

is a solution of

$$i\,\Phi_t + \sum_{k=1}^{N} \alpha_k\,\Phi_{x_k x_k} + a_2\,|\Phi|^2\,\Phi = 0, \tag{6.2}$$

with the replacements

$$x \to \sum_{k=1}^{N} c_k\,(x_k - x_{0k}),$$
$$t \to t,$$
$$a_1 \to \sum_{k=1}^{N} c_k^2\,\alpha_k,$$
$$a_2 \to a_2,$$

where α_k, a_2, c_k, and x_{0k} are arbitrary real constants.

Example* 1. sech(x_1, x_2) *2D bright soliton
(Figure 6.1)
Given

$$\psi(x, t) = A_0\,\sqrt{\frac{2a_1}{a_2}}\,\operatorname{sech}[A_0\,(x - x_0)]\,e^{i\left[a_1\,A_0^2\,(t - t_0) + \phi_0\right]}$$

is a solution of (2.1), then

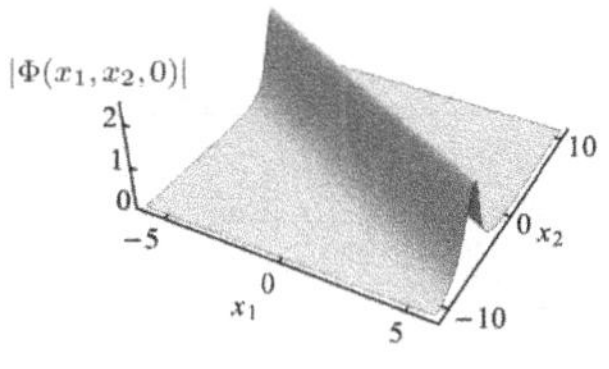

Figure 6.1. 2D bright soliton (6.3) at $t = 0$, with $\alpha_1 = \alpha_2 = A_0 = c_1 = c_2 = a_2 = 1$ and $x_{01} = x_{02} = t_0 = \phi_0 = 0$.

$$\Phi(x_1, x_2, t) = A_0 \sqrt{\frac{2\left(c_1^2\,\alpha_1 + c_2^2\,\alpha_2\right)}{a_2}}\ \operatorname{sech}\{A_0\,[c_1\,(x_1 - x_{01}) + c_2\,(x_2 - x_{02})]\} \times e^{i\left[A_0^2\left(c_1^2\,\alpha_1 + c_2^2\,\alpha_2\right)(t-t_0)+\phi_0\right]} \tag{6.3}$$

is a solution of

$$i\,\Phi_t + \alpha_1\,\Phi_{x_1x_1} + \alpha_2\,\Phi_{x_2x_2} + a_2\,|\Phi|^2\,\Phi = 0, \tag{6.4}$$

where
$a_2\,(c_1^2\,\alpha_1 + c_2^2\,\alpha_2) > 0$,
A_0, t_0, and ϕ_0 are arbitrary real constants.

***Example* 2. sech(x_1, x_2, x_3)** *3D bright soliton*
Given

$$\psi(x, t) = A_0\sqrt{\frac{2\,a_1}{a_2}}\ \operatorname{sech}[A_0\,(x - x_0)]\ e^{i\left[a_1\,A_0^2\,(t-t_0)+\phi_0\right]}$$

is a solution of (2.1), then

$$\Phi(x_1, x_2, x_3, t) = A_0\sqrt{\frac{2\left(c_1^2\,\alpha_1 + c_2^2\,\alpha_2 + c_3^2\,\alpha_3\right)}{a_2}} \times \operatorname{sech}\{A_0\,[c_1\,(x_1 - x_{01}) + c_2\,(x_2 - x_{02}) + c_3\,(x_3 - x_{03})]\}\ e^{i\left[A_0^2\left(c_1^2\,\alpha_1 + c_2^2\,\alpha_2 + c_3^2\,\alpha_3\right)(t-t_0)+\phi_0\right]} \tag{6.5}$$

is a solution of

$$i\,\Phi_t + \alpha_1\,\Phi_{x_1x_1} + \alpha_2\,\Phi_{x_2x_2} + \alpha_3\,\Phi_{x_3x_3} + a_2\,|\Phi|^2\,\Phi = 0, \tag{6.6}$$

where
$a_2\,(c_1^2\,\alpha_1 + c_2^2\,\alpha_2 + c_3^2\,\alpha_3) > 0$,
A_0, t_0, and ϕ_0 are arbitrary real constants.

***Example* 3. tanh(x_1, x_2)** *2D dark soliton*
(Figure 6.2)
Given

$$\psi(x, t) = A_0\sqrt{\frac{-2\,a_1}{a_2}}\ \tanh[A_0\,(x - x_0)]\ e^{-i\left[2\,a_1\,A_0^2\,(t-t_0)+\phi_0\right]}$$

is a solution of (2.1), then

$$\Phi(x_1, x_2, t) = A_0\sqrt{\frac{-2\left(c_1^2\,\alpha_1 + c_2^2\,\alpha_2\right)}{a_2}} \times \tanh\{A_0\,[c_1\,(x_1 - x_{01}) + c_2\,(x_2 - x_{02})]\} \times e^{-i\left[2\,A_0^2\left(c_1^2\,\alpha_1 + c_2^2\,\alpha_2\right)(t-t_0)+\phi_0\right]} \tag{6.7}$$

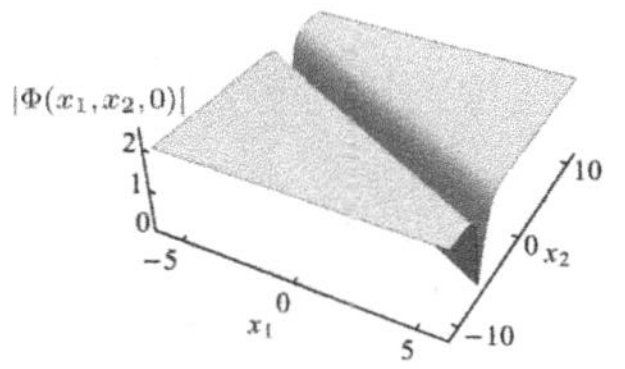

Figure 6.2. 2D dark soliton (6.7) at $t = 0$, with $\alpha_1 = \alpha_2 = A_0 = c_1 = c_2 = 1$, $a_2 = -1$ and $x_{01} = x_{02} = t_0 = \phi_0 = 0$.

is a solution of (6.4), where

$a_2\,(c_1^2\,\alpha_1 + c_2^2\,\alpha_2) < 0$,

A_0, t_0, and ϕ_0 are arbitrary real constants.

***Example* 4. tanh(x_1, x_2, x_3)** *3D dark soliton*

Given

$$\psi(x,\,t) = A_0\,\sqrt{\frac{-2\,a_1}{a_2}}\;\tanh[A_0\,(x - x_0)]\,e^{-i\left[2\,a_1\,A_0^2\,(t-t_0)+\phi_0\right]}$$

is a solution of (2.1), then

$$\begin{aligned}\Phi(x_1,\,x_2,\,x_3,\,t) = A_0\,&\sqrt{\frac{-2\left(c_1^2\,\alpha_1 + c_2^2\,\alpha_2 + c_3^2\,\alpha_3\right)}{a_2}}\\ &\times \tanh\{A_0\,[c_1\,(x_1 - x_{01}) + c_2\,(x_2 - x_{02})\\ &+ c_3\,(x_3 - x_{03})]\}\;e^{-i\left[2\,A_0^2\left(c_1^2\,\alpha_1+c_2^2\,\alpha_2+c_3^2\,\alpha_3\right)(t-t_0)+\phi_0\right]}\end{aligned} \tag{6.8}$$

is a solution of (6.6), where

$a_2\,(c_1^2\,\alpha_1 + c_2^2\,\alpha_2 + c_3^2\,\alpha_3) < 0$,

A_0, t_0, and ϕ_0 are arbitrary real constants.

***Example* 5. Periodicity in *t* and Localization in x_1 and x_2** *2D Kuznetsov–Ma breather*

(Figure 6.3)

Given

$$\begin{aligned}\psi(x,\,t) = \frac{1}{\sqrt{a_2}}&\left\{\frac{-p^2\,\cos[\omega\,(t - t_0)] - 2\,i\,p\,\nu\,\sin[\omega\,(t - t_0)]}{2\,\cos[\omega\,(t - t_0)] - 2\,\nu\,\cosh\left[\frac{p}{\sqrt{2\,a_1}}\,(x - x_0)\right]} - 1\right\}\\ &\times e^{i\,[t-t_0+\phi_0]}\end{aligned}$$

is a solution of (2.1), then

$$\begin{aligned}\Phi(x_1,\,x_2,\,t) = \frac{-1}{\sqrt{a_2}}&\left(\frac{p^2\,\cos[\omega\,(t - t_0)] + 2\,i\,p\,\nu\,\sin[\omega\,(t - t_0)]}{2\,\cos[\omega\,(t - t_0)] - q(x_1,\,x_2)} + 1\right)\\ &\times e^{i\,[(t-t_0)+\phi_0]}\end{aligned} \tag{6.9}$$

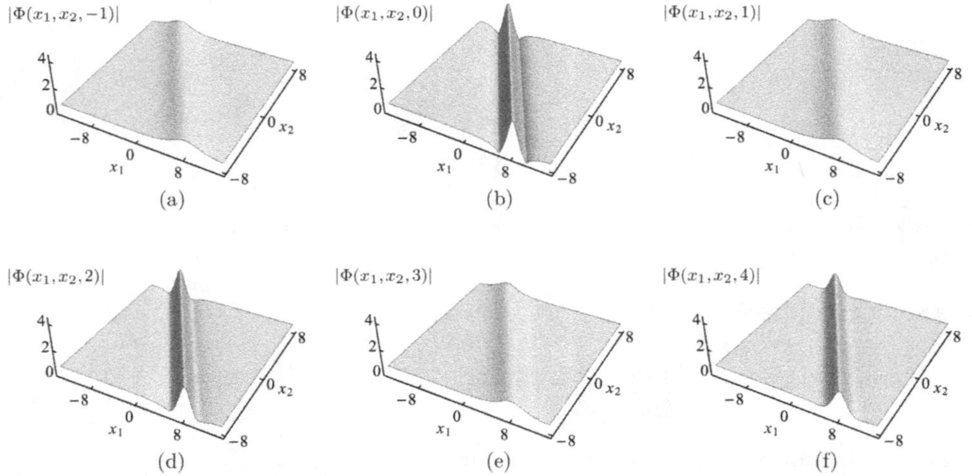

Figure 6.3. 2D Kuznetsov–Ma breather (6.9). (a) at $t = -1$, (b) at $t = 0$, (c) at $t = 1$, (d) at $t = 2$, (e) at $t = 3$, and (f) at $t = 4$. Used parameters: $\alpha_1 = \alpha_2 = c_1 = c_2 = 3$, $a_2 = 2$, $x_{01} = x_{02} = t_0 = \phi_0 = 0$, and $\nu = 1.5$. Animation available online at https://iopscience.iop.org/book/978-0-7503-2428-1.

is a solution of (6.4), where

$$q(x_1, x_2) = 2\,\nu\,\cosh\left\{\frac{p}{\sqrt{2\,(c_1^2\,\alpha_1 + c_2^2\,\alpha_2)}}\,[c_1\,(x_1 - x_{01}) + c_2\,(x_2 - x_{02})]\right\},$$

$p = 2\sqrt{\nu^2 - 1}$,
$\omega = p\,\nu$,
$\nu > 1$,
$a_2 > 0$,
$(c_1^2\,\alpha_1 + c_2^2\,\alpha_2) > 0$,
t_0 and ϕ_0 are arbitrary real constants.

***Example* 6. Periodicity in *t* and Localization in x_1x_2, and x_3** *3D Kuznetsov–Ma breather*

Given

$$\psi(x, t) = \frac{1}{\sqrt{a_2}}\left\{\frac{-p^2\cos[\omega\,(t - t_0)] - 2\,i\,p\,\nu\,\sin[\omega\,(t - t_0)]}{2\cos[\omega\,(t - t_0)] - 2\,\nu\,\cosh\left[\frac{p}{\sqrt{2\,a_1}}\,(x - x_0)\right]} - 1\right\} \times e^{i\,[t - t_0 + \phi_0]}$$

is a solution of (2.1), then

$$\Phi(x_1, x_2, x_3, t) = \frac{-1}{\sqrt{a_2}}\left(\frac{p^2\cos[\omega\,(t - t_0)] + 2\,i\,p\,\nu\,\sin[\omega\,(t - t_0)]}{2\cos[\omega\,(t - t_0)] - q(x_1, x_2, x_3)} + 1\right) \times e^{i\,[(t - t_0) + \phi_0]} \tag{6.10}$$

is a solution of (6.6), where

$$q(x_1, x_2, x_3) = 2\,\nu \cosh\left\{\frac{p}{\sqrt{2\,(c_1^2\,\alpha_1 + c_2^2\,\alpha_2 + c_3^2\,\alpha_3)}}\right.$$
$$\left.\times\,[c_1\,(x_1 - x_{01}) + c_2\,(x_2 - x_{02}) + c_3\,(x_3 - x_{03})]\right\},$$

$p = 2\,\sqrt{\nu^2 - 1}$,
$\omega = p\,\nu$,
$\nu > 1$,
$a_2 > 0$,
$(c_1^2\,\alpha_1 + c_2^2\,\alpha_2 + c_3^2\,\alpha_3) > 0$,
t_0 and ϕ_0 are arbitrary real constants.

***Example* 7. Periodicity in x_1 and x_2 and Localization in t** *2D Akhmediev breather* (Figure 6.4)
Given

$$\psi(x, t) = \frac{1}{\sqrt{a_2}}\left\{\frac{\kappa^2 \cosh[\delta\,(t - t_0)] + 2\,i\,\kappa\,\nu\,\sinh[\delta\,(t - t_0)]}{2\cosh[\delta\,(t - t_0)] - 2\,\nu\cos\left[\frac{\kappa}{\sqrt{2\,a_1}}\,(x - x_0)\right]} - 1\right\} e^{i\,[t - t_0 + \phi_0]}$$

is a solution of (2.1), then

$$\Phi(x_1, x_2, t) = \frac{1}{\sqrt{a_2}}\left(\frac{\kappa^2 \cosh[\delta\,(t - t_0)] + 2\,i\,\kappa\,\nu\,\sinh[\delta\,(t - t_0)]}{2\cosh[\delta\,(t - t_0)] - q(x_1, x_2)} - 1\right)$$
$$\times\, e^{i\,[(t - t_0) + \phi_0]} \tag{6.11}$$

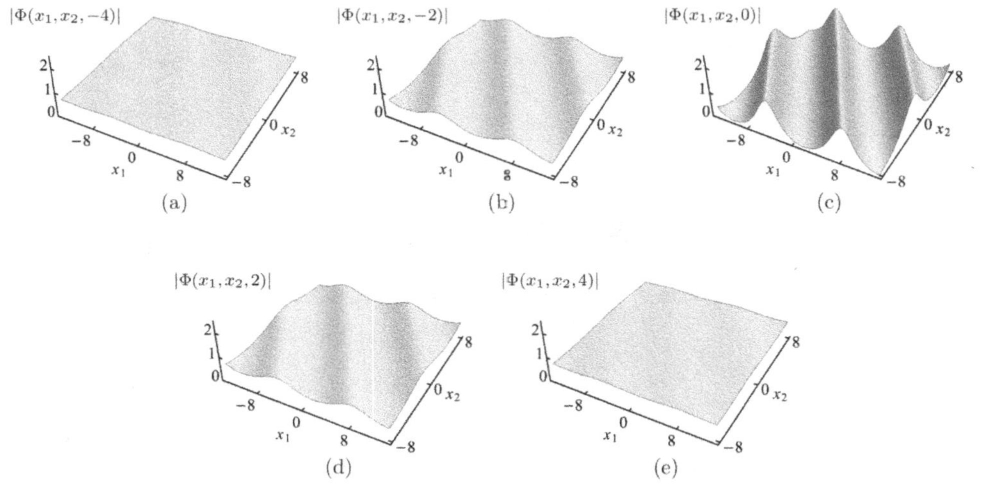

Figure 6.4. 2D Akhmediev breather (6.11). (a) at $t = -4$, (b) at $t = -2$, (c) at $t = 0$, (d) at $t = 2$, and (e) at $t = 4$. Used parameters: $\alpha_1 = \alpha_2 = c_1 = c_2 = 3$, $a_2 = 2$, $x_{01} = x_{02} = t_0 = \phi_0 = 0$, and $\nu = 0.5$. Animation available online at https://iopscience.iop.org/book/978-0-7503-2428-1.

is a solution of (6.4), where

$q(x_1, x_2) = 2\nu \cos\left\{\frac{\kappa}{\sqrt{2\,(c_1^2\,\alpha_1 + c_2^2\,\alpha_2)}}\,[c_1\,(x_1 - x_{01}) + c_2\,(x_2 - x_{02})]\right\},$
$\kappa = 2\sqrt{1 - \nu^2},$
$\delta = \kappa\,\nu,$
$\nu < 1,$
$a_2 > 0,$
$(c_1^2\,\alpha_1 + c_2^2\,\alpha_2) > 0,$
t_0 and ϕ_0 are arbitrary real constants.

***Example* 8. Periodicity in x_1, x_2, and x_3 and Localization in t** *3D Akhmediev breather.*
Given

$$\psi(x, t) = \frac{1}{\sqrt{a_2}}\left\{\frac{\kappa^2\cosh[\delta\,(t - t_0)] + 2\,i\,\kappa\,\nu\,\sinh[\delta\,(t - t_0)]}{2\cosh[\delta\,(t - t_0)] - 2\,\nu\,\cos\left[\frac{\kappa}{\sqrt{2\,a_1}}\,(x - x_0)\right]} - 1\right\}e^{i\,[t - t_0 + \phi_0]}$$

is a solution of (2.1), then

$$\Phi(x_1, x_2, x_3, t) = \frac{1}{\sqrt{a_2}}\left(\frac{\kappa^2\cosh[\delta\,(t - t_0)] + 2\,i\,\kappa\,\nu\,\sinh[\delta\,(t - t_0)]}{2\cosh[\delta\,(t - t_0)] - q(x_1, x_2, x_3)} - 1\right)\times e^{i\,[(t - t_0) + \phi_0]} \tag{6.12}$$

is a solution of (6.6), where

$q(x_1, x_2, x_3)$
$= 2\,\nu\,\cos\left\{\frac{\kappa}{\sqrt{2\,(c_1^2\,\alpha_1 + c_2^2\,\alpha_2 + c_3^2\,\alpha_3)}}\,[c_1\,(x_1 - x_{01}) + c_2\,(x_2 - x_{02}) + c_3\,(x_3 - x_{03})]\right\},$
$\kappa = 2\,\sqrt{1 - \nu^2},$
$\delta = \kappa\,\nu,$
$\nu < 1,$
$a_2 > 0,$
$(c_1^2\,\alpha_1 + c_2^2\,\alpha_2 + c_3^2\,\alpha_3) > 0,$
t_0 and ϕ_0 are arbitrary real constants.

***Example* 9. Localization in t, x_1, and x_2** *2D Peregrine soliton*
(Figure 6.5)
Given

$$\psi(x, t) = \frac{1}{\sqrt{a_2}}\left[\frac{4 + i\,8\,(t - t_0)}{1 + 4\,(t - t_0)^2 + \frac{2}{a_1}\,(x - x_0)^2} - 1\right]e^{i\,[t - t_0 + \phi_0]}$$

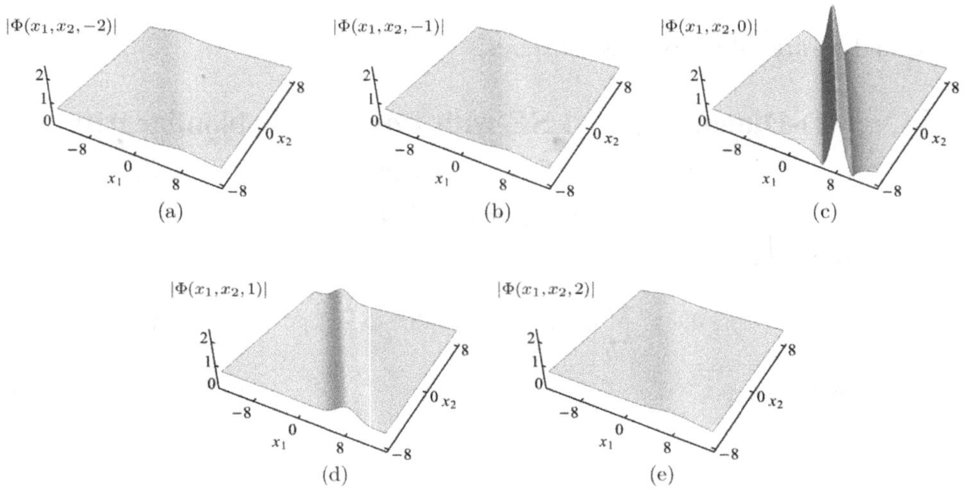

Figure 6.5. 2D Peregrine soliton (6.13). (a) at $t = -2$, (b) at $t = -1$, (c) at $t = 0$, (d) at $t = 1$, and (e) at $t = 2$. Used parameters: $\alpha_1 = \alpha_2 = c_1 = c_2 = a_2 = 2$, and $x_{01} = x_{02} = t_0 = \phi_0 = 0$. Animation available online at https://iopscience.iop.org/book/978-0-7503-2428-1.

is a solution of (2.1), then

$$\Phi(x_1, x_2, t) = \frac{1}{\sqrt{a_2}}\left\{\frac{4 + i\,8\,(t - t_0)}{1 + 4\,(t - t_0)^2 + \frac{2}{c_1^2\,\alpha_1 + c_2^2\,\alpha_2}\,[c_1\,(x_1 - x_{01}) + c_2\,(x_2 - x_{02})]^2} - 1\right\} \times e^{i\,[(t - t_0) + \phi_0]}, \tag{6.13}$$

is a solution of (6.4), where

$a_2 > 0$,

t_0 and ϕ_0 are arbitrary real constants.

Example* 10. Localization in t, x_1, x_2, and x_3 *3D Peregrine soliton

Given

$$\psi(x, t) = \frac{1}{\sqrt{a_2}}\left[\frac{4 + i\,8\,(t - t_0)}{1 + 4\,(t - t_0)^2 + \frac{2}{a_1}\,(x - x_0)^2} - 1\right] e^{i\,[t - t_0 + \phi_0]}$$

is a solution of (2.1), then

$$\Phi(x_1, x_2, x_3, t) = \frac{1}{\sqrt{a_2}}\left\{\frac{4 + i\,8\,(t - t_0)}{1 + 4\,(t - t_0)^2 + q(x_1, x_2, x_3)} - 1\right\} e^{i\,[(t - t_0) + \phi_0]} \tag{6.14}$$

is a solution of (6.6), where

$q(x_1, x_2, x_3) = \frac{2}{c_1^2\,\alpha_1 + c_2^2\,\alpha_2 + c_3^2\,\alpha_3}\,[c_1\,(x_1 - x_{01}) + c_2\,(x_2 - x_{02}) + c_3\,(x_3 - x_{03})]^2$,

$a_2 > 0$,
t_0 and ϕ_0 are arbitrary real constants.

6.2 (*N* + 1)-Dimensional NLSE with Power Law Nonlinearity

If $\psi(x, t; a_1, a_2)$ is a solution to the NLSE with power law nonlinearity, (3.1),

$$i\,\psi_t + a_1\,\psi_{xx} + a_2\,|\psi|^n\,\psi = 0,$$

then (6.1) is a solution of

$$i\,\Phi_t + \sum_{k=1}^{N} \alpha_k\,\Phi_{x_k x_k} + a_2\,|\Phi|^n\,\Phi = 0, \tag{6.15}$$

with the replacements

$x \to \sum_{k=1}^{N} c_k\,(x_k - x_{0k})$,
$t \to t$,
$a_1 \to \sum_{k=1}^{N} c_k^2\,\alpha_k$,
$a_2 \to a_2$,
where α_k, a_2, and n are real constants.

***Example* 1. sech(x_1, x_2)** *2D bright soliton.*
Given

$$\psi(x, t) = \left\{\frac{2\,A_0^2\,a_1\,(n+2)}{a_2\,n^2}\,\text{sech}^2[A_0\,(x - x_0)]\right\}^{\frac{1}{n}} e^{i\left[\frac{4\,a_1\,A_0^2}{n^2}(t-t_0)+\phi_0\right]}$$

is a solution of (3.1), then

$$\Phi(x_1, x_2, t) = \left(\frac{2\,A_0^2\,(n+2)\left(c_1^2\,\alpha_1 + c_2^2\,\alpha_2\right)}{a_2\,n^2} \times \text{sech}^2\{A_0\,[\,c_1\,(x_1 - x_{01}) + c_2\,(x_2 - x_{02})]\}\right)^{\frac{1}{n}} \times e^{i\left[\frac{4\,A_0^2\left(c_1^2\,\alpha_1 + c_2^2\,\alpha_2\right)}{n^2}(t-t_0)+\phi_0\right]} \tag{6.16}$$

is a solution of

$$i\,\Phi_t + \alpha_1\,\Phi_{x_1x_1} + \alpha_2\,\Phi_{x_2x_2} + a_2\,|\Phi|^n\,\Phi = 0, \tag{6.17}$$

where
$a_2\,(n+2)\,(c_1^2\,\alpha_1 + c_2^2\,\alpha_2) > 0$,
A_0, t_0, and ϕ_0 are arbitrary real constants.

***Example* 2. sech(x_1, x_2, x_3)** *3D bright soliton.*
Given

$$\psi(x,\, t) = \left\{\frac{2\, A_0^2\, a_1\, (n+2)}{a_2\, n^2}\, \mathrm{sech}^2[A_0\, (x - x_0)]\right\}^{\frac{1}{n}}\, e^{i\left[\frac{4\, a_1\, A_0^2}{n^2}\, (t - t_0) + \phi_0\right]}$$

is a solution of (3.1), then

$$\begin{aligned}\Phi(x_1,\, x_2,\, x_3,\, t) = &\left(\frac{2\, A_0^2\, (n+2)\left(c_1^2\, \alpha_1 + c_2^2\, \alpha_2 + c_3^2\, \alpha_3\right)}{a_2\, n^2}\right.\\ &\times \mathrm{sech}^2\{A_0\, [c_1\, (x_1 - x_{01}) + c_2\, (x_2 - x_{02})\\ &\left. +\; c_3\, (x_3 - x_{03})]\}\right)^{\frac{1}{n}}\, e^{i\left[\frac{4\, A_0^2\left(c_1^2\, \alpha_1 + c_2^2\, \alpha_2 + c_3^2\, \alpha_3\right)}{n^2}\, (t - t_0) + \phi_0\right]}\end{aligned} \tag{6.18}$$

is a solution of

$$i\, \Phi_t + \alpha_1\, \Phi_{x_1 x_1} + \alpha_2\, \Phi_{x_2 x_2} + \alpha_3\, \Phi_{x_3 x_3} + a_2\, |\Phi|^n\, \Phi = 0, \tag{6.19}$$

where
$a_2\, (n+2)\, (c_1^2\, \alpha_1 + c_2^2\, \alpha_2 + c_3^2\, \alpha_3) > 0$,
A_0, t_0, and ϕ_0 are arbitrary real constants.

6.3 (N + 1)-Dimensional NLSE with Dual Power Law Nonlinearity

If $\psi(x,\, t;\, a_1,\, a_2,\, a_3)$ is a solution of the NLSE with dual power law nonlinearity, (3.19),

$$i\, \psi_t + a_1\, \psi_{xx} + a_2\, |\psi|^n\, \psi + a_3\, |\psi|^m\, \psi = 0,$$

then (6.1) is a solution of

$$i\, \Phi_t + \sum_{k=1}^{N} \alpha_k\, \Phi_{x_k x_k} + a_2\, |\Phi|^n\, \Phi + a_3\, |\Phi|^m\, \Phi = 0, \tag{6.20}$$

with the replacements
$x \to \sum_{k=1}^{N} c_k\, (x_k - x_{0k})$,
$t \to t$,
$a_1 \to \sum_{k=1}^{N} c_k^2\, \alpha_k$,
$a_2 \to a_2$,
$a_3 \to a_3$,
where α_k, a_2, a_3, n, and m are real constants.

***Example* 1. sech(x_1, x_2)** *2D flat-top soliton*
(Figure 6.6)
Given

$$\psi(x, t) = \left(\frac{A_0 \, A_1 \, (n+2)}{a_2 \left\{ A_0 + 2 \cosh\left[n \sqrt{\frac{A_1}{a_1}} \, (x - x_0) \right] \right\}} \right)^{\frac{1}{n}} e^{i \, [A_1 \, (t - t_0) + \phi_0]}$$

is a solution of (3.19), then

$$\Phi(x_1, x_2, t) = \left[\frac{A_0 \, A_1 \, (n+2)}{a_2 \, A_0 + 2 \, a_2 \cosh\left\{ n \sqrt{\frac{A_1}{c_1^2 \, \alpha_1 + c_2^2 \, \alpha_2}} \, [c_1 \, (x_1 - x_{01}) + c_2 \, (x_2 - x_{02})] \right\}} \right]^{\frac{1}{n}} \times e^{i \, [A_1 \, (t - t_0) + \phi_0]} \tag{6.21}$$

is a solution of

$$i \, \Phi_t + \alpha_1 \, \Phi_{x_1 x_1} + \alpha_2 \, \Phi_{x_2 x_2} + a_2 \, |\Phi|^n \, \Phi + a_3 \, |\Phi|^m \, \Phi = 0, \tag{6.22}$$

where

$A_0 = \frac{2a_2}{n+2} \sqrt{\frac{n+1}{A_1 \, \delta}}$,
$\delta = a_3 + \frac{a_2^2 \, (n+1)}{A_1 (n+2)^2}$,
$A_1 \, (c_1^2 \, \alpha_1 + c_2^2 \, \alpha_2) > 0$,
$A_1 \, \delta \, (n+1) > 0$,
$m = 2n$,
t_0, A_1, and ϕ_0 are arbitrary real constants.

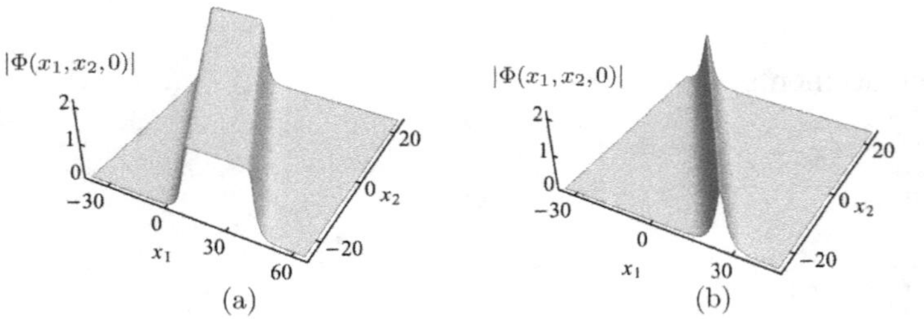

Figure 6.6. Plot of solution (6.21) at $t = 0$. (a) 2D flat-top soliton with $a_3 = -0.0694444$, (b) 2D bright soliton with $a_3 = 0.0305556$. Values of the other parameters are: $A_1 = 2$, $n = 4$, $\alpha_1 = \alpha_2 = 5$, $c_1 = c_2 = a_2 = 1$ and $x_{01} = x_{02} = t_0 = \phi_0 = 0$.

***Example* 2. sech(x_1, x_2, x_3)** *3D flat-top soliton.*
Given

$$\psi(x, t) = \left(\frac{A_0\, A_1\, (n + 2)}{a_2 \left\{ A_0 + 2 \cosh\left[n \sqrt{\frac{A_1}{\alpha_1}}\, (x - x_0) \right] \right\}} \right)^{\frac{1}{n}} e^{i\, [A_1\, (t - t_0) + \phi_0]}$$

is a solution of (3.19), then

$$\Phi(x_1, x_2, x_3, t) = \left[\frac{A_0\, A_1\, (n + 2)}{a_2\, A_0 + q(x_1, x_2, x_3)} \right]^{\frac{1}{n}} e^{i\, [A_1\, (t - t_0) + \phi_0]} \tag{6.23}$$

is a solution of

$$i\Phi_t + \alpha_1\, \Phi_{x_1 x_1} + \alpha_2\, \Phi_{x_2 x_2} + \alpha_3\, \Phi_{x_3 x_3} + a_2\, |\Phi|^n\, \Phi + a_3\, |\Phi|^m\, \Phi = 0, \tag{6.24}$$

where

$q(x_1, x_2, x_3)=$
$2a_2 \cosh\left\{ n \sqrt{\frac{A_1}{(c_1^2\, \alpha_1 + c_2^2\, \alpha_2 + c_3^2\, \alpha_3)}}\, [c_1\, (x_1 - x_{01}) + c_2\, (x_2 - x_{02}) + c_3\, (x_3 - x_{03})] \right\},$

$A_0 = \frac{2a_2}{n+2} \sqrt{\frac{n+1}{A_1\, \delta}},$
$\delta = a_3 + \frac{a_2^2\, (n+1)}{A_1\, (n+2)^2},$
$A_1\, (c_1^2\, \alpha_1 + c_2^2\, \alpha_2 + c_3^2\, \alpha_3) > 0,$
$A_1\, \delta\, (n + 1) > 0,$
$m = 2n,$
t_0, A_1, and ϕ_0 are arbitrary real constants.

***Example* 3. tanh(x_1, x_2)** *2D dark soliton*
(Figure 6.7)
Given

$$\psi(x, t) = \left(\frac{2a_1\, A_0^2\, (n + 2)}{a_2\, n^2}\, \{1 - \tanh[A_0\, (x - x_0)]\} \right)^{\frac{1}{n}} e^{i\left[\frac{4\, a_1\, A_0^2}{n^2}\, (t - t_0) + \phi_0 \right]}$$

is a solution of (3.19), then

$$\Phi(x_1, x_2, t) = \left[\frac{2A_0^2 (n + 2)\left(c_1^2\, \alpha_1 + c_2^2\, \alpha_2 \right)}{a_2\, n^2} (1 - \tanh\{A_0[c_1(x_1 - x_{01}) + c_2(x_2 - x_{02})]\}) \right]^{\frac{1}{n}} \times e^{i\left[\frac{4\, A_0^2\, \left(c_1^2\, \alpha_1 + c_2^2\, \alpha_2 \right)}{n^2}\, (t - t_0) + \phi_0 \right]} \tag{6.25}$$

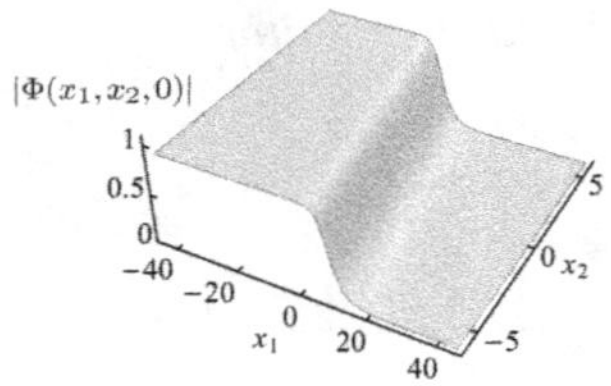

Figure 6.7. 2D dark soliton (6.25) at $t = 0$, with $\alpha_1 = \alpha_2 = c_1 = c_2 = a_2 = 1$, $a_3 = -1$, $x_{01} = x_{02} = t_0 = \phi_0 = 0$, and $n = 2.5$.

is a solution of (6.22), where

$A_0 = \frac{a_2\, n}{2\,(n+2)} \sqrt{\frac{-(n+1)}{a_3\,(c_1^2\,\alpha_1 + c_2^2\,\alpha_2)}}$,

$a_2\,(n+2)\,(c_1^2\,\alpha_1 + c_2^2\,\alpha_2) > 0$,

$a_3\,(n+1)\,(c_1^2\,\alpha_1 + c_2^2\,\alpha_2) < 0$,

$m = 2\,n$,

t_0 and ϕ_0 are arbitrary real constants.

***Example* 4. tanh(x_1, x_2, x_3)** *3D dark soliton*

Given

$$\psi(x,\, t) = \left(\frac{2a_1\, A_0^2\,(n+2)}{a_2\, n^2}\,\{1 - \tanh[A_0\,(x - x_0)]\}\right)^{\frac{1}{n}} e^{i\left[\frac{4\,\alpha_1\, A_0^2}{n^2}\,(t-t_0)+\phi_0\right]}$$

is a solution of (3.19), then

$$\begin{aligned}\Phi(x_1,\, x_2,\, x_3,\, t) = \Bigg[&\frac{2A_0^2\,(n+2)\left(c_1^2\,\alpha_1 + c_2^2\,\alpha_2 + c_3^2\,\alpha_3\right)}{a_2\, n^2}\\ &\times (1 - \tanh\{A_0\,[c_1\,(x_1 - x_{01}) + c_2\,(x_2 - x_{02})\\ &+ c_3\,(x_3 - x_{03})]\})\Bigg]^{\frac{1}{n}} e^{i\left[\frac{4\,A_0^2\left(c_1^2\,\alpha_1+c_2^2\,\alpha_2+c_3^2\,\alpha_3\right)}{n^2}\,(t-t_0)+\phi_0\right]}\end{aligned} \tag{6.26}$$

is a solution of (6.24), where

$A_0 = \frac{a_2\, n}{2\,(n+2)} \sqrt{\frac{-(n+1)}{a_3\,(c_1^2\,\alpha_1 + c_2^2\,\alpha_2 + c_3^2\,\alpha_3)}}$,

$a_2\,(n+2)\,(c_1^2\,\alpha_1 + c_2^2\,\alpha_2 + c_3^2\,\alpha_3) > 0$,

$a_3\,(n+1)\,(c_1^2\,\alpha_1 + c_2^2\,\alpha_2 + c_3^2\,\alpha_3) < 0$,

$m = 2n$,

t_0 and ϕ_0 are arbitrary real constants.

6.4 Galilean Transformation in $(N + 1)$-Dimensions (Movable Solutions)

If $\psi(x, t; a_1)$ is a solution of one of the three equations, fundamental NLSE, (2.1), NLSE with power law nonlinearity, (3.1), and NLSE with dual power law nonlinearity, (3.19),

$$i\,\psi_t + a_1\,\psi_{xx} + a_2\,|\psi|^2\,\psi = 0,$$
$$i\,\psi_t + a_1\,\psi_{xx} + a_2\,|\psi|^n\,\psi = 0,$$
$$i\,\psi_t + a_1\,\psi_{xx} + a_2\,|\psi|^n\,\psi + a_3\,|\psi|^m\,\psi = 0,$$

then

$$\begin{aligned}
&\Phi(x_1, x_2, \dots, x_N, t; \alpha_1, \alpha_2, \dots, \alpha_N) = \\
&\psi\Big\{ c_1\,[x_1 - (x_{01} + v_1\,t)] + c_2\,[x_2 - (x_{02} + v_2\,t)] + \cdots \\
&+ c_N\,[x_N - (x_{0N} + v_N\,t)],\ t;\ c_1^2\,\alpha_1 + c_1^2\,\alpha_2 + \cdots + c_N^2\,\alpha_N \Big\} \\
&\times e^{i\left[\frac{v_1}{2\,\alpha_1}(x_1 - x_{01}) + \frac{v_2}{2\,\alpha_2}(x_2 - x_{02}) + \cdots + \frac{v_N}{2\,\alpha_N}(x_N - x_{0N}) - \left(\frac{v_1^2}{4\,\alpha_1} + \frac{v_2^2}{4\,\alpha_2} + \cdots + \frac{v_N^2}{4\,\alpha_N}\right)(t - t_0)\right]}
\end{aligned} \tag{6.27}$$

is a movable solution of the

$$i\,\Phi_t + \sum_{k=1}^{N} \alpha_k\,\Phi_{x_k x_k} + a_2\,|\Phi|^2\,\Phi = 0,$$
$$i\,\Phi_t + \sum_{k=1}^{N} \alpha_k\,\Phi_{x_k x_k} + a_2\,|\Phi|^n\,\Phi = 0,$$
$$i\,\Phi_t + \sum_{k=1}^{N} \alpha_k\,\Phi_{x_k x_k} + a_2\,|\Phi|^n\,\Phi + a_3\,|\Phi|^m\,\Phi = 0,$$

respectively, with the replacements

$$\begin{aligned}
&x \to \sum_{k=1}^{N} c_k\,[x_k - (x_{0k} + v_k\,t)], \\
&t \to t, \\
&a_1 \to \sum_{k=1}^{N} c_k^2\,\alpha_k, \\
&a_2 \to a_2, \\
&a_3 \to a_3, \\
&\psi \to \psi\,e^{i\sum_{k=1}^{N}\frac{v_k}{2\alpha_k}(x_k - x_{0k}) - \frac{v_k^2}{4\alpha_k}(t - t_0)},
\end{aligned}$$

where α_k, c_k, x_{0k}, v_k, a_2, a_3, n, m, and t_0 are real constants.

***Example* 1. sech(x_1, x_2, t)** *moving 2D bright soliton*
(Figure 6.8)
Given

$$\psi(x,\, t) = A_0 \sqrt{\frac{2\, a_1}{a_2}}\, \text{sech}[A_0\, (x - x_0)]\, e^{i\left[a_1\, A_0^2\, (t-t_0)+\phi_0\right]}$$

is a static solution of (2.1), then

$$\begin{aligned}\Phi(x_1,\, x_2,\, t) = {} & A_0 \sqrt{\frac{2\left(c_1^2\, \alpha_1 + c_2^2\, \alpha_2\right)}{a_2}} \\ & \times \text{sech}(A_0\, \{c_1\, [x_1 - (x_{01} + v_1\, t)] + c_2\, [x_2 - (x_{02} + v_2\, t)]\}) \\ & \times e^{i\left\{\frac{v_1\, (x_1-x_{01})}{2\, \alpha_1}+\frac{v_2\, (x_2-x_{02})}{2\, \alpha_2}+\left[A_0^2\, \left(c_1^2\, \alpha_1+c_2^2\, \alpha_2\right)-\frac{v_1^2\, \alpha_2+v_2^2\, \alpha_1}{4\, \alpha_1\, \alpha_2}\right](t-t_0)+\phi_0\right\}}\end{aligned} \tag{6.28}$$

is a movable solution of (6.4), where
$a_2\, (c_1^2\, \alpha_1 + c_2^2\, \alpha_2) > 0$,
A_0, t_0, and ϕ_0 are arbitrary real constants.

***Example* 2. sech(x_1, x_2, x_3, t)** *moving 3D bright soliton*
Given

$$\psi(x,\, t) = A_0 \sqrt{\frac{2a_1}{a_2}}\, \text{sech}[A_0\, (x - x_0)]\, e^{i\left[a_1\, A_0^2\, (t-t_0)+\phi_0\right]}$$

is a static solution of (2.1), then

$$\begin{aligned}\Phi(x_1,\, x_2,\, x_3,\, t) = {} & A_0 \sqrt{\frac{2\left(c_1^2\, \alpha_1 + c_2^2\, \alpha_2 + c_3^2\, \alpha_3\right)}{a_2}}\, \text{sech}(A_0\, \{c_1\, [x_1 - (x_{01} + v_1\, t)] \\ & + c_2\, [x_2 - (x_{02} + v_2\, t)] + c_3\, [x_3 - (x_{03} + v_3\, t)]\})\, e^{i\, \phi(x_1, x_2, x_3, t)}\end{aligned} \tag{6.29}$$

is a movable solution of (6.6), where

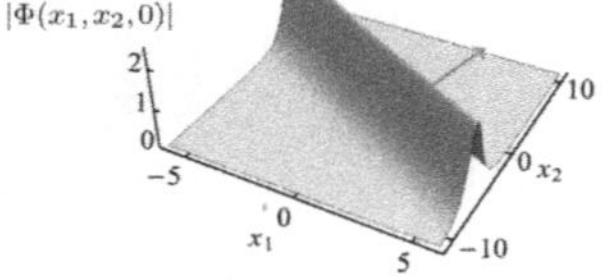

Figure 6.8. Moving 2D bright soliton (6.28) at $t = 0$, with $a_2 = \alpha_1 = \alpha_2 = A_0 = c_1 = c_2 = 1$, $v_1 = v_2 = 1/2$, and $x_{01} = x_{02} = t_0 = \phi_0 = 0$. The arrow shows the direction of motion. Animation available online at https://iopscience.iop.org/book/978-0-7503-2428-1.

$$\phi(x_1, x_2, x_3, t) = \frac{v_1\,(x_1 - x_{01})}{2\alpha_1} + \frac{v_2\,(x_2 - x_{02})}{2\alpha_2} + \frac{v_3\,(x_3 - x_{03})}{2\alpha_3} + \left[A_0^2\left(c_1^2\,\alpha_1 + c_2^2\,\alpha_2 + c_3^2\,\alpha_3\right) - \frac{v_1^2\,\alpha_2\,\alpha_3 + v_2^2\,\alpha_1\,\alpha_3 + v_3^2\,\alpha_1\,\alpha_2}{4\,\alpha_1\,\alpha_2\,\alpha_3}\right] \times (t - t_0) + \phi_0,$$

$a_2\,(c_1^2\,\alpha_1 + c_2^2\,\alpha_2 + c_3^2\,\alpha_3) > 0$,
A_0, t_0, and ϕ_0 are arbitrary real constants.

***Example* 3. tanh(x_1, x_2, t)** *moving 2D dark soliton*
(Figure 6.9)
Given

$$\psi(x, t) = A_0\,\sqrt{\frac{-2a_1}{a_2}}\;\tanh[A_0\,(x - x_0)]\;e^{-i\left[2a_1\,A_0^2\,(t-t_0)+\phi_0\right]}$$

is a static solution of (2.1), then

$$\Phi(x_1, x_2, t) = A_0\,\sqrt{\frac{-2\left(c_1^2\,\alpha_1 + c_2^2\,\alpha_2\right)}{a_2}} \times \tanh(A_0\,\{c_1\,[x_1 - (x_{01} + v_1\,t)] + c_2\,[x_2 - (x_{02} + v_2\,t)]\}) \times e^{i\left\{\frac{v_1\,(x_1-x_{01})}{2\alpha_1}+\frac{v_2\,(x_2-x_{02})}{2\alpha_2}-\left[2\,A_0^2\left(c_1^2\,\alpha_1+c_2^2\,\alpha_2\right)+\frac{v_1^2\,\alpha_2+v_2^2\,\alpha_1}{4\,\alpha_1\,\alpha_2}\right](t-t_0)+\phi_0\right\}} \tag{6.30}$$

is a movable solution of (6.4), where
$a_2\,(c_1^2\,\alpha_1 + c_2^2\,\alpha_2) < 0$,
A_0, t_0, and ϕ_0 are arbitrary real constants.

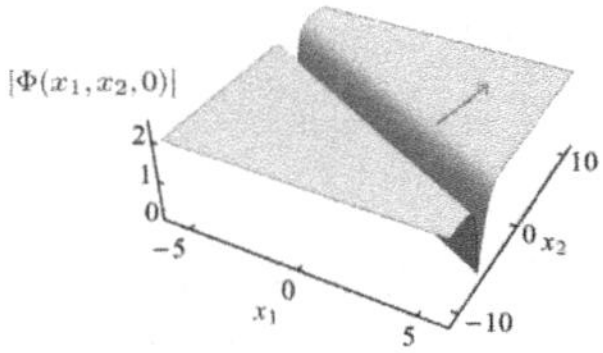

Figure 6.9. Moving 2D dark soliton (6.30) at $t = 0$, with $\alpha_1 = \alpha_2 = A_0 = c_1 = c_2 = 1$, $a_2 = -1$, $v_1 = v_2 = 1/2$, and $x_{01} = x_{02} = t_0 = \phi_0 = 0$. The arrow shows the direction of motion. Animation available online at https://iopscience.iop.org/book/978-0-7503-2428-1.

***Example* 4. tanh(x_1, x_2, x_3, t)** *moving 3D dark soliton*
Given

$$\psi(x, t) = A_0 \sqrt{\frac{-2\, a_1}{a_2}}\; \tanh[A_0\, (x - x_0)]\; e^{-i\left[2\, a_1\, A_0^2\, (t-t_0)+\phi_0\right]}$$

is a static solution of (2.1), then

$$\Phi(x_1, x_2, x_3, t) = A_0 \sqrt{\frac{-2\left(c_1^2\, \alpha_1 + c_2^2\, \alpha_2 + c_3^2\, \alpha_3\right)}{a_2}}\tanh(A_0\{c_1[x_1 - (x_{01} + v_1\, t)] + c_2\, [x_2 - (x_{02} + v_2\, t)] + c_3\, [x_3 - (x_{03} + v_3\, t)]\})\; e^{i\, \phi(x_1, x_2, x_3, t)} \tag{6.31}$$

is a movable solution of (6.6), where

$$\phi(x_1, x_2, x_3, t) = \phi_0 + \frac{v_1\, (x_1 - x_{01})}{2\alpha_1} + \frac{v_2\, (x_2 - x_{02})}{2\alpha_2} + \frac{v_3\, (x_3 - x_{03})}{2\alpha_3} - \left[2A_0^2\, (c_1^2\, \alpha_1 + c_2^2\, \alpha_2 + c_3^2\, \alpha_3) + \frac{v_1^2\, \alpha_2\, \alpha_3 + v_2^2\, \alpha_1\, \alpha_3 + v_3^2\, \alpha_1\, \alpha_2}{4\, \alpha_1\, \alpha_2\, \alpha_3}\right] \times (t - t_0),$$

$a_2\, (c_1^2\, \alpha_1 + c_2^2\, \alpha_2 + c_3^2\, \alpha_3) < 0$,
A_0, t_0, and ϕ_0 are arbitrary real constants.

***Example* 5. sech(x_1, x_2, t)** *moving 2D bright soliton*
Given

$$\psi(x, t) = \left\{\frac{2\, A_0^2\, a_1\, (n + 2)}{a_2\, n^2}\; \mathrm{sech}^2[A_0\, (x - x_0)]\right\}^{\frac{1}{n}}\; e^{i\left[\frac{4\, a_1\, A_0^2}{n^2}\, (t-t_0)+\phi_0\right]}$$

is a static solution of (3.1), then

$$\Phi(x_1, x_2, t) = \left(\frac{2A_0^2\, (n + 2)\left(c_1^2\, \alpha_1 + c_2^2\, \alpha_2\right)}{a_2\, n^2} \times \mathrm{sech}^2(A_0\, \{c_1\, [x_1 - (x_{01} + v_1\, t)] + c_2\, [x_2 - (x_{02} + v_2\, t)]\})\right)^{\frac{1}{n}} \tag{6.32}$$
$$\times\, e^{i\left\{\frac{v_1\, (x_1 - x_{01})}{2\, \alpha_1} + \frac{v_2\, (x_2 - x_{02})}{2\, \alpha_2} + \left[\frac{4\, A_0^2\left(c_1^2\, \alpha_1 + c_2^2\, \alpha_2\right)}{n^2} - \frac{v_1^2\, \alpha_2 + v_2^2\, \alpha_1}{4\, \alpha_1\, \alpha_2}\right](t-t_0)+\phi_0\right\}}$$

is a movable solution of (6.17), where

$a_2\,(n+2)\,(c_1^2\,\alpha_1 + c_2^2\,\alpha_2) > 0$,
A_0, t_0, and ϕ_0 are arbitrary real constants.

***Example* 6. sech(x_1, x_2, x_3, t)** *moving 3D bright soliton*
Given

$$\psi(x,\,t) = \left\{\frac{2\,A_0^2\,a_1\,(n+2)}{a_2\,n^2}\,\mathrm{sech}^2[A_0\,(x - x_0)]\right\}^{\frac{1}{n}} e^{i\left[\frac{4\,a_1\,A_0^2}{n^2}(t-t_0)+\phi_0\right]}$$

is a static solution of (3.1), then

$$\Phi(x_1,\,x_2,\,x_3,\,t) = \left(\frac{2\,A_0^2\,(n+2)\left(c_1^2\,\alpha_1 + c_2^2\,\alpha_2 + c_3^2\,\alpha_3\right)}{a_2\,n^2}\,\mathrm{sech}^2[q(x_1,\,x_2,\,x_3,\,t)]\right)^{\frac{1}{n}} \times e^{i\,\phi(x_1,x_2,x_3,t)} \tag{6.33}$$

is a movable solution of (6.19),where

$$\begin{aligned}\phi(x_1,\,x_2,\,x_3,\,t) = \phi_0 &+ \frac{v_1\,(x_1 - x_{01})}{2\,\alpha_1} + \frac{v_2\,(x_2 - x_{02})}{2\,\alpha_2} + \frac{v_3\,(x_3 - x_{03})}{2\,\alpha_3}\\ &+ \left[\frac{4\,A_0^2\left(c_1^2\,\alpha_1 + c_2^2\,\alpha_2 + c_3^2\,\alpha_3\right)}{n^2}\right.\\ &\left.- \frac{v_1^2\,\alpha_2\,\alpha_3 + v_2^2\,\alpha_1\,\alpha_3 + v_3^2\,\alpha_1\,\alpha_2}{4\,\alpha_1\,\alpha_2\,\alpha_3}\right] \times (t - t_0),\end{aligned}$$

$$\begin{aligned}q(x_1,\,x_2,\,x_3,\,t) = A_0\,\{&c_1\,[x_1 - (x_{01} + v_1\,t)]\\ &+ c_2\,[x_2 - (x_{02} + v_2\,t)] + c_3\,[x_3 - (x_{03} + v_3\,t)]\}\text{'}\end{aligned}$$

$a_2\,(n+2)\,(c_1^2\,\alpha_1 + c_2^2\,\alpha_2 + c_3^2\,\alpha_3) > 0$,
A_0, t_0, and ϕ_0 are arbitrary real constants.

***Example* 7. sech(x_1, x_2, t)** *moving 2D flat-top soliton*
(Figure 6.10)
Given

$$\psi(x,\,t) = \left(\frac{A_0\,A_1\,(n+2)}{a_2\left\{A_0 + 2\cosh\left[n\,\sqrt{\frac{A_1}{a_1}}\,(x - x_0)\right]\right\}}\right)^{\frac{1}{n}} e^{i\,[A_1\,(t-t_0)+\phi_0]}$$

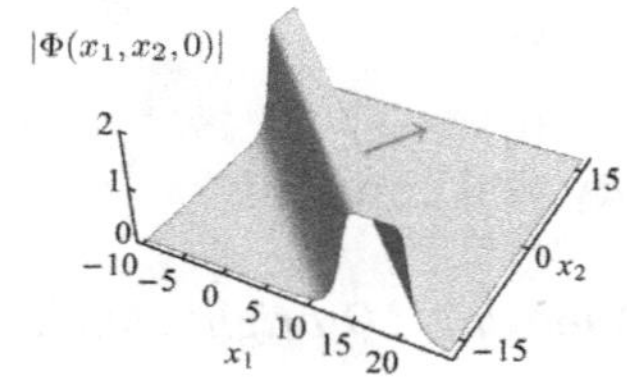

Figure 6.10. Moving 2D flat-top soliton (6.34) at $t = 0$, with $a_2 = 1$, $A_1 = 2$, $c_1 = c_2 = \alpha_1 = \alpha_2 = 1$, $a_3 = -0.06944444444$, $n = 4$, $v_1 = v_2 = 1/2$, and $x_{01} = x_{02} = t_0 = \phi_0 = 0$. The arrow shows the direction of motion. Animation available online at https://iopscience.iop.org/book/978-0-7503-2428-1.

is a static solution of (3.19), then

$$\Phi(x_1, x_2, t) = \left[\frac{A_0\, A_1\,(n+2)}{a_2\, A_0 + q(x_1, x_2, t)}\right]^{\frac{1}{n}} \times e^{i\left[\frac{v_1\,(x_1 - x_{01})}{2\,\alpha_1} + \frac{v_2\,(x_2 - x_{02})}{2\,\alpha_2} + \left(A_1 - \frac{v_1^2\,\alpha_2 + v_2^2\,\alpha_1}{4\,\alpha_1\,\alpha_2}\right)(t - t_0) + \phi_0\right]} \tag{6.34}$$

is a movable solution of (6.22), where

$$q(x_1, x_2, t) = 2\,a_2 \cosh\left(n\,\sqrt{\frac{A_1}{c_1^2\,\alpha_1 + c_2^2\,\alpha_2}}\,\{c_1\,[x_1 - (x_{01} + v_1\,t)] + c_2\,[x_2 - (x_{02} + v_2\,t)]\}\right),$$

$A_0 = \frac{2\,a_2}{n+2}\sqrt{\frac{n+1}{A_1\,\delta}}$,
$\delta = a_3 + \frac{a_2^2\,(n+1)}{A_1\,(n+2)^2}$,
$A_1\,(c_1^2\,\alpha_1 + c_2^2\,\alpha_2) > 0$,
$A_1\,\delta\,(n+1) > 0$,
$m = 2\,n$,
t_0, A_1, and ϕ_0 are arbitrary real constants.

***Example* 8. sech(x_1, x_2, x_3, t)** *moving 3D flat-top soliton*
Given

$$\psi(x, t) = \left(\frac{A_0\, A_1\,(n+2)}{a_2\left\{A_0 + 2\cosh\left[n\,\sqrt{\frac{A_1}{a_1}}\,(x - x_0)\right]\right\}}\right)^{\frac{1}{n}} e^{i\,[A_1\,(t - t_0) + \phi_0]}$$

is a static solution of (3.19), then

$$\Phi(x_1, x_2, x_3, t) = \left[\frac{A_0\, A_1\,(n+2)}{a_2\, A_0 + q(x_1, x_2, x_3)}\right]^{\frac{1}{n}} e^{i\,\phi(x_1, x_2, x_3, t)} \tag{6.35}$$

is a movable solution of (6.24), where

$$\phi(x_1, x_2, x_3, t) = \frac{v_1 (x_1 - x_{01})}{2 \alpha_1} + \frac{v_2 (x_2 - x_{02})}{2 \alpha_2} + \frac{v_3 (x_3 - x_{03})}{2 \alpha_3} + \left(A_1 - \frac{v_1^2 \alpha_2 \alpha_3 + v_2^2 \alpha_1 \alpha_3 + v_3^2 \alpha_1 \alpha_2}{4 \alpha_1 \alpha_2 \alpha_3}\right)(t - t_0) + \phi_0,$$

$$q(x_1, x_2, x_3, t) = 2 a_2 \cosh\left(n \sqrt{\frac{A_1}{(c_1^2 \alpha_1 + c_2^2 \alpha_2 + c_3^2 \alpha_3)}} \{c_1 [x_1 - (x_{01} + v_1 t)] + c_2 [x_2 - (x_{02} + v_2 t)] + c_3 [x_3 - (x_{03} + v_3 t)]\}\right),$$

$A_0 = \frac{2 a_2}{n+2} \sqrt{\frac{n+1}{A_1 \delta}}$,
$\delta = a_3 + \frac{a_2^2 (n+1)}{A_1 (n+2)^2}$,
$A_1 (c_1^2 \alpha_1 + c_2^2 \alpha_2 + c_3^2 \alpha_3) > 0$,
$A_1 \delta (n + 1) > 0$,
$m = 2 n$,
t_0, A_1, and ϕ_0 are arbitrary real constants.

6.5 NLSE in (2 + 1)-Dimensions with $\Phi_{x_1 x_2}$ Term

If $\psi(x, t)$ is a solution to the fundamental NLSE, (2.1),

$$i \psi_t + a_1 \psi_{xx} + a_2 |\psi|^2 \psi = 0,$$

then

$$\Phi(x_1, x_2, t) = \sqrt{\frac{a_2}{b_4}} \psi[X(x_1, x_2), t], \tag{6.36}$$

is a solution of

$$i \Phi_t + b \Phi_{x_1 x_1} + b_2 \Phi_{x_2 x_2} + b_3 \Phi_{x_1 x_2} + b_4 |\Phi|^2 \Phi = 0, \tag{6.37}$$

where

$$X(x_1, x_2) = c_0 - \frac{1}{2 b_1}\left(b_3 c_1 - \sqrt{4 a_1 b_1 - 4 b_1 b_2 c_1^2 + b_3^2 c_1^2}\right) x_1 + c_1 x_2, \tag{6.38}$$

$a_2 b_4 > 0$,
$4 a_1 b_1 (1 - b_2 c_1^2) > b_3^2 c_1^2$,
b_1, b_2, b_3, b_4, c_0 and c_1 are arbitrary real constants.

***Example* 1. sech(x_1, x_2)** *bright soliton*
Given

$$\psi(x, t) = A_0 \sqrt{\frac{2\, a_1}{a_2}}\ \text{sech}[A_0\,(x - x_0)]\, e^{i\left[a_1 A_0^2\,(t-t_0)+\phi_0\right]}$$

is a solution of (2.1), then

$$\Phi(x_1, x_2, t) = A_0\sqrt{\frac{2a_1}{b_4}}\,\text{sech}\left\{A_0\left[c_0 - \frac{\left(b_3\, c_1 - \sqrt{4\,a_1\,b_1 - 4\,b_1\,b_2\,c_1^2 + b_3^2\,c_1^2}\right)(x_1 - x_{01})}{2b_1} + c_1\,(x_2 - x_{02})\right]\right\} e^{i\left[a_1 A_0^2\,(t-t_0)+\phi_0\right]} \tag{6.39}$$

is a solution of (6.37), where
$a_1\, b_4 > 0$,
$(4\,a_1\,b_1 - 4\,b_1\,b_2\,c_1^2 + b_3^2\,c_1^2) > 0$,
A_0, t_0, and ϕ_0 are arbitrary real constants.

***Example* 2. tanh(x_1, x_2)** *dark soliton*
Given

$$\psi(x, t) = A_0 \sqrt{\frac{-2\, a_1}{a_2}}\ \tanh[A_0\,(x - x_0)]\, e^{-i\left[2\,a_1 A_0^2\,(t-t_0)+\phi_0\right]}$$

is a solution of (2.1), then

$$\Phi(x_1, x_2, t) = A_0\sqrt{\frac{-2a_1}{b_4}}\,\tanh\left\{A_0\left[c_0 + \frac{\left(\sqrt{4\,a_1\,b_1 - 4\,b_1\,b_2\,c_1^2 + b_3^2\,c_1^2} - b_3\,c_1\right)(x_1 - x_{01})}{2\,b_1} + c_1\,(x_2 - x_{02})\right]\right\} e^{-i\left[2\,a_1 A_0^2\,(t-t_0)+\phi_0\right]} \tag{6.40}$$

is a solution of (6.37), where
$a_1\, b_4 < 0$,
$(4\,a_1\,b_1 - 4\,b_1\,b_2\,c_1^2 + b_3^2\,c_1^2) > 0$,
A_0, t_0, and ϕ_0 are arbitrary real constants.

6.6 Summary of Sections 6.1–6.5

$(N+1)$-Dimensional NLSE with Cubic Nonlinearity

Transformation: $\Phi(x_1, x_2, \dots, x_N, t; \alpha_1, \alpha_2, \dots, \alpha_N) = \Phi[c_1(x_1 - x_{01}) + c_2(x_2 - x_{02}) + \cdots + c_N(x_N - x_{0N}), t; c_1^2\alpha_1 + c_1^2\alpha_2 + \cdots + c_N^2\alpha_N, a_2]$

Equation: $i\,\Phi_t + \sum_{k=1}^{N}\alpha_k\,\Phi_{x_k x_k} + a_2|\Phi|^2\Phi = 0$

#	Example	Conditions	Name	Eq. #
1.	$\Phi(x_1, x_2, t) = A_0\sqrt{\frac{2(c_1^2\alpha_1 + c_2^2\alpha_2)}{a_2}} \times \operatorname{sech}\{A_0[c_1(x_1 - x_{01}) + c_2(x_2 - x_{02})]\} \times e^{i[A_0^2(c_1^2\alpha_1 + c_2^2\alpha_2)(t-t_0)+\phi_0]}$	$a_2(c_1^2\alpha_1 + c_2^2\alpha_2) > 0$, A_0, t_0, and ϕ_0 are arbitrary real constants	2D bright soliton	(6.3)
2.	$\Phi(x_1, x_2, x_3, t) = A_0\sqrt{\frac{2(c_1^2\alpha_1 + c_2^2\alpha_2 + c_3^2\alpha_3)}{a_2}} \times \operatorname{sech}\{A_0[c_1(x_1 - x_{01}) + c_2(x_2 - x_{02}) + c_3(x_3 - x_{03})]\} \times e^{i[A_0^2(c_1^2\alpha_1 + c_2^2\alpha_2 + c_3^2\alpha_3)(t-t_0)+\phi_0]}$	$a_2(c_1^2\alpha_1 + c_2^2\alpha_2 + c_3^2\alpha_3) > 0$, A_0, t_0, and ϕ_0 are arbitrary real constants	3D bright soliton	(6.5)
3.	$\Phi(x_1, x_2, t) = A_0\sqrt{\frac{-2(c_1^2\alpha_1 + c_2^2\alpha_2)}{a_2}} \times \tanh\{A_0[c_1(x_1 - x_{01}) + c_2(x_2 - x_{02})]\} \times e^{-i[2A_0^2(c_1^2\alpha_1 + c_2^2\alpha_2)(t-t_0)+\phi_0]}$	$a_2(c_1^2\alpha_1 + c_2^2\alpha_2) < 0$, A_0, t_0, and ϕ_0 are arbitrary real constants	2D dark soliton	(6.7)
4.	$\Phi(x_1, x_2, x_3, t) = A_0\sqrt{\frac{-2(c_1^2\alpha_1 + c_2^2\alpha_2 + c_3^2\alpha_3)}{a_2}} \times \tanh\{A_0[c_1(x_1 - x_{01}) + c_2(x_2 - x_{02}) + c_3(x_3 - x_{03})]\} \times e^{-i[2A_0^2(c_1^2\alpha_1 + c_2^2\alpha_2 + c_3^2\alpha_3)(t-t_0)+\phi_0]}$	$a_2(c_1^2\alpha_1 + c_2^2\alpha_2 + c_3^2\alpha_3) < 0$, A_0, t_0, and ϕ_0 are arbitrary real constants	3D dark soliton	(6.8)

5.	$\Phi(x_1, x_2, t) = \frac{-1}{\sqrt{a_2}}\left(\frac{p^2 \cos[\omega\,(t-t_0)] + 2\,i\,p\,\nu \sin[\omega\,(t-t_0)]}{2\cos[\omega\,(t-t_0)] - q(x_1,x_2)} + 1\right) e^{i\,[(t-t_0)+\phi_0]}$, $q(x_1, x_2) = 2\,\nu \cosh\left\{\frac{p}{\sqrt{2\left(c_1^2\,\alpha_1 + c_2^2\,\alpha_2\right)}}\,[c_1\,(x_1 - x_{01}) + c_2\,(x_2 - x_{02})]\right\}$	$p = 2\sqrt{\nu^2 - 1}$, $\omega = p\,\nu$, $\nu > 1$, $a_2 > 0$, $(c_1^2\,\alpha_1 + c_2^2\,\alpha_2) > 0$, t_0 and ϕ_0 are arbitrary real constants	2D Kuznetsov–Ma breather (6.9)
6.	$\Phi(x_1, x_2, x_3, t) = \frac{-1}{\sqrt{a_2}}\left(\frac{p^2 \cos[\omega\,(t-t_0)] + 2\,i\,p\,\nu \sin[\omega\,(t-t_0)]}{2\cos[\omega\,(t-t_0)] - q(x_1,x_2,x_3)} + 1\right) \times e^{i\,[(t-t_0)+\phi_0]}$, $q(x_1, x_2, x_3) = 2\,\nu \cosh \left\{\frac{p}{\sqrt{2\left(c_1^2\,\alpha_1 + c_2^2\,\alpha_2 + c_3^2\,\alpha_3\right)}}\,[c_1\,(x_1 - x_{01}) + c_2\,(x_2 - x_{02}) + c_3\,(x_3 - x_{03})]\right\}$	$p = 2\sqrt{\nu^2 - 1}$, $\omega = p\,\nu$, $\nu > 1$, $a_2 > 0$, $(c_1^2\,\alpha_1 + c_2^2\,\alpha_2 + c_3^2\,\alpha_3) > 0$, t_0 and ϕ_0 are arbitrary real constants	3D Kuznetsov–Ma breather (6.10)
7.	$\Phi(x_1, x_2, t) = \frac{1}{\sqrt{a_2}}\left(\frac{\kappa^2 \cosh[\delta\,(t-t_0)] + 2\,i\,\kappa\,\nu \sinh[\delta\,(t-t_0)]}{2\cosh[\delta\,(t-t_0)] - q(x_1,x_2)} - 1\right) \times e^{i\,[(t-t_0)+\phi_0]}$, $q(x_1, x_2) = 2\,\nu \cos\left\{\frac{\kappa}{\sqrt{2\left(c_1^2\,\alpha_1 + c_2^2\,\alpha_2\right)}}\,[c_1\,(x_1 - x_{01}) + c_2\,(x_2 - x_{02})]\right\}$	$\kappa = 2\sqrt{1 - \nu^2}$, $\delta = \kappa\,\nu$, $\nu < 1$, $a_2 > 0$ $(c_1^2\,\alpha_1 + c_2^2\,\alpha_2) > 0$, t_0 and ϕ_0 are arbitrary real constants	2D Akhmediev breather (6.11)
8.	$\Phi(x_1, x_2, x_3, t) = \frac{1}{\sqrt{a_2}}\left(\frac{\kappa^2 \cosh[\delta\,(t-t_0)] + 2\,i\,\kappa\,\nu \sinh[\delta\,(t-t_0)]}{2\cosh[\delta\,(t-t_0)] - q(x_1,x_2,x_3)} - 1\right) \times e^{i\,[(t-t_0)+\phi_0]}$, $q(x_1, x_2, x_3) = 2\,\nu \cos \left\{\frac{\kappa}{\sqrt{2\left(c_1^2\,\alpha_1 + c_2^2\,\alpha_2 + c_3^2\,\alpha_3\right)}}\,[c_1\,(x_1 - x_{01}) + c_2\,(x_2 - x_{02}) + c_3\,(x_3 - x_{03})]\right\}$	$\kappa = 2\sqrt{1 - \nu^2}$, $\delta = \kappa\,\nu$, $\nu < 1$, $a_2 > 0$, $(c_1^2\,\alpha_1 + c_2^2\,\alpha_2 + c_3^2\,\alpha_3) > 0$, t_0 and ϕ_0 are arbitrary real constants	3D Akhmediev breather (6.12)

(*Continued*)

#	Example	Conditions	Name	Eq. #
9.	$\Phi(x_1, x_2, t) = \frac{1}{\sqrt{a_2}}\left(\frac{4 + i\,8\,(t - t_0)}{1 + 4\,(t - t_0)^2 + q(x_1, x_2)} - 1\right) e^{i\,[(t - t_0) + \phi_0]}$, $q(x_1, x_2) = \frac{2}{c_1^2\,\alpha_1 + c_2^2\,\alpha_2}\,[c_1\,(x_1 - x_{01}) + c_2\,(x_2 - x_{02})]^2$	$a_2 > 0$, t_0 and ϕ_0 are arbitrary real constants	2D Peregrine soliton	(6.13)
10.	$\Phi(x_1, x_2, x_3, t) = \frac{1}{\sqrt{a_2}}\left(\frac{4 + i\,8\,(t - t_0)}{1 + 4\,(t - t_0)^2 + q(x_1, x_2, x_3)} - 1\right) e^{i\,[(t - t_0) + \phi_0]}$, $q(x_1, x_2, x_3) = \frac{2}{c_1^2\,\alpha_1 + c_2^2\,\alpha_2 + c_3^2\,\alpha_3}\,[c_1\,(x_1 - x_{01}) + c_2\,(x_2 - x_{02}) + c_3\,(x_3 - x_{03})]^2$	$a_2 > 0$, t_0 and ϕ_0 are arbitrary real constants	3D Peregrine soliton	(6.14)

$(N + 1)$-Dimensional NLSE with Power Law Nonlinearity

Transformation: $\Phi(x_1, x_2, \ldots, x_N, t; \alpha_1, \alpha_2, \ldots, \alpha_N) = \psi[c_1\,(x_1 - x_{01}) + c_2\,(x_2 - x_{02}) + \cdots + c_N\,(x_N - x_{0N}), t; c_1^2\,\alpha_1 + c_1^2\,\alpha_2 + \cdots + c_N^2\,\alpha_N, a_2]$

Equation: $i\,\Phi_t + \sum_{k=1}^{N} \alpha_k\,\Phi_{x_k x_k} + a_2\,|\Phi|^n\,\Phi = 0$

#	Example	Conditions	Name	Eq. #
1.	$\Phi(x_1, x_2, t) = \left(\frac{2\,A_0^2\,(n+2)\,(c_1^2\,\alpha_1 + c_2^2\,\alpha_2)}{a_2\,n^2}\,\mathrm{sech}^2\{A_0\,[\,c_1\,(x_1 - x_{01}) + c_2\,(x_2 - x_{02})]\}\right)^{\frac{1}{n}} e^{i\left[\frac{4\,A_0^2\,(c_1^2\,\alpha_1 + c_2^2\,\alpha_2)}{n^2}\,(t - t_0) + \phi_0\right]}$	$a_2\,(n + 2)\,(c_1^2\,\alpha_1 + c_2^2\,\alpha_2) > 0$, $N = 2$, A_0, t_0, and ϕ_0 are arbitrary real constants	2D bright soliton	(6.16)
2.	$\Phi(x_1, x_2, x_3, t) = \left(\frac{2\,A_0^2\,(n+2)\,(c_1^2\,\alpha_1 + c_2^2\,\alpha_2 + c_3^2\,\alpha_3)}{a_2\,n^2} \times \mathrm{sech}^2\{A_0\,[c_1\,(x_1 - x_{01}) + c_2\,(x_2 - x_{02}) + c_3\,(x_3 - x_{03})]\}\right)^{\frac{1}{n}} \times e^{i\left[\frac{4\,A_0^2\,(c_1^2\,\alpha_1 + c_2^2\,\alpha_2 + c_3^2\,\alpha_3)}{n^2}\,(t - t_0) + \phi_0\right]}$	$N = 3$, $a_2\,(n + 2)\,(c_1^2\,\alpha_1 + c_2^2\,\alpha_2 + c_3^2\,\alpha_3) > 0$, A_0, t_0, and ϕ_0 are arbitrary real constants	3D bright soliton	(6.18)

$(N+1)$-Dimensional NLSE with Dual Power Law Nonlinearity

Transformation: $\Phi(x_1, x_2, \ldots, x_N, t; \alpha_1, \alpha_2, \ldots, \alpha_N) = \psi[c_1(x_1 - x_{01}) + c_2(x_2 - x_{02}) + \cdots + c_N(x_N - x_{0N}), t; c_1^2\alpha_1 + c_1^2\alpha_2 + \cdots + c_N^2\alpha_N, a_2]$

Equation: $i\,\Phi_t + \sum_{k=1}^{N}\alpha_k\,\Phi_{x_k x_k} + a_2\,|\Phi|^n\,\Phi + a_3\,|\Phi|^m\,\Phi = 0$

#	Example	Conditions	Name	Eq. #
1.	$\Phi(x_1, x_2, t) = \left[\frac{A_0 A_1 (n+2)}{a_2 A_0 + q(x_1, x_2)}\right]^{\frac{1}{n}} e^{i\,[A_1(t-t_0)+\phi_0]}$, $q(x_1, x_2) = 2a_2 \cosh\left\{n\sqrt{\frac{A_1}{c_1^2\alpha_1 + c_2^2\alpha_2}}\,[c_1(x_1 - x_{01}) + c_2(x_2 - x_{02})]\right\}$	$A_0 = \frac{2a_2}{n+2}\sqrt{\frac{n+1}{A_1\delta}}$, $\delta = a_3 + \frac{a_2^2(n+1)}{A_1(n+2)^2}$, $A_1(c_1^2\alpha_1 + c_2^2\alpha_2) > 0$, $A_1\delta(n+1) > 0$, $m = 2n$, $N = 2$, t_0, A_1, and ϕ_0 are arbitrary real constants	2D flat-top soliton	(6.21)
2.	$\Phi(x_1, x_2, x_3, t) = \left[\frac{A_0 A_1 (n+2)}{a_2 A_0 + q(x_1, x_2, x_3)}\right]^{\frac{1}{n}} e^{i\,[A_1(t-t_0)+\phi_0]}$, $q(x_1, x_2, x_3) = 2a_2 \cosh\left\{n\sqrt{\frac{A_1}{(c_1^2\alpha_1 + c_2^2\alpha_2 + c_3^2\alpha_3)}}\,[c_1(x_1 - x_{01}) + c_2(x_2 - x_{02}) + c_3(x_3 - x_{03})]\right\}$	$A_0 = \frac{2a_2}{n+2}\sqrt{\frac{n+1}{A_1\delta}}$, $N = 3$, $\delta = a_3 + \frac{a_2^2(n+1)}{A_1(n+2)^2}$, $A_1(c_1^2\alpha_1 + c_2^2\alpha_2 + c_3^2\alpha_3) > 0$, $A_1\delta(n+1) > 0$, $m = 2n$, t_0, A_1, and ϕ_0 are arbitrary real constants	3D flat-top soliton	(6.23)
3.	$\Phi(x_1, x_2, t) = \left[\frac{2A_0^2(n+2)(c_1^2\alpha_1 + c_2^2\alpha_2)}{a_2 n^2}\left(1 - \tanh\{A_0[c_1(x_1 - x_{01}) + c_2(x_2 - x_{02})]\}\right)\right]^{\frac{1}{n}} \times e^{i\left[\frac{4A_0^2(c_1^2\alpha_1 + c_2^2\alpha_2)}{n^2}(t-t_0)+\phi_0\right]}$	$A_0 = \frac{a_2 n}{2(n+2)}\sqrt{\frac{-(n+1)}{a_3(c_1^2\alpha_1 + c_2^2\alpha_2)}}$, $a_2(n+2)(c_1^2\alpha_1 + c_2^2\alpha_2) > 0$, $N = 2$, $a_3(n+1)(c_1^2\alpha_1 + c_2^2\alpha_2) < 0$, $m = 2n$, t_0 and ϕ_0 are arbitrary real constants	2D dark soliton	(6.25)

(Continued)

#	Example	Conditions	Name	Eq. #
4.	$\Phi(x_1, x_2, x_3, t) = \left[\frac{2 A_0^2 (n+2)(c_1^2 \alpha_1 + c_2^2 \alpha_2 + c_3^2 \alpha_3)}{a_2 n^2} (1 - \tanh\{A_0 [c_1 (x_1 - x_{01}) + c_2 (x_2 - x_{02}) + c_3 (x_3 - x_{03})]\})\right]^{\frac{1}{n}} e^{i\left[\frac{4 A_0^2 (c_1^2 \alpha_1 + c_2^2 \alpha_2 + c_3^2 \alpha_3)}{n^2} (t - t_0) + \phi_0\right]}$	$A_0 = \frac{a_2 n}{2(n+2)} \sqrt{\frac{-(n+1)}{a_3 (c_1^2 \alpha_1 + c_2^2 \alpha_2 + c_3^2 \alpha_3)}}$, $a_2 (n+2)(c_1^2 \alpha_1 + c_2^2 \alpha_2 + c_3^2 \alpha_3) > 0$, $a_3 (n+1)(c_1^2 \alpha_1 + c_2^2 \alpha_2 + c_3^2 \alpha_3) < 0$, $m = 2n$, $N = 3$, t_0 and ϕ_0 are arbitrary real constants	3D dark soliton	(6.26)

Galilean Transformation in (N + 1)-Dimensions (Movable Solutions)

Transformation:

$$\Phi(x_1, x_2, \ldots, x_N, t; \alpha_1, \alpha_2, \ldots, \alpha_N, a_2) = \Phi\{c_1 [x_1 - (x_{01} + v_1 t)] + c_2 [(x_2 - x_{02} + v_2 t)] + \cdots + c_N [x_N - (x_{0N} + v_N t)], t; c_1^2 \alpha_1 + c_1^2 \alpha_2 + \cdots + c_N^2 \alpha_N, a_2\} \times e^{i\left[\frac{v_1}{2\alpha_1}(x_1 - x_{01}) + \frac{v_2}{2\alpha_2}(x_2 - x_{02}) + \cdots + \frac{v_N}{2\alpha_N}(x_N - x_{0N}) - \left(\frac{v_1^2}{4\alpha_1} + \frac{v_2^2}{4\alpha_2} + \cdots + \frac{v_N^2}{4\alpha_N}\right)(t - t_0)\right]}$$

Equation: $i\,\Phi_t + \sum_{k=1}^{N} \alpha_k \Phi_{x_k x_k} + a_2 |\Phi|^2 \Phi = 0$

#	Example	Conditions	Name	Eq. #
1.	$\Phi(x_1, x_2, t) = A_0 \sqrt{\frac{2(c_1^2 \alpha_1 + c_2^2 \alpha_2)}{a_2}} \times \text{sech}(A_0 \{c_1 [x_1 - (x_{01} + v_1 t)] + c_2 [x_2 - (x_{02} + v_2 t)]\}) \times e^{i\left\{\frac{v_1 (x_1 - x_{01})}{2\alpha_1} + \frac{v_2 (x_2 - x_{02})}{2\alpha_2} + \left[A_0^2 (c_1^2 \alpha_1 + c_2^2 \alpha_2) - \frac{v_1^2 \alpha_2 + v_2^2 \alpha_1}{4\alpha_1 \alpha_2}\right](t - t_0) + \phi_0\right\}}$	$a_2 (c_1^2 \alpha_1 + c_2^2 \alpha_2) > 0$, $N = 2$, A_0, t_0, and ϕ_0 are arbitrary real constants	moving 2D bright soliton	(6.28)

2.	$\Phi(x_1, x_2, x_3, t) = A_0 \sqrt{\frac{2 (c_1^2 \alpha_1 + c_2^2 \alpha_2 + c_3^2 \alpha_3)}{a_2}} \times \operatorname{sech}(A_0 \{c_1 [x_1 - (x_{01} + v_1 t)] + c_2 [x_2 - (x_{02} + v_2 t)] + c_3 [x_3 - (x_{03} + v_3 t)]\}) e^{i \phi(x_1, x_2, x_3, t)}$, $\phi(x_1, x_2, x_3, t) = \phi_0 + \frac{v_1 (x_1 - x_{01})}{2 \alpha_1} + \frac{v_2 (x_2 - x_{02})}{2 \alpha_2} + \frac{v_3 (x_3 - x_{03})}{2 \alpha_3} + \left[A_0^2 (c_1^2 \alpha_1 + c_2^2 \alpha_2 + c_3^2 \alpha_3) - \frac{v_1^2 \alpha_2 \alpha_3 + v_2^2 \alpha_1 \alpha_3 + v_3^2 \alpha_1 \alpha_2}{4 \alpha_1 \alpha_2 \alpha_3}\right] (t - t_0)$	$a_2 (c_1^2 \alpha_1 + c_2^2 \alpha_2 + c_3^2 \alpha_3) > 0$, $N = 3$, A_0, t_0, and ϕ_0 are arbitrary real constants	moving 3D bright soliton	(6.29)
3.	$\Phi(x_1, x_2, t) = A_0 \sqrt{\frac{-2 (c_1^2 \alpha_1 + c_2^2 \alpha_2)}{a_2}} \times \tanh(A_0 \{c_1 [x_1 - (x_{01} + v_1 t)] + c_2 [x_2 - (x_{02} + v_2 t)]\}) \times e^{i \left\{\frac{v_1 (x_1 - x_{01})}{2 \alpha_1} + \frac{v_2 (x_2 - x_{02})}{2 \alpha_2} - \left[2 A_0^2 (c_1^2 \alpha_1 + c_2^2 \alpha_2) + \frac{v_1^2 \alpha_2 + v_2^2 \alpha_1}{4 \alpha_1 \alpha_2}\right] (t - t_0) + \phi_0\right\}}$	$a_2 (c_1^2 \alpha_1 + c_2^2 \alpha_2) < 0$, $N = 2$, A_0, t_0, and ϕ_0 are arbitrary real constants	moving 2D dark soliton	(6.30)
4.	$\Phi(x_1, x_2, x_3, t) = A_0 \sqrt{\frac{-2 (c_1^2 \alpha_1 + c_2^2 \alpha_2 + c_3^2 \alpha_3)}{a_2}} \times \tanh(A_0 \{c_1 [x_1 - (x_{01} + v_1 t)] + c_2 [x_2 - (x_{02} + v_2 t)] + c_3 [x_3 - (x_{03} + v_3 t)]\}) e^{i \phi(x_1, x_2, x_3, t)}$, $\phi(x_1, x_2, x_3, t) = \frac{v_1 (x_1 - x_{01})}{2 \alpha_1} + \frac{v_2 (x_2 - x_{02})}{2 \alpha_2} + \frac{v_3 (x_3 - x_{03})}{2 \alpha_3} - \left[2 A_0^2 (c_1^2 \alpha_1 + c_2^2 \alpha_2 + c_3^2 \alpha_3) + \frac{v_1^2 \alpha_2 \alpha_3 + v_2^2 \alpha_1 \alpha_3 + v_3^2 \alpha_1 \alpha_2}{4 \alpha_1 \alpha_2 \alpha_3}\right] \times (t - t_0) + \phi_0$,	$a_2 (c_1^2 \alpha_1 + c_2^2 \alpha_2 + c_3^2 \alpha_3) < 0$, $N = 3$, A_0, t_0, and ϕ_0 are arbitrary real constants	moving 3D dark soliton	(6.31)

Equation: $i \Phi_t + \sum_{k=1}^{N} \alpha_k \Phi_{x_k x_k} + a_2 |\Phi|^n \Phi = 0$

(*Continued*)

#	Example	Conditions	Name	Eq. #
1.	$\Phi(x_1, x_2, t) = \left(\frac{2 A_0^2 (n+2)(c_1^2 \alpha_1 + c_2^2 \alpha_2)}{a_2 n^2} \times \text{sech}^2(A_0 \{c_1 [x_1 - (x_{01} + v_1 t)] + c_2 [x_2 - (x_{02} + v_2 t)]\}) \right)^{\frac{1}{n}} \times e^{i \left\{ \frac{v_1 (x_1 - x_{01})}{2 \alpha_1} + \frac{v_2 (x_2 - x_{02})}{2 \alpha_2} + \left[\frac{4 A_0^2 (c_1^2 \alpha_1 + c_2^2 \alpha_2)}{n^2} - \frac{v_1^2 \alpha_2 + v_2^2 \alpha_1}{4 \alpha_1 \alpha_2} \right] (t - t_0) + \phi_0 \right\}}$	$a_2 (c_1^2 \alpha_1 + c_2^2 \alpha_2) > 0$, $N = 2$, A_0, t_0, and ϕ_0 are arbitrary real constants	moving 2D bright soliton	(6.32)
2.	$\Phi(x_1, x_2, x_3, t) = \left(\frac{2 A_0^2 (n+2)(c_1^2 \alpha_1 + c_2^2 \alpha_2 + c_3^2 \alpha_3)}{a_2 n^2} \text{sech}^2[q(x_1, x_2, x_3, t)] \right)^{\frac{1}{n}} \times e^{i \phi(x_1, x_2, x_3, t)}$, $q(x_1, x_2, x_3, t) = A_0 \{c_1 [x_1 - (x_{01} + v_1 t)] + c_2 [x_2 - (x_{02} + v_2 t)] + c_3 [x_3 - (x_{03} + v_3 t)]\}$, $\phi(x_1, x_2, x_3, t) = \phi_0 + \frac{v_1 (x_1 - x_{01})}{2 \alpha_1} + \frac{v_2 (x_2 - x_{02})}{2 \alpha_2} + \frac{v_3 (x_3 - x_{03})}{2 \alpha_3} + \left[\frac{4 A_0^2 (c_1^2 \alpha_1 + c_2^2 \alpha_2 + c_3^2 \alpha_3)}{n^2} - \frac{v_1^2 \alpha_2 \alpha_3 + v_2^2 \alpha_1 \alpha_3 + v_3^2 \alpha_1 \alpha_2}{4 \alpha_1 \alpha_2 \alpha_3} \right] (t - t_0)$	$a_2 (n + 2)(c_1^2 \alpha_1 + c_2^2 \alpha_2 + c_3^2 \alpha_3) > 0$, $N = 3$, A_0, t_0, and ϕ_0 are arbitrary real constants	moving 3D bright soliton	(6.33)

Equation: $i \Phi_t + \sum_{k=1}^{N} \alpha_k \Phi_{x_k x_k} + a_2 |\Phi|^n \Phi + a_3 |\Phi|^m \Phi = 0$

#	Example	Conditions	Name	Eq. #

1.	$\Phi(x_1, x_2, t) = \left[\frac{A_0\, A_1\,(n+2)}{a_2\, A_0 + q(x_1, x_2, t)}\right]^{\frac{1}{n}}$ $\times\, e^{i\left[\frac{v_1\,(x_1 - x_{01})}{2\,\alpha_1} + \frac{v_2\,(x_2 - x_{02})}{2\,\alpha_2} + \left(A_1 - \frac{v_1^2\,\alpha_2 + v_2^2\,\alpha_1}{4\,\alpha_1\,\alpha_2}\right)(t - t_0) + \phi_0\right]}$, $q(x_1, x_2, t) = 2a_2 \cosh\left(n\sqrt{\frac{A_1}{c_1^2\,\alpha_1 + c_2^2\,\alpha_2}}\,\{c_1\,[x_1 - (x_{01} + v_1\,t)] + c_2\,[x_2 - (x_{02} + v_2\,t)]\}\right)$	$A_0 = \frac{2\,a_2}{n+2}\sqrt{\frac{n+1}{A_1\,\delta}}$, $\delta = a_3 + \frac{a_2^2\,(n+1)}{A_1(n+2)^2}$, $A_1\,(c_1^2\,\alpha_1 + c_2^2\,\alpha_2) > 0$, $N = 2$, $A_1\,\delta\,(n+1) > 0$, $m = 2\,n$, t_0, A_1, and ϕ_0 are arbitrary real constants	moving 2D flat-top soliton	(6.34)
2.	$\Phi(x_1, x_2, x_3, t) = \left[\frac{A_0\, A_1\,(n+2)}{a_2\, A_0 + q(x_1, x_2, x_3)}\right]^{\frac{1}{n}} e^{i\,\phi(x_1, x_2, x_3, t)}$, $\phi(x_1, x_2, x_3, t) = \frac{v_1\,(x_1 - x_{01})}{2\alpha_1} + \frac{v_2\,(x_2 - x_{02})}{2\alpha_2} + \frac{v_3\,(x_3 - x_{03})}{2\alpha_3} + \left(A_1 - \frac{v_1^2\,\alpha_2\,\alpha_3 + v_2^2\,\alpha_1\,\alpha_3 + v_3^2\,\alpha_1\,\alpha_2}{4\,\alpha_1\,\alpha_2\,\alpha_3}\right)(t - t_0) + \phi_0$, $q(x_1, x_2, x_3, t) = 2\,a_2 \cosh\left(n\sqrt{\frac{A_1}{(c_1^2\,\alpha_1 + c_2^2\,\alpha_2 + c_3^2\,\alpha_3)}}\,\{c_1\,[x_1 - (x_{01} + v_1\,t)] + c_2\,[x_2 - (x_{02} + v_2\,t)] + c_3\,[x_3 - (x_{03} + v_3\,t)]\}\right)$	$A_0 = \frac{2\,a_2}{n+2}\sqrt{\frac{n+1}{A_1\,\delta}}$, $\delta = a_3 + \frac{a_2^2\,(n+1)}{A_1(n+2)^2}$, $A_1\,(c_1^2\,\alpha_1 + c_2^2\,\alpha_2 + c_3^2\,\alpha_3) > 0$, $A_1\,\delta\,(n+1) > 0$, $m = 2\,n$, $N = 3$, t_0, A_1, and ϕ_0 are arbitrary real constants	moving 3D flat-top soliton	(6.35)

NLSE in (2 + 1)-Dimensions with $\Phi_{x_1 x_2}$ Term

Transformation: $\Phi(x_1, x_2, t) = \sqrt{\frac{a_2}{b_4}}\,\Phi[X(x_1, x_2), t]$, $\quad X(x_1, x_2) = c_0 - \frac{1}{2\,b}\left(b_3\,c_1 - \sqrt{4\,a_1\,b - 4\,b\,b_2\,c_1^2 + b_3^2\,c_1^2}\right)x_1 + c_1\,x_2$

Equation: $i\,\Phi_t + b\,\Phi_{x_1 x_1} + b_2\,\Phi_{x_2 x_2} + b_3\,\Phi_{x_1 x_2} + b_4\,|\Phi|^2\,\Phi = 0$

(Continued)

#	Example	Conditions	Name	Eq. #
1.	$\Phi(x_1, x_2, t) = A_0 \sqrt{\frac{2\,a_1}{b_4}} \times \operatorname{sech}\left\{ A_0 \left[c_0 - \frac{\left(b_3\,c_1 - \sqrt{4\,a_1\,b_1 - 4\,b_1\,b_2\,c_1^2 + b_3^2\,c_1^2}\right)(x_1 - x_{01})}{2\,b_1} + c_1\,(x_2 - x_{02}) \right] \right\} e^{i\left[a_1\,A_0^2\,(t-t_0)+\phi_0\right]}$	$a_1\,b_4 > 0$, $(4\,a_1\,b_1 - 4\,b_1\,b_2\,c_1^2 + b_3^2\,c_1^2) > 0$,	bright soliton	(6.39)
2.	$\Phi(x_1, x_2, t) = A_0 \sqrt{\frac{-2a_1}{b_4}} \times \tanh\left\{ A_0 \left[c_0 + \frac{\left(\sqrt{4a_1\,b_1 - 4\,b_1\,b_2\,c_1^2 + b_3^2\,c_1^2} - b_3\,c_1\right)(x_1 - x_{01})}{2\,b_1} + c_1\,(x_2 - x_{02}) \right] \right\} e^{i\left[2a_1\,A_0^2\,(t-t_0)+\phi_0\right]}$	$a_1\,b_4 < 0$, $(4\,a_1\,b_1 - 4\,b_1\,b_2\,c_1^2 + b_3^2\,c_1^2) > 0$, A_0, t_0, and ϕ_0 are arbitrary real constants	dark soliton	(6.40)

6.7 ($N + 1$)-Dimensional Isotropic NLSE with Cubic Nonlinearity in Polar Coordinate System

If $\psi(x, t)$ is a solution of the fundamental NLSE in 1D, (2.1),

$$i\,\psi_t + a_1\,\psi_{xx} + a_2\,|\psi|^2\,\psi = 0,$$

then

$$\Phi(r, t) = A(r, t)\,e^{i\,B(r,t)}\,\psi[R(r, t),\, T(r, t)] \tag{6.41}$$

is a solution of

$$i\,\Phi_t + b_1(r, t)\left(\Phi_{rr} + \frac{N-1}{r}\Phi_r\right) + b_2(r, t)\,|\Phi|^2\,\Phi + [b_{3r}(r, t) + i\,b_{3i}(r, t)]\Phi = 0, \tag{6.42}$$

where

$$T(r, t) = g_1(t), \tag{6.43}$$

$$A(r, t) = \frac{g_2(t)\,r^{\frac{1-N}{2}}}{\sqrt{R_r(r, t)}}, \tag{6.44}$$

$$B(r, t) = g_3(t) - \int \frac{R_t(r, t)\,R_r(r, t)\,dr}{2a_1\,g_1'(t)}, \tag{6.45}$$

$$b_1(r, t) = \frac{a_1\,g_1'(t)}{R_r^2(r, t)}, \tag{6.46}$$

$$b_2(r, t) = \frac{a_2\,g_1'(t)}{A^2(r, t)}, \tag{6.47}$$

$$b_{3i}(r, t) = \frac{R_{rt}(r, t)}{R_r(r, t)} - \frac{g_2'(t)}{g_2(t)}, \tag{6.48}$$

$$b_{3r}(r, t) = \frac{1}{4a_1\,g_1'^2(t)}\Big\{2g_1''(t)\int R_t(r, t)\,R_r(r, t)\,dr \tag{6.49}$$

$$- 2g_1'(t)\int [R_{tt}(r, t)\,R_r(r, t) + R_t(r, t)\,R_{rt}(r, t)]\,dr + g_1'(t)\,R_t^2(r, t)$$
$$+ \frac{a_1^2\,g_1'^3(t)}{r^2\,R_r^4(r, t)}\Big[(N-1)(N-3)\,R_r^2(r, t) \tag{6.50}$$
$$- 3r^2\,R_{rr}^2(r, t) + 2r^2\,R_r(r, t)\,R_{rrr}(r, t)\Big]\Big\},$$

$g_1(t)$, $g_2(t)$, and $R(r, t)$ are arbitrary real functions.

6.7.1 Angular Dependence

If $\psi(x, t)$ is a solution to the fundamental NLSE, (2.1), then

$$\Phi(r, \theta, t) = r^{(1-N)/2}\, e^{i\,\frac{1}{2}\sqrt{-(N-1)(N-3)}\,\theta}\psi(r, t) \tag{6.51}$$

is a solution to the NLSE

$$i\,\Phi_t + a_1\left(\Phi_{rr} + \frac{N-1}{r}\,\Phi_r + \frac{1}{r^2}\,\Phi_{\theta\theta}\right) + a_2\,r^{N-1}\,|\Phi|^2\,\Phi = 0. \tag{6.52}$$

This is obtained from the previous section with the special choices: $g_1 = t,\ g_2 = 1,\ g_3 = 0,\ R = r$.

Remarks:

1. The angular term $\Phi_{\theta\theta}/r^2$ vanishes in 1D and 3D.
2. The prefactor $r^{(1-N)/2}$ in Φ diverges at $r = 0$ as $r^{-1/2}$ and r^{-1} for 2D and 3D, respectively. This divergence may be removed with certain solutions of the fundamental NLSE, ψ, such as the tanh(r) solution.

***Example* 1. General Case**

Given

$$\psi(x, t) = A_0\sqrt{\frac{-2\,a_1}{a_2}}\,\tanh[A_0\,(x - x_0)]\,e^{-i\,[2\,a_1\,A_0^2\,(t-t_0)+\phi_0]}$$

is a solution of (2.1), then

$$\Phi(r, \theta, t) = r^{(1-N)/2}\,e^{i\,\frac{1}{2}\sqrt{-(N-1)(N-3)}\,\theta}\left[A_0\sqrt{\frac{-2a_1}{a_2}}\,e^{-i\left(2a_1\,A_0^2\,t+\phi_0\right)}\tanh(A_0\,r)\right] \tag{6.53}$$

is a solution of (6.52), where

$a_1\,a_2 < 0$,

A_0 and ϕ_0 are arbitrary real constants.

***Example* 2. 2D** *vortex soliton*

(Figure 6.11)

$$\Phi(r, \theta, t) = r^{-1/2}\,e^{i\,\sqrt{a_1}\,\theta}\left[A_0\sqrt{\frac{-2a_1}{a_2}}\,e^{-i\left(2a_1\,A_0^2\,t+\phi_0\right)}\tanh(A_0\,r)\right] \tag{6.54}$$

is a solution of

$$i\,\Phi_t + a_1\left(\Phi_{rr} + \frac{1}{r}\,\Phi_r - \frac{1}{4r^2}\right) + a_2\,r\,|\Phi|^2\,\Phi = 0, \tag{6.55}$$

where

$a_1\,a_2 < 0$,

A_0 and ϕ_0 are arbitrary real constants.

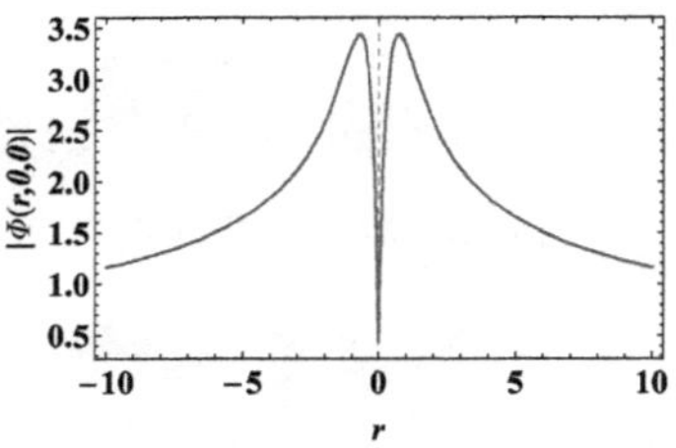

Figure 6.11. Vortex soliton (6.54) at $\theta = t = 0$, with $a_1 = A_0 = 3/2$, $a_2 = -1/2$, and $\phi_0 = 0$.

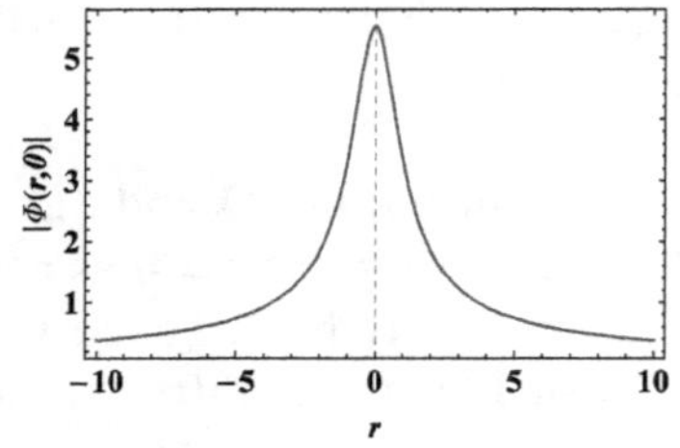

Figure 6.12. Plot of solution (6.56) at $t = 0$ with $a_1 = A_0 = 3/2$, $a_2 = -1/2$, and $\phi_0 = 0$.

***Example* 3. 3D**

(Figure 6.12)

$$\Phi(r,\, t) = r^{-1}\left[A_0 \sqrt{\frac{-2a_1}{a_2}}\; e^{-i\left(2a_1\, A_0^2\, t + \phi_0\right)} \tanh(A_0\, r)\right] \tag{6.56}$$

is a solution to

$$i\,\Phi_t + a_1\left(\Phi_{rr} + \frac{2}{r}\,\Phi_r\right) + a_2\, r^2\, |\Phi|^2\, \Phi = 0, \tag{6.57}$$

where

$a_1\, a_2 < 0$,

A_0 and ϕ_0 are arbitrary real constants.

6.7.2 Constant Dispersion and Real Potential

If $\psi(x,\, t)$ is a solution of the fundamental NLSE, (2.1),

$$i\,\psi_t + a_1\,\psi_{xx} + a_2\,|\psi|^2\,\psi = 0,$$

then

$$\begin{aligned}\Phi(r,\, t) = {} & c_0\, g_1'^{1/4}(t)\, r^{\frac{1-N}{2}}\, e^{i\left\{g_4(t) - \frac{\left[4\sqrt{b_{10}\, g_1'(t)}\; g_2'(t) + \sqrt{a_1}\; g_1''(t)\, r\right] r}{8\, b_{10}\, \sqrt{a_1}\; g_1'(t)}\right\}} \\ & \times \psi\left[g_2(t) + \sqrt{\frac{a_1\, g_1'(t)}{b_{10}}}\; r,\; g_1(t)\right]\end{aligned} \tag{6.58}$$

is a solution of

$$i\,\Phi_t + b_{10}\,\Phi_{rr} + \frac{b_{10}\,(N-1)}{r}\,\Phi_r + \frac{a_2\,\sqrt{g_1'(t)}}{c_0^2\,r^{1-N}}\,|\Phi|^2\,\Phi$$
$$+\left\{\frac{b_{10}\,(N-1)\,(N-3)}{4r^2} + \frac{g_2'^2(t)}{4a_1\,g_1'(t)} + g_4'(t) + \frac{\left[g_2'(t)\,g_1''(t) - g_1'(t)\,g_2''(t)\right]r}{2\sqrt{a_1\,b_{10}}\,g_1'^{3/2}(t)}\right. \tag{6.59}$$
$$\left. + \frac{\left[3g_1''^2(t) - 2g_1'(t)\,g_1'''(t)\right]r^2}{16b_{10}\,g_1'^2(t)}\right\}\Phi = 0,$$

where

$$T(r,\,t) = g_1(t), \tag{6.60}$$

$$R(r,\,t) = g_2(t) + \sqrt{\frac{a_1\,g_1'(t)}{b_{10}}}\;r, \tag{6.61}$$

$$A(r,\,t) = g_3(t)\,r^{\frac{1-N}{2}}, \tag{6.62}$$

$$B(r,\,t) = g_4(t) - \frac{\left[4\sqrt{b_{10}\,g_1'(t)}\;g_2'(t) + \sqrt{a_1}\;g_1''(t)\,r\right]r}{8b_{10}\,\sqrt{a_1}\;g_1'(t)}, \tag{6.63}$$

$$b_1(r,\,t) = b_{10}, \tag{6.64}$$

$$b_2(r,\,t) = \frac{a_2\,g_1'(t)}{A^2(r,\,t)}, \tag{6.65}$$

$$b_{3r}(r,\,t) = \frac{b_{10}\,(N-1)\,(N-3)}{4\,r^2} + \frac{g_2'^2(t)}{4\,a_1\,g_1'(t)} + g_4'(t) + \frac{\left[g_2'(t)\,g_1''(t) - g_1'(t)\,g_2''(t)\right]r}{2\,\sqrt{a_1\,b_{10}}\;g_1'^{3/2}(t)} \tag{6.66}$$
$$+ \frac{\left[3\,g_1''^2(t) - 2\,g_1'(t)\,g_1'''(t)\right]r^2}{16\,b_{10}\,g_1'^2(t)},$$

$$g_3(t) = c_0\,g_1'^{1/4}(t), \tag{6.67}$$

$g_1(t)$, $g_2(t)$, and $g_4(t)$ are arbitrary real functions,
c_0 and b_{10} are arbitrary real constants.

***Example* 1. sech(r, t)** *bright soliton*

Given

$$\psi(x,\,t) = A_0 \sqrt{\frac{2\,a_1}{a_2}}\ \mathrm{sech}[A_0\,(x - x_0)]\, e^{i\left[a_1 A_0^2\,(t-t_0)+\phi_0\right]}$$

is a solution of (2.1), then

$$\Phi(r,\,t) = c_0\,A_0 \sqrt{\frac{2a_1}{a_2}}\ g_1'^{1/4}(t)\, r^{\frac{1-N}{2}}\ \mathrm{sech}\left\{A_0\left[g_2(t) + \sqrt{\frac{a_1\,g_1'(t)}{b_{10}}}\ r\right]\right\}$$
$$\times\, e^{i\left\{A_0^2\,a_1\,[g_1(t)-t_0]+g_4(t)-\frac{g_2'(t)}{2\sqrt{a_1\,b_{10}\,g_1'(t)}}\,r-\frac{g_1''(t)}{8\,b_{10}\,g_1'(t)}\,r^2+\phi_0\right\}} \qquad (6.68)$$

is a solution of (6.59), where

$a_1\,a_2 > 0$,
$a_1\,b_{10} > 0$,
A_0, t_0, and ϕ_0 are arbitrary real constants.

***Example* 2. tanh(r, t)** *dark soliton*

Given

$$\psi(x,\,t) = A_0 \sqrt{\frac{-2a_1}{a_2}}\ \tanh[A_0\,(x - x_0)]\, e^{-i\left[2\,a_1 A_0^2\,(t-t_0)+\phi_0\right]}$$

is a solution of (2.1), then

$$\Phi(r,\,t) = c_0\,A_0 \sqrt{\frac{-2a_1}{a_2}}\ g_1'^{1/4}(t)\, r^{\frac{1-N}{2}}\ \tanh\left\{A_0\left[g_2(t) + \sqrt{\frac{a_1\,g_1'(t)}{b_{10}}}\ r\right]\right\}$$
$$\times\, e^{-i\left\{2\,A_0^2\,a_1\,[g_1(t)-t_0]-g_4(t)+\frac{g_2'(t)}{2\sqrt{a_1\,b_{10}\,g_1'(t)}}\,r+\frac{g_1''(t)}{8\,b_{10}\,g_1'(t)}\,r^2+\phi_0\right\}} \qquad (6.69)$$

is a solution of (6.59), where

$a_1\,a_2 < 0$,
$a_1\,b_{10} > 0$,
A_0, t_0, and ϕ_0 are arbitrary real constants.

6.8 Summary of Section 6.7

$(N+1)$-Dimensional NLSE with Cubic Nonlinearity in a Polar Coordinate System

Transformation: $\Phi(r, t) = A(r, t)\, e^{i\, B(r, t)}\, \psi[R(r, t),\, T(r, t)]$

Equation: $i\, \Phi_t + b_1(r, t)\left(\Phi_{rr} + \frac{N-1}{r}\, \Phi_r\right) + b_2(r, t)\, |\Phi|^2\, \Phi + [b_{3r}(r, t) + i\, b_{3i}(r, t)]\Phi = 0$

Angular Dependence

Transformation: $\Phi(r, \theta, t) = r^{(1-N)/2}\, e^{i\, \frac{1}{2} \sqrt{-(N-1)(N-3)}\, \theta} \psi(r, t)$

Equation (1): $i\, \Phi_t + a_1\left(\Phi_{rr} + \frac{N-1}{r}\, \Phi_r + \frac{1}{r^2}\, \Phi_{\theta\theta}\right) + a_2\, r^{N-1}\, |\Phi|^2\, \Phi = 0$

# Example: General Case	Conditions	Name	Eq. #
1. $\Phi(r, \theta, t) = r^{(1-N)/2}\, e^{i\, \frac{1}{2} \sqrt{-(N-1)(N-3)}\, \theta} \times\left[A_0 \sqrt{\frac{-2a_1}{a_2}}\, e^{-i\left(2a_1 A_0^2\, t + \phi_0\right)} \tanh(A_0\, r)\right]$	$a_1\, a_2 < 0$, A_0 and ϕ_0 are arbitrary real constants	dark soliton	(6.53)

Equation (2): $i\, \Phi_t + a_1\left(\Phi_{rr} + \frac{1}{r}\, \Phi_r - \frac{1}{4r^2}\right) + a_2\, r\, |\Phi|^2\, \Phi = 0$

# Example: 2D	Conditions	Name	Eq. #
1. $\Phi(r, \theta, t) = r^{-1/2}\, e^{i\, \sqrt{a_1}\, \theta}\left[A_0 \sqrt{\frac{-2a_1}{a_2}}\, e^{-i\left(2a_1 A_0^2\, t + \phi_0\right)} \tanh(A_0\, r)\right]$	$a_1\, a_2 < 0$, A_0 and ϕ_0 are arbitrary real constants	dark(vortex) soliton	(6.54)

Equation (3): $i\,\Phi_t + a_1\left(\Phi_{rr} + \frac{2}{r}\,\Phi_r\right) + a_2\,r^2\,|\Phi|^2\,\Phi = 0$

#	Example: 3D	Conditions	Name	Eq.#
1.	$\Phi(r,t) = r^{-1}\left[A_0\sqrt{\frac{-2a_1}{a_2}}\,e^{-i\left(2a_1 A_0^2\,t+\phi_0\right)}\tanh(A_0\,r)\right]$	$a_1\,a_2 < 0$, A_0 and ϕ_0 are arbitrary real constants	dark soliton	(6.56)

Constant Dispersion and Real Potential

Transformation: $\Phi(r,t) = c_0\,g_1'^{1/4}(t)\,r^{\frac{1-N}{2}}\,e^{i\left\{g_4(t)-\frac{\left[4\sqrt{b_{10}\,g_1'(t)}\,g_2'(t)+\sqrt{a_1}\,g_1''(t)\,r\right]r}{8\,b_{10}\sqrt{a_1}\,g_1'(t)}\right\}}\,\psi\left[g_2(t)+\sqrt{\frac{a_1\,g_1'(t)}{b_{10}}}\,r,\ g_1(t)\right]$

Equation:

$$i\,\Phi_t + b_{10}\,\Phi_{rr} + \frac{b_{10}\,(N-1)}{r}\,\Phi_r + \frac{a_2\sqrt{g_1'(t)}}{c_0^2\,r^{1-N}}\,|\Phi|^2\,\Phi + \left\{\frac{b_{10}\,(N-1)\,(N-3)}{4\,r^2} + \frac{g_2'^2(t)}{4\,a_1\,g_1'(t)} + g_4'(t) + \frac{\left[g_2'(t)\,g_1''(t) - g_1'(t)\,g_2''(t)\right]r}{2\sqrt{a_1\,b_{10}}\,g_1'^{3/2}(t)} + \frac{\left[3\,g_1''^2(t) - 2\,g_1'(t)\,g_1'''(t)\right]r^2}{16\,b_{10}\,g_1'^2(t)}\right\}\Phi = 0$$

#	Example	Conditions	Name	Eq.#
1.	$\Phi(r,t) = c_0\,A_0\sqrt{\frac{2\,a_1}{a_2}}\,g_1'^{1/4}(t)\,r^{\frac{1-N}{2}}\,\mathrm{sech}\left\{A_0\left[g_2(t)+\sqrt{\frac{a_1\,g_1'(t)}{b_{10}}}\,r\right]\right\} \times e^{i\left\{A_0^2\,a_1\,[g_1(t)-t_0]+g_4(t)-\frac{g_2'(t)}{2\sqrt{a_1\,b_{10}\,g_1'(t)}}\,r-\frac{g_1''(t)}{8\,b_{10}\,g_1'(t)}\,r^2+\phi_0\right\}}$	$a_1\,a_2 > 0$, $a_1\,b_{10} > 0$, A_0, t_0, and ϕ_0 are arbitrary real constants	bright soliton	(6.68)

(Continued)

2. $\Phi(r, t) = c_0\, A_0 \sqrt{\frac{-2\, a_1}{a_2}}\, g_1'^{1/4}(t)\, r^{\frac{1-N}{2}} \tanh\left\{A_0\left[g_2(t) + \sqrt{\frac{a_1\, g_1'(t)}{b_{10}}}\, r\right]\right\} \times e^{-i\left\{2\, A_0^2\, a_1\, [g_1(t)-t_0]-g_4(t)+\frac{g_2'(t)}{2\sqrt{a_1\, b_{10}\, g_1'(t)}}\, r+\frac{g_1''(t)}{8\, b_{10}\, g_1'(t)}\, r^2+\phi_0\right\}}$	$a_1\, a_2 < 0$, $a_1\, b_{10} > 0$, A_0, t_0, and ϕ_0 are arbitrary real constants	dark soliton	(6.69)
2. $\Phi(x_1, x_2, t) = A_0 \sqrt{\frac{-2\, a_1}{b_4}} \times \tanh\left\{A_0\left[c_0 + \frac{\left(\sqrt{4\, a_1\, b_1 - 4\, b_1\, b_2\, c_1^2 + b_3^2\, c_1^2} - b_3\, c_1\right)(x_1 - x_{01})}{2\, b_1} + c_1\, (x_2 - x_{02})\right]\right\} e^{i\left[2\, a_1\, A_0^2\, (t-t_0)+\phi_0\right]}$	$a_1\, b_4 < 0$, $(4\, a_1\, b_1 - 4\, b_1\, b_2\, c_1^2 + b_3^2\, c_1^2) > 0$, A_0, t_0, and ϕ_0 are arbitrary real constants	dark soliton	(6.40)

6.9 Power Series Solutions to (2 + 1)-Dimensional NLSE with Cubic Nonlinearity in a Polar Coordinate System

The function

$$\Phi(r, \theta, t) = Z(r)\, e^{i\,(\lambda\, t+\alpha\, \theta)} \tag{6.70}$$

with

$$\begin{aligned} Z(r) &= a_0 + \left(\frac{5\, b_0}{4}\right) r + \left(\frac{25\, [-a_2\, (a_0)^3 - a_1\, b_0 + a_1\, a_0\, \alpha^2 + a_0\, \lambda]}{32\, a_1}\right) r^2 \\ &+ \left(\frac{15\, [a_2\, (a_0)^2\, (a_0 - 3\, b_0) - 3\, a_1\, a_0\, \alpha^2 + a_1\, b_0\, (2 + \alpha^2) + (b_0 - a_0)\, \lambda]}{64\, a_1}\right) r^3 \\ &+ \frac{5}{512\, a_1^2}\Big(a_1^2\, [a_0\, \alpha^2\, (68 + 7\, \alpha^2) - 3\, b_0\, (12 + 13\, \alpha^2)] \\ &+ a_1 \Big\{-a_2\, a_0 \left[-33\, a_0\, b_0 + 42\, (b_0)^2 + 2\, (a_0)^2\, (9 + 14\, \alpha^2)\right] \\ &+ \left[-11\, b_0 + 2\, a_0\, (9 + 7\, \alpha^2)\right] \lambda\Big\} \\ &+ 7\, a_0 \left[3 a_2^2\, (a_0)^4 - 4\, a_2\, (a_0)^2\, \lambda + \lambda^2\right]\Big)\, r^4 + O^5(r) \end{aligned} \tag{6.71}$$

is a stationary solution of

$$i\, \Phi_t + a_1 \left[\Phi_{rr} + \frac{1}{r}\, \Phi_r + \frac{1}{r^2}\, \Phi_{\theta\theta}\right] + a_2\, |\Phi|^2\, \Phi = 0, \tag{6.72}$$

where λ, α, a_0, and b_0 are arbitrary real constants.

The solution is obtained using an *Iterative Power Series* (IPS) method [1, 2], which is briefly described as follows: The function $Z(r)$ obeys the ordinary differential equation

$$-\lambda\, Z(r) + a_1 \left[Z''(r) + \frac{1}{r}\, Z'(r) - \frac{\alpha^2}{r^2}\, Z(r)\right] + a_2\, Z^3(r) = 0. \tag{6.73}$$

A convergent power series solution is obtained by the following algorithm:

1. Set initial values a_0 and b_0.
2. Expand $Z(r)$ in power series around the arbitrary real r_0: $Z(r) = a_0 + b_0\, (r - r_0) + \sum_{j=2}^{n_{\max}} c_j (r - r_0)^j$.
3. Substitute in (6.71) to obtain the recursion relation for c_n in terms of a_0 and b_0.
4. Calculate $Z(\Delta)$ and $Z'(\Delta)$, where $\Delta = (r - r_0)/I$, and I is an integer larger than 1.
5. Assign: $a_0 = Z(\Delta)$ and $b_0 = Z'(\Delta)$.
6. Obtain c_n in terms of a_0 and b_0.
7. Repeat steps 2–6 I times.
8. At the Ith step, a_0 will correspond to the power series of Z.

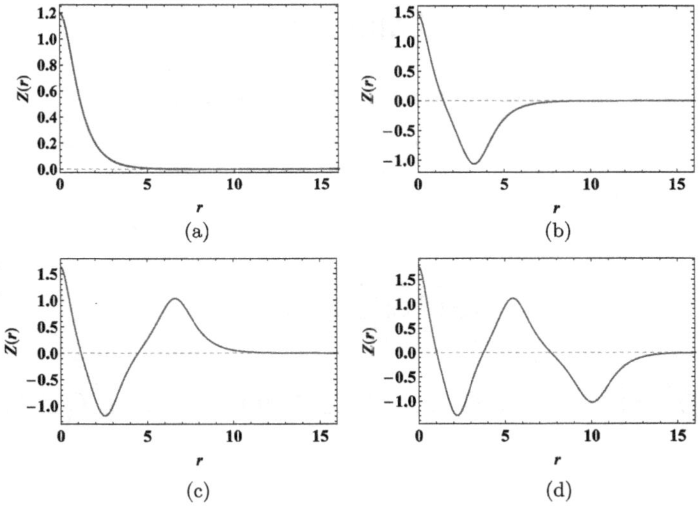

Figure 6.13. Stationary power series solutions of (6.73) with a different number of nodes. (a) nodeless solution with $a_0 = 1.1986644033$, (b) single-node solution with $a_0 = 1.4605146251$, root at $r = 1.46$, (c) double-node solution with $a_0 = 1.6326670699$, roots at $r = 1.17, 4.46$, and (d) triple-node solution with $a_0 = 1.7642003085$, roots at $r = 1.02, 3.75, 7.64$. IPS parameters used are: $n_{\max} = 2$, $I = 1700$, and $\Delta = 0.01$.

Solution (6.71) is obtained with $I = 4$, $n_{\max} = 2$, and for arbitrary a_1, a_2, λ, α, a_0, and b_0.

6.9.1 Family of Infinite Number of Localized Solutions

Using the IPS method, described above, we tune the parameter a_0 to obtain a family of infinite number of solutions differing by the number of nodes. In Figure 6.13, we present some plots showing the nodeless, single-node, double-node, and triple-node solutions of (6.73). Other parameters are fixed to $b_0 = 0$, $a_1 = 1$, $a_2 = 2$, $\lambda = 1$, and $\alpha = 0$.

References

[1] Al Khawaja U and Al-Mdallal Q M 2018 Convergent power series of and solutions to nonlinear differential equations *Int. J. Differ. Equ.* **2018** 1–10

[2] Al Sakkaf L Y, Al-Mdallal Q M and Al Khawaja U 2018 A Numerical algorithm for solving higher-order nonlinear BVPs with an application on fluid flow over a shrinking permeable infinite long cylinder *Complexity* **2018** 1–11

IOP Publishing

Handbook of Exact Solutions to the Nonlinear Schrödinger Equations

Usama Al Khawaja and Laila Al Sakkaf

Chapter 7

Coupled Nonlinear Schrödinger Equations

A Glance at Chapter 7

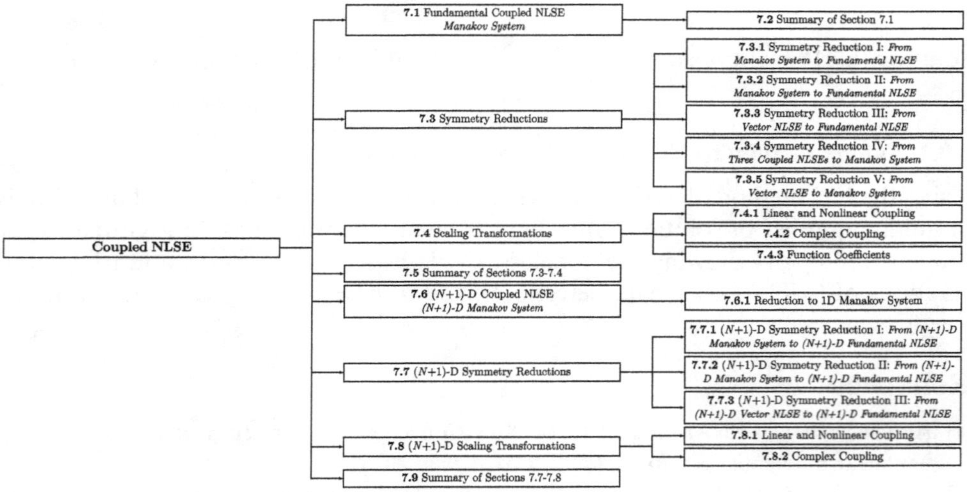

doi:10.1088/978-0-7503-2428-1ch7

A Statistical View of Chapter 7

	Equation	Solutions
1	$i\,\psi_{1_t} + b_0\,\psi_{1xx} + (b_1\,|\psi_1|^2 + b_2\,|\psi_2|^2)\,\psi_1 = 0,$ $i\,\psi_{2_t} + c_0\,\psi_{2xx} + (c_1\,|\psi_1|^2 + c_2\,|\psi_2|^2)\,\psi_2 = 0$	13
2	$i\,\psi_{1_t} + b_0\,\psi_{1xx} + (c_1 + c_2|\sigma|^2)\,|\psi_1|^2\,\psi_1 = 0$	0
3	$i\,\psi_{1_t} + b_0\,\psi_{1xx} - (c_1 + c_2)\,|\psi_1|^2\,\psi_1 = 0$	0
4	$i\,\psi_{1_t} + b_{0j}\,\psi_{1xx} + \sum_{k=1}^{N} b_{1j}|\sigma_j|^2\,|\psi_1|^2\,\psi_1 = 0, \quad j = 1, 2, \ldots, N$	0
5	$i\,\Phi_{1_t} + \Phi_{1xx} + (g_1\,|\Phi_1|^2 - g_2\,|\Phi_2|^2)\,\Phi_1 + g_0\,(g_1 + g_2)\,\Phi_1 - 2\,g_0\,g_2\,\Phi_2 = 0,$ $i\,\Phi_{2_t} + \Phi_{2xx} + (g_1\,|\Phi_1|^2 - g_2\,|\Phi_2|^2)\,\Phi_2 - g_0\,(g_1 + g_2)\,\Phi_2 + 2\,g_0\,g_1\,\Phi_1 = 0$	0
6	$i\,\psi_{1_t} + a_1\,\psi_{1xx} + (b_1\,|\psi_1|^2 + b_3\,|\psi_3|^2)\,\psi_1 = 0,$ $i\,\psi_{3_t} + a_3\,\psi_{3xx} + (d_1\,|\psi_1|^2 + d_3\,|\psi_3|^2)\,\psi_3 = 0$	2
7	$i\,\psi_{1_t} + a_1\,\psi_{1xx} + (b_1\,|\psi_1|^2 + b_2\,|\psi_2|^2)\,\psi_1 = 0,$ $i\,\psi_{2_t} + a_2\,\psi_{2xx} + (c_1\,|\psi_1|^2 + c_2\,|\psi_2|^2)\,\psi_2 = 0$	0
8	$i\,\Phi_{1_t} + \Phi_{1xx} + (g_1\,|\Phi_1|^2 - g_2\,|\Phi_2|^2)\,\Phi_1 = 0,$ $i\,\Phi_{2_t} + \Phi_{2xx} + (g_1\,|\Phi_1|^2 - g_2\,|\Phi_2|^2)\,\Phi_2 = 0$	1
9	$i\,\Phi_{1_t} + \Phi_{1xx} + 2\,(a_{11}\,|\Phi_1|^2 + a_{12}\,|\Phi_2|^2)\,\Phi_1 + 2\,(b_{11}\,\Phi_1\,\Phi_2^* + b_{12}\,\Phi_2\,\Phi_1^*)\,\Phi_1 = 0,$ $i\,\Phi_{2_t} + \Phi_{2xx} + 2\,(a_{21}\,|\Phi_1|^2 + a_{22}\,|\Phi_2|^2)\,\Phi_2 + 2\,(b_{21}\,\Phi_1\,\Phi_2^* + b_{22}\,\Phi_2\,\Phi_1^*)\,\Phi_2 = 0$	0

10	$i\,\Phi_{1t} + \Phi_{1xx} - 2(a+b)\,(\lvert\Phi_1\rvert^2 + \lvert\Phi_2\rvert^2)\,\Phi_1 + 2\,((a+i\,b)\,\Phi_1\,\Phi_2^* + (a-i\,b)\,\Phi_2\,\Phi_1^*)\,\Phi_1 = 0,$ $i\,\Phi_{2t} + \Phi_{2xx} - 2(a+b)\,(\lvert\Phi_1\rvert^2 + \lvert\Phi_2\rvert^2)\,\Phi_2 + 2\,((a+i\,b)\,\Phi_1\,\Phi_2^* + (a-i\,b)\,\Phi_2\,\Phi_1^*)\,\Phi_2 = 0$	0
11	$i\,\psi_{1t} + \sum_{k=1}^N b_{0k}\,\psi_{1x_kx_k} + (b_1\,\lvert\psi_1\rvert^2 + b_2\,\lvert\psi_2\rvert^2)\,\psi_1 = 0,$ $i\,\psi_{2t} + \sum_{k=1}^N c_{0k}\,\psi_{2x_kx_k} + (c_1\,\lvert\psi_1\rvert^2 + c_2\,\lvert\psi_2\rvert^2)\,\psi_2 = 0$	1
12	$i\,\psi_{1t} + \sum_{k=1}^{N} c_{0k}\,\psi_{1x_kx_k} + (c_1 + c_2\lvert\sigma\rvert^2)\,\lvert\psi_1\rvert^2\,\psi_1 = 0$	0
13	$i\,\psi_{1t} - \sum_{k=1}^{N} c_{0k}\,\psi_{1x_kx_k} - 2\,(c_1 + c_2)\,\lvert\psi_1\rvert^2\,\psi_1 = 0$	0
14	$i\,\psi_{1t} + \sum_{k=1}^{N} b_{10k}\,\psi_{1x_kx_k} + \sum_{k=1}^{N} b_{1j}\lvert\sigma_{j+1}\rvert^2\,\lvert\psi_1\rvert^2\,\psi_1 = 0$	0
15	$i\,\Phi_{1t} + \sum_{k=1}^N \Phi_{1x_kx_k} + (g_1\,\lvert\Phi_1\rvert^2 - g_2\,\lvert\Phi_2\rvert^2)\,\Phi_1 + g_0\,(g_1 + g_2)\,\Phi_1 - 2\,g_0\,g_2\,\Phi_2 = 0,$ $i\,\Phi_{2t} + \sum_{k=1}^N \Phi_{2x_kx_k} + (g_1\,\lvert\Phi_1\rvert^2 - g_2\,\lvert\Phi_2\rvert^2)\,\Phi_2 - g_0\,(g_1 + g_2)\,\Phi_2 + 2\,g_0\,g_1\,\Phi_1 = 0$	0
16	$i\,\Phi_{1t} + \sum_{k=1}^N \Phi_{1x_kx_k} + (g_1\,\lvert\Phi_1\rvert^2 - g_2\,\lvert\Phi_2\rvert^2)\,\Phi_1 = 0,$ $i\,\Phi_{2t} + \sum_{k=1}^N \Phi_{2x_kx_k} + (g_1\,\lvert\Phi_1\rvert^2 - g_2\,\lvert\Phi_2\rvert^2)\,\Phi_2 = 0$	0
17	$i\,\Phi_{1t} + \sum_{k=1}^N \Phi_{1x_kx_k} + 2\,(a_{11}\,\lvert\Phi_1\rvert^2 + a_{12}\,\lvert\Phi_2\rvert^2)\,\Phi_1 + 2\,(b_{11}\,\Phi_1\,\Phi_2^* + b_{12}\,\Phi_2\,\Phi_1^*)\,\Phi_1 = 0,$ $i\,\Phi_{2t} + \sum_{k=1}^N \Phi_{2x_kx_k} + 2\,(a_{21}\,\lvert\Phi_1\rvert^2 + a_{22}\,\lvert\Phi_2\rvert^2)\,\Phi_2 + 2\,(b_{21}\,\Phi_1\,\Phi_2^* + b_{22}\,\Phi_2\,\Phi_1^*)\,\Phi_2 = 0$	0
18	$i\,\Phi_{1t} + \sum_{k=1}^N \Phi_{1x_kx_k} - 2(a+b)\,(\lvert\Phi_1\rvert^2 + \lvert\Phi_2\rvert^2)\,\Phi_1 + 2\,((a+i\,b)\,\Phi_1\,\Phi_2^* + (a-i\,b)\,\Phi_2\,\Phi_1^*)\,\Phi_1 = 0,$ $i\,\Phi_{2t} + \sum_{k=1}^N \Phi_{2x_kx_k} - 2(a+b)\,(\lvert\Phi_1\rvert^2 + \lvert\Phi_2\rvert^2)\,\Phi_2 + 2\,((a+i\,b)\,\Phi_1\,\Phi_2^* + (a-i\,b)\,\Phi_2\,\Phi_1^*)\,\Phi_2 = 0$	0
Total	18	17

7.1 Fundamental Coupled NLSE *Manakov System*

Equation:

$$
\begin{aligned}
& i\,\psi_{1_t} + b_0\,\psi_{1_{xx}} + (b_1\,|\psi_1|^2 + b_2\,|\psi_2|^2)\,\psi_1 = 0,\\
& i\,\psi_{2_t} + c_0\,\psi_{2_{xx}} + (c_1\,|\psi_1|^2 + c_2\,|\psi_2|^2)\,\psi_2 = 0,
\end{aligned}
\tag{7.1}
$$

where

$\psi_j = \psi_j(x,\,t)$ is the complex function profile,
x and t are its two independent variables,
b_0, c_0, b_1, c_1, b_2, and c_2 are real constants.

Solutions:

***Solution* 1. Constant Amplitude I** *continuous wave (CW), t- and x-independent phase*

$$
\begin{aligned}
\psi_1(x,\,t) &= A_0\,e^{i\phi_1},\\
\psi_2(x,\,t) &= B_0\,e^{i\phi_2},
\end{aligned}
\tag{7.2}
$$

where

$b_1 = -\frac{b_2\,B_0^2}{A_0^2}$,
$c_2 = -\frac{c_1\,A_0^2}{B_0^2}$,
A_0, B_0, ϕ_1, and ϕ_2 are arbitrary real constants.

***Solution* 2. Constant Amplitude II** *CW, t-dependent phase*

$$
\begin{aligned}
\psi_1(x,\,t) &= A_0\,e^{i\,[A_1\,(t-t_0)+\phi_1]},\\
\psi_2(x,\,t) &= B_0\,e^{i\,[B_1\,(t-t_0)+\phi_2]},
\end{aligned}
\tag{7.3}
$$

where

$A_1 = A_0^2\,b_1 + B_0^2\,b_2$,
$B_1 = A_0^2\,c_1 + B_0^2\,c_2$,
A_0, B_0, t_0, ϕ_1, and ϕ_2 are arbitrary real constants.

***Solution* 3. Constant Amplitude III** *CW, x-dependent phase*

$$
\begin{aligned}
\psi_1(x,\,t) &= A_0\,e^{i\,[A_1\,(x-x_0)+\phi_1]},\\
\psi_2(x,\,t) &= B_0\,e^{i\,[B_1\,(x-x_0)+\phi_2]},
\end{aligned}
\tag{7.4}
$$

where

$A_1 = \pm\sqrt{\frac{A_0^2\,b_1 + B_0^2\,b_2}{b_0}}$,

$B_1 = \pm\sqrt{\frac{A_0^2\,c_1 + B_0^2\,c_2}{c_0}}$,

$b_0\,(A_0^2\,b_1 + B_0^2\,b_2) > 0$,
$c_0\,(A_0^2\,c_1 + B_0^2\,c_2) > 0$,
A_0, B_0, x_0, ϕ_1, and ϕ_2 are arbitrary real constants.

***Solution* 4. Constant Amplitude IV** *CW, t- and x-dependent phase*

$$\begin{aligned}\psi_1(x,\,t) &= A_0\,e^{i\,[A_1\,(t-t_0)+A_2\,(x-x_0)+\phi_1]},\\ \psi_2(x,\,t) &= B_0\,e^{i\,[B_1\,(t-t_0)+B_2\,(x-x_0)+\phi_2]},\end{aligned} \tag{7.5}$$

where
$A_1 = -A_2^2\,b_0 + A_0^2\,b_1 + B_0^2\,b_2$,
$B_1 = -B_2^2\,c_0 + A_0^2\,c_1 + B_0^2\,c_2$,
A_0, B_0, A_2, B_2, x_0, t_0, ϕ_1, and ϕ_2 are arbitrary real constants.

***Solution* 5. Rational Solution** *decaying wave (DW)*

$$\begin{aligned}\psi_1(x,\,t) &= \frac{A_0}{\sqrt{2\,[A_1 + t - t_0]}}\,e^{i\,\phi_1(x,t)},\\ \psi_2(x,\,t) &= \frac{B_0}{\sqrt{2\,[B_1 + t - t_0]}}\,e^{i\,\phi_2(x,t)},\end{aligned} \tag{7.6}$$

where

$$\begin{aligned}\phi_1(x,\,t) &= \tfrac{(b_0\,A_2 + x - x_0)^2}{4\,b_0\,(A_1 + t - t_0)} + \tfrac{b_1\,A_0^2}{2}\ln[2(A_1 + t - t_0)]\\ &\quad + \tfrac{b_2\,B_0^2}{2}\ln[2\,(B_1 + t - t_0)] + \phi_{01},\\ \phi_2(x,\,t) &= \tfrac{(c_0\,B_2 + x - x_0)^2}{4\,c_0\,(B_1 + t - t_0)} + \tfrac{c_1\,A_0^2}{2}\ln[2(A_1 + t - t_0)]\\ &\quad + \tfrac{c_2\,B_0^2}{2}\ln[2\,(B_1 + t - t_0)] + \phi_{02},\end{aligned}$$

A_0, A_1, A_2, B_0, B_1, B_2, x_0, t_0, ϕ_{01}, and ϕ_{02} are arbitrary real constants.

***Solution* 6. tanh(*x*, *t*)-sech(*x*, *t*)** *dark-bright soliton*
(Figure 7.1)

$$\begin{aligned}\psi_1(x,\,t) &= (A_0\tanh\{A_1\,[x - x_0 - v\,(t - t_0)]\} + i\,A_2)\,e^{i\,[b_1\,(t-t_0)+\phi_1]},\\ \psi_2(x,\,t) &= B_0\,\mathrm{sech}\{A_1\,[x - x_0 - v\,(t - t_0)]\}\,e^{i\,[(A_1^2+b_2)\,(t-t_0)+\theta(x,t)+\phi_2]},\end{aligned} \tag{7.7}$$

where

$$A_0 = \cos\left[\tan^{-1}\left(\tfrac{-v}{2\,A_1}\right)\right],$$

$$A_1 = \sqrt{\tfrac{b_1\,(1-b_2^2)}{2\,(1+b_1\,b_2)} - \tfrac{v^2}{4}},$$

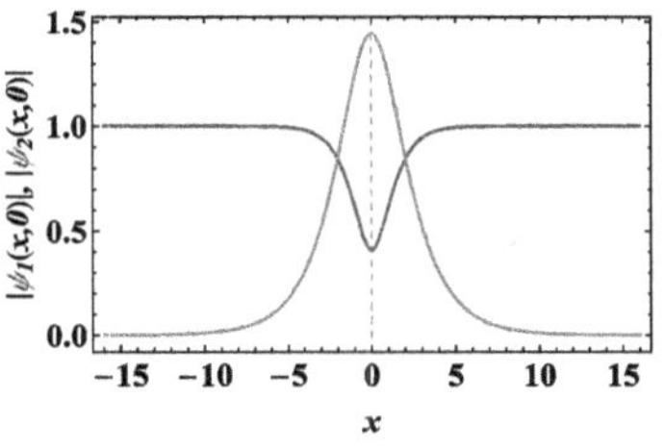

Figure 7.1. Dark-bright soliton (7.7) at $t = 0$. Blue is ψ_1 and red is ψ_2 with $b_2 = 1/2$, $v = 1/2$, and $x_0 = t_0 = \phi_1 = \phi_2 = 0$. Animation available online at https://iopscience.iop.org/book/978-0-7503-2428-1.

$$A_2 = \sin\left[\tan^{-1}\left(\frac{-v}{2A_1}\right)\right],$$
$$B_0 = \sqrt{\frac{2A_1^2(b_1+b_2)}{1-b_2^2}},$$
$$\theta(x, t) = \frac{v}{2}\left[x - x_0 - \frac{v}{2}(t - t_0)\right],$$
$$b_0 = -1,$$
$$c_0 = 1,$$
$$c_1 = 1/b_1,$$
$$c_2 = b_2,$$
$$b_1 b_2 = 1,$$
$$\frac{b_1(1-b_2^2)}{2(1+b_1 b_2)} > \frac{v^2}{4},$$
$$2A_1^2(b_1 + b_2)(1 - b_2^2) > 0,$$
x_0, t_0, ϕ_1, ϕ_2, and v are arbitrary real constants.

- *Reference*: [2].

***Solution* 7.**
(Figure 7.2)

$$\begin{aligned}\psi_1(x, t) = {} & A_0\Big(1 - 4\left[\alpha_1^2(x, t) + \alpha_2^2(x, t) - \alpha_1(x, t) + i\,\alpha_2(x, t)\right]e^{-\delta(x,t)}\\ & \times\Big\{\left[2\,\alpha_1^2(x, t) + 2\,\alpha_2^2(x, t) - 2\,\alpha_1(x, t) + 1\right]e^{-\delta(x,t)}\\ & + \left(\beta_3^2 + \beta_4^2\right)e^{2\,\delta(x,t)}\Big\}^{-1}\Big) \times e^{i\,[A_1 x + (2A_0^2 - A_1^2)\,t]},\\ \psi_2(x, t) = {} & -4A_0\,\{\beta_3\,[\alpha_1(x, t) - 1] - \beta_4\,\alpha_2(x, t) + i\,[\beta_3\,\alpha_2(x, t) + \beta_4\,\alpha_1(x, t) - \beta_4]\}\\ & \times\left(\left\{2\left[\alpha_1^2(x, t) + \alpha_2^2(x, t)\right] - 2\,\alpha_1(x, t) + 1\right\}e^{-\delta(x,t)} + \left(\beta_3^2 + \beta_4^2\right)e^{2\,\delta(x,t)}\right)^{-1}\\ & \times e^{i\,[A_1 x + (3A_0^2 - A_1^2)] + \frac{\delta(x,t)}{2}},\end{aligned} \tag{7.8}$$

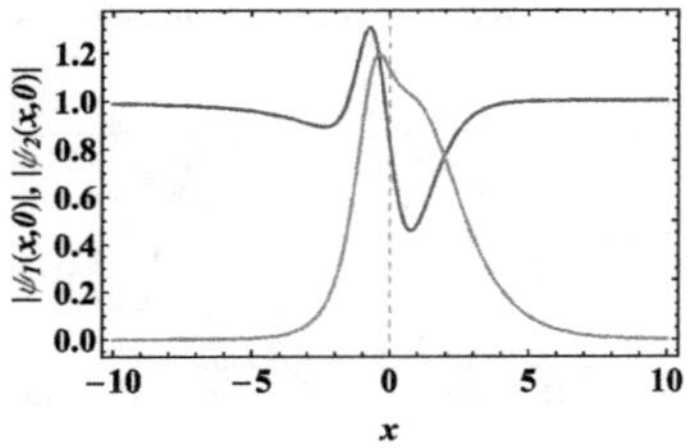

Figure 7.2. Plot of solution (7.8) at $t = 0$. Blue is ψ_1 and red is ψ_2 with $A_0 = A_1 = \beta_1 = \beta_2 = \beta_3 = \beta_4 = 1$. Animation available online at https://iopscience.iop.org/book/978-0-7503-2428-1.

where

$\delta(x, t) = \frac{2 A_0 (x - 2 A_1 t)}{3}$,
$\alpha_1(x, t) = \beta_1 A_0 + A_0 (x - 2 A_1 t)$,
$\alpha_2(x, t) = \beta_2 A_0 - 2 A_0^2 t$,
$b_0 = c_0 = 1$,
$b_1 = b_2 = c_1 = c_2 = 2$,
A_0, A_1, β_1, β_2, β_3, and β_4 are arbitrary real constants.

- *Reference*: [3].

***Solution* 8.**

(Figure 7.3)

$$\psi_1(x, t) = A_0 \left(\frac{-6 \sqrt{3} A_0 \delta(x, t) - 36 \sqrt{3} A_0^2 t + i \left[36 A_0^2 t + 6 A_0 \delta(x, t) + 5 \sqrt{3} \right] - 3}{12 A_0^2 \delta^2(x, t) + 8 \sqrt{3} A_0 \delta(x, t) + 144 A_0^4 t^2 + 5} - (1 + i \sqrt{3}) \right) \times e^{i [A_1 x + (16 A_0^2 - A_1^2) t]},$$

$$\psi_2(x, t) = A_0 \left(\frac{-6 \sqrt{3} A_0 \delta(x, t) + 36 \sqrt{3} A_0^2 t + i \left[36 A_0^2 t - 6 A_0 \delta(x, t) - 5 \sqrt{3} \right] - 3}{12 A_0^2 \delta^2(x, t) + 8 \sqrt{3} A_0 \delta(x, t) + 144 A_0^4 t^2 + 5} - (1 - i \sqrt{3}) \right) \times e^{i [A_2 x + (16 A_0^2 - A_2^2) t]}, \quad (7.9)$$

where

$\delta(x, t) = x + 6 A_3 t$,
$b_0 = c_0 = 1$,
$b_1 = b_2 = c_1 = c_2 = 2$,
$A_0 = A_2 + 3 A_3$,
$A_1 = A_2 - 2 A_0$,
A_2 and A_3 are arbitrary real constants.

- *Reference*: [3].

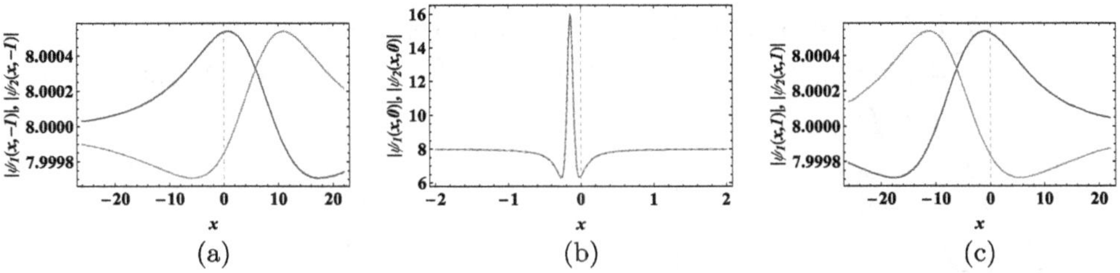

Figure 7.3. Plot of solution (7.9). Blue is ψ_1 and red is ψ_2 with $A_2 = A_3 = 1$. (a) at $t = -1$, (b) at $t = 0$, and (c) at $t = 1$.

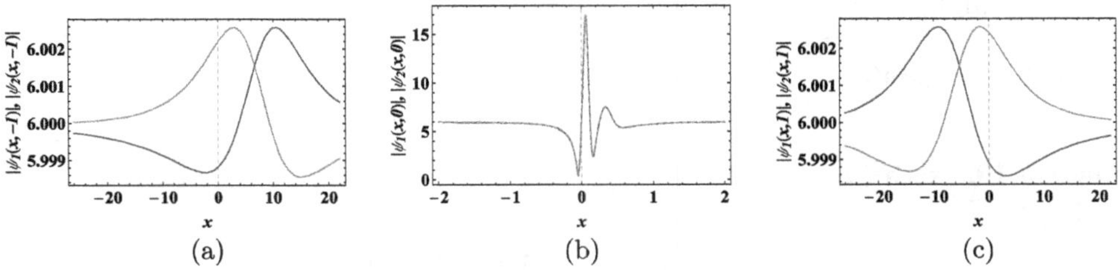

Figure 7.4. Plot of solution (7.10). Blue is ψ_1 and red is ψ_2 with $A_2 = A_3 = 1$. (a) at $t = -1$, (b) at $t = 0$, and (c) at $t = 1$.

***Solution* 9.**

(Figure 7.4)

$$\begin{aligned}
\psi_1(x,t) &= A_0\left[\frac{\alpha_1(x,t) + i\,\beta_1(x,t)}{\gamma(x,t)} - (1 + i\sqrt{3})\right] e^{i\left[A_1 x + \left(16A_0^2 - A_1^2\right)t\right]},\\
\psi_2(x,t) &= A_0\left[\frac{\alpha_2(x,t) + i\,\beta_2(x,t)}{\gamma(x,t)} - (1 - i\sqrt{3})\right] e^{i\left[A_2 x + \left(16A_0^2 - A_2^2\right)t\right]},
\end{aligned} \tag{7.10}$$

where

$$\begin{aligned}
\alpha_1(x,t) = &-864\sqrt{3}A_0^6 t^3 - 144\sqrt{3}A_0^5\,\delta(x,t)\,t^2 - 72\sqrt{3}A_0^4\,\delta^2(x,t)\,t\\
&-216A_0^4 t^2 - 12\sqrt{3}A_0^3\,\delta^3(x,t) - 144A_0^3\,\delta(x,t)\,t\\
&-18A_0^2\,\delta^2(x,t) - 12\sqrt{3}A_0^2 t + 3,\\
\alpha_2(x,t) = &\;864A_0^6\sqrt{3}t^3 - 144\sqrt{3}A_0^5\,\delta(x,t)\,t^2 + 72\sqrt{3}A_0^4\,\delta^2(x,t)\,t\\
&-216A_0^4 t^2 - 12\sqrt{3}A_0^3\,\delta^3(x,t) + 144A_0^3\,\delta(x,t)\,t\\
&-18A_0^2\,\delta^2(x,t) + 12\sqrt{3}A_0^2 t + 3,\\
\beta_1(x,t) = &\;864A_0^6 t^3 + 144A_0^5\,\delta(x,t)\,t^2 + 72A_0^4\,\delta^2(x,t)\,t + 312\sqrt{3}A_0^4 t^2\\
&+12A_0^3\,\delta^3(x,t) + 96\sqrt{3}A_0^3\,\delta(x,t)\,t + 18\sqrt{3}A_0^2\,\delta^2(x,t)\\
&+108A_0^2 t + 12A_0\,\delta(x,t) + \sqrt{3},\\
\beta_2(x,t) = &\;864A_0^6 t^3 - 144A_0^5\,\delta(x,t)\,t^2 + 72A_0^4\,\delta^2(x,t)\,t - 312\sqrt{3}A_0^4 t^2\\
&-12A_0^3\,\delta^3(x,t) + 96\sqrt{3}A_0^3\,\delta(x,t)\,t - 18\sqrt{3}A_0^2\,\delta^2(x,t)\\
&+108A_0^2 t - 12A_0\,\delta(x,t) - \sqrt{3},
\end{aligned}$$

$$\gamma(x,t) = 1728\,A_0^8\,t^4 + 288\,A_0^6\,\delta^2(x,t)\,t^2 + 384\sqrt{3}\,A_0^5\,\delta(x,t)\,t^2 + 12\,A_0^4\,\delta^4(x,t) + 432\,A_0^4\,t^2 + 16\sqrt{3}\,A_0^3\,\delta^3(x,t) + 24\,A_0^2\,\delta^2(x,t) + 4\sqrt{3}\,A_0\,\delta(x,t) + 1,$$

$A_0 = A_2 + 3\,A_3,$
$A_1 = A_2 - 2\,A_0,$
$A_2 = -6A_3,$
$\delta(x,t) = x + 6\,A_3\,t,$
$b_0 = c_0 = 1,$
$b_1 = b_2 = c_1 = c_2 = 2,$
A_3 is an arbitrary real constant.

- *Reference*: [3].

***Solution* 10. $\mathbf{sech^2(x)}$**
(Figure 7.5)

$$\psi_1(x,t) = \{A_0\,\mathrm{sech}^2[A_1\,(x-x_0)] + A_3\}\,e^{-i\,[\omega_1\,(t-t_0)+\phi_1]},$$
$$\psi_2(x,t) = B_0\,\mathrm{sech}^2[A_1\,(x-x_0)]\,e^{-i\,[\omega_2\,(t-t_0)+\phi_2]}, \tag{7.11}$$

where
$A_3 = \frac{-2\,A_0}{3},$
$\omega_1 = 2\,A_1^2,$
$\omega_2 = -2\,A_1^2,$
$b_1 = \frac{-9\,A_1^2}{2\,A_0^2},$
$b_2 = \frac{9\,A_1^2}{2\,B_0^2},$
$b_0 = c_0 = 1,$
$b_1 = c_1,$
$b_2 = c_2,$
$A_0 \neq 0,$
$B_0 \neq 0,$
A_1, x_0, t_0, ϕ_1, and ϕ_2 are arbitrary real constants.

- *Reference*: [4], *taken from the nonlocal case.*

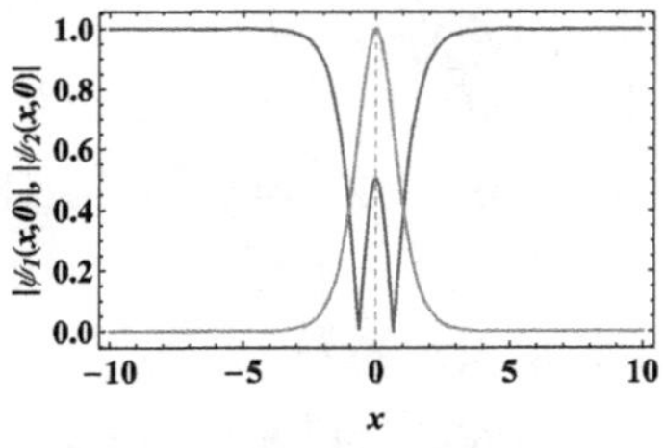

Figure 7.5. Plot of solution (7.11) at $t = 0$. Blue is ψ_1 and red is ψ_2 with $A_0 = 3/2$, $B_0 = A_1 = 1$, $b_1 = b_2 = 2$, and $x_0 = t_0 = \phi_1 = \phi_2 = 0$.

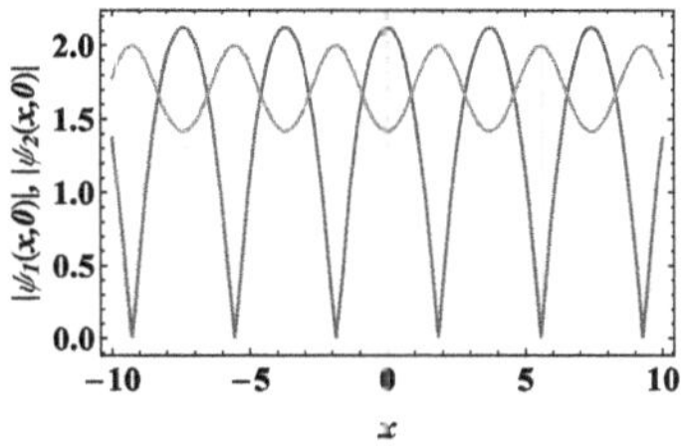

Figure 7.6. Solitary wave (7.12) at $t = 0$. Blue is ψ_1 and red is ψ_2 with $A_0 = 3$, $B_0 = 2$, $A_1 = B_1 = 1$, $b_2 = c_2 = 2$, $m = 1/2$, and $x_0 = t_0 = \phi_1 = \phi_2 = 0$.

***Solution* 11. cd(x, m)-nd(x, m)** *solitary wave (SW)* (Figure 7.6)

$$\begin{aligned} \psi_1(x,t) &= A_0\,\sqrt{m}\,\mathrm{cd}[A_1\,(x-x_0),\,m]\,e^{-i\,[\omega_1\,(t-t_0)+\phi_1]},\\ \psi_2(x,t) &= B_0\,\sqrt{1-m}\,\mathrm{nd}[A_1\,(x-x_0),\,m]\,e^{-i\,[\omega_2\,(t-t_0)+\phi_2]}, \end{aligned} \tag{7.12}$$

where

$\omega_1 = -(1-m)\,A_1^2 - b_1\,A_0^2$,
$\omega_2 = -(2-m)\,A_1^2 - c_1\,A_0^2$,
$b_0 = c_0 = 1$,
$b_1 = \frac{-2\,A_1^2 + b_2\,B_0^2}{A_0^2}$,
$c_1 = \frac{-2\,A_1^2 + c_2\,B_0^2}{A_0^2}$,
$0 < m \leqslant 1$,
$A_0 \neq 0$,
B_0, x_0, t_0, ϕ_1, and ϕ_2 are arbitrary real constants.

- *Reference*: [4], *taken from the nonlocal case.*

***Solution* 12. Weierstrass Elliptic Function I: $\wp$(z, g$_2$, g$_3$)** *SW*

$$\begin{aligned} \psi_1(x,t) &= \Phi_1(z)\,e^{i\,\phi_1(z,t)},\\ \psi_2(x,t) &= \Phi_2(z)\,e^{i\,\phi_2(z,t)}, \end{aligned} \tag{7.13}$$

where

$$\begin{aligned} \Phi_1(z) &= \sqrt{F\,\wp^2(z,g_2,g_3) + A_0\,\wp(z,g_2,g_3) + A_1}\,,\\ \Phi_2(z) &= \sqrt{G\,\wp^2(z,g_2,g_3) + B_0\,\wp(z,g_2,g_3) + B_1}, \end{aligned} \tag{7.14}$$

which satisfy

$$\begin{aligned} &-A_2^2\,b_0 + \left[b_1\,\Phi_1^2(z) + b_2\,\Phi_2^2(z) - \gamma\right]\Phi_1^4(z) + b_0\,\Phi_1^3(z)\,\Phi_1''(z) = 0,\\ &-B_2^2\,c_0 + \left[b_2\,\Phi_1^2(z) + b_1^{-1}\,\Phi_2^2(z) - \beta\right]\Phi_2^4(z) + c_0\,\Phi_2^3(z)\,\Phi_2''(z) = 0, \end{aligned} \tag{7.15}$$

with
$\phi_1(z, t) = \frac{b_0\,\alpha}{2}\,z - b_0\,A_2\left[\int\frac{dz}{\Phi_1^2(z)}\right] + \phi_{01}(t),$
$\phi_2(z, t) = \frac{c_0\,\alpha}{2}\,z - c_0\,B_2\left[\int\frac{dz}{\Phi_2^2(z)}\right] + \phi_{02}(t),$
$c_1 = b_2,$
$c_2 = b_1^{-1},$
$z = x - c\,t,$
$\phi_{01}(t) = \left(\gamma + \frac{b_0\,\alpha^2}{4}\right)t + \phi_{011},$
$\phi_{02}(t) = \left(\beta + \frac{c_0\,\alpha^2}{4}\right)t + \phi_{022},$
$\gamma = b_1\,A_1 + b_2\,B_1 - 3\,b_0\,\frac{A_0}{F},$
$\beta = b_2\,A_1 + b_1^{-1}\,B_1 - 3\,c_0\,\frac{B_0}{G},$
$A_1 = \frac{4\,A_0^2 - F^2\,g_2}{4\,F},$
$A_2 = -\sqrt{\frac{3\,F^2}{4}\left(\frac{A_0^2}{F^2} - \frac{g_2}{3}\right)\left(4\,\frac{A_0^3}{F^3} - \frac{g_2\,A_0}{F} - g_3\right)},$
$B_0 = \frac{-6\,b_0 - b_1\,A_0}{b_2},$
$B_2 = -\sqrt{\frac{3\,G^2}{4}\left(\frac{B_0^2}{G^2} - \frac{g_2}{3}\right)\left(4\,\frac{B_0^3}{G^3} - \frac{g_2\,B_0}{G} - g_3\right)},$
$g_2 = \frac{4\,(B_0^2 - B_1\,G)}{G^2},$
$G = \frac{-b_1\,F}{b_2},$
$F \neq 0,$
$b_1 = \frac{b_0\,b_2}{c_0},$
$b_2 = 1,$
b_0, c_0, A_0, B_1, g_3, α, c, ϕ_{011}, and ϕ_{022} are arbitrary real constants.

- *Reference*: [4].

***Solution* 13. Weierstrass Elliptic Function II: $\wp$(z, g$_2$, g$_3$)** *SW*

$$\psi_1(x, t) = \Phi_1(z)\,e^{i\,\phi_1(z,t)}, \qquad \psi_2(x, t) = \Phi_2(z)\,e^{i\,\phi_2(z,t)}, \tag{7.16}$$

where

$$\Phi_1(z) = \sqrt{A_0\,\wp(z, g_2, g_3) + A_1}, \qquad \Phi_2(z) = \sqrt{B_0\,\wp(z, g_2, g_3) + B_1}, \tag{7.17}$$

which satisfy (7.15), with
$\phi_1(z, t) = \frac{b_0\,\alpha}{2}\,z - b_0\,A_2\left[\int\frac{dz}{\Phi_1^2(z)}\right] + \phi_{01}(t),$

$\phi_2(z, t) = \frac{c_0 \alpha}{2} z - c_0 B_2 \left[\int \frac{dz}{\Phi_2^2(z)}\right] + \phi_{02}(t),$
$c_1 = b_2,$
$c_2 = b_1^{-1},$
$z = x - c\,t,$
$\phi_{01}(t) = \left(\gamma + \frac{b_0 \alpha^2}{4}\right) t + \phi_{011},$
$\phi_{02}(t) = \left(\beta + \frac{c_0 \alpha^2}{4}\right) t + \phi_{022},$
$\gamma = \frac{b_2 (B_1 A_0 - A_1 B_0) - 3 b_0 A_1}{A_0},$
$\beta = \frac{b_2 (A_1 B_0 - B_1 A_0) - 3 c_0 B_1}{B_0},$
$A_0 = \frac{2 (b_0 c_0 b_1 - b_0)}{b_1 (1 - b_2^2)},$
$A_2 = -\sqrt{\frac{A_0^2}{4}\left(\frac{-A_1 g_2}{A_0} + g_3 + \frac{4 A_1^3}{A_0^3}\right)},$
$B_0 = \frac{2 (b_0 b_2 - c_0 b_1)}{1 - b_2^2},$
$B_2 = -\sqrt{\frac{B_0^2}{4}\left(\frac{-B_1 g_2}{B_0} + g_3 + \frac{4 B_1^3}{B_0^3}\right)},$
$b_0, b_1, b_2, c_0, A_1, B_1, g_2, g_3, \alpha, c, \phi_{011},$ and ϕ_{022} are arbitrary real constants.

- *Reference*: [4].

7.2 Summary of Section 7.1

Note: For lengthy conditions, the reader is referred to the solutions in section 7.1.

Equation (1): $i\,\psi_{1_t} + b_0\,\psi_{1xx} + (b_1\,|\psi_1|^2 + b_2\,|\psi_2|^2)\,\psi_1 = 0,\ i\,\psi_{2_t} + c_0\,\psi_{2xx} + (c_1\,|\psi_1|^2 + c_2\,|\psi_2|^2)\,\psi_2 = 0$

#	Solution	Conditions	Name	Eq. #
1.	$\psi_1(x, t) = A_0\,e^{i\,\phi_1}$, $\psi_2(x, t) = B_0\,e^{i\,\phi_2}$	$b_1 = -\frac{b_2\,B_0^2}{A_0^2}$, $c_2 = -\frac{c_1\,A_0^2}{B_0^2}$, A_0, B_0, ϕ_1, and ϕ_2 are arbitrary real constants	continuous wave, t- and x-independent phase	(7.2)
2.	$\psi_1(x, t) = A_0\,e^{i\,[A_1\,(t-t_0)+\phi_1]}$, $\psi_2(x, t) = B_0\,e^{i\,[B_1\,(t-t_0)+\phi_2]}$	$A_1 = A_0^2\,b_1 + B_0^2\,b_2$, $B_1 = A_0^2\,c_1 + B_0^2\,c_2$, A_0, B_0, t_0, ϕ_1, and ϕ_2 are arbitrary real constants	continuous wave, t-dependent phase	(7.3)
3.	$\psi_1(x, t) = A_0\,e^{i\,[A_1\,(x-x_0)+\phi_1]}$, $\psi_2(x, t) = B_0\,e^{i\,[B_1\,(x-x_0)+\phi_2]}$	$A_1 = \pm\sqrt{\frac{A_0^2\,b_1 + B_0^2\,b_2}{b_0}}$, $B_1 = \pm\sqrt{\frac{A_0^2\,c_1 + B_0^2\,c_2}{c_0}}$, $b_0\,(A_0^2\,b_1 + B_0^2\,b_2) > 0$, $c_0\,(A_0^2\,c_1 + B_0^2\,c_2) > 0$, A_0, B_0, x_0, ϕ_1, and ϕ_2 are arbitrary real constants	continuous wave, x-dependent phase	(7.4)
4.	$\psi_1(x, t) = A_0\,e^{i\,[A_1\,(t-t_0)+A_2\,(x-x_0)+\phi_1]}$, $\psi_2(x, t) = B_0\,e^{i\,[B_1\,(t-t_0)+B_2\,(x-x_0)+\phi_2]}$	$A_1 = -A_2^2\,b_0 + A_0^2\,b_1 + B_0^2\,b_2$, $B_1 = -B_2^2\,c_0 + A_0^2\,c_1 + B_0^2\,c_2$, A_0, B_0, A_2, B_2, x_0, t_0, ϕ_1, and ϕ_2 are arbitrary real constants	continuous wave, t- and x-dependent phase	(7.5)

(Continued)

5. $\psi_1(x,\,t) = \frac{A_0}{\sqrt{2\,[A_1 + t - t_0]}}\, e^{i\,\phi_1(x,t)}$, $\psi_2(x,\,t) = \frac{B_0}{\sqrt{2\,[B_1 + t - t_0]}}\, e^{i\,\phi_2(x,t)}$	$\phi_1(x,\,t) = \frac{[b_0\,A_2 + (x - x_0)]^2}{4\,b_0\,[A_1 + t - t_0]} + \frac{b_1\,A_0^2}{2}\ln[2(A_1$ $+ t - t_0)] + \frac{b_2\,B_0^2}{2}\ln[2\,(B_1 + t - t_0)]$ $+ \phi_{01}$, $\phi_2(x,\,t) = \frac{[c_0\,B_2 + (x - x_0)]^2}{4\,c_0\,[B_1 + t - t_0]} + \frac{c_1\,A_0^2}{2}\ln[2(A_1$ $+ t - t_0)] + \frac{c_2\,B_0^2}{2}\ln[2\,(B_1 + t - t_0)]$ $+ \phi_{02}$, A_0, A_1, A_2, B_0, B_1, B_2, x_0, t_0, ϕ_{01} and ϕ_{02} are arbitrary real constants	decaying wave	(7.6)
6. $\psi_1(x,\,t) = (A_0\tanh\{A_1\,[(x - x_0) - v\,(t - t_0)]\} + i\,A_2)\,e^{i\,[b_1\,(t - t_0) + \phi_1]}$, $\psi_2(x,\,t) = B_0\,\mathrm{sech}\{A_1\,[(x - x_0) - v\,(t - t_0)]\}$ $\times\, e^{i\,[(A_1^2 + b_2)\,(t - t_0) + \theta(x,t) + \phi_2]}$	$A_0 = \cos[\tan^{-1}(\frac{-v}{2\,A_1})]$, $A_1 = \sqrt{\frac{b_1\,(1 - b_2^2)}{2\,(1 + b_1\,b_2)} - \frac{v^2}{4}}$, $A_2 = \sin[\tan^{-1}(\frac{-v}{2\,A_1})]$, $B_0 = \sqrt{\frac{2\,A_1^2\,(b_1 + b_2)}{1 - b_2^2}}$, $\theta(x,\,t) = \frac{v}{2}\,[x - x_0 - \frac{v}{2}\,(t - t_0)]$, $b_0 = -1$, $\frac{b_1\,(1 - b_2^2)}{2\,(1 + b_1\,b_2)} > \frac{v^2}{4}$, $2\,A_1^2\,(b_1 + b_2)\,(1 - b_2^2) > 0$, $c_0 = 1$, $c_1 = 1/b_1$, $c_2 = b_2$, $b_1\,b_2 = 1$, x_0, t_0, ϕ_1, ϕ_2, and v are arbitrary real constants	dark-bright soliton	(7.7)
7. $\psi_1(x,\,t) = (1 - 4\,[\alpha_1^2(x,\,t) + \alpha_2^2(x,\,t) - \alpha_1(x,\,t) + i\,\alpha_2(x,\,t)]$ $\times\, e^{-\delta(x,t)}\,\{[2\,\alpha_1^2(x,\,t) + 2\,\alpha_2^2(x,\,t) - 2\,\alpha_1(x,\,t) + 1]\,e^{-\delta(x,t)}$ $+ (\beta_3^2 + \beta_4^2)\,e^{2\,\delta(x,t)}\}^{-1})\,e^{i\,[A_1\,x + (2\,A_0^2 - A_1^2)\,t]}$, $\psi_2(x,\,t) = -4\,A_0\,\{\beta_3\,[\alpha_1(x,\,t) - 1] - \beta_4\,\alpha_2(x,\,t) + i\,[\beta_3\,\alpha_2(x,\,t)$ $+ \beta_4\,\alpha_1(x,\,t) - \beta_4]\}\,(\{2\,[\alpha_1^2(x,\,t) + \alpha_2^2(x,\,t)] - 2\,\alpha_1(x,\,t) + 1\}$ $\times\, e^{-\delta(x,t)} + (\beta_3^2 + \beta_4^2)\,e^{2\,\delta(x,t)})^{-1}\,e^{i\,[A_1\,x + (3\,A_0^2 - A_1^2)] + \frac{\delta(x,t)}{2}}$	$\delta(x,\,t) = \frac{2\,A_0\,(x - 2\,A_1\,t)}{3}$, $\alpha_1(x,\,t) = \beta_1\,A_0 + A_0\,(x - 2\,A_1\,t)$, $\alpha_2(x,\,t) = \beta_2\,A_0 - 2\,A_0^2\,t$, $b_0 = c_0 = 1$, $b_1 = b_2 = c_1 = c_2 = 2$, A_0, A_1, β_1, β_2, β_3, and β_4 are arbitrary real constants	—	(7.8)

8. $\psi_1(x,\,t) = A_0 \left(\frac{-6\sqrt{3}\,A_0\,\delta(x,t) - 36\sqrt{3}\,A_0^2\,t + i\,[36\,A_0^2\,t + 6\,A_0\,\delta(x,t) + 5\sqrt{3}] - 3}{12\,A_0^2\,\delta^2(x,t) + 8\sqrt{3}\,A_0\,\delta(x,t) + 144\,A_0^4\,t^2 + 5} - (1 + i\sqrt{3}) \right) e^{i\,[A_1\,x + (16\,A_0^2 - A_1^2)\,t]}$, $\psi_2(x,\,t) = A_0 \left(\frac{-6\sqrt{3}\,A_0\,\delta(x,t) + 36\sqrt{3}\,A_0^2\,t + i\,[36\,A_0^2\,t - 6\,A_0\,\delta(x,t) - 5\sqrt{3}] - 3}{12\,A_0^2\,\delta^2(x,t) + 8\sqrt{3}\,A_0\,\delta(x,t) + 144\,A_0^4\,t^2 + 5} - (1 - i\sqrt{3}) \right) e^{i\,[A_2\,x + (16\,A_0^2 - A_2^2)\,t]}$	$\delta(x,\,t) = x + 6\,A_3\,t$, $b_0 = c_0 = 1$, $b_1 = b_2 = c_1 = c_2 = 2$, $A_0 = A_2 + 3\,A_3$, $A_1 = A_2 - 2\,A_0$, A_2 and A_3 are arbitrary real constants	–	(7.9)
9. $\psi_1(x,\,t) = A_0 \left[\frac{\alpha_1(x,t) + i\,\beta_1(x,t)}{\gamma(x,t)} - (1 + i\sqrt{3}) \right] e^{i\,[A_1\,x + (16\,A_0^2 - A_1^2)\,t]}$, $\psi_2(x,\,t) = A_0 \left[\frac{\alpha_2(x,t) + i\,\beta_2(x,t)}{\gamma(x,t)} - (1 - i\sqrt{3}) \right] e^{i\,[A_2\,x + (16\,A_0^2 - A_2^2)\,t]}$	See text.	–	(7.10)
10. $\psi_1(x,\,t) = \{A_0\,\mathrm{sech}^2[A_1\,(x - x_0)] + A_3\}\,e^{-i\,[\omega_1\,(t - t_0) + \phi_1]}$, $\psi_2(x,\,t) = B_0\,\mathrm{sech}^2[A_1\,(x - x_0)]\,e^{-i\,[\omega_2\,(t - t_0) + \phi_2]}$	$A_3 = \frac{-2\,A_0}{3}$, $\omega_1 = 2\,A_1^2$, $\omega_2 = -2\,A_1^2$, $b_1 = \frac{-9\,A_1^2}{2\,A_0^2}$, $b_2 = \frac{9\,A_1^2}{2\,B_0^2}$, $b_0 = c_0 = 1$, $b_1 = c_1$, $b_2 = c_2$, $A_0 \neq 0$, $B_0 \neq 0$, A_1, B_1, x_0, t_0, ϕ_1, and ϕ_2 are arbitrary real constants	–	(7.11)
11. $\psi_1(x,\,t) = A_0\,\sqrt{m}\,\mathrm{cd}[A_1\,(x - x_0),\,m]\,e^{-i\,[\omega_1\,(t - t_0) + \phi_1]}$, $\psi_2(x,\,t) = B_0\,\sqrt{1 - m}\,\mathrm{nd}[A_1\,(x - x_0),\,m]\,e^{-i\,[\omega_2\,(t - t_0) + \phi_2]}$	$\omega_1 = -(1 - m)\,A_1^2 - b_1\,A_0^2$, $\omega_2 = -(2 - m)\,A_1^2 - c_1\,A_0^2$, $b_0 = c_0 = 1$, $b_1 = \frac{-2\,A_1^2 + b_2\,B_0^2}{A_0^2}$, $c_1 = \frac{-2\,A_1^2 + c_2\,B_0^2}{A_0^2}$, $0 < m \leqslant 1$, A_0, B_0, x_0, t_0, ϕ_1, and ϕ_2 are arbitrary real constants	solitary wave	(7.12)

(*Continued*)

Equation (2): $-A_2^2\, b_0 + [b_1\, \Phi_1^2(z) + b_2\, \Phi_2^2(z) - \gamma]\, \Phi_1^4(z) + b_0\, \Phi_1^3(z)\, \Phi_1''(z) = 0,\ -B_2^2\, c_0 + [b_2\, \Phi_1^2(z) + b_1^{-1}\, \Phi_2^2(z) - \beta]\, \Phi_2^4(z) + c_0\, \Phi_2^3(z)\, \Phi_2''(z) = 0$

12. $\Phi_1(z) = \sqrt{F\, \wp^2(z, g_2, g_3) + A_0\, \wp(z, g_2, g_3) + A_1}$, $\Phi_2(z) = \sqrt{G\, \wp^2(z, g_2, g_3) + B_0\, \wp(z, g_2, g_3) + B_1}$	$\phi_{01}(t) = (\gamma + \frac{b_0\, \alpha^2}{4})\, t + \phi_{011}$, $\phi_{02}(t) = \left(\beta + \frac{c_0\, \alpha^2}{4}\right) t + \phi_{022}$, $\gamma = b_1\, A_1 + b_2\, B_1 - 3\, b_0\, \frac{A_0}{F}$, $\beta = b_2\, A_1 + b_1^{-1}\, B_1 - 3\, c_0\, \frac{B_0}{G}$, $A_1 = \frac{4\, A_0^2 - F^2\, g_2}{4\, F}$, $A_2 = -\sqrt{\frac{3\, F^2}{4} \left(\frac{A_0^2}{F^2} - \frac{g_2}{3}\right) \left(4\, \frac{A_0^3}{F^3} - \frac{g_2\, A_0}{F} - g_3\right)}$, $B_0 = \frac{-6\, b_0 - b_1\, A_0}{b_2}$, $B_2 = -\sqrt{\frac{3\, G^2}{4} \left(\frac{B_0^2}{G^2} - \frac{g_2}{3}\right) \left(4\, \frac{B_0^3}{G^3} - \frac{g_2\, B_0}{G} - g_3\right)}$, $g_2 = \frac{4\, (B_0^2 - B_1\, G)}{G^2}$, $G = \frac{-b_1\, F}{b_2}$, $F \neq 0$, $b_1 = \frac{b_0\, b_2}{c_0}$, $b_2 = 1$, b_0, c_0, c, A_0, B_1, g_3, α, ϕ_{011}, and ϕ_{022} are arbitrary real constants	Weierstrass elliptic function I	(7.14)
13. $\Phi_1(z) = \sqrt{A_0\, \wp(z, g_2, g_3) + A_1}$, $\Phi_2(z) = \sqrt{B_0\, \wp(z, g_2, g_3) + B_1}$	$\phi_{01}(t) = (\gamma + \frac{b_0\, \alpha^2}{4})\, t + \phi_{011}$, $\phi_{02}(t) = (\beta + \frac{c_0\, \alpha^2}{4})\, t + \phi_{022}$, $\gamma = \frac{b_2\, (B_1\, A_0 - A_1\, B_0) - 3\, b_0\, A_1}{A_0}$, $\beta = \frac{b_2\, (A_1\, B_0 - B_1\, A_0) - 3\, c_0\, B_1}{B_0}$, $A_0 = \frac{2\, (b_0\, c_0\, b_1 - b_0)}{b_1\, (1 - b_2^2)}$, $A_2 = -\sqrt{\frac{A_0^2}{4} (\frac{-A_1\, g_2}{A_0} + g_3 + \frac{4\, A_1^3}{A_0^3})}$, $B_0 = \frac{2\, (b_0\, b_2 - c_0\, b_1)}{1 - b_2^2}$, $B_2 = -\sqrt{\frac{B_0^2}{4} (\frac{-B_1\, g_2}{B_0} + g_3 + \frac{4\, B_1^3}{B_0^3})}$, b_0, b_1, b_2, c_0, c, A_1, B_1, g_2, g_3, α, ϕ_{011}, and ϕ_{022} are arbitrary real constants	Weierstrass elliptic function II	(7.17)

7.3 Symmetry Reductions

7.3.1 Symmetry Reduction I

From Manakov System to Fundamental NLSE

The CNLSE, (7.1),

$$i\,\psi_{1_t} + b_0\,\psi_{1_{xx}} + (b_1\,|\psi_1|^2 + b_2\,|\psi_2|^2)\,\psi_1 = 0,$$
$$i\,\psi_{2_t} + b_0\,\psi_{2_{xx}} + (c_1\,|\psi_1|^2 + c_2\,|\psi_2|^2)\,\psi_2 = 0,$$

transforms to the scalar NLSE

$$i\,\psi_{1_t} + b_0\,\psi_{1_{xx}} + (c_1 + c_2|\sigma|^2)\,|\psi_1|^2\,\psi_1 = 0, \tag{7.18}$$

with the replacements

$\psi_2(x, t) = \sigma\,\psi_1(x, t)$,

$b_1 = c_1 + (c_2 - b_2)\,|\sigma|^2$,

where σ is an arbitrary complex constant.

Conclusion:

If $\psi_1(x, t)$ is a solution of the fundamental NLSE

$$i\,\psi_{1_t} + a_1\,\psi_{1_{xx}} + a_2\,|\psi_1|^2\,\psi_1 = 0,$$

then

$$(\psi_1, \psi_2) = (\psi_1, \sigma\,\psi_1) \tag{7.19}$$

is a solution of the CNLSE

$$i\,\psi_{1_t} + b_0\,\psi_{1_{xx}} + (b_1\,|\psi_1|^2 + b_2\,|\psi_2|^2)\,\psi_1 = 0,$$
$$i\,\psi_{2_t} + b_0\,\psi_{2_{xx}} + (c_1\,|\psi_1|^2 + c_2\,|\psi_2|^2)\,\psi_2 = 0,$$

with

$a_1 = b_0$,

$a_2 = c_1 + c_2|\sigma|^2$,

$b_1 = c_1 + (c_2 - b_2)\,|\sigma|^2$.

7.3.2 Symmetry Reduction II

From Manakov System to Fundamental NLSE

The CNLSE, (7.1),

$$i\,\psi_{1_t} + b_0\,\psi_{1_{xx}} + (b_1\,|\psi_1|^2 + b_2\,|\psi_2|^2)\,\psi_1 = 0,$$
$$i\,\psi_{2_t} - b_0\,\psi_{2_{xx}} + (c_1\,|\psi_1|^2 + c_2\,|\psi_2|^2)\,\psi_2 = 0,$$

transforms to the scalar NLSE

$$i\,\psi_{1_t} + b_0\,\psi_{1_{xx}} - (c_1 + c_2)\,|\psi_1|^2\,\psi_1 = 0, \tag{7.20}$$

with the replacements

$$\psi_2(x, t) = e^{i\phi}\, \psi_1^*(x, t),$$
$$b_1 = -(c_1 + c_2 + b_2),$$

where ϕ is an arbitrary real constant.

Conclusion:
If $\psi_1(x, t)$ is a solution of the fundamental NLSE

$$i\, \psi_{1_t} + b_0\, \psi_{1xx} + a_2\, |\psi_1|^2\, \psi_1 = 0,$$

then

$$(\psi_1, \psi_2) = \left(\psi_1, e^{i\phi}\, \psi_1^*\right) \tag{7.21}$$

is a solution of the CNLSE

$$i\, \psi_{1_t} + b_0\, \psi_{1xx} + (b_1\, |\psi_1|^2 + b_2\, |\psi_2|^2)\, \psi_1 = 0,$$
$$i\, \psi_{2_t} - b_0\, \psi_{2xx} + (c_1\, |\psi_1|^2 + c_2\, |\psi_2|^2)\, \psi_2 = 0,$$

with $a_2 = -(c_1 + c_2)$.

7.3.3 Symmetry Reduction III

From Vector NLSE to Fundamental NLSE
The vector CNLSE

$$i\, \psi_{j_t} + b_{0j}\, \psi_{jxx} + \left(\sum_{k=1}^{N} b_{1k}\, |\psi_k|^2\right) \psi_j = 0, \quad j = 1, 2, \ldots, N, \tag{7.22}$$

transforms to the scalar NLSE

$$i\, \psi_{1_t} + b_{0j}\, \psi_{1xx} + \sum_{k=1}^{N} b_{1j} |\sigma_j|^2\, |\psi_1|^2\, \psi_1 = 0,$$

with the replacement: $\psi_j(x, t) = \sigma_j\, \psi_1(x, t)$,
where
σ_j are arbitrary complex constants,
$\sigma_1 = 1$.

Conclusion:
If $\psi_1(x, t)$ is a solution of the fundamental NLSE

$$i\, \psi_{1_t} + a_1\, \psi_{1xx} + a_2\, |\psi_1|^2\, \psi_1 = 0,$$

then

$$(\psi_1, \psi_2, \psi_3, \ldots) = (\psi_1, \sigma_2\, \psi_1, \sigma_3\, \psi_1, \ldots) \tag{7.23}$$

is a solution of the vector CNLSE

$$i \, \psi_{j_t} + b_{0j} \, \psi_{j_{xx}} + \left(\sum_{k=1}^{N} b_{1k} \, |\psi_k|^2 \right) \psi_j = 0, \quad j = 1, 2, \dots, N,$$

with

$b_{0j} = a_1,$

$$a_2 = \sum_{j=1}^{N} b_{1j} |\sigma_j|^2.$$

7.3.4 Symmetry Reduction IV

From Three Coupled NLSEs to Manakov System

The three CNLSEs,

$$\begin{aligned}
i \, \psi_{1_t} + a_1 \, \psi_{1_{xx}} + (b_1 \, |\psi_1|^2 + b_2 \, |\psi_2|^2 + b_3 \, |\psi_3|^2) \, \psi_1 &= 0, \\
i \, \psi_{2_t} + a_2 \, \psi_{2_{xx}} + (c_1 \, |\psi_1|^2 + c_2 \, |\psi_2|^2 + c_3 \, |\psi_3|^2) \, \psi_2 &= 0, \\
i \, \psi_{3_t} + a_3 \, \psi_{3_{xx}} + (d_1 \, |\psi_1|^2 + d_2 \, |\psi_2|^2 + d_3 \, |\psi_3|^2) \, \psi_3 &= 0,
\end{aligned} \tag{7.24}$$

transform to the CNLSE

$$\begin{aligned}
i \, \psi_{1_t} + a_1 \, \psi_{1_{xx}} + (b_1 \, |\psi_1|^2 + b_3 \, |\psi_3|^2) \, \psi_1 &= 0, \\
i \, \psi_{3_t} + a_3 \, \psi_{3_{xx}} + (d_1 \, |\psi_1|^2 + d_3 \, |\psi_3|^2) \, \psi_3 &= 0,
\end{aligned} \tag{7.25}$$

with the replacements

$\psi_2(x, t) = \sigma \, \psi_1(x, t),$
$c_1 = b_1 + (b_2 - c_2) \, |\sigma|^2,$
$a_1 = a_2,$
$c_3 = b_3,$

where

$a_{1,2,3}$, $b_{1,2,3}$, $c_{1,2,3}$, and $d_{1,2,3}$ are real constants, σ is an arbitrary complex constant.

Example 1. **tanh(x, t)-tanh(x, t)-sech(x, t)** *dark-dark-bright soliton* (Figure 7.7)

$$\begin{aligned}
\psi_1(x, t) &= (A_0 \tanh\{A_1 \, [x - x_0 - v \, (t - t_0)]\} + i \, A_2) \, e^{i \, [b_1 \, (t - t_0) + \phi_1]}, \\
\psi_2(x, t) &= \sigma_2 \, \psi_1(x, t), \\
\psi_3(x, t) &= B_0 \, \mathrm{sech}\{A_1 \, [x - x_0 - v \, (t - t_0)]\} \, e^{i \, [(A_1^2 + b_2) \, (t - t_0) + \theta(x, \, t) + \phi_2]},
\end{aligned} \tag{7.26}$$

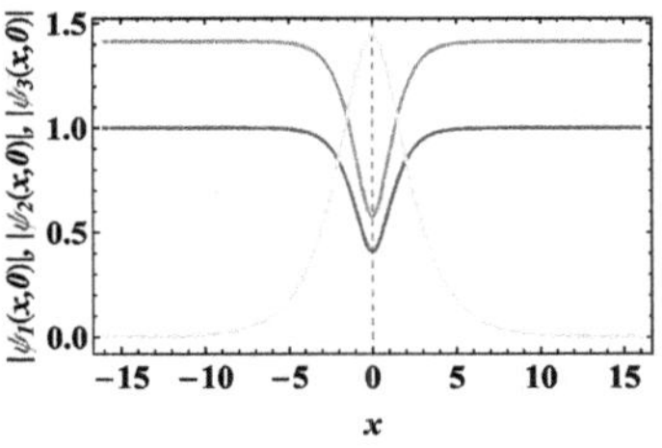

Figure 7.7. Dark-dark-bright soliton (7.26) at $t = 0$. Blue is ψ_1, red is ψ_2, and green is ψ_3 with $b_2 = 1/2$, $v = 1/2$, $\sigma_{2r} = \sigma_{2i} = 1$, and $x_0 = t_0 = \phi_1 = \phi_2 = 0$. Animation available online at https://iopscience.iop.org/book/978-0-7503-2428-1.

are a solution of

$$\begin{aligned} &i\,\psi_{1_t} + a_1\,\psi_{1xx} + (b_{11}\,|\psi_1|^2 + b_{12}\,|\psi_2|^2 + b_{13}\,|\psi_3|^2)\,\psi_1 = 0,\\ &i\,\psi_{2_t} + a_2\,\psi_{2xx} + (c_{11}\,|\psi_1|^2 + c_{12}\,|\psi_2|^2 + c_{13}\,|\psi_3|^2)\,\psi_2 = 0,\\ &i\,\psi_{3_t} + a_3\,\psi_{3xx} + (d_{11}\,|\psi_1|^2 + d_{12}\,|\psi_2|^2 + d_{13}\,|\psi_3|^2)\,\psi_3 = 0, \end{aligned} \tag{7.27}$$

where

$$\begin{aligned} &A_0 = \cos\left[\tan^{-1}\left(\frac{-v}{2\,A_1}\right)\right],\\ &A_1 = \sqrt{\frac{b_1\,(1-b_2^2)}{2\,(1+b_1\,b_2)} - \frac{v^2}{4}},\\ &A_2 = \sin\left[\tan^{-1}\left(\frac{-v}{2\,A_1}\right)\right],\\ &B_0 = \sqrt{\frac{2\,A_1^2\,(b_1+b_2)}{1-b_2^2}},\\ &\theta(x,\,t) = \frac{v}{2}\,[x - x_0 - \frac{v}{2}\,(t - t_0)],\\ &b_0 = -1,\\ &c_0 = 1,\\ &c_1 = 1/b_1,\\ &c_2 = b_2,\\ &b_1\,b_2 = 1,\\ &a_1 = a_2 = b_0,\\ &a_3 = c_0,\\ &b_{11} = b_1 - b_{12}\,|\sigma_2|^2,\\ &b_{13} = b_2,\\ &d_{11} = c_1 - d_{12}\,|\sigma_2|^2,\\ &d_{13} = c_2,\\ &c_{13} = b_{13},\\ &c_{11} = b_{11} + (b_{12} - c_{12})\,|\sigma_2|^2,\\ &\sigma_2 = \sigma_{2r} + i\,\sigma_{2i},\\ &\frac{b_1\,(1-b_2^2)}{2\,(1+b_1\,b_2)} > \frac{v^2}{4},\\ &(b_1 + b_2)\,(1 - b_2^2) > 0, \end{aligned}$$

x_0, t_0, ϕ_1, ϕ_2, σ_{2r}, σ_{2i}, and v are arbitrary real constants.

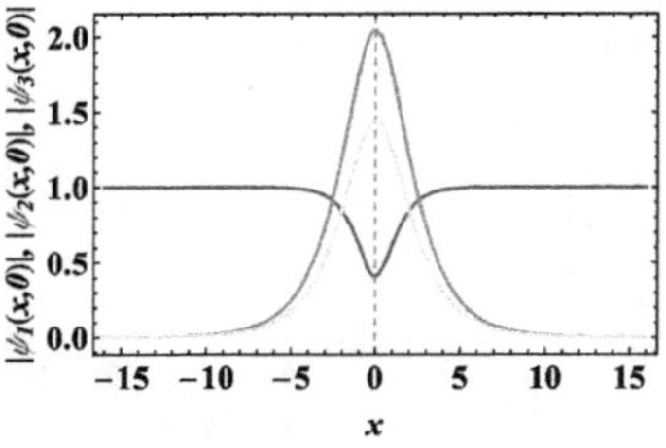

Figure 7.8. Dark-bright-bright soliton (7.28) at $t = 0$. Blue is ψ_1, red is ψ_2, and green is ψ_3 with $b_2 = 1/2$, $v = 1/2$, $\sigma_{2r} = \sigma_{2i} = 1$, and $x_0 = t_0 = \phi_1 = \phi_2 = 0$. Animation available online at https://iopscience.iop.org/book/978-0-7503-2428-1.

***Example* 2. tanh(*x*, *t*)-sech(*x*, *t*)-sech(*x*, *t*)** *dark-bright-bright soliton* (Figure 7.8)

$$\begin{aligned}
&\psi_1(x, t) = (A_0 \tanh\{A_1 [x - x_0 - v (t - t_0)]\} + i A_2) e^{i [b_1 (t-t_0)+\phi_1]},\\
&\psi_2(x, t) = \sigma_2 \psi_3(x, t),\\
&\psi_3(x, t) = B_0 \operatorname{sech}\{A_1 [x - x_0 - v (t - t_0)]\} e^{i [(A_1^2+b_2) (t-t_0)+\theta(x, t)+\phi_2]},
\end{aligned} \tag{7.28}$$

are a solution of (7.27), where

$A_0 = \cos\left[\tan^{-1}\left(\frac{-v}{2 A_1}\right)\right]$,
$A_1 = \sqrt{\frac{b_1 (1 - b_2^2)}{2 (1 + b_1 b_2)} - \frac{v^2}{4}}$,
$A_2 = \sin\left[\tan^{-1}\left(\frac{-v}{2 A_1}\right)\right]$,
$B_0 = \sqrt{\frac{2 A_1^2 (b_1 + b_2)}{1 - b_2^2}}$,
$\theta(x, t) = \frac{v}{2} [(x - x_0) - \frac{v}{2} (t - t_0)]$,
$b_0 = -1$,
$c_0 = 1$,
$c_1 = 1/b_1$,
$c_2 = b_2$,
$b_1 b_2 = 1$,
$a_1 = a_2 = b_0$,
$a_3 = c_0$,
$b_{11} = b_1 - b_{12} |\sigma_2|^2$,
$b_{13} = b_2$,
$d_{11} = c_1 - d_{12} |\sigma_2|^2$,
$d_{13} = c_2$,
$c_{13} = b_{13}$,
$c_{11} = b_{11} + (b_{12} - c_{12}) |\sigma_2|^2$,
$\sigma = \sigma_{2r} + i \sigma_{2i}$,
$\frac{b_1 (1 - b_2^2)}{2 (1 + b_1 b_2)} > \frac{v^2}{4}$,

$(b_1 + b_2)(1 - b_2^2) > 0$,
x_0, t_0, ϕ_1, ϕ_2, σ_{2r}, σ_{2i}, and v are arbitrary real constants.

7.3.5 Symmetry Reduction V

From Vector NLSE to Manakov System
The vector CNLSE

$$i\,\psi_{j_t} + a_{1j}\,\psi_{j_{xx}} + \left(\sum_{k=1}^{N} b_{j_k}\,|\psi_k|^2\right)\psi_j = 0, \quad j = 1, 2, \ldots, N, \tag{7.29}$$

transforms to the CNLSE

$$\begin{aligned} i\,\psi_{1_t} + a_1\,\psi_{1_{xx}} + (b_1\,|\psi_1|^2 + b_2\,|\psi_2|^2)\,\psi_1 = 0,\\ i\,\psi_{2_t} + a_2\,\psi_{2_{xx}} + (c_1\,|\psi_1|^2 + c_2\,|\psi_2|^2)\,\psi_2 = 0, \end{aligned} \tag{7.30}$$

with the replacements:

$$\psi_k(x, t) = \begin{cases} \sigma_k\,\psi_1(x, t), & k = 1, 2, \ldots, m, \sigma_1 = 1,\\ \sigma_k\,\psi_{m+1}(x, t), & k = m+1, \ldots, N, \sigma_{m+1} = 1, \end{cases}$$

$$b_1 = \sum_{k=1}^{m} b_{j_k}\,|\sigma_k|^2, \; j = 1, 2, \ldots, m,$$

$$b_2 = \sum_{k=m+1}^{N} b_{j_k}\,|\sigma_k|^2, \; j = 1, 2, \ldots, m,$$

$$c_1 = \sum_{k=1}^{m} b_{j_k}\,|\sigma_k|^2, \; j = m, m+1, \ldots, N,$$

$$c_2 = \sum_{k=m+1}^{N} b_{j_k}\,|\sigma_k|^2, \; j = m, m+1, \ldots, N,$$

$$a_{1j} = \begin{cases} a_1, & j = 1, 2, \ldots, m,\\ a_2, & j = m+1, m+2, \ldots, N, \end{cases}$$

where
a_1 and b_{1k} are real constants,
σ_k is an arbitrary complex constant.

7.4 Scaling Transformations

7.4.1 Linear and Nonlinear Coupling

7.4.1.1 General Case
If (ψ_1, ψ_2) is a solution of

$$\begin{aligned} i\,\psi_{1_t} + \psi_{1_{xx}} + (b_1\,|\psi_1|^2 + b_2\,|\psi_2|^2)\,\psi_1 = 0,\\ i\,\psi_{2_t} + \psi_{2_{xx}} + (b_1\,|\psi_1|^2 + b_2\,|\psi_2|^2)\,\psi_2 = 0, \end{aligned} \tag{7.31}$$

then

$$
\begin{aligned}
\Phi_1(x,t) &= \sqrt{\frac{b_1}{g_1-g_2}}\,\psi_1(x,t)\,e^{i\,g_0\,(g_1-g_2)\,t} + \sqrt{\frac{g_2\,b_2}{g_1\,(g_2-g_1)}}\,\psi_2(x,t)\,e^{-i\,g_0\,(g_1-g_2)\,t},\\
\Phi_2(x,t) &= \sqrt{\frac{b_1}{g_1-g_2}}\,\psi_1(x,t)\,e^{i\,g_0\,(g_1-g_2)\,t} + \sqrt{\frac{g_1\,b_2}{g_2\,(g_2-g_1)}}\,\psi_2(x,t)\,e^{-i\,g_0\,(g_1-g_2)\,t}
\end{aligned}
\tag{7.32}
$$

is a solution of

$$
\begin{aligned}
&i\,\Phi_{1t} + \Phi_{1xx} + (g_1\,|\Phi_1|^2 - g_2\,|\Phi_2|^2)\,\Phi_1 + g_0\,(g_1+g_2)\,\Phi_1 - 2\,g_0\,g_2\,\Phi_2 = 0,\\
&i\,\Phi_{2t} + \Phi_{2xx} + (g_1\,|\Phi_1|^2 - g_2\,|\Phi_2|^2)\,\Phi_2 - g_0\,(g_1+g_2)\,\Phi_2 + 2\,g_0\,g_1\,\Phi_1 = 0,
\end{aligned}
\tag{7.33}
$$

where

$b_1\,(g_1 - g_2) > 0$,
$b_2\,g_1\,g_2\,(g_2 - g_1) > 0$,
b_1, b_2, g_0, g_1, and g_2 are real constants.

7.4.1.2 Specific Case I: Manakov System to Another Manakov System

If (ψ_1, ψ_2) is a solution of

$$
\begin{aligned}
&i\,\psi_{1t} + \psi_{1xx} + (b_1\,|\psi_1|^2 + b_2\,|\psi_2|^2)\,\psi_1 = 0,\\
&i\,\psi_{2t} + \psi_{2xx} + (b_1\,|\psi_1|^2 + b_2\,|\psi_2|^2)\,\psi_2 = 0,
\end{aligned}
\tag{7.34}
$$

then

$$
\begin{aligned}
\Phi_1(x,t) &= \sqrt{\frac{b_1}{g_1-g_2}}\,\psi_1(x,t) + \sqrt{\frac{g_2\,b_2}{g_1\,(g_2-g_1)}}\,\psi_2(x,t),\\
\Phi_2(x,t) &= \sqrt{\frac{b_1}{g_1-g_2}}\,\psi_1(x,t) + \sqrt{\frac{g_1\,b_2}{g_2\,(g_2-g_1)}}\,\psi_2(x,t)
\end{aligned}
\tag{7.35}
$$

is a solution of

$$
\begin{aligned}
&i\,\Phi_{1t} + \Phi_{1xx} + (g_1\,|\Phi_1|^2 - g_2\,|\Phi_2|^2)\,\Phi_1 = 0,\\
&i\,\Phi_{2t} + \Phi_{2xx} + (g_1\,|\Phi_1|^2 - g_2\,|\Phi_2|^2)\,\Phi_2 = 0,
\end{aligned}
\tag{7.36}
$$

where

$b_1\,(g_1 - g_2) > 0$,
$b_2\,g_1\,g_2\,(g_2 - g_1) > 0$,
b_1, b_2, g_1 and g_2 are real constants.

7.4.1.3 Specific Case II: Manakov System to the Same Manakov System Superposition Principle for a Nonlinear System

If (ψ_1, ψ_2) is a solution of

$$
\begin{aligned}
&i\,\psi_{1t} + \psi_{1xx} + (b_1\,|\psi_1|^2 + b_2\,|\psi_2|^2)\,\psi_1 = 0,\\
&i\,\psi_{2t} + \psi_{2xx} + (b_1\,|\psi_1|^2 + b_2\,|\psi_2|^2)\,\psi_2 = 0,
\end{aligned}
\tag{7.37}
$$

then

$$\Phi_1(x, t) = \sqrt{\frac{b_1}{b_1 + b_2}} \left[\psi_1(x, t) - \frac{b_2}{b_1} \psi_2(x, t)\right],$$
$$\Phi_2(x, t) = \sqrt{\frac{b_1}{b_1 + b_2}} [\psi_1(x, t) + \psi_2(x, t)] \quad (7.38)$$

is also a solution of (7.37), where
$b_1 + b_2 \neq 0$,
$b_1 (b_1 + b_2) > 0$.

***Example* 1. $\text{sech}^2(x)$**
(Figure 7.9)
Given

$$\psi_1(x, t) = \{A_0 \text{ sech}^2[A_1 (x - x_0)] + A_3\} e^{-i [\omega_1 (t-t_0)+\phi_1]},$$
$$\psi_2(x, t) = B_0 \text{ sech}^2[A_1 (x - x_0)] e^{-i [\omega_2 (t-t_0)+\phi_2]}$$

is a solution of (7.1), then

$$\Phi_1(x, t) = \sqrt{\frac{b_1}{b_1 + b_2}} \left(\{A_0 \text{ sech}^2[A_1 (x - x_0)] + A_3\} e^{-i [\omega_1 (t-t_0)+\phi_1]} - \frac{b_2}{b_1} B_0 \text{ sech}^2[A_1 (x - x_0)] e^{-i [\omega_2 (t-t_0)+\phi_2]}\right),$$
$$\Phi_2(x, t) = \sqrt{\frac{b_1}{b_1 + b_2}} \left(\{A_0 \text{ sech}^2[A_1 (x - x_0)] + A_3\} e^{-i [\omega_1 (t-t_0)+\phi_1]} + B_0 \text{ sech}^2[A_1 (x - x_0)] e^{-i [\omega_2 (t-t_0)+\phi_2]}\right) \quad (7.39)$$

is a solution of (7.37), where
$A_3 = \frac{-2 A_0}{3}$,

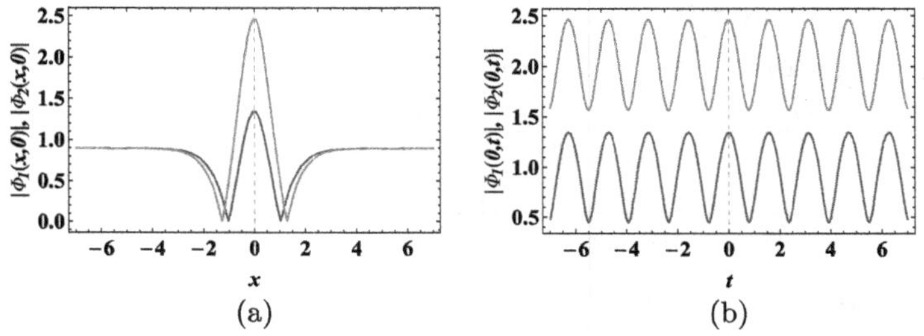

Figure 7.9. Plot of solution (7.39). (a) $t = 0$, (b) $x = 0$, with $A_0 = A_1 = B_1 = 1$, $B_0 = 3/2$, and $x_0 = t_0 = \phi_1 = \phi_2 = 0$. Blue is ϕ_1 and red is ϕ_2.

$\omega_1 = 2\,A_1^2,$
$\omega_2 = -2\,A_1^2,$
$b_1 = \frac{-9\,A_1^2}{2\,A_0^2},$
$b_2 = \frac{9\,A_1^2}{2\,B_0^2},$
$b_0 = c_0 = 1,$
$b_1 = c_1,$
$b_2 = c_2,$
$b_1 + b_2 \neq 0,$
$b_1\,(b_1 + b_2) > 0,$
$A_0 \neq 0,$
$B_0 \neq 0,$
A_1, B_1, x_0, t_0, ϕ_1, and ϕ_2 are arbitrary real constants.

7.4.2 Complex Coupling

7.4.2.1 General Case

If (ψ_1, ψ_2) is a solution of

$$\begin{aligned} i\,\psi_{1_t} + \psi_{1xx} + q_1\,(q_2\,|\psi_1|^2 + q_3\,|\psi_2|^2)\,\psi_1 = 0, \\ i\,\psi_{2_t} + \psi_{2xx} + q_1\,(q_2\,|\psi_1|^2 + q_3\,|\psi_2|^2)\,\psi_2 = 0, \end{aligned} \tag{7.40}$$

then

$$\begin{aligned} \Phi_1(x,\,t) = c_1\,\psi_1(x,\,t) + c_2\,\psi_2(x,\,t), \\ \Phi_2(x,\,t) = c_3\,\psi_1(x,\,t) + c_4\,\psi_2(x,\,t) \end{aligned} \tag{7.41}$$

is a solution of

$$\begin{aligned} i\,\Phi_{1_t} + \Phi_{1xx} + 2\,(a_{11}\,|\Phi_1|^2 + a_{12}\,|\Phi_2|^2)\,\Phi_1 + 2\,(b_{11}\,\Phi_1\,\Phi_2^* + b_{12}\,\Phi_2\,\Phi_1^*)\,\Phi_1 = 0, \\ i\,\Phi_{2_t} + \Phi_{2xx} + 2\,(a_{21}\,|\Phi_1|^2 + a_{22}\,|\Phi_2|^2)\,\Phi_2 + 2\,(b_{21}\,\Phi_1\,\Phi_2^* + b_{22}\,\Phi_2\,\Phi_1^*)\,\Phi_2 = 0, \end{aligned} \tag{7.42}$$

where

$q_1 = \frac{2\,(c_1\,c_4 - c_2\,c_3)\,(c_2^*\,c_3^* - c_1^*\,c_4^*)}{c_1^*\,c_2\,c_3\,c_4^* - c_1\,c_2^*\,c_3^*\,c_4},$
$q_2 = (a - i\,b)\,c_1^*\,c_3 - (a + i\,b)\,c_1\,c_3^*,$
$q_3 = a\,(c_2\,c_4^* - c_2^*\,c_4) + i\,b\,(c_2^*\,c_4 + c_2\,c_4^*),$
$a_{12} = \frac{b_{12}\,c_1^*\,c_2^*\,(c_2\,c_3 - c_1\,c_4) + b_{11}\,c_1\,c_2\,(c_1^*\,c_4^* - c_2^*\,c_3^*)}{c_1\,c_2^*\,c_3^*\,c_4 - c_1^*\,c_2\,c_3\,c_4^*},$
$a_{11} = \frac{b_{11}\,c_3^*\,c_4^*\,(c_2\,c_3 - c_1\,c_4) + b_{12}\,c_3\,c_4\,(c_1^*\,c_4^* - c_2^*\,c_3^*)}{c_1\,c_2^*\,c_3^*\,c_4 - c_1^*\,c_2\,c_3\,c_4^*},$
$a_{22} = a_{12},$
$b_{12} = a - i\,b,$
$b_{21} = a + i\,b,$
$b_{11} = b_{21},$
$b_{22} = b_{12},$
$c_1\,c_2^*\,c_3^*\,c_4$ should not be pure real or pure imaginary,

$c_2\, c_3 - c_1\, c_4 \neq 0$,
c_j, $j = 1, 2, 3, 4$, are complex constants,
a and b are real constants.

7.4.2.2 Specific Case
If (ψ_1, ψ_2) is a solution of

$$\begin{aligned} &i\, \psi_{1_t} + \psi_{1_{xx}} - 4\,(b\, |\psi_1|^2 + a\, |\psi_2|^2)\, \psi_1 = 0, \\ &i\, \psi_{2_t} + \psi_{2_{xx}} - 4\,(b\, |\psi_1|^2 + a\, |\psi_2|^2)\, \psi_2 = 0, \end{aligned} \tag{7.43}$$

then

$$\begin{aligned} &\Phi_1(x, t) = \psi_1(x, t) + \psi_2(x, t), \\ &\Phi_2(x, t) = \psi_1(x, t) + i\, \psi_2(x, t) \end{aligned} \tag{7.44}$$

is a solution of

$$\begin{aligned} &i\, \Phi_{1t} + \Phi_{1xx} - 2(a + b)\, (|\Phi_1|^2 + |\Phi_2|^2)\, \Phi_1 \\ &+ 2\left[(a + i\, b)\, \Phi_1\, \Phi_2^* + (a - i\, b)\, \Phi_2\, \Phi_1^*\right] \Phi_1 = 0, \\ &i\, \Phi_{2t} + \Phi_{2xx} - 2(a + b)\, (|\Phi_1|^2 + |\Phi_2|^2)\, \Phi_2 \\ &+ 2\left[(a + i\, b)\, \Phi_1\, \Phi_2^* + (a - i\, b)\, \Phi_2\, \Phi_1^*\right] \Phi_2 = 0, \end{aligned} \tag{7.45}$$

where a and b are real constants.

7.4.3 Function Coefficients

7.4.3.1 General Case
If (ψ_1, ψ_2) is a solution of

$$\begin{aligned} &i\, \psi_{1_t} + a_{11}\, \psi_{1_{xx}} + (a_{12}\, |\psi_1|^2 + a_{13}\, |\psi_2|^2)\, \psi_1 = 0, \\ &i\, \psi_{2_t} + a_{21}\, \psi_{2_{xx}} + (a_{22}\, |\psi_1|^2 + a_{23}\, |\psi_2|^2)\, \psi_2 = 0, \end{aligned} \tag{7.46}$$

then

$$\begin{aligned} &\Phi_1(x, t) = A(x, t)\, e^{i\, B_1(x,t)}\, \psi_1[X(x, t),\, T(x, t)], \\ &\Phi_2(x, t) = A(x, t)\, e^{i\, B_2(x,t)}\, \psi_2[X(x, t),\, T(x, t)] \end{aligned} \tag{7.47}$$

is a solution of

$$\begin{aligned} &i\Phi_{1t} + b_{11}(x, t)\Phi_{1xx} + [b_{12}(x, t)|\Phi_1|^2 + b_{13}(x, t)|\Phi_2|^2]\Phi_1 \\ &\quad + [b_{14r}(x, t) + ib_{14i}(x, t)]\Phi_1 = 0, \\ &i\Phi_{2t} + b_{21}(x, t)\Phi_{2xx} + [b_{22}(x, t)|\Phi_1|^2 + b_{23}(x, t)|\Phi_2|^2]\Phi_2 \\ &\quad + [b_{24r}(x, t) + ib_{24i}(x, t)]\Phi_2 = 0, \end{aligned} \tag{7.48}$$

where

$$T(x, t) = g_1(t), \tag{7.49}$$

$$A(x, t) = \frac{g_3(t)}{\sqrt{X_x(x, t)}}, \tag{7.50}$$

$$B_1(x, t) = g_2(t) - \int \frac{X_t(x, t)}{2\, b_{11}(x, t)\, X_x(x, t)}\, dx, \tag{7.51}$$

$$b_{11}(x, t) = \frac{a_{11}\, g_1'(t)}{X_x^2(x, t)}, \tag{7.52}$$

$$b_{12}(x, t) = \frac{a_{12}\, g_1'(t)}{A^2(x, t)}, \tag{7.53}$$

$$b_{13}(x, t) = \frac{a_{13}\, g_1'(t)\, X_x(x, t)}{g_3^2(t)}, \tag{7.54}$$

$$B_2(x, t) = g_2(t) - \int \frac{X_t(x, t)}{2\, b_{21}(x, t)\, X_x(x, t)}\, dx, \tag{7.55}$$

$$b_{21}(x, t) = \frac{a_{21}\, g_1'(t)}{X_x^2(x, t)}, \tag{7.56}$$

$$b_{22}(x, t) = \frac{a_{22}\, g_1'(t)}{A^2(x, t)}, \tag{7.57}$$

$$b_{23}(x, t) = \frac{a_{23}\, g_1'(t)\, X_x(x, t)}{g_3^2(t)}, \tag{7.58}$$

$$\begin{aligned} b_{14r}(x, t) = g_2'(t) + \frac{1}{4\, a_{11}\, g_1'^2(t)} \Bigg(& 2\, g_1''(t) \int X_t(x, t)\, X_x(x, t)\, dx \\ & + g_1'(t) \Bigg\{ -2 \int [X_{tt}(x, t)\, X_x(x, t) + X_t(x, t)\, X_{xt}(x, t)]\, dx \\ & + X_t^2(x, t) + \frac{a_{11}^2\, g_1'^2(t)}{X_x^4(x, t)} \left[-3\, X_{xx}^2(x, t) + 2\, X_x(x, t)\, X_{xxx}(x, t) \right] \Bigg\} \Bigg), \end{aligned} \tag{7.59}$$

$$b_{14i}(x, t) = \frac{X_{xt}(x, t)}{X_x(x, t)} - \frac{g_3'(t)}{g_3(t)}, \tag{7.60}$$

$$\begin{aligned} b_{24r}(x, t) = g_2'(t) + \frac{1}{4\, a_{21}\, g_1'^2(t)} \Bigg(2\, g_1''(t) \int X_t(x, t)\, X_x(x, t)\, dx \\ + g_1'(t) \Bigg\{ -2 \int [X_{tt}(x, t)\, X_x(x, t) + X_t(x, t)\, X_{xt}(x, t)]\, dx \\ + X_t^2(x, t) + \frac{a_{21}^2\, g_1'^2(t)}{X_x^4(x, t)} \left[-3\, X_{xx}^2(x, t) + 2\, X_x(x, t)\, X_{xxx}(x, t) \right] \Bigg\} \Bigg), \end{aligned} \tag{7.61}$$

$$b_{24i}(x, t) = \frac{X_{xt}(x, t)}{X_x(x, t)} - \frac{g_3'(t)}{g_3(t)}, \tag{7.62}$$

a_{11}, a_{12}, a_{13}, a_{21}, a_{22}, and a_{23} are arbitrary real constants.

7.4.3.2 Specific Case: Constant Dispersion and Real Quadratic Potential
If (ψ_1, ψ_2) is a solution of

$$\begin{aligned} i\, \psi_{1_t} + a_{11}\, \psi_{1_{xx}} + (a_{12}\, |\psi_1|^2 + a_{13}\, |\psi_2|^2)\, \psi_1 = 0, \\ i\, \psi_{2_t} + a_{21}\, \psi_{2_{xx}} + (a_{22}\, |\psi_1|^2 + a_{23}\, |\psi_2|^2)\, \psi_2 = 0, \end{aligned} \tag{7.63}$$

then

$$\begin{aligned} \Phi_1(x, t) = e^{\frac{\gamma(t)}{2}}\, e^{-\frac{i\, x^2\, \gamma'(t)}{a_{11}}}\, \psi_1\!\left[e^{\gamma(t)}\, x, \frac{1}{4} \int e^{2\,\gamma(t)}\, dt \right], \\ \Phi_2(x, t) = e^{\frac{\gamma(t)}{2}}\, e^{-\frac{i\, x^2\, \gamma'(t)}{a_{21}}}\, \psi_2\!\left[e^{\gamma(t)}\, x, \frac{1}{4} \int e^{2\,\gamma(t)}\, dt \right] \end{aligned} \tag{7.64}$$

is a solution of

$$\begin{aligned} i\, \Phi_{1t} + a_{11}\, \Phi_{1xx} + e^{\gamma(t)}\, (a_{12}\, |\Phi_1|^2 + a_{13}\, |\Phi_2|^2)\, \Phi_1 \\ + \frac{4}{a_{11}} [\gamma'^2(t) - \gamma''(t)]\, x^2\, \Phi_1 = 0, \\ i\, \Phi_{2t} + a_{21}\, \Phi_{2xx} + e^{\gamma(t)}\, (a_{22}\, |\Phi_1|^2 + a_{23}\, |\Phi_2|^2)\, \Phi_2 \\ + \frac{4}{a_{21}} [\gamma'^2(t) - \gamma''(t)]\, x^2\, \Phi_2 = 0, \end{aligned} \tag{7.65}$$

where $\gamma(t)$ is an arbitrary real function.

Remark: This case is obtained from the general case with the specifications:
$X(x, t) = c(t)\,x,$
$g_1(t) = \int c^2(t)\,dt,$
$g_2(t) = 0,$
$g_3(t) = c(t),$
$c(t) = e^{\gamma(t)}.$

7.5 Summary of Sections 7.3–7.4

Symmetry Reductions

Symmetry Reduction I: *From Manakov system to fundamental NLSE*

Transformation: $\psi_2(x, t) = \sigma\, \psi_1(x, t)$, $b_1 = c_1 + (c_2 - b_2)\,|\sigma|^2$, (ψ_1, ψ_2) is a solution of the fundamental CNLSE

Equation: $i\,\psi_{1_t} + b_0\,\psi_{1_{xx}} + (c_1 + c_2|\sigma|^2)\,|\psi_1|^2\,\psi_1 = 0$

Symmetry Reduction II: *From Manakov system to fundamental NLSE*

Transformation: $\psi_2(x, t) = e^{i\,\phi}\,\psi_1^*(x, t)$, $b_1 = -(c_1 + c_2 + b_2)$

Equation: $i\,\psi_{1_t} + b_0\,\psi_{1_{xx}} - (c_1 + c_2)\,|\psi_1|^2\,\psi_1 = 0$

Symmetry Reduction III: *From vector NLSE to fundamental NLSE*

Transformation: $\psi_j(x, t) = \sigma_j\,\psi_1(x, t)$

Equation: $i\,\psi_{1_t} + b_{0\,j}\,\psi_{1_{xx}} + \sum_{k=1}^{N} b_{1\,j}|\sigma_j|^2\,|\psi_1|^2\,\psi_1 = 0$

Symmetry Reduction IV: *From three coupled NLSEs to Manakov system*

Transformation: $\psi_2(x, t) = \sigma\,\psi_1(x, t)$, $c_1 = b_1 + (b_2 - c_2)\,|\sigma|^2$, $a_1 = a_2$, $c_3 = b_3$

Equation: $i\,\psi_{1_t} + a_1\,\psi_{1_{xx}} + (b_1\,|\psi_1|^2 + b_3\,|\psi_3|^2)\,\psi_1 = 0, i\,\psi_{3_t} + a_3\,\psi_{3_{xx}} + (d_1\,|\psi_1|^2 + d_3\,|\psi_3|^2)\,\psi_3 = 0$

Symmetry Reduction V: *From vector NLSE to Manakov system*

Transformation:

$$\psi_k(x, t) = \begin{cases} \sigma_k\,\psi_1(x, t), & k = 1, 2, \ldots, m,\ \sigma_1 = 1, \\ \sigma_k\,\psi_{m+1}(x, t), & k = m+1, \ldots, N,\ \sigma_{m+1} = 1, \end{cases}$$

$$b_1 = \sum_{k=1}^{m} b_{j_k}\,|\sigma_k|^2,\ j = 1, 2, \ldots, m,$$

$$b_2 = \sum_{k=m+1}^{N} b_{j_k}\,|\sigma_k|^2,\ j = 1, 2, \ldots, m,$$

$$c_1 = \sum_{k=1}^{m} b_{j_k}\,|\sigma_k|^2,\ j = m, m+1, \ldots, N,$$

$$c_2 = \sum_{k=m+1}^{N} b_{j_k}\,|\sigma_k|^2,\ j = m, m+1, \ldots, N,$$

$$a_{1\,j} = \begin{cases} a_1, & j = 1, 2, \ldots, m, \\ a_2, & j = m+1, m+2, \ldots, N \end{cases}$$

Equation: $i\,\psi_{1_t} + a_1\,\psi_{1_{xx}} + (b_1\,|\psi_1|^2 + b_2\,|\psi_2|^2)\,\psi_1 = 0,\ i\,\psi_{2_t} + a_2\,\psi_{2_{xx}} + (c_1\,|\psi_1|^2 + c_2\,|\psi_2|^2)\,\psi_2 = 0$

Scaling Transformations

Linear and Nonlinear Coupling

General Case

Transformation: $\Phi_1(x, t) = \sqrt{\frac{b_1}{g_1 - g_2}}\, \psi_1(x, t)\, e^{i\, g_0\, (g_1 - g_2)\, t} + \sqrt{\frac{g_2\, b_2}{g_1\, (g_2 - g_1)}}\, \psi_2(x, t)\, e^{-i\, g_0\, (g_1 - g_2)\, t}$,
$\Phi_2(x, t) = \sqrt{\frac{b_1}{g_1 - g_2}}\, \psi_1(x, t)\, e^{i\, g_0\, (g_1 - g_2)\, t} + \sqrt{\frac{g_1\, b_2}{g_2\, (g_2 - g_1)}}\, \psi_2(x, t)\, e^{-i\, g_0\, (g_1 - g_2)\, t}$

Equation: $i\, \Phi_{1t} + \Phi_{1xx} + (g_1\, |\Phi_1|^2 - g_2\, |\Phi_2|^2)\, \Phi_1 + g_0\, (g_1 + g_2)\, \Phi_1 - 2\, g_0\, g_2\, \Phi_2 = 0$,
$i\, \Phi_{2t} + \Phi_{2xx} + (g_1\, |\Phi_1|^2 - g_2\, |\Phi_2|^2)\, \Phi_2 - g_0\, (g_1 + g_2)\, \Phi_2 + 2\, g_0\, g_1\, \Phi_1 = 0$

Specific Case I: Manakov System to Another Manakov System

Transformation: $\Phi_1(x, t) = \sqrt{\frac{b_1}{g_1 - g_2}}\, \psi_1(x, t) + \sqrt{\frac{g_2\, b_2}{g_1\, (g_2 - g_1)}}\, \psi_2(x, t)$,
$\Phi_2(x, t) = \sqrt{\frac{b_1}{g_1 - g_2}}\, \psi_1(x, t) + \sqrt{\frac{g_1\, b_2}{g_2\, (g_2 - g_1)}}\, \psi_2(x, t)$

Equation: $i\, \Phi_{1t} + \Phi_{1xx} + (g_1\, |\Phi_1|^2 - g_2\, |\Phi_2|^2)\, \Phi_1 = 0$, $i\, \Phi_{2t} + \Phi_{2xx} + (g_1\, |\Phi_1|^2 - g_2\, |\Phi_2|^2)\, \Phi_2 = 0$

Specific Case II: Manakov System to the Same Manakov System *Superposition Principle for a Nonlinear System*

Transformation: $\Phi_1(x, t) = \sqrt{\frac{b_1}{b_1 + b_2}}\left[\psi_1(x, t) - \frac{b_2}{b_1}\, \psi_2(x, t)\right]$, $\Phi_2(x, t) = \sqrt{\frac{b_1}{b_1 + b_2}}$
$\times\, [\psi_1(x, t) + \psi_2(x, t)]$

Complex Coupling

General Case

Transformation: $\Phi_1(x, t) = c_1\, \psi_1(x, t) + c_2\, \psi_2(x, t)$, $\Phi_2(x, t) = c_3\, \psi_1(x, t) + c_4\, \psi_2(x, t)$

Equation: $i\, \Phi_{1t} + \Phi_{1xx} + 2\, (a_{11}\, |\Phi_1|^2 + a_{12}\, |\Phi_2|^2)\, \Phi_1 + 2\, (b_{11}\, \Phi_1\, \Phi_2^* + b_{12}\, \Phi_2\, \Phi_1^*)\, \Phi_1 = 0$,
$i\, \Phi_{2t} + \Phi_{2xx} + 2\, (a_{21}\, |\Phi_1|^2 + a_{22}\, |\Phi_2|^2)\, \Phi_2 + 2\, (b_{21}\, \Phi_1\, \Phi_2^* + b_{22}\, \Phi_2\, \Phi_1^*)\, \Phi_2 = 0$

Specific Case

Transformation: $\Phi_1(x, t) = \psi_1(x, t) + \psi_2(x, t)$, $\Phi_2(x, t) = \psi_1(x, t) + i\, \psi_2(x, t)$

Equation:
$i\, \Phi_{1t} + \Phi_{1xx} - 2(a + b)\, (|\Phi_1|^2 + |\Phi_2|^2)\, \Phi_1 + 2\, [(a + i\, b)\, \Phi_1\, \Phi_2^* + (a - i\, b)\, \Phi_2\, \Phi_1^*]\, \Phi_1 = 0$,

$i\, \Phi_{2t} + \Phi_{2xx} - 2(a + b)\, (|\Phi_1|^2 + |\Phi_2|^2)\, \Phi_2 + 2\, [(a + i\, b)\, \Phi_1\, \Phi_2^* + (a - i\, b)\, \Phi_2\, \Phi_1^*]\, \Phi_2 = 0$

Function Coefficients

General Case

Transformation: $\Phi_1(x, t) = A(x, t)\, e^{i\, B_1(x,t)}\, \psi_1[X(x, t),\, T(x, t)]$,
$\Phi_2(x, t) = A(x, t)\, e^{i\, B_2(x,t)}\, \psi_2[X(x, t),\, T(x, t)]$

(Continued)

(Continued)

Symmetry Reductions
Equation: $i\,\Phi_{1t} + b_{11}(x,t)\,\Phi_{1xx} + [b_{12}(x,t)\,\lvert\Phi_1\rvert^2 + b_{13}(x,t)\,\lvert\Phi_2\rvert^2]\,\Phi_1 + [b_{14r}(x,t) + i\,b_{14i}(x,t)]\,\Phi_1 = 0,$
$i\,\Phi_{2t} + b_{21}(x,t)\Phi_{2xx} + [b_{22}(x,t)\,\lvert\Phi_1\rvert^2 + b_{23}(x,t)\,\lvert\Phi_2\rvert^2]\,\Phi_2 + [b_{24r}(x,t) + i\,b_{24i}(x,t)_i]\,\Phi_2 = 0,$
Specific Case: Constant Dispersion and Real Quadratic Potential
Transformation: $\Phi_1(x,t) = e^{\frac{\gamma(t)}{2}}\, e^{-\frac{i\,x^2\gamma'(t)}{a_{11}}}\, \psi_1\left[e^{\gamma(t)}\,x,\ \frac{1}{4}\int e^{2\,\gamma(t)}\,dt\right],$ $\Phi_2(x,t) = e^{\frac{\gamma(t)}{2}}\, e^{-\frac{i\,x^2\gamma'(t)}{a_{21}}}\, \psi_2\left[e^{\gamma(t)}\,x,\ \frac{1}{4}\int e^{2\,\gamma(t)}\,dt\right]$
Equation: $i\,\Phi_{1t} + a_{11}\,\Phi_{1xx} + e^{\gamma(t)}\,[a_{12}\,\lvert\Phi_1\rvert^2 + a_{13}\,\lvert\Phi_2\rvert^2]\,\Phi_1 + \frac{4}{a_{11}}\,[\gamma'^2(t) - \gamma''(t)]\,x^2\,\Phi_1 = 0,$ $i\,\Phi_{2t} + a_{21}\,\Phi_{2xx} + e^{\gamma(t)}\,[a_{22}\,\lvert\Phi_1\rvert^2 + a_{23}\,\lvert\Phi_2\rvert^2]\,\Phi_2 + \frac{4}{a_{21}}\,[\gamma'^2(t) - \gamma''(t)]\,x^2\,\Phi_2 = 0$

7.6 $(N + 1)$-Dimensional Coupled NLSE

$(N + 1)$-Dimensional Manakov System

7.6.1 Reduction to 1D Manakov System

If (ψ_1, ψ_2) is a solution of the Manakov system

$$\begin{aligned} &i\,\psi_{1_t} + a_1\,\psi_{1_{xx}} + (b_1\,\lvert\psi_1\rvert^2 + b_2\,\lvert\psi_2\rvert^2)\,\psi_1 = 0, \quad \psi_1 = \psi_1(x,t;\,a_1,b_1,b_2),\\ &i\,\psi_{2_t} + a_2\,\psi_{2_{xx}} + (c_1\,\lvert\psi_1\rvert^2 + c_2\,\lvert\psi_2\rvert^2)\,\psi_2 = 0, \quad \psi_2 = \psi_2(x,t;\,a_2,c_1,c_2), \end{aligned} \tag{7.66}$$

then

$$\begin{aligned} \Phi_1(x_1, x_2, \ldots, x_N, t) &= \psi_1\left[\sum_{k=1}^{N} d_k\,(x_k - x_{k0}),\ t;\ \sum_{k=1}^{N} d_k^2\,b_{0k},\ b_1,\ b_2\right],\\ \Phi_2(x_1, x_2, \ldots, x_N, t) &= \psi_2\left[\sum_{k=1}^{N} d_k\,(x_k - x_{k0}),\ t;\ \sum_{k=1}^{N} d_k^2\,c_{0k},\ c_1,\ c_2\right] \end{aligned} \tag{7.67}$$

is a solution of

$$\begin{aligned} &i\,\Phi_{1t} + \sum_{k=1}^{N} b_{0k}\,\Phi_{1x_kx_k} + \left(b_1\,\lvert\Phi_1\rvert^2 + b_2\,\lvert\Phi_2\rvert^2\right)\Phi_1 = 0,\\ &i\,\Phi_{2t} + \sum_{k=1}^{N} c_{0k}\,\Phi_{2x_kx_k} + \left(c_1\,\lvert\Phi_1\rvert^2 + c_2\,\lvert\Phi_2\rvert^2\right)\Phi_2 = 0, \end{aligned} \tag{7.68}$$

where

$\Phi_j = \Phi_j(x_1, x_2, \ldots, x_N, t)$ is the complex function profile, $j = 1, 2$,
b_{0k}, c_{0k}, d_k, b_1, c_1, b_2, and c_2 are real constants.

***Example* 1. tanh(x_1, x_2)-sech(x_1, x_2)** *2D dark-bright soliton*
(Figure 7.10)
Given

$$\psi_1(x, t) = (A_0 \tanh\{A_1 [x - x_0 - v\,(t - t_0)]\} + i\,A_2)\, e^{i\,[b_1 (t-t_0)+\phi_1]},$$
$$\psi_2(x, t) = B_0 \operatorname{sech}\{A_1 [x - x_0 - v\,(t - t_0)]\}\, e^{i\,[(A_1^2+b_2)(t-t_0)+\theta(x,t)+\phi_2]}$$

is a solution of (7.1), then

$$\begin{aligned}\Phi_1(x_1, x_2, t) &= (A_0 \tanh\{A_1 [d_1 (x_1 - x_{01}) + d_2 (x_2 - x_{02}) - v\,(t - t_0)]\} + i\,A_2)\\ &\quad\times e^{i\,[b_1 (t-t_0)+\Phi_1]},\\ \Phi_2(x_1, x_2, t) &= B_0 \operatorname{sech}\{A_1 [d_1 (x_1 - x_{01}) + d_2 (x_2 - x_{02}) - v\,(t - t_0)]\}\\ &\quad\times e^{i\,[(A_1^2+b_2)(t-t_0)+\theta(x_1,x_2,t)+\phi_2]}\end{aligned} \tag{7.69}$$

is a solution of

$$\begin{aligned}&i\,\Phi_{1t} + b_{01}\,\Phi_{1x_1x_1} + b_{02}\,\Phi_{1x_2x_2} + (b_1\,|\Phi_1|^2 + b_2\,|\Phi_2|^2)\,\Phi_1 = 0,\\ &i\,\Phi_{2t} + c_{01}\,\Phi_{2x_1x_1} + c_{02}\,\Phi_{2x_2x_2} + (c_1\,|\Phi_1|^2 + c_2\,|\Phi_2|^2)\,\Phi_2 = 0,\end{aligned} \tag{7.70}$$

where

$$A_0 = \cos\left[\tan^{-1}\left(\frac{-v}{2\,A_1}\right)\right],$$
$$A_1 = \sqrt{\frac{b_1\,(1-b_2^2)}{2\,(1+b_1\,b_2)} - \frac{v^2}{4}},$$
$$A_2 = \sin\left[\tan^{-1}\left(\frac{-v}{2\,A_1}\right)\right],$$
$$B_0 = \sqrt{\frac{2\,A_1^2\,(b_1+b_2)}{1-b_2^2}},$$
$$\theta(x_1, x_2, t) = \frac{v}{2}\left[d_1 (x_1 - x_{01}) + d_2 (x_2 - x_{02}) - \frac{v}{2}\,(t - t_0)\right],$$

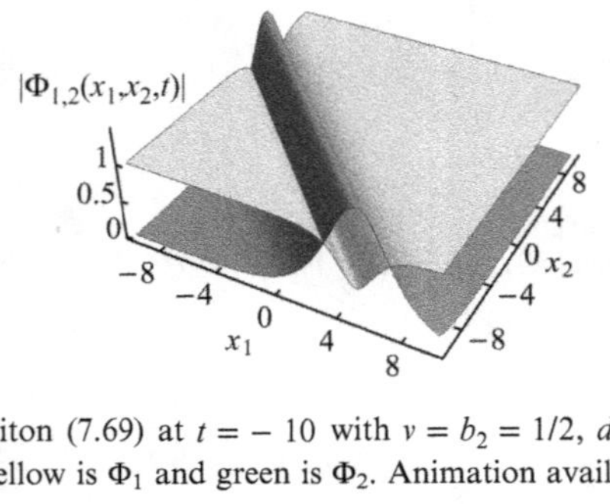

Figure 7.10. 2D dark-bright soliton (7.69) at $t = -10$ with $v = b_2 = 1/2$, $d_1 = d_2 = 1$, $b_{02} = c_{02} = 1/4$, and $x_{01} = x_{02} = t_0 = \phi_1 = \phi_2 = 0$. Yellow is Φ_1 and green is Φ_2. Animation available online at https://iopscience.iop.org/book/978-0-7503-2428-1.

$b_{01} = \frac{-1 - d_2^2\, b_{02}}{d_1^2}$,
$c_{01} = \frac{1 - d_2^2\, c_{02}}{d_1^2}$,
$c_1 = 1/b_1$,
$c_2 = b_2$,
$b_1\, b_2 = 1$,
$\frac{b_1\,(1 - b_2^2)}{2\,(1 + b_1\, b_2)} > \frac{v^2}{4}$,
$(b_1 + b_2)\,(1 - b_2^2) > 0$,
$d_1 \neq 0$,
x_{01}, x_{02}, t_0, ϕ_1, ϕ_2, d_2, and v are arbitrary real constants.

7.7 Symmetry Reductions of (N + 1)-Dimensional CNLSE to Scalar NLSE

7.7.1 Symmetry Reduction I

From (N + 1)-Dimensional Manakov System to (N + 1)-Dimensional Fundamental NLSE

The (N + 1)-dimensional CNLSE, (7.68),

$$i\,\psi_{1_t} + \sum_{k=1}^{N} b_{0k}\,\psi_{1x_kx_k} + (b_1\,|\psi_1|^2 + b_2\,|\psi_2|^2)\,\psi_1 = 0,$$
$$i\,\psi_{2_t} + \sum_{k=1}^{N} c_{0k}\,\psi_{2x_kx_k} + (c_1\,|\psi_1|^2 + c_2\,|\psi_2|^2)\,\psi_2 = 0,$$

transforms to the scalar (N + 1)-dimensional NLSE

$$i\,\psi_{1_t} + \sum_{k=1}^{N} c_{0k}\,\psi_{1x_kx_k} + (c_1 + c_2|\sigma|^2)\,|\psi_1|^2\,\psi_1 = 0, \tag{7.71}$$

with the replacements:
$\psi_2(x_1, x_2, \ldots, x_N, t) = \sigma\,\psi_1(x_1, x_2, \ldots, x_N, t)$,
$b_1 = c_1 + (c_2 - b_2)\,|\sigma|^2$,
$b_{0k} = c_{0k}$,

where σ is an arbitrary complex constant.

Conclusion:
If $\psi_1(x_1, x_2, \ldots, x_N, t)$ is a solution of the (N + 1)-dimensional NLSE

$$i\,\psi_{1_t} + \sum_{k=1}^{N} a_k\,\psi_{1x_kx_k} + a_2\,|\psi_1|^2\,\psi_1 = 0, \tag{7.72}$$

then

$$(\psi_1, \psi_2) = (\psi_1, \sigma\,\psi_1) \tag{7.73}$$

is a solution of the $(N+1)$-dimensional CNLSE

$$i\,\psi_{1_t} + \sum_{k=1}^{N} b_{0k}\,\psi_{1_{x_k x_k}} + (b_1\,|\psi_1|^2 + b_2\,|\psi_2|^2)\,\psi_1 = 0,$$
$$i\,\psi_{2_t} + \sum_{k=1}^{N} c_{0k}\,\psi_{2_{x_k x_k}} + (c_1\,|\psi_1|^2 + c_2\,|\psi_2|^2)\,\psi_2 = 0,$$

with
$\alpha_k = b_{0k} = c_{0k}$,
$a_2 = c_1 + c_2|\sigma|^2$,
$b_1 = c_1 + (c_2 - b_2)\,|\sigma|^2$.

7.7.2 Symmetry Reduction II

From $(N+1)$-Dimensional Manakov System to $(N+1)$-Dimensional Fundamental NLSE

The $(N+1)$-dimensional CNLSE, (7.68),

$$i\,\psi_{1_t} + \sum_{k=1}^{N} b_{0k}\,\psi_{1_{x_k x_k}} + (b_1\,|\psi_1|^2 + b_2\,|\psi_2|^2)\,\psi_1 = 0,$$
$$i\,\psi_{2_t} + \sum_{k=1}^{N} c_{0k}\,\psi_{2_{x_k x_k}} + (c_1\,|\psi_1|^2 + c_2\,|\psi_2|^2)\,\psi_2 = 0,$$

transforms to the scalar $(N+1)$-dimensional NLSE

$$i\,\psi_{1_t} - \sum_{k=1}^{N} c_{0k}\,\psi_{1_{x_k x_k}} - 2\,(c_1 + c_2)\,|\psi_1|^2\,\psi_1 = 0, \tag{7.74}$$

with the replacements:
$\psi_2(x_1, x_2, \ldots, x_N, t) = e^{i\,\phi}\,\psi_1^*(x_1, x_2, \ldots, x_N, t)$,
$b_1 = -(c_1 + c_2 + b_2)$,
$b_{0k} = -c_{0k}$,

where ϕ is an arbitrary real constant.

Conclusion:
If $\psi_1(x_1, x_2, \ldots, x_N, t)$ is a solution of the $(N+1)$-dimensional NLSE

$$i\,\psi_{1_t} + \sum_{k=1}^{N} \alpha_k\,\psi_{1_{x_k x_k}} + a_2\,|\psi_1|^2\,\psi_1 = 0, \tag{7.75}$$

then

$$(\psi_1, \psi_2) = (\psi_1, e^{i\,\phi}\,\psi_1^*) \tag{7.76}$$

is a solution of the $(N+1)$-dimensional CNLSE

$$i\,\psi_{1_t} + \sum_{k=1}^{N} b_{0k}\,\psi_{1_{x_k x_k}} + (b_1\,|\psi_1|^2 + b_2\,|\psi_2|^2)\,\psi_1 = 0,$$

$$i\,\psi_{2_t} + \sum_{k=1}^{N} c_{0k}\,\psi_{2_{x_k x_k}} + (c_1\,|\psi_1|^2 + c_2\,|\psi_2|^2)\,\psi_2 = 0,$$

with

$\alpha_k = b_{0k} = -c_{0k}$,
$a_2 = -2\,(c_1 + c_2)$.

7.7.3 Symmetry Reduction III

From (N + 1)-Dimensional Vector NLSE to (N + 1)-Dimensional Fundamental NLSE

The generalized $(N + 1)$-dimensional CNLSE

$$i\,\psi_{j_t} + \sum_{k=1}^{N} b_{j0k}\,\psi_{j_{x_k x_k}} + \left(\sum_{k=1}^{M} b_{1k}\,|\psi_k|^2\right)\psi_j = 0, \quad j = 1, 2, \ldots, M, \tag{7.77}$$

transforms to the scalar $(N + 1)$-dimensional NLSE

$$i\,\psi_{1_t} + \sum_{k=1}^{N} b_{10k}\,\psi_{1_{x_k x_k}} + \sum_{k=1}^{N} b_{1j}|\sigma_{j+1}|^2\,|\psi_1|^2\,\psi_1 = 0, \tag{7.78}$$

with the replacement:

$$\psi_j(x_1, x_2, \ldots, x_N, t) = \sigma_j\,\psi_1(x_1, x_2, \ldots, x_N, t),$$

where

σ_j are arbitrary complex constants,
$\sigma_1 = 1$.

Conclusion:

If $\psi_1(x_1, x_2, \ldots, x_N, t)$ is a solution of the $(N + 1)$-dimensional NLSE

$$i\,\psi_{1_t} + \sum_{k=1}^{N} \alpha_k\,\psi_{1_{x_k x_k}} + a_2\,|\psi_1|^2\,\psi_1 = 0,$$

then

$$(\psi_1, \psi_2, \psi_3, \ldots) = (\psi_1, \sigma_2\,\psi_1, \sigma_3\,\psi_1, \ldots) \tag{7.79}$$

is a solution of the generalized $(N + 1)$-dimensional CNLSE

$$i\,\psi_{j_t} + \sum_{k=1}^{N} b_{0k\,j}\,\psi_{j_{x_k x_k}} + \left(\sum_{k=1}^{M} b_{1k}\,|\psi_k|^2\right)\psi_j = 0, \quad j = 1, 2, \ldots, M,$$

with

$$\alpha_k = b_{10k},$$
$$a_2 = \sum_{j=1}^{N} b_{1j}|\sigma_{j+1}|^2.$$

Notes:

The $(N + 1)$-dimensional three coupled NLSEs reduce to $(N + 1)$-dimensional Manakov system in a similar manner to the above described symmetry reductions, (7.7.1–7.7.3).

The $(N + 1)$-dimensional vector NLSE reduces to $(N + 1)$-dimensional Manakov system in a similar manner to the above described symmetry reductions, (7.7.1–7.7.3).

7.8 $(N + 1)$-Dimensional Scaling Transformations

7.8.1 Linear and Nonlinear Coupling

7.8.1.1 General Case

If (ψ_1, ψ_2) is a solution of

$$\begin{aligned} i\,\psi_{1_t} + \sum_{k=1}^{N} \psi_{1x_kx_k} + (b_1\,|\psi_1|^2 + b_2\,|\psi_2|^2)\,\psi_1 = 0,\\ i\,\psi_{2_t} + \sum_{k=1}^{N} \psi_{2x_kx_k} + (b_1\,|\psi_1|^2 + b_2\,|\psi_2|^2)\,\psi_2 = 0, \end{aligned} \tag{7.80}$$

then

$$\begin{aligned} \Phi_1(x_1, x_2, \ldots, x_N, t) &= \sqrt{\frac{b_1}{g_1 - g_2}}\;\psi_1(x_1, x_2, \ldots, x_N, t)\, e^{i\,g_0\,(g_1-g_2)\,t}\\ &\quad + \sqrt{\frac{g_2\,b_2}{g_1\,(g_2 - g_1)}}\;\psi_2(x_1, x_2, \ldots, x_N, t)\, e^{-i\,g_0\,(g_1-g_2)\,t},\\ \Phi_2(x_1, x_2, \ldots, x_N, t) &= \sqrt{\frac{b_1}{g_1 - g_2}}\;\psi_1(x_1, x_2, \ldots, x_N, t)\, e^{i\,g_0\,(g_1-g_2)\,t}\\ &\quad + \sqrt{\frac{g_1\,b_2}{g_2\,(g_2 - g_1)}}\;\psi_2(x_1, x_2, \ldots, x_N, t)\, e^{-i\,g_0\,(g_1-g_2)\,t} \end{aligned} \tag{7.81}$$

is a solution of

$$\begin{aligned} i\,\Phi_{1_t} + \sum_{k=1}^{N} \Phi_{1x_kx_k} + (g_1\,|\Phi_1|^2 - g_2\,|\Phi_2|^2)\,\Phi_1 + g_0\,(g_1 + g_2)\,\Phi_1 - 2\,g_0\,g_2\,\Phi_2 = 0,\\ i\,\Phi_{2_t} + \sum_{k=1}^{N} \Phi_{2x_kx_k} + (g_1\,|\Phi_1|^2 - g_2\,|\Phi_2|^2)\,\Phi_2 - g_0\,(g_1 + g_2)\,\Phi_2 + 2\,g_0\,g_1\,\Phi_1 = 0, \end{aligned} \tag{7.82}$$

where

$b_1\,(g_1 - g_2) > 0,$
$b_2\,g_1\,g_2\,(g_2 - g_1) > 0,$
b_1, b_2, g_0, g_1, and g_2 are real constants.

7.8.1.2 Specific Case I: (N + 1)-Dimensional Manakov System to Another (N + 1)-Dimensional Manakov System

If (ψ_1, ψ_2) is a solution of

$$i\,\psi_{1_t} + \sum_{k=1}^{N}\psi_{1_{x_k x_k}} + (b_1\,|\psi_1|^2 + b_2\,|\psi_2|^2)\,\psi_1 = 0,$$
$$i\,\psi_{2_t} + \sum_{k=1}^{N}\psi_{2_{x_k x_k}} + (b_1\,|\psi_1|^2 + b_2\,|\psi_2|^2)\,\psi_2 = 0,$$

then

$$\begin{aligned}\Phi_1(x_1, x_2, \ldots, x_N, t) &= \sqrt{\frac{b_1}{g_1 - g_2}}\;\psi_1(x_1, x_2, \ldots, x_N, t)\\ &\quad + \sqrt{\frac{g_2\,b_2}{g_1\,(g_2 - g_1)}}\;\psi_2(x_1, x_2, \ldots, x_N, t),\\ \Phi_2(x_1, x_2, \ldots, x_N, t) &= \sqrt{\frac{b_1}{g_1 - g_2}}\;\psi_1(x_1, x_2, \ldots, x_N, t)\\ &\quad + \sqrt{\frac{g_1\,b_2}{g_2\,(g_2 - g_1)}}\;\psi_2(x_1, x_2, \ldots, x_N, t)\end{aligned} \tag{7.83}$$

is a solution of

$$\begin{aligned}i\,\Phi_{1_t} + \sum_{k=1}^{N}\Phi_{1_{x_k x_k}} + (g_1\,|\Phi_1|^2 - g_2\,|\Phi_2|^2)\,\Phi_1 = 0,\\ i\,\Phi_{2_t} + \sum_{k=1}^{N}\Phi_{2_{x_k x_k}} + (g_1\,|\Phi_1|^2 - g_2\,|\Phi_2|^2)\,\Phi_2 = 0,\end{aligned} \tag{7.84}$$

where

$b_1\,(g_1 - g_2) > 0,$
$b_2\,g_1\,g_2\,(g_2 - g_1) > 0,$
b_1, b_2, g_1, and g_2 are real constants.

7.8.1.3 Specific Case II: (N + 1)-Dimensional Manakov System to the Same (N + 1)-Dimensional Manakov System

If (ψ_1, ψ_2) is a solution of

$$i\,\psi_{1_t} + \sum_{k=1}^{N}\psi_{1_{x_k x_k}} + (b_1\,|\psi_1|^2 + b_2\,|\psi_2|^2)\,\psi_1 = 0,$$

$$i\,\psi_{2t} + \sum_{k=1}^{N} \psi_{2x_kx_k} + (b_1\,|\psi_1|^2 + b_2\,|\psi_2|^2)\,\psi_2 = 0,$$

then

$$\begin{aligned}\Phi_1(x_1, x_2, \dots, x_N, t) &= \sqrt{\frac{b_1}{b_1+b_2}}\left[\psi_1(x_1, x_2, \dots, x_N, t) - \frac{b_2}{b_1}\,\psi_2(x_1, x_2, \dots, x_N, t)\right],\\ \Phi_2(x_1, x_2, \dots, x_N, t) &= \sqrt{\frac{b_1}{b_1+b_2}}\,[\psi_1(x_1, x_2, \dots, x_N, t) + \psi_2(x_1, x_2, \dots, x_N, t)]\end{aligned} \tag{7.85}$$

is also a solution of (7.34), where $b_1 + b_2 \neq 0$.

7.8.2 Complex Coupling

7.8.2.1 General Case

If (ψ_1, ψ_2) is a solution of

$$\begin{aligned}&i\,\psi_{1t} + \sum_{k=1}^{N} \psi_{1x_kx_k} + q_1\,\{q_2\,|\psi_1|^2 + q_3\,|\psi_2|^2\}\,\psi_1 = 0,\\ &i\,\psi_{2t} + \sum_{k=1}^{N} \psi_{2x_kx_k} + q_1\,\{q_2\,|\psi_1|^2 + q_3\,|\psi_2|^2\}\,\psi_2 = 0,\end{aligned} \tag{7.86}$$

then

$$\begin{aligned}\Phi_1(x_1, x_2, \dots, x_N, t) &= c_1\,\psi_1(x_1, x_2, \dots, x_N, t) + c_2\,\psi_2(x_1, x_2, \dots, x_N, t),\\ \Phi_2(x_1, x_2, \dots, x_N, t) &= c_3\,\psi_1(x_1, x_2, \dots, x_N, t) + c_4\,\psi_2(x_1, x_2, \dots, x_N, t)\end{aligned} \tag{7.87}$$

is a solution of

$$\begin{aligned}&i\,\Phi_{1t} + \sum_{k=1}^{N} \Phi_{1x_kx_k} + 2\,(a_{11}\,|\Phi_1|^2 + a_{12}\,|\Phi_2|^2)\,\Phi_1\\ &+ 2\,(b_{11}\,\Phi_1\,\Phi_2^* + b_{12}\,\Phi_2\,\Phi_1^*)\,\Phi_1 = 0,\\ &i\,\Phi_{2t} + \sum_{k=1}^{N} \Phi_{2x_kx_k} + 2\,(a_{21}\,|\Phi_1|^2 + a_{22}\,|\Phi_2|^2)\,\Phi_2\\ &+ 2\,(b_{21}\,\Phi_1\,\Phi_2^* + b_{22}\,\Phi_2\,\Phi_1^*)\,\Phi_2 = 0,\end{aligned} \tag{7.88}$$

where

$$q_1 = \frac{2\,(c_1\,c_4 - c_2\,c_3)\,(c_2^*\,c_3^* - c_1^*\,c_4^*)}{c_1^*\,c_2\,c_3\,c_4^* - c_1\,c_2^*\,c_3^*\,c_4},$$

$$q_2 = (a - i\,b)\,c_1^*\,c_3 - (a + i\,b)\,c_1\,c_3^*,$$

$$q_3 = a\,(c_2\,c_4^* - c_2^*\,c_4) + i\,b\,(c_2^*\,c_4 + c_2\,c_4^*),$$

$$a_{12} = \frac{b_{12}\,c_1^*\,c_2^*\,(c_2\,c_3 - c_1\,c_4) + b_{11}\,c_1\,c_2\,(c_1^*\,c_4^* - c_2^*\,c_3^*)}{c_1\,c_2^*\,c_3^*\,c_4 - c_1^*\,c_2\,c_3\,c_4^*},$$

$$a_{11} = \frac{b_{11}\,c_3^*\,c_4^*\,(c_2\,c_3 - c_1\,c_4) + b_{12}\,c_3\,c_4\,(c_1^*\,c_4^* - c_2^*\,c_3^*)}{c_1\,c_2^*\,c_3^*\,c_4 - c_1^*\,c_2\,c_3\,c_4^*},$$

$a_{22} = a_{12}$,
$b_{12} = a - i\, b$,
$b_{21} = a + i\, b$,
$b_{11} = b_{21}$,
$b_{22} = b_{12}$,
$c_1\, c_2^*\, c_3^*\, c_4$ should not be pure real or pure imaginary,
$c_2\, c_3 - c_1\, c_4 \neq 0$,
c_j, $j = 1, 2, 3, 4$ are complex constants,
a and b are real constants.

7.8.2.2 Specific Case
If (ψ_1, ψ_2) is a solution of

$$\begin{aligned} &i\,\psi_{1_t} + \sum_{k=1}^{N} \psi_{1_{x_k x_k}} - 4\,(b\,|\psi_1|^2 + a\,|\psi_2|^2)\,\psi_1 = 0,\\ &i\,\psi_{2_t} + \sum_{k=1}^{N} \psi_{2_{x_k x_k}} - 4\,(b\,|\psi_1|^2 + a\,|\psi_2|^2)\,\psi_2 = 0, \end{aligned} \tag{7.89}$$

then

$$\begin{aligned} &\Phi_1(x_1, x_2, \ldots, x_N, t) = \psi_1(x_1, x_2, \ldots, x_N, t) + \psi_2(x_1, x_2, \ldots, x_N, t),\\ &\Phi_2(x_1, x_2, \ldots, x_N, t) = \psi_1(x_1, x_2, \ldots, x_N, t) + i\,\psi_2(x_1, x_2, \ldots, x_N, t) \end{aligned} \tag{7.90}$$

is a solution of

$$\begin{aligned} &i\,\Phi_{1_t} + \sum_{k=1}^{N} \Phi_{1_{x_k x_k}} - 2(a+b)\,(|\Phi_1|^2 + |\Phi_2|^2)\,\Phi_1\\ &\qquad + 2\,[(a + i\,b)\,\Phi_1\,\Phi_2^* + (a - i\,b)\,\Phi_2\,\Phi_1^*]\,\Phi_1 = 0,\\ &i\,\Phi_{2_t} + \sum_{k=1}^{N} \Phi_{2_{x_k x_k}} - 2(a+b)\,(|\Phi_1|^2 + |\Phi_2|^2)\,\Phi_2\\ &\qquad + 2\,[(a + i\,b)\,\Phi_1\,\Phi_2^* + (a - i\,b)\,\Phi_2\,\Phi_1^*]\,\Phi_2 = 0, \end{aligned} \tag{7.91}$$

where a and b are real constants.

7.9 Summary of Sections 7.7–7.8

$(N + 1)$-Dimensional Symmetry Reductions
$(N + 1)$-Dimensional Symmetry Reduction I: *From $(N + 1)$-dimensional Manakov system to $(N + 1)$-dimensional fundamental NLSE*
Transformation: $\psi_2(x_1, x_2, \ldots, x_N, t) = \sigma\,\psi_1(x_1, x_2, \ldots, x_N, t)$, $b_1 = c_1 + (c_2 - b_2)\,|\sigma|^2$, $b_{0k} = c_{0k}$, (ψ_1, ψ_2) is a solution of the $(N + 1)$-dimensional CNLSE

Equation: $i\, \psi_{1_t} + \sum_{k=1}^{N} c_{0k}\, \psi_{1_{x_k x_k}} + (c_1 + c_2|\sigma|^2)\, |\psi_1|^2\, \psi_1 = 0$

(*N* +1)-Dimensional Symmetry Reduction II: *From (N + 1)-dimensional Manakov system to (N + 1)-dimensional fundamental NLSE*

Transformation: $\psi_2(x_1, x_2, \ldots, x_N, t) = e^{i\,\phi}\, \psi_1^*(x_1, x_2, \ldots, x_N, t), b_1 = -(c_1 + c_2 + b_2), b_{0k} = -c_{0k}$

Equation: $i\, \psi_{1_t} - \sum_{k=1}^{N} c_{0k}\, \psi_{1_{x_k x_k}} - 2\,(c_1 + c_2)\, |\psi_1|^2\, \psi_1 = 0$

(*N* +1)-Dimensional Symmetry Reduction III: *From (N + 1)-dimensional vector NLSE to (N + 1)-dimensional fundamental NLSE*

Transformation: $\psi_j(x_1, x_2, \ldots, x_N, t) = \sigma_j\, \psi_1(x_1, x_2, \ldots, x_N, t)$

Equation: $i\, \psi_{1_t} + \sum_{k=1}^{N} b_{10k}\, \psi_{1_{x_k x_k}} + \sum_{k=1}^{N} b_{1j}|\sigma_{j+1}|^2\, |\psi_1|^2\, \psi_1 = 0$

(*N* +1)-Dimensional Scaling Transformations

Linear and Nonlinear Coupling

General Case

Transformation:

$$\Phi_1(x_1, x_2, \ldots, x_N, t) = \sqrt{\frac{b_1}{g_1 - g_2}}\, \psi_1(x_1, x_2, \ldots, x_N, t)\, e^{i\, g_0\,(g_1 - g_2)\, t} + \sqrt{\frac{g_2\, b_2}{g_1\,(g_2 - g_1)}}\, \psi_2(x_1, x_2, \ldots, x_N, t)\, e^{-i\, g_0\,(g_1 - g_2)\, t},$$

$$\Phi_2(x_1, x_2, \ldots, x_N, t) = \sqrt{\frac{b_1}{g_1 - g_2}}\, \psi_1(x_1, x_2, \ldots, x_N, t)\, e^{i\, g_0\,(g_1 - g_2)\, t} + \sqrt{\frac{g_1\, b_2}{g_2\,(g_2 - g_1)}}\, \psi_2(x_1, x_2, \ldots, x_N, t)\, e^{-i\, g_0\,(g_1 - g_2)\, t}$$

Equation: $i\, \Phi_{1_t} + \sum_{k=1}^{N} \Phi_{1_{x_k x_k}} + (g_1\, |\Phi_1|^2 - g_2\, |\Phi_2|^2)\, \Phi_1 + g_0\,(g_1 + g_2)\, \Phi_1 - 2\, g_0\, g_2\, \Phi_2 = 0,$

$$i\, \Phi_{2t} + \sum_{k=1}^{N} \Phi_{2_{x_k x_k}} + (g_1\, |\Phi_1|^2 - g_2\, |\Phi_2|^2)\, \Phi_2 - g_0\,(g_1 + g_2)\, \Phi_2 + 2\, g_0\, g_1\, \Phi_1 = 0$$

Specific Case I: (*N* +1)-Dimensional Manakov System to Another (*N* + 1)-Dimensional Manakov System

Transformation:

$$\Phi_1(x_1, x_2, \ldots, x_N, t) = \sqrt{\frac{b_1}{g_1 - g_2}}\, \psi_1(x_1, x_2, \ldots, x_N, t) + \sqrt{\frac{g_2\, b_2}{g_1\,(g_2 - g_1)}}\, \psi_2(x_1, x_2, \ldots, x_N, t),$$

$$\Phi_2(x_1, x_2, \ldots, x_N, t) = \sqrt{\frac{b_1}{g_1 - g_2}}\, \psi_1(x_1, x_2, \ldots, x_N, t) + \sqrt{\frac{g_1\, b_2}{g_2\,(g_2 - g_1)}}\, \psi_2(x_1, x_2, \ldots, x_N, t)$$

(*Continued*)

(Continued)

(N + 1)-Dimensional Symmetry Reductions
Equation: $i\,\Phi_{1t} + \sum_{k=1}^{N}\Phi_{1x_kx_k} + (g_1\,\lvert\Phi_1\rvert^2 - g_2\,\lvert\Phi_2\rvert^2)\,\Phi_1 = 0,$ $i\,\Phi_{2t} + \sum_{k=1}^{N}\Phi_{2x_kx_k} + (g_1\,\lvert\Phi_1\rvert^2 - g_2\,\lvert\Phi_2\rvert^2)\,\Phi_2 = 0$
Specific Case II: (N + 1)-Dimensional Manakov System to the Same (N + 1)-Dimensional Manakov System
Superposition Principle to a Nonlinear System
Transformation: $\Phi_1(x_1, x_2, \ldots, x_N, t) = \sqrt{\frac{b_1}{b_1 + b_2}}\left[\psi_1(x_1, x_2, \ldots, x_N, t) + \frac{b_2}{b_1}\psi_2(x_1, x_2, \ldots, x_N, t)\right],$ $\Phi_2(x_1, x_2, \ldots, x_N, t) = \sqrt{\frac{b_1}{b_1 + b_2}}\,[\psi_1(x_1, x_2, \ldots, x_N, t) + \psi_2(x_1, x_2, \ldots, x_N, t)]$
Complex Coupling
General Case
Transformation: $\Phi_1(x_1, x_2, \ldots, x_N, t) = c_1\,\psi_1(x_1, x_2, \ldots, x_N, t) + c_2\,\psi_2(x_1, x_2, \ldots, x_N, t),$ $\Phi_2(x_1, x_2, \ldots, x_N, t) = c_3\,\psi_1(x_1, x_2, \ldots, x_N, t) + c_4\,\psi_2(x_1, x_2, \ldots, x_N, t)$
Equation: $i\,\Phi_{1t} + \sum_{k=1}^{N}\Phi_{1x_kx_k} + 2\,(a_{11}\,\lvert\Phi_1\rvert^2 + a_{12}\,\lvert\Phi_2\rvert^2)\,\Phi_1 + 2\,(b_{11}\,\Phi_1\,\Phi_2^* + b_{12}\,\Phi_2\,\Phi_1^*)\,\Phi_1 = 0,$ $i\,\Phi_{2t} + \sum_{k=1}^{N}\Phi_{2x_kx_k} + 2\,(a_{21}\,\lvert\Phi_1\rvert^2 + a_{22}\,\lvert\Phi_2\rvert^2)\,\Phi_2 + 2\,(b_{21}\,\Phi_1\,\Phi_2^* + b_{22}\,\Phi_2\,\Phi_1^*)\,\Phi_2 = 0$
Specific Case
Transformation: $\Phi_1(x_1, x_2, \ldots, x_N, t) = \psi_1(x_1, x_2, \ldots, x_N, t) + \psi_2(x_1, x_2, \ldots, x_N, t),$ $\Phi_2(x_1, x_2, \ldots, x_N, t) = \psi_1(x_1, x_2, \ldots, x_N, t) + i\,\psi_2(x_1, x_2, \ldots, x_N, t)$
Equation: $i\,\Phi_{1t} + \sum_{k=1}^{N}\Phi_{1x_kx_k} - 2(a+b)(\lvert\Phi_1\rvert^2 + \lvert\Phi_2\rvert^2)\,\Phi_1 + 2((a+i\,b)\,\Phi_1\,\Phi_2^* + (a-i\,b)\,\Phi_2\,\Phi_1^*)\,\Phi_1 = 0,$ $i\,\Phi_{2t} + \sum_{k=1}^{N}\Phi_{2x_kx_k} - 2(a+b)(\lvert\Phi_1\rvert^2 + \lvert\Phi_2\rvert^2)\,\Phi_2 + 2((a+i\,b)\,\Phi_1\,\Phi_2^* + (a-i\,b)\,\Phi_2\,\Phi_1^*)\,\Phi_2 = 0$

References

[1] Gordon J P 1983 Interaction forces among solitons in optical fibers *Opt. Lett.* **8** 596–8

[2] Buryak A V, Kivshar Y S and Parker D F 1996 Coupling between dark and bright solitons *Phys. Lett.* A **215** 57–62

[3] Guo B L and Ling L M 2011 Rogue wave, breathers and bright-dark-rogue solutions for the coupled Schrödinger equations *Chin. Phys. Lett.* **28** 110202
[4] Khare A and Saxena A 2015 Periodic and hyperbolic soliton solutions of a number of nonlocal nonlinear equations *J. Math. Phys.* **56** 032104–27
[5] Porubov A V and Parker D F 1999 Some general periodic solutions to coupled nonlinear Schrödinger equations *Wave Motion* **29** 97–109

IOP Publishing

Handbook of Exact Solutions to the Nonlinear Schrödinger Equations

Usama Al Khawaja and Laila Al Sakkaf

Chapter 8

Discrete Nonlinear Schrödinger Equation

A Glance at Chapter 8

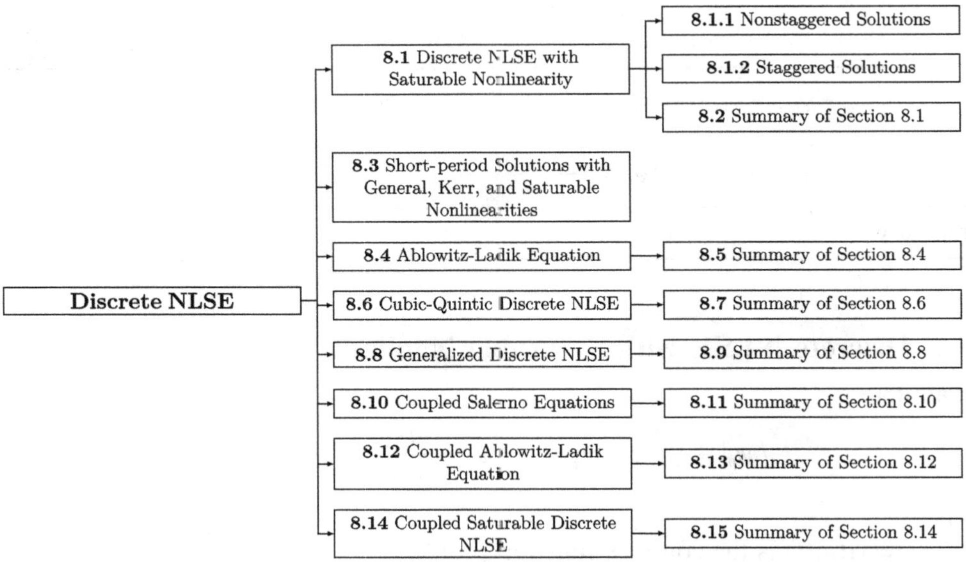

doi:10.1088/978-0-7503-2428-1ch8

A Statistical View of Chapter 8

	Equation	Solutions
1	$i\,\psi_{nt} + \psi_{n+1} + \psi_{n-1} - 2\,\psi_n + \frac{a_2\,\|\psi_n\|^2\,\psi_n}{1+\mu\,\|\psi_n\|^2} = 0$	35
2	$i\,\psi_{nt} + \psi_{n+1} + \psi_{n-1} - 2\,\psi_n + a_2\,F[\|\psi_n\|^2]\,\psi_n = 0$	6
3	$i\,\psi_{nt} + \psi_{n+1} + \psi_{n-1} - 2\,\psi_n + a_2\,\|\psi_n\|^2\,\psi_n = 0$	6
4	$i\,\psi_{nt} + \psi_{n+1} + \psi_{n-1} - 2\,\psi_n + a_2\,(\psi_{n+1} + \psi_{n-1})\,\|\psi_n\|^2 = 0$	12
5	$i\,\psi_{nt} + a_1\,(\psi_{n+1} + \psi_{n-1} - 2\,\psi_n) + a_2\,\|\psi_n\|^2\,\psi_n + (a_3\,\|\psi_n\|^2 + a_4\,\|\psi_n\|^4)(\psi_{n+1} + \psi_{n-1}) = 0$	5
6	$i\,\psi_{nt} + a_1\,(\psi_{n+1} + \psi_{n-1} - 2\,\psi_n) + f[\psi_{n-1}, \psi_n, \psi_{n+1}] = 0$	7
7	$i\,\psi_{1nt} + \psi_{1n+1} + \psi_{1n-1} - 2\,\psi_{1n} + (\mu_1\,\|\psi_{1n}\|^2 + \mu_2\,\|\psi_{2n}\|^2)\,(\psi_{1n+1} + \psi_{1n-1} + \frac{\nu_1 - 2\,\mu_1}{\mu_1}\,\psi_{1n}) = 0,$ $i\,\psi_{2nt} + [\psi_{2n+1} + \psi_{2n-1} - (2 + \frac{\nu_1\,\mu_2}{\mu_1^2} - \frac{\nu_2}{\mu_2})\,\psi_{2n}] + (\mu_1\,\|\psi_{1n}\|^2 + \mu_2\,\|\psi_{2n}\|^2)\,[\psi_{2n+1}$ $+ \psi_{2n-1} + (\frac{\nu_2 - 2\,\mu_2}{\mu_2})\,\psi_{2n}] = 0$	14
8	$i\,\psi_{1nt} + \psi_{1n+1} + \psi_{1n-1} - 2\,\psi_{1n} + (\mu_1\,\|\psi_{1n}\|^2 + \mu_2\,\|\psi_{2n}\|^2)\,(\psi_{1n+1} + \psi_{1n-1}) = 0,$ $i\,\psi_{2nt} + \psi_{2n+1} + \psi_{2n-1} - \frac{2\,\mu_2}{\mu_1}\,\psi_{2n} + (\mu_1\,\|\psi_{1n}\|^2 + \mu_2\,\|\psi_{2n}\|^2)\,(\psi_{2n+1} + \psi_{2n-1}) = 0$	18
9	$i\,\psi_{1nt} + \psi_{1n+1} + \psi_{1n-1} - 2\,\psi_{1n} + \frac{\nu_1\,(\mu_1\,\|\psi_{1n}\|^2 + \mu_2\,\|\psi_{2n}\|^2)\,\psi_{1n}}{\mu_1\,(1 + \mu_1\,\|\psi_{1n}\|^2 + \mu_2\,\|\psi_{2n}\|^2)} = 0,$ $i\,\psi_{2nt} + \psi_{2n+1} + \psi_{2n-1} - 2\,\psi_{2n} + \frac{\left[\nu_2 - \frac{\nu_1\,\mu_2^2}{\mu_1^2} + \nu_2\,(\mu_1\,\|\psi_{1n}\|^2 + \mu_2\,\|\psi_{2n}\|^2)\right]\psi_{2n}}{\mu_2\,(1 + \mu_1\,\|\psi_{1n}\|^2 + \mu_2\,\|\psi_{2n}\|^2)} = 0$	3
Total	9	106

8.1 Discrete NLSE with Saturable Nonlinearity

Equation:

$$i\,\psi_{nt} + \psi_{n+1} + \psi_{n-1} - 2\,\psi_n + \frac{a_2\,|\psi_n|^2\,\psi_n}{1 + \mu\,|\psi_n|^2} = 0, \tag{8.1}$$

where

$\psi_n = \psi(n, t)$ is the complex function profile,
the integer site index, n, and t are its two independent variables,
a_2 and μ are real constants.

Solutions:

8.1.1 Nonstaggered Solutions

***Solution* 1. Constant Amplitude** *discrete continuous wave (CW), t- and n-dependent phase*

$$\psi(n,\,t) = A_0\, e^{i[A_1\,(n-n_0)-A_2\,(t-t_0)+\phi_0]}, \tag{8.2}$$

where

$A_2 = 4\,\sin^2(A_1/2) - \frac{a_2\,A_0^2}{1+\mu\,A_0^2}$,

A_0, A_1, t_0, n_0, and ϕ_0 are arbitrary real constants.

- *Reference*: [2].

***Solution* 2. sec(*n*)**

$$\psi(n,\,t) = A_0\,\sec[A_1\,(n-n_0)]\,e^{-i\,[A_2\,(t-t_0)+\phi_0]}, \tag{8.3}$$

where

$A_0 = \frac{\sin(A_1)}{\sqrt{-\mu}}$,

$A_2 = \frac{2\,\mu - a_2}{\mu}$,

$a_2 = 2\,\mu\,\cos(A_1)$,

$\mu < 0$,

A_1, t_0, n_0, and ϕ_0 are arbitrary real constants.

- *Reference*: [3].

***Solution* 3. tan(*n*)**

$$\psi(n,\,t) = A_0\,\tan[A_1\,(n-n_0)]\,e^{-i\,[A_2\,(t-t_0)+\phi_0]}, \tag{8.4}$$

where

$A_0 = \frac{\tan(A_1)}{\sqrt{-\mu}}$,

$A_2 = 2 - 2\,\sec^2(A_1)$,

$a_2 = 2\,\mu\,\sec^2(A_1)$,

$\mu < 0$,

A_1, t_0, n_0, and ϕ_0 are arbitrary real constants.

- *Reference*: [3], *we corrected the expression of* A_0.

***Solution* 4. sech(*n*)** *discrete bright soliton*
(Figure 8.1)

$$\psi(n,\,t) = A_0\,\mathrm{sech}[A_1\,(n-n_0)]\,e^{-i\,[A_2\,(t-t_0)+\phi_0]}, \tag{8.5}$$

where

$A_0 = \frac{\sinh(A_1)}{\sqrt{\mu}}$,

$A_2 = \frac{2\,\mu - a_2}{\mu}$,

$\mu = \frac{a_2\,\mathrm{sech}(A_1)}{2} > 0$,

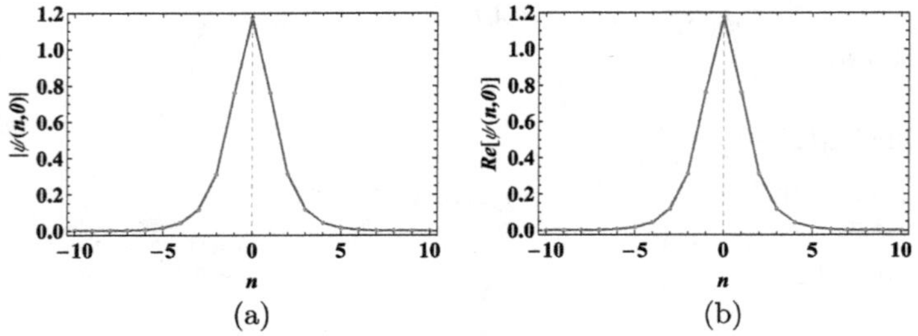

Figure 8.1. Discrete bright soliton (8.5) at $t = 0$. (a) Absolute value, (b) real part, with $a_2 = A_1 = 1$ and $n_0 = t_0 = \phi_0 = 0$. The lines are guides for the eye.

A_1, t_0, n_0, and ϕ_0 are arbitrary real constants.

- *Reference*: [2].

***Solution* 5. csch(*n*)**

$$\psi(n,\, t) = A_0 \operatorname{csch}[A_1\, (n - n_0)]\, e^{-i\, [A_2\, (t - t_0) + \phi_0]}, \tag{8.6}$$

where

$A_0 = \frac{\sinh(A_1)}{\sqrt{-\mu}}$,
$A_2 = \frac{2\,\mu - a_2}{\mu}$,
$a_2 = 2\,\mu \cosh(A_1)$,
$\mu < 0$, A_1, t_0, n_0, and ϕ_0 are arbitrary real constants.

- *Reference*: [3].

***Solution* 6. tanh(*n*)** *discrete dark soliton*
(Figure 8.2)

$$\psi(n,\, t) = A_0 \tanh[A_1\, (n - n_0)]\, e^{-i\, [A_2\, (t - t_0) + \phi_0]}, \tag{8.7}$$

where

$A_0 = \frac{\tanh(A_1)}{\sqrt{-\mu}}$,
$A_2 = \frac{2\,\mu - a_2}{\mu}$,
$a_2 = 2\,\mu\, \operatorname{sech}^2(A_1)$,
$\mu < 0$,
A_1, t_0, n_0, and ϕ_0 are arbitrary real constants.

- *Reference*: [3].

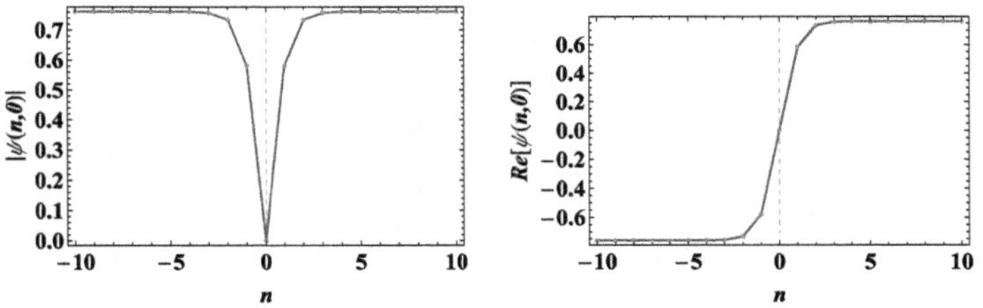

(a) (b)

Figure 8.2. Discrete dark soliton (8.7) at $t = 0$. (a) Absolute value, (b) Real part, with $A_1 = 1$, $\mu = -1$, and $n_0 = t_0 = \phi_0 = 0$. The lines are guide for the eye.

***Solution* 7. coth(*n*)**

$$\psi(n,\, t) = A_0 \coth[A_1\,(n - n_0)]\, e^{-i\,[A_2\,(t-t_0)+\phi_0]}, \tag{8.8}$$

where

$A_0 = \frac{\tanh(A_1)}{\sqrt{-\mu}}$,
$A_2 = \frac{2\,\mu - a_2}{\mu}$,
$a_2 = 2\,\mu\,\mathrm{sech}^2(A_1)$,
$\mu < 0$, A_1, t_0, n_0, and ϕ_0 are arbitrary real constants.

- *Reference*: [3].

***Solution* 8. sn(*n*,*m*)** *discrete solitary wave (SW)* (Figure 8.3)

$$\psi(n,\, t) = A_0\, \mathrm{sn}[A_1\,(n - n_0),\, m]\, e^{-i\,[A_2\,(t-t_0)+\phi_0]}, \tag{8.9}$$

where

$A_0 = \frac{\sqrt{m}\,\mathrm{sn}(A_1, m)}{\sqrt{-\mu}}$,
$A_2 = \frac{2\,\mu - a_2}{\mu}$,
$a_2 = 2\,\mu\,\mathrm{cn}(A_1,\, m)\,\mathrm{dn}(A_1,\, m)$,
$\mu < 0$,
$0 \leqslant m \leqslant 1$,
A_1, t_0, n_0, and ϕ_0 are arbitrary real constants.

- *Reference*: [3], *we corrected the expression of* A_0.

***Solution* 9. cn(*n*,*m*)** *discrete SW* (Figure 8.4)

$$\psi(n,\, t) = A_0\, \mathrm{cn}[A_1\,(n - n_0),\, m]\, e^{-i\,[A_2\,(t-t_0)+\phi_0]}, \tag{8.10}$$

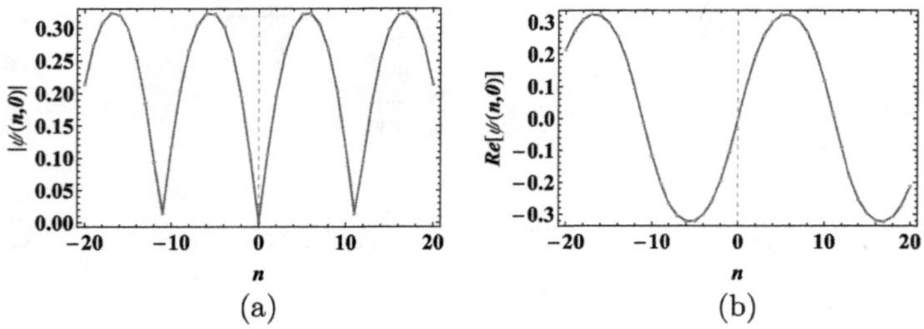

Figure 8.3. Discrete solitary wave (8.9) at $t = 0$. (a) Absolute value, (b) real part, with $A_1 = 1/3$, $\mu = -1/2$, $m = 1/2$, and $n_0 = t_0 = \phi_0 = 0$. The lines are guides for the eye.

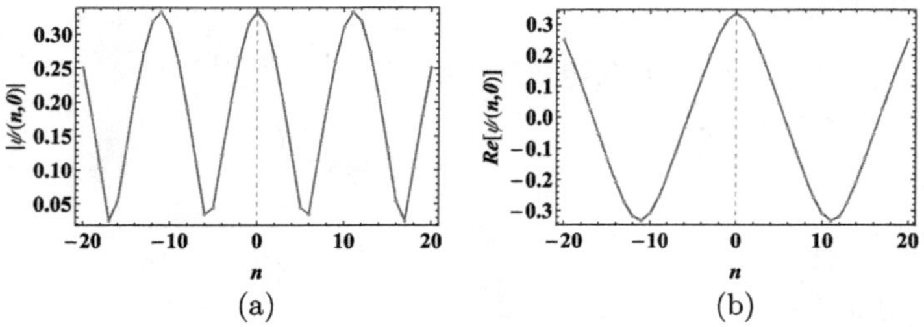

Figure 8.4. Discrete solitary wave (8.10) at $t = 0$. (a) Absolute value, (b) real part, with $a_2 = 1$, $A_1 = 1/3$, $m = 1/2$, and $n_0 = t_0 = \phi_0 = 0$. The lines are guides for the eye.

where

$$A_0 = \frac{\sqrt{m}\,\mathrm{sn}(A_1, m)}{\sqrt{\mu}\,\mathrm{dn}(A_1, m)},$$

$$A_2 = \frac{2\mu - a_2}{\mu},$$

$$\mu = \frac{a_2\,\mathrm{dn}^2(A_1, m)}{2\,\mathrm{cn}(A_1, m)} > 0,$$

$0 \leqslant m \leqslant 1$,

A_1, t_0, n_0, and ϕ_0 are arbitrary real constants.

- *Reference*: [2].

***Solution* 10. dn(*n*,*m*)** *discrete SW*
(Figure 8.5)

$$\psi(n, t) = A_0\,\mathrm{dn}[A_1(n - n_0), m]\, e^{-i\,[A_2(t-t_0)+\phi_0]}, \tag{8.11}$$

where

$$A_0 = \frac{\mathrm{sn}(A_1, m)}{\sqrt{\mu}\,\mathrm{cn}(A_1, m)},$$

$$A_2 = \frac{2\mu - a_2}{\mu},$$

$$\mu = \frac{a_2\,\mathrm{cn}^2(A_1, m)}{2\,\mathrm{dn}(A_1, m)} > 0,$$

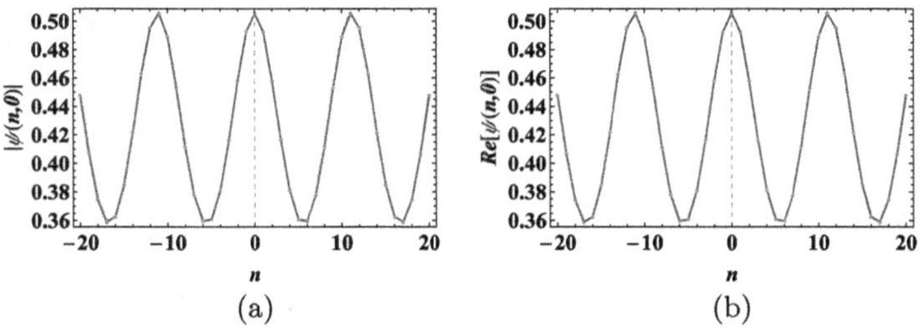

Figure 8.5. Discrete solitary wave (8.11) at $t = 0$. (a) Absolute value, (b) real part, with $a_2 = 1$, $A_1 = 1/3$, $m = 1/2$, and $n_0 = t_0 = \phi_0 = 0$. The lines are guides for the eye.

$0 \leqslant m \leqslant 1$,
A_1, t_0, n_0, and ϕ_0 are arbitrary real constants.

- *Reference*: [2].

***Solution* 11. ns(*n*,*m*)** *discrete SW*

$$\psi(n,\, t) = A_0 \,\mathrm{ns}[A_1\,(n - n_0),\, m]\, e^{-i\,[A_2\,(t-t_0)+\phi_0]}, \tag{8.12}$$

where

$A_0 = \dfrac{\mathrm{sn}(A_1, m)}{\sqrt{-\mu}}$,
$A_2 = \dfrac{2\,\mu - a_2}{\mu}$,
$a_2 = 2\,\mu\, \mathrm{dn}(A_1,\, m)\, \mathrm{cn}(A_1,\, m)$,
$\mu < 0$,
$0 \leqslant m \leqslant 1$,
A_1, t_0, n_0, and ϕ_0 are arbitrary real constants.

- *Reference*: [3].

***Solution* 12. cs(*n*,*m*)** *discrete SW*

$$\psi(n,\, t) = A_0 \,\mathrm{cs}[A_1\,(n - n_0),\, m]\, e^{-i\,[A_2\,(t-t_0)+\phi_0]}, \tag{8.13}$$

where

$A_0 = \dfrac{\mathrm{sn}(A_1, m)}{\sqrt{-\mu}\;\mathrm{cn}(A_1, m)}$,
$A_2 = \dfrac{2\,\mu - a_2}{\mu}$,
$a_2 = \dfrac{2\,\mu\, \mathrm{dn}(A_1, m)}{\mathrm{cn}^2(A_1, m)}$,
$\mu < 0$,
$0 \leqslant m \leqslant 1$,
A_1, t_0, n_0, and ϕ_0 are arbitrary real constants.

- *Reference*: [3].

***Solution* 13. ds(*n*,*m*)** *discrete SW*

$$\psi(n,\ t) = A_0\ \mathrm{ds}[A_1\ (n - n_0),\ m]\ e^{-i\ [A_2\ (t-t_0)+\phi_0]}, \tag{8.14}$$

where

$A_0 = \frac{\mathrm{sn}(A_1, m)}{\sqrt{-\mu}\ \mathrm{dn}(A_1, m)}$,
$A_2 = \frac{2\,\mu - a_2}{\mu}$,
$a_2 = \frac{2\,\mu\ \mathrm{cn}(A_1, m)}{\mathrm{dn}^2(A_1, m)}$,
$\mu < 0$,
$0 \leqslant m \leqslant 1$,
A_1, t_0, n_0, and ϕ_0 are arbitrary real constants.

- *Reference*: [3].

***Solution* 14. cd(*n*,*m*)** *discrete SW*

$$\psi(n,\ t) = A_0\ \mathrm{cd}[A_1\ (n - n_0),\ m]\ e^{-i\ [A_2\ (t-t_0)+\phi_0]}, \tag{8.15}$$

where

$A_0 = \frac{\sqrt{m}\ \mathrm{sn}(A_1, m)}{\sqrt{-\mu}}$,
$A_2 = \frac{2\,\mu - a_2}{\mu}$,
$a_2 = 2\,\mu\ \mathrm{cn}(A_1,\ m)\ \mathrm{dn}(A_1,\ m)$,
$\mu < 0$,
$0 \leqslant m \leqslant 1$,
A_1, t_0, n_0, and ϕ_0 are arbitrary real constants.

- *Reference*: [3], *we corrected the expression of* A_0.

***Solution* 15. dc(*n*,*m*)** *discrete SW*

$$\psi(n,\ t) = A_0\ \mathrm{dc}[A_1\ (n - n_0),\ m]\ e^{-i\ [A_2\ (t-t_0)+\phi_0]}, \tag{8.16}$$

where

$A_0 = \frac{\mathrm{sn}(A_1, m)}{\sqrt{-\mu}}$,
$A_2 = \frac{2\,\mu - a_2}{\mu}$,
$a_2 = 2\,\mu\ \mathrm{cn}(A_1,\ m)\ \mathrm{dn}(A_1,\ m)$,
$\mu < 0$,
$0 \leqslant m \leqslant 1$,
A_1, t_0, n_0, and ϕ_0 are arbitrary real constants.

- *Reference*: [3].

***Solution* 16. dn(*n*,*m*) + cn(*n*,*m*)** *discrete SW*

$$\psi(n,\, t) = \left\{\frac{A_0}{2}\,\mathrm{dn}[A_1\,(n - n_0),\, m] + \frac{B_0}{2}\,\sqrt{m}\;\mathrm{cn}[A_1\,(n - n_0),\, m]\right\} \times e^{-i\,[A_2\,(t-t_0)+\phi_0]}, \quad (8.17)$$

where

$A_0 = \frac{2}{\mathrm{cs}(A_1, m) + \mathrm{ds}(A_1, m)}$,
$B_0 = \pm A_0$,
$A_2 = 2 - a_2$,
$a_2 = \frac{4}{\mathrm{cn}(A_1, m) + \mathrm{dn}(A_1, m)}$,
$\mu = 1$,
$0 \leqslant m \leqslant 1$,
A_1, t_0, and ϕ_0 are arbitrary real constants,
n_0 is an arbitrary real integer.

- *Reference*: [1], *taken from the nonlocal case.*

8.1.2 Staggered Solutions

If $\psi(n,\, t;\, a_2)$ is a nonstaggered solution of (8.1), then

$$\psi_s(n,\, t,\, a_2) = (-1)^n\, \psi^*(n,\, t,\, -a_2)\, e^{-4\,i\,(t-t_0)} \quad (8.18)$$

is a staggered solution of the same equation, where ψ^* is the complex conjugate of the nonstaggered solution.

***Solution* 1. Constant Amplitude** *staggered discrete CW, t- and n-dependent phase*

$$\psi_s(n,\, t) = (-1)^n\, A_0\, e^{-i[A_1\,(n-n_0)-(A_2-4)\,(t-t_0)+\phi_0]}, \quad (8.19)$$

where

$A_2 = 4\sin^2(A_1/2) + \frac{a_2\, A_0^2}{1+\mu\, A_0^2}$,
A_0, A_1, t_0, n_0, and ϕ_0 are arbitrary real constants.

***Solution* 2. cos(*n*) I**

$$\psi_s(n,\, t) = A_0\cos[\pi\,(n - n_0)]\, e^{-i\,[A_2\,(t-t_0)+\phi_0]}, \quad (8.20)$$

where

$A_2 = 4 - \frac{a_2\, A_0^2}{1+\mu\, A_0^2}$,
A_0, t_0, n_0, and ϕ_0 are arbitrary real constants.

- *Reference*: [4].

***Solution* 3. cos(*n*) II**

$$\psi_s(n, t) = A_0 \cos\left[\frac{\pi}{2}(n - n_0)\right] e^{\pm i\,[A_2\,(t-t_0)+\phi_0]}, \tag{8.21}$$

where

$A_2 = -2 + \frac{A_0^2\, a_2}{A_0^2\, \mu + \sec^2[\pi\,(n - n_0)/2]}$,

A_0, t_0, n_0, and ϕ_0 are arbitrary real constants.

- *Reference*: [4].

***Solution* 4. cos(*n*)-sin(*n*)**

$$\psi_s(n, t) = A_0 \left\{\cos\left[\frac{\pi}{2}(n - n_0)\right] - \sin\left[\frac{\pi}{2}(n - n_0)\right]\right\} e^{-i\,[A_2\,(t-t_0)+\phi_0]}, \tag{8.22}$$

where

$A_2 = 2 - \frac{a_2\, A_0^2}{1 + \mu\, A_0^2}$,

A_0, t_0, and ϕ_0 are arbitrary real constants,

n_0 is an arbitrary real integer.

- *Reference*: [4].

***Solution* 5. sec(*n*)**

$$\psi_s(n, t) = (-1)^n A_0 \sec[A_1\,(n - n_0)]\, e^{i\,[A_2\,(t-t_0)-4\,(t-t_0)+\phi_0]}, \tag{8.23}$$

where

$A_0 = \frac{\sin(A_1)}{\sqrt{-\mu}}$,

$A_2 = \frac{2\,\mu + a_2}{\mu}$,

$a_2 = -2\,\mu \cos(A_1)$,

$\mu < 0$,

A_1, t_0, n_0, and ϕ_0 are arbitrary real constants.

***Solution* 6. tan(*n*)**

$$\psi_s(n, t) = (-1)^n A_0 \tan[A_1\,(n - n_0)]\, e^{i\,[A_2\,(t-t_0)-4\,(t-t_0)+\phi_0]}, \tag{8.24}$$

where

$A_0 = \frac{\tan(A_1)}{\sqrt{-\mu}}$,

$A_2 = 2 - 2\sec^2(A_1)$,

$a_2 = -2\,\mu \sec^2(A_1)$,

$\mu < 0$,

A_1, t_0, n_0, and ϕ_0 are arbitrary real constants.

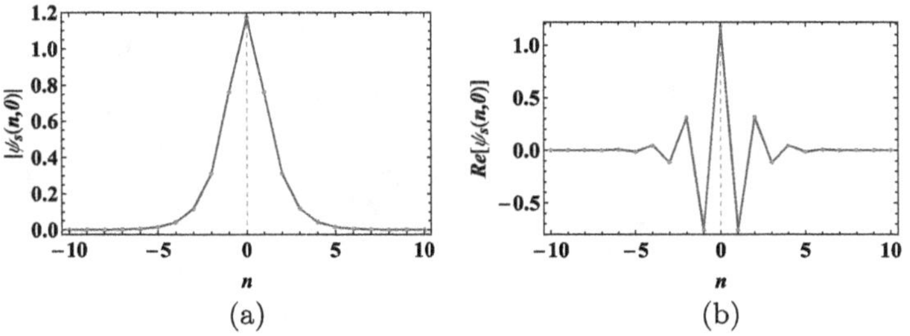

Figure 8.6. Staggered discrete bright soliton (8.25) at $t = 0$. (a) Absolute value, (b) real part, with $a_2 = -1$, $A_1 = 1$, and $n_0 = t_0 = \phi_0 = 0$. The lines are guides for the eye.

***Solution* 7. sech(n)** *staggered discrete bright soliton*
(Figure 8.6)

$$\psi_s(n, t) = (-1)^n A_0 \operatorname{sech}[A_1 (n - n_0)] e^{i [A_2 (t-t_0)-4 (t-t_0)+\phi_0]}, \tag{8.25}$$

where

$A_0 = \frac{\sinh(A_1)}{\sqrt{\mu}}$,
$A_2 = \frac{2 \mu + a_2}{\mu}$,
$\mu = -\frac{a_2 \operatorname{sech}(A_1)}{2} > 0$,
A_1, t_0, n_0, and ϕ_0 are arbitrary real constants.

***Solution* 8. csch(n)**

$$\psi_s(n, t) = (-1)^n A_0 \operatorname{csch}[A_1 (n - n_0)] e^{i [A_2 (t-t_0)-4 (t-t_0)+\phi_0]}, \tag{8.26}$$

where

$A_0 = \frac{\sinh(A_1)}{\sqrt{-\mu}}$,
$A_2 = \frac{2 \mu + a_2}{\mu}$,
$a_2 = -2 \mu \cosh(A_1)$,
$\mu < 0$,
A_1, t_0, n_0, and ϕ_0 are arbitrary real constants.

***Solution* 9. tanh(n)** *staggered discrete dark soliton*
(Figure 8.7)

$$\psi_s(n, t) = (-1)^n A_0 \tanh[A_1 (n - n_0)] e^{i [A_2 (t-t_0)-4 (t-t_0)+\phi_0]}, \tag{8.27}$$

where

$A_0 = \frac{\tanh(A_1)}{\sqrt{-\mu}}$,
$A_2 = \frac{2 \mu + a_2}{\mu}$,
$a_2 = -2 \mu \operatorname{sech}^2(A_1)$,

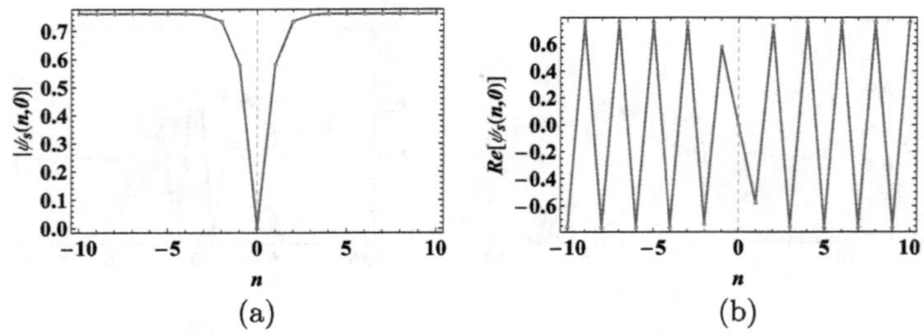

Figure 8.7. Staggered discrete dark soliton (8.27) at $t = 0$. (a) Absolute value, (b) real part, with $A_1 = 1$, $\mu = -1$, and $n_0 = t_0 = \phi_0 = 0$. The lines are guides for the eye.

$\mu < 0$,
A_1, t_0, n_0, and ϕ_0 are arbitrary real constants.

***Solution* 10. coth(*n*)**

$$\psi_s(n, t) = (-1)^n A_0 \coth[A_1 (n - n_0)]\, e^{i\,[A_2 (t-t_0) - 4 (t-t_0) + \phi_0]}, \tag{8.28}$$

where
$A_0 = \frac{\tanh(A_1)}{\sqrt{-\mu}}$,
$A_2 = \frac{2\mu + a_2}{\mu}$,
$a_2 = -2\mu \operatorname{sech}^2(A_1)$,
$\mu < 0$,
A_1, t_0, n_0, and ϕ_0 are arbitrary real constants.

***Solution* 11. sn(*n*,*m*)** *staggered discrete SW*
(Figure 8.8)

$$\psi_s(n, t) = (-1)^n A_0 \operatorname{sn}[A_1 (n - n_0), m]\, e^{i\,[A_2 (t-t_0) - 4 (t-t_0) + \phi_0]}, \tag{8.29}$$

where
$A_0 = \frac{\sqrt{m}\, \operatorname{sn}(A_1, m)}{\sqrt{-\mu}}$,
$A_2 = \frac{2\mu + a_2}{\mu}$,
$a_2 = -2\mu \operatorname{cn}(A_1, m) \operatorname{dn}(A_1, m)$,
$\mu < 0$,
$0 \leqslant m \leqslant 1$,
A_1, t_0, n_0, and ϕ_0 are arbitrary real constants.

***Solution* 12. cn(*n*,*m*)** *staggered discrete SW*
(Figure 8.9)

$$\psi_s(n, t) = (-1)^n A_0 \operatorname{cn}[A_1 (n - n_0), m]\, e^{i\,[A_2 (t-t_0) - 4 (t-t_0) + \phi_0]}, \tag{8.30}$$

where
$A_0 = \frac{\sqrt{m}\, \operatorname{sn}(A_1, m)}{\sqrt{\mu}\, \operatorname{dn}(A_1, m)}$,

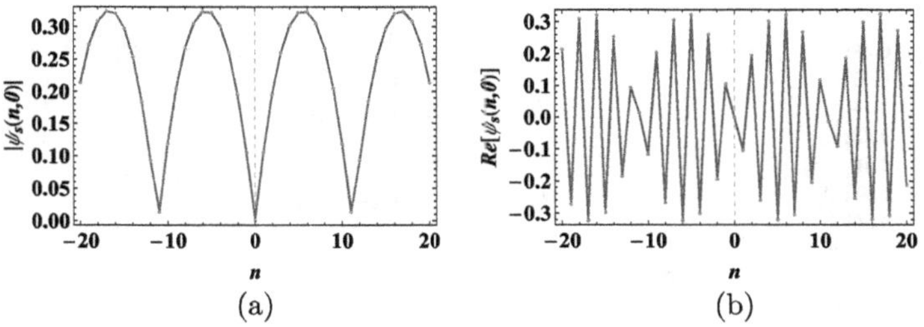

Figure 8.8. Staggered discrete solitary wave (8.29) at $t = 0$. (a) Absolute value, (b) real part, with $A_1 = 1/3$, $\mu = -1/2$, $m = 1/2$, and $n_0 = t_0 = \phi_0 = 0$. The lines are guides for the eye.

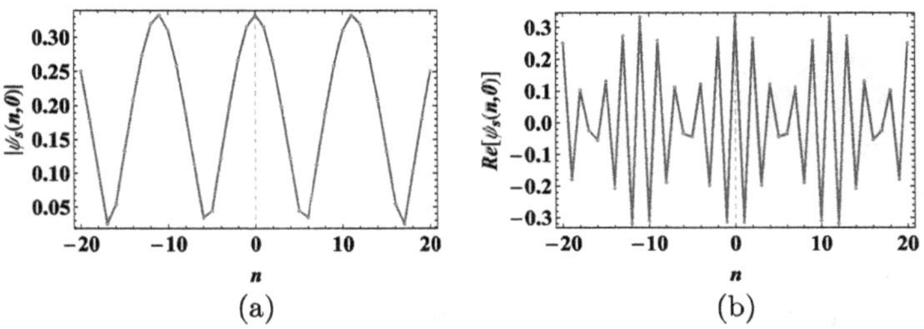

Figure 8.9. Staggered discrete solitary wave (8.30) at $t = 0$. (a) Absolute value, (b) real part, with $a_2 = -1$, $A_1 = 1/3$, $m = 1/2$, and $n_0 = t_0 = \phi_0 = 0$. The lines are guides for the eye.

$$A_2 = \frac{2\mu + a_2}{\mu},$$
$$\mu = -\frac{a_2\,\mathrm{dn}^2(A_1, m)}{2\,\mathrm{cn}(A_1, m)} > 0,$$
$$0 \leqslant m \leqslant 1,$$
A_1, t_0, n_0, and ϕ_0 are arbitrary real constants.

***Solution* 13. dn(*n*,*m*)** *staggered discrete SW* (Figure 8.10)

$$\psi_s(n, t) = (-1)^n A_0\,\mathrm{dn}[A_1(n - n_0), m]\, e^{i\,[A_2(t-t_0) - 4(t-t_0) + \phi_0]}, \tag{8.31}$$

where
$$A_0 = \frac{\mathrm{sn}(A_1, m)}{\sqrt{\mu}\,\mathrm{cn}(A_1, m)},$$
$$A_2 = \frac{2\mu + a_2}{\mu},$$
$$\mu = -\frac{a_2\,\mathrm{cn}^2(A_1, m)}{2\,\mathrm{dn}(A_1, m)} > 0,$$
$$0 < m \leqslant 1,$$
A_1, t_0, n_0, and ϕ_0 are arbitrary real constants.

***Solution* 14. ns(*n*,*m*)** *staggered discrete SW*

$$\psi_s(n, t) = (-1)^n A_0\,\mathrm{ns}[A_1(n - n_0), m]\, e^{i\,[A_2(t-t_0) - 4(t-t_0) + \phi_0]}, \tag{8.32}$$

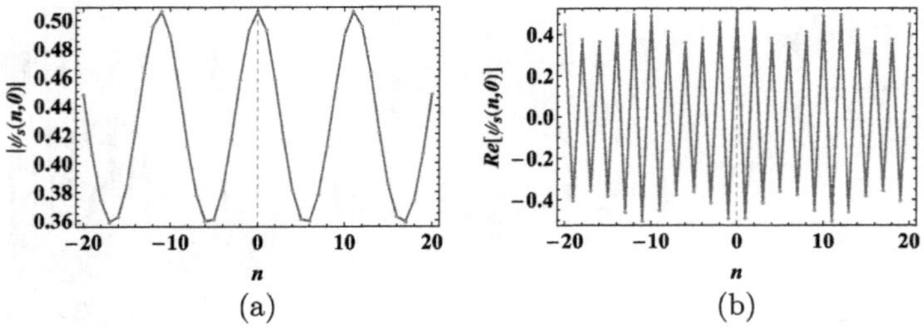

Figure 8.10. Staggered discrete solitary wave (8.31) at $t = 0$. (a) Absolute value, (b) real part, with $a_2 = -1$, $A_1 = 1/3$, $m = 1/2$, and $n_0 = t_0 = \phi_0 = 0$. The lines are guides for the eye.

where

$$A_0 = \frac{\mathrm{sn}(A_1, m)}{\sqrt{-\mu}},$$
$$A_2 = \frac{2\mu + a_2}{\mu},$$
$$a_2 = -2\mu\, \mathrm{dn}(A_1, m)\, \mathrm{cn}(A_1, m),$$
$$\mu < 0,$$
$$0 \leqslant m \leqslant 1,$$

A_1, t_0, n_0, and ϕ_0 are arbitrary real constants.

***Solution* 15. cs(*n*,*m*)** *staggered discrete SW*

$$\psi_s(n, t) = (-1)^n A_0\, \mathrm{cs}[A_1 (n - n_0), m]\, e^{i\,[A_2 (t-t_0) - 4(t-t_0) + \phi_0]}, \tag{8.33}$$

where

$$A_0 = \frac{\mathrm{sn}(A_1, m)}{\sqrt{-\mu}\, \mathrm{cn}(A_1, m)},$$
$$A_2 = \frac{2\mu + a_2}{\mu},$$
$$a_2 = -\frac{2\mu\, \mathrm{dn}(A_1, m)}{\mathrm{cn}^2(A_1, m)},$$
$$\mu < 0,$$
$$0 \leqslant m \leqslant 1,$$

A_1, t_0, n_0, and ϕ_0 are arbitrary real constants.

***Solution* 16. ds(*n*,*m*)** *staggered discrete SW*

$$\psi_s(n, t) = (-1)^n A_0\, \mathrm{ds}[A_1 (n - n_0), m]\, e^{i\,[A_2 (t-t_0) - 4(t-t_0) + \phi_0]}, \tag{8.34}$$

where

$$A_0 = \frac{\mathrm{sn}(A_1, m)}{\sqrt{-\mu}\, \mathrm{dn}(A_1, m)},$$
$$A_2 = \frac{2\mu + a_2}{\mu},$$
$$a_2 = -\frac{2\mu\, \mathrm{cn}(A_1, m)}{\mathrm{dn}^2(A_1, m)},$$
$$\mu < 0,$$

$0 \leqslant m \leqslant 1$,
A_1, t_0, n_0, and ϕ_0 are arbitrary real constants.

***Solution* 17. cd(*n*,*m*)** *staggered discrete SW*

$$\psi_s(n,\, t) = (-1)^n\, A_0\, \mathrm{cd}[A_1\,(n - n_0),\, m]\, e^{i\,[A_2\,(t-t_0)-4\,(t-t_0)+\phi_0]}, \tag{8.35}$$

where

$A_0 = \frac{\sqrt{m}\ \mathrm{sn}(A_1, m)}{\sqrt{-\mu}}$,
$A_2 = \frac{2\,\mu + a_2}{\mu}$,
$a_2 = -2\,\mu\ \mathrm{cn}(A_1,\, m)\ \mathrm{dn}(A_1,\, m)$,
$\mu < 0$,
$0 \leqslant m \leqslant 1$,
A_1, t_0, n_0, and ϕ_0 are arbitrary real constants.

***Solution* 18. dc(*n*,*m*)** *staggered discrete SW*

$$\psi_s(n,\, t) = (-1)^n\, A_0\, \mathrm{dc}[A_1\,(n - n_0),\, m]\, e^{i\,[A_2\,(t-t_0)-4\,(t-t_0)+\phi_0]}, \tag{8.36}$$

where

$A_0 = \frac{\mathrm{sn}(A_1, m)}{\sqrt{-\mu}}$,
$A_2 = \frac{2\,\mu + a_2}{\mu}$,
$a_2 = -2\,\mu\ \mathrm{cn}(A_1,\, m)\ \mathrm{dn}(A_1,\, m)$,
$\mu < 0$,
$0 \leqslant m \leqslant 1$,
A_1, t_0, n_0, and ϕ_0 are arbitrary real constants.

8.2 Summary of Section 8.1

Nonstaggered Solutions

Equation: $i\,\psi_{n_t} + \psi_{n+1} + \psi_{n-1} - 2\,\psi_n + \frac{a_2\,|\psi_n|^2\,\psi_n}{1+\mu\,|\psi_n|^2} = 0$

#	Solution	Conditions	Name	Eq. #
1.	$\psi(n,\,t) = A_0\,e^{i[A_1\,(n-n_0)-A_2\,(t-t_0)+\phi_0]}$	$A_2 = 4\,\sin^2(A_1/2) - \frac{a_2\,A_0^2}{1+\mu\,A_0^2}$, A_0, A_1, t_0, and ϕ_0 are arbitrary real constants, n_0 is an arbitrary real integer	discrete constant wave, t- and n-dependent phase	(8.2)
2.	$\psi(n,\,t) = A_0\,\sec[A_1\,(n-n_0)]\,e^{-i\,[A_2\,(t-t_0)+\phi_0]}$	$A_0 = \frac{\sin(A_1)}{\sqrt{-\mu}}$, $A_2 = \frac{2\,\mu-a_2}{\mu}$, $a_2 = 2\,\mu\,\cos(A_1)$, $\mu < 0$, A_1, t_0, and ϕ_0 are arbitrary real constants, n_0 is an arbitrary real integer	–	(8.3)
3.	$\psi(n,\,t) = A_0\,\tan[A_1\,(n-n_0)]\,e^{-i\,[A_2\,(t-t_0)+\phi_0]}$	$A_0 = \frac{\tan(A_1)}{\sqrt{-\mu}}$, $A_2 = 2 - 2\,\sec^2(A_1)$, $a_2 = 2\,\mu\,\sec^2(A_1)$, $\mu < 0$, A_1, t_0, ϕ_0 are arbitrary real constants, n_0 is an arbitrary real integer	–	(8.4)
4.	$\psi(n,\,t) = A_0\,\mathrm{sech}[A_1\,(n-n_0)]\,e^{-i\,[A_2\,(t-t_0)+\phi_0]}$	$A_0 = \frac{\sinh(A_1)}{\sqrt{\mu}}$, $A_2 = \frac{2\,\mu-a_2}{\mu}$, $\mu = \frac{a_2\,\mathrm{sech}(A_1)}{2} > 0$, A_1, t_0, and ϕ_0 are arbitrary real constants, n_0 is an arbitrary real integer	discrete bright soliton	(8.5)
5.	$\psi(n,\,t) = A_0\,\mathrm{csch}[A_1\,(n-n_0)]\,e^{-i\,[A_2\,(t-t_0)+\phi_0]}$	$A_0 = \frac{\sinh(A_1)}{\sqrt{-\mu}}$, $A_2 = \frac{2\,\mu-a_2}{\mu}$, $a_2 = 2\,\mu\,\cosh(A_1)$, $\mu < 0$, A_1, t_0, and ϕ_0 are arbitrary real constants, n_0 is an arbitrary real integer	–	(8.6)

6. $\psi(n, t) = A_0 \tanh[A_1 (n - n_0)] \, e^{-i \, [A_2 (t-t_0)+\phi_0]}$	$A_0 = \frac{\tanh(A_1)}{\sqrt{-\mu}}$, $A_2 = \frac{2\mu - a_2}{\mu}$, $a_2 = 2 \mu \operatorname{sech}^2(A_1)$, $\mu < 0$, A_1, t_0, and ϕ_0 are arbitrary real constants, n_0 is an arbitrary real integer	discrete dark soliton	(8.7)
7. $\psi(n, t) = A_0 \coth[A_1 (n - n_0)] \, e^{-i \, [A_2 (t-t_0)+\phi_0]}$	$A_0 = \frac{\tanh(A_1)}{\sqrt{-\mu}}$, $A_2 = \frac{2\mu - a_2}{\mu}$, $a_2 = 2 \mu \operatorname{sech}^2(A_1)$, $\mu < 0$, A_1, t_0, and ϕ_0 are arbitrary real constants, n_0 is an arbitrary real integer	–	(8.8)
8. $\psi(n, t) = A_0 \operatorname{sn}[A_1 (n - n_0), m] \, e^{-i \, [A_2 (t-t_0)+\phi_0]}$	$A_0 = \frac{\sqrt{m} \operatorname{sn}(A_1, m)}{\sqrt{-\mu}}$, $A_2 = \frac{2\mu - a_2}{\mu}$, $a_2 = 2 \mu \operatorname{cn}(A_1, m) \operatorname{dn}(A_1, m)$, $\mu < 0, 0 \leqslant m \leqslant 1$, A_1, t_0, and ϕ_0 are arbitrary real constants, n_0 is an arbitrary real integer	discrete solitary wave	(8.9)
9. $\psi(n, t) = A_0 \operatorname{cn}[A_1 (n - n_0), m] \, e^{-i \, [A_2 (t-t_0)+\phi_0]}$	$A_0 = \frac{\sqrt{m} \operatorname{sn}(A_1, m)}{\sqrt{\mu} \operatorname{dn}(A_1, m)}$, $A_2 = \frac{2\mu - a_2}{\mu}$, $\mu = \frac{a_2 \operatorname{dn}^2(A_1, m)}{2 \operatorname{cn}(A_1, m)} > 0, 0 \leqslant m \leqslant 1$, A_1, t_0, and ϕ_0 are arbitrary real constants, n_0 is an arbitrary real integer	discrete solitary wave	(8.10)
10. $\psi(n, t) = A_0 \operatorname{dn}[A_1 (n - n_0), m] \, e^{-i \, [A_2 (t-t_0)+\phi_0]}$	$A_0 = \frac{\operatorname{sn}(A_1, m)}{\sqrt{\mu} \operatorname{cn}(A_1, m)}$, $A_2 = \frac{2\mu - a_2}{\mu}$, $\mu = \frac{a_2 \operatorname{cn}^2(A_1, m)}{2 \operatorname{dn}(A_1, m)} > 0, 0 \leqslant m \leqslant 1$, A_1, t_0, and ϕ_0 are arbitrary real constants, n_0 is an arbitrary real integer	discrete solitary wave	(8.11)
11. $\psi(n, t) = A_0 \operatorname{ns}[A_1 (n - n_0), m] \, e^{-i \, [A_2 (t-t_0)+\phi_0]}$	$A_0 = \frac{\operatorname{sn}(A_1, m)}{\sqrt{-\mu}}$, $A_2 = \frac{2\mu - a_2}{\mu}$, $a_2 = 2 \mu \operatorname{dn}(A_1, m) \operatorname{cn}(A_1, m)$, $\mu < 0, 0 \leqslant m \leqslant 1$, A_1, t_0, and ϕ_0 are arbitrary real constants, n_0 is an arbitrary real integer	discrete solitary wave	(8.12)

(*Continued*)

12. $\psi(n,t) = A_0\,\mathrm{cs}[A_1(n-n_0), m]\, e^{-i[A_2(t-t_0)+\phi_0]}$	$A_0 = \frac{\mathrm{sn}(A_1,m)}{\sqrt{-\mu}\,\mathrm{cn}(A_1,m)}$, $A_2 = \frac{2\mu - a_2}{\mu}$, $a_2 = \frac{2\mu\,\mathrm{dn}(A_1,m)}{\mathrm{cn}^2(A_1,m)}$, $\mu < 0,\ 0 \leqslant m \leqslant 1$, A_1, t_0, and ϕ_0 are arbitrary real constants, n_0 is an arbitrary real integer	discrete solitary wave	(8.13)
13. $\psi(n,t) = A_0\,\mathrm{ds}[A_1(n-n_0), m]\, e^{-i[A_2(t-t_0)+\phi_0]}$	$A_0 = \frac{\mathrm{sn}(A_1,m)}{\sqrt{-\mu}\,\mathrm{dn}(A_1,m)}$, $A_2 = \frac{2\mu - a_2}{\mu}$, $a_2 = \frac{2\mu\,\mathrm{cn}(A_1,m)}{\mathrm{dn}^2(A_1,m)}$, $\mu < 0,\ 0 \leqslant m \leqslant 1$, A_1, t_0, and ϕ_0 are arbitrary real constants, n_0 is an arbitrary real integer	discrete solitary wave	(8.14)
14. $\psi(n,t) = A_0\,\mathrm{cd}[A_1(n-n_0), m]\, e^{-i[A_2(t-t_0)+\phi_0]}$	$A_0 = \frac{\sqrt{m}\,\mathrm{sn}(A_1,m)}{\sqrt{-\mu}}$, $A_2 = \frac{2\mu - a_2}{\mu}$, $a_2 = 2\mu\,\mathrm{cn}(A_1,m)\,\mathrm{dn}(A_1,m)$, $\mu < 0,\ 0 \leqslant m \leqslant 1$, A_1, t_0, and ϕ_0 are arbitrary real constants, n_0 is an arbitrary real integer	discrete solitary wave	(8.15)
15. $\psi(n,t) = A_0\,\mathrm{dc}[A_1(n-n_0), m]\, e^{-i[A_2(t-t_0)+\phi_0]}$	$A_0 = \frac{\mathrm{sn}(A_1,m)}{\sqrt{-\mu}}$, $A_2 = \frac{2\mu - a_2}{\mu}$, $a_2 = 2\mu\,\mathrm{cn}(A_1,m)\,\mathrm{dn}(A_1,m)$, $\mu < 0,\ 0 \leqslant m \leqslant 1$, A_1, t_0, and ϕ_0 are arbitrary real constants, n_0 is an arbitrary real integer	discrete solitary wave	(8.16)
16. $\psi(n,t) = \left\{\frac{A_0}{2}\,\mathrm{dn}[A_1(n-n_0), m] + \frac{B_0}{2}\sqrt{m}\,\mathrm{cn}[A_1(n-n_0), m]\right\} \times e^{-i[A_2(t-t_0)+\phi_0]}$	$A_0 = \frac{2}{\mathrm{cs}(A_1,m)+\mathrm{ds}(A_1,m)}$, $B_0 = \pm A_0$, $A_2 = 2 - a_2$, $a_2 = \frac{4}{\mathrm{cn}(A_1,m)+\mathrm{dn}(A_1,m)}$, $\mu = 1,\ 0 \leqslant m \leqslant 1$, A_1, t_0, and ϕ_0 are arbitrary real constants, n_0 is an arbitrary real integer	discrete solitary wave	(8.17)

Staggered Solutions: $\psi_s(n, t, a_2) = (-1)^n \psi^*(n, t, -a_2)\, e^{-4i(t-t_0)}$

#	Solution	Conditions	Name	Eq. #
1.	$\psi_s(n, t) = (-1)^n A_0\, e^{-i[A_1(n-n_0)-(A_2-4)(t-t_0)+\phi_0]}$	$A_2 = 4\sin^2(A_1/2) + \frac{a_2 A_0^2}{1+\mu A_0^2}$, A_0, A_1, t_0, ϕ_0 are arbitrary real constants, n_0 is an arbitrary real integer	staggered discrete constant wave, t- and n-dependent phase	(8.19)
2.	$\psi_s(n, t) = A_0 \cos[\pi(n-n_0)]\, e^{-i[A_2(t-t_0)+\phi_0]}$	$A_2 = 4 - \frac{a_2 A_0^2}{1+\mu A_0^2}$, A_0, t_0, and ϕ_0 are arbitrary real constants, n_0 is an arbitrary real integer	–	(8.20)
3.	$\psi_s(n, t) = A_0 \cos\left[\frac{\pi}{2}(n-n_0)\right] e^{\pm i[A_2(t-t_0)+\phi_0]}$	$A_2 = -2 + \frac{A_0^2 a_2}{A_0^2\mu + \sec^2[\pi(n-n_0)/2]}$, A_0, t_0, and ϕ_0 are arbitrary real constants, n_0 is an arbitrary real integer	–	(8.21)
4.	$\psi_s(n, t) = A_0 \left\{\cos\left[\frac{\pi}{2}(n-n_0)\right] - \sin\left[\frac{\pi}{2}(n-n_0)\right]\right\} e^{-i[A_2(t-t_0)+\phi_0]}$	$A_2 = 2 - \frac{a_2 A_0^2}{1+\mu A_0^2}$, A_0, t_0, and ϕ_0 are arbitrary real constants, n_0 is an arbitrary real integer	–	(8.22)
5.	$\psi_s(n, t) = (-1)^n A_0 \sec[A_1(n-n_0)]\, e^{i[A_2(t-t_0)-4(t-t_0)+\phi_0]}$	$A_0 = \frac{\sin(A_1)}{\sqrt{-\mu}}$, $A_2 = \frac{2\mu + a_2}{\mu}$, $a_2 = -2\mu\cos(A_1)$, $\mu < 0$, A_1, t_0, and ϕ_0 are arbitrary real constants, n_0 is an arbitrary real integer	–	(8.23)
6.	$\psi_s(n, t) = (-1)^n A_0 \tan[A_1(n-n_0)]\, e^{i[A_2(t-t_0)-4(t-t_0)+\phi_0]}$	$A_0 = \frac{\tan(A_1)}{\sqrt{-\mu}}$, $A_2 = 2 - 2\sec^2(A_1)$, $a_2 = -2\mu\sec^2(A_1)$, $\mu < 0$, A_1, t_0, and ϕ_0 are arbitrary real constants, n_0 is an arbitrary real integer	–	(8.24)

(Continued)

7. $\psi_s(n, t) = (-1)^n A_0 \operatorname{sech}[A_1 (n - n_0)] e^{i [A_2 (t-t_0) - 4 (t-t_0) + \phi_0]}$	$A_0 = \frac{\sinh(A_1)}{\sqrt{\mu}}$, $A_2 = \frac{2\mu + a_2}{\mu}$, $\mu = -\frac{a_2 \operatorname{sech}(A_1)}{2} > 0$, A_1, t_0, and ϕ_0 are arbitrary real constants, n_0 is an arbitrary real integer	staggered discrete bright soliton	(8.25)
8. $\psi_s(n, t) = (-1)^n A_0 \operatorname{csch}[A_1 (n - n_0)] e^{i [A_2 (t-t_0) - 4 (t-t_0) + \phi_0]}$	$A_0 = \frac{\sinh(A_1)}{\sqrt{-\mu}}$, $A_2 = \frac{2\mu + a_2}{\mu}$, $a_2 = -2\mu \cosh(A_1)$, $\mu < 0$, A_1, t_0, and ϕ_0 are arbitrary real constants, n_0 is an arbitrary real integer	–	(8.26)
9. $\psi_s(n, t) = (-1)^n A_0 \tanh[A_1 (n - n_0)] e^{i [A_2 (t-t_0) - 4 (t-t_0) + \phi_0]}$	$A_0 = \frac{\tanh(A_1)}{\sqrt{-\mu}}$, $A_2 = \frac{2\mu + a_2}{\mu}$, $a_2 = -2\mu \operatorname{sech}^2(A_1)$, $\mu < 0$, A_1, t_0, and ϕ_0 are arbitrary real constants, n_0 is an arbitrary real integer	staggered discrete dark soliton	(8.27)
10. $\psi_s(n, t) = (-1)^n A_0 \coth[A_1 (n - n_0)] e^{i [A_2 (t-t_0) - 4 (t-t_0) + \phi_0]}$	$A_0 = \frac{\tanh(A_1)}{\sqrt{-\mu}}$, $A_2 = \frac{2\mu + a_2}{\mu}$, $a_2 = -2\mu \operatorname{sech}^2(A_1)$, $\mu < 0$, A_1, t_0, and ϕ_0 are arbitrary real constants, n_0 is an arbitrary real integer	–	(8.28)
11. $\psi_s(n, t) = (-1)^n A_0 \operatorname{sn}[A_1 (n - n_0), m] e^{i [A_2 (t-t_0) - 4 (t-t_0) + \phi_0]}$	$A_0 = \frac{\sqrt{m}\, \operatorname{sn}(A_1, m)}{\sqrt{-\mu}}$, $A_2 = \frac{2\mu + a_2}{\mu}$, $a_2 = -2\mu \operatorname{cn}(A_1, m) \operatorname{dn}(A_1, m)$, $\mu < 0, 0 \leqslant m \leqslant 1$, A_1, t_0, and ϕ_0 are arbitrary real constants, n_0 is an arbitrary real integer	staggered discrete solitary wave	(8.29)
12. $\psi_s(n, t) = (-1)^n A_0 \operatorname{cn}[A_1 (n - n_0), m] e^{i [A_2 (t-t_0) - 4 (t-t_0) + \phi_0]}$	$A_0 = \frac{\sqrt{m}\, \operatorname{sn}(A_1, m)}{\sqrt{\mu}\, \operatorname{dn}(A_1, m)}$, $A_2 = \frac{2\mu + a_2}{\mu}$, $\mu = -\frac{a_2 \operatorname{dn}^2(A_1, m)}{2 \operatorname{cn}(A_1, m)} > 0, 0 \leqslant m \leqslant 1$, A_1, t_0, and ϕ_0 are arbitrary real constants, n_0 is an arbitrary real integer	staggered discrete solitary wave	(8.30)

13. $\psi_s(n,t) = (-1)^n A_0 \,\mathrm{dn}[A_1(n-n_0), m]\, e^{i\,[A_2(t-t_0)-4(t-t_0)+\phi_0]}$	$A_0 = \frac{\mathrm{sn}(A_1,m)}{\sqrt{\mu}\,\mathrm{cn}(A_1,m)}$, $A_2 = \frac{2\mu + a_2}{\mu}$, $\mu = -\frac{a_2\,\mathrm{cn}^2(A_1,m)}{2\,\mathrm{dn}(A_1,m)} > 0$, $0 < m \leqslant 1$, A_1, t_0, and ϕ_0 are arbitrary real constants, n_0 is an arbitrary real integer	staggered discrete solitary wave	(8.31)
14. $\psi_s(n,t) = (-1)^n A_0 \,\mathrm{ns}[A_1(n-n_0), m]\, e^{i\,[A_2(t-t_0)-4(t-t_0)+\phi_0]}$	$A_0 = \frac{\mathrm{sn}(A_1,m)}{\sqrt{-\mu}}$, $A_2 = \frac{2\mu + a_2}{\mu}$, $a_2 = -2\mu\,\mathrm{dn}(A_1,m)\,\mathrm{cn}(A_1,m)$, $\mu < 0$, $0 \leqslant m \leqslant 1$, A_1, t_0, and ϕ_0 are arbitrary real constants, n_0 is an arbitrary real integer	staggered discrete solitary wave	(8.32)
15. $\psi_s(n,t) = (-1)^n A_0 \,\mathrm{cs}[A_1(n-n_0), m]\, e^{i\,[A_2(t-t_0)-4(t-t_0)+\phi_0]}$	$A_0 = \frac{\mathrm{sn}(A_1,m)}{\sqrt{-\mu}\,\mathrm{cn}(A_1,m)}$, $A_2 = \frac{2\mu + a_2}{\mu}$, $a_2 = -\frac{2\mu\,\mathrm{dn}(A_1,m)}{\mathrm{cn}^2(A_1,m)}$, $\mu < 0$, $0 \leqslant m \leqslant 1$, A_1, t_0, and ϕ_0 are arbitrary real constants, n_0 is an arbitrary real integer	staggered discrete solitary wave	(8.33)
16. $\psi_s(n,t) = (-1)^n A_0 \,\mathrm{ds}[A_1(n-n_0), m]\, e^{i\,[A_2(t-t_0)-4(t-t_0)+\phi_0]}$	$A_0 = \frac{\mathrm{sn}(A_1,m)}{\sqrt{-\mu}\,\mathrm{dn}(A_1,m)}$, $A_2 = \frac{2\mu + a_2}{\mu}$, $a_2 = -\frac{2\mu\,\mathrm{cn}(A_1,m)}{\mathrm{dn}^2(A_1,m)}$, $\mu < 0$, $0 \leqslant m \leqslant 1$, A_1, t_0, and ϕ_0 are arbitrary real constants, n_0 is an arbitrary real integer	staggered discrete solitary wave	(8.34)
17. $\psi_s(n,t) = (-1)^n A_0 \,\mathrm{cd}[A_1(n-n_0), m]\, e^{i\,[A_2(t-t_0)-4(t-t_0)+\phi_0]}$	$A_0 = \frac{\sqrt{m}\,\mathrm{sn}(A_1,m)}{\sqrt{-\mu}}$, $A_2 = \frac{2\mu + a_2}{\mu}$, $a_2 = -2\mu\,\mathrm{cn}(A_1,m)\,\mathrm{dn}(A_1,m)$, $\mu < 0$, $0 \leqslant m \leqslant 1$, A_1, t_0, and ϕ_0 are arbitrary real constants, n_0 is an arbitrary real integer	staggered discrete solitary wave	(8.35)
18. $\psi_s(n,t) = (-1)^n A_0 \,\mathrm{dc}[A_1(n-n_0), m]\, e^{i\,[A_2(t-t_0)-4(t-t_0)+\phi_0]}$	$A_0 = \frac{\mathrm{sn}(A_1,m)}{\sqrt{-\mu}}$, $A_2 = \frac{2\mu + a_2}{\mu}$, $a_2 = -2\mu\,\mathrm{cn}(A_1,m)\,\mathrm{dn}(A_1,m)$, $\mu < 0$, $0 \leqslant m \leqslant 1$, A_1, t_0, and ϕ_0 are arbitrary real constants, n_0 is an arbitrary real integer	staggered discrete solitary wave	(8.36)

8.3 Short-period Solutions with General, Kerr, and Saturable Nonlinearities

Equations:

Case I: DNLS with General Nonlinearity (GN)

$$i\,\psi_{nt} + \psi_{n+1} + \psi_{n-1} - 2\,\psi_n + a_2\,F[|\psi_n|^2]\,\psi_n = 0, \tag{8.37}$$

Case II: DNLS with Kerr Nonlinearity (KN)

$$i\,\psi_{nt} + \psi_{n+1} + \psi_{n-1} - 2\,\psi_n + a_2\,|\psi_n|^2\,\psi_n = 0, \tag{8.38}$$

Case III: DNLS with Saturable Nonlinearity (SN)

$$i\,\psi_{nt} + \psi_{n+1} + \psi_{n-1} - 2\,\psi_n + a_2\,\frac{|\psi_n|^2\,\psi_n}{1 + \mu\,|\psi_n|^2} = 0, \tag{8.39}$$

where

a_2 and μ are real constants,
$\psi_n = \psi(n, t)$ is the complex function profile,
the integer site index, n, and t are its two independent variables,
F is a general real function.
For other solutions of Case III, see section 8.1.

General Solutions:

$$\psi(n, t) = A_0\,(\ldots,\, c_0,\, c_1,\, c_2,\, c_3,\, \ldots)\,e^{i\,[A_2\,(t-t_0)+\phi_0]}, \tag{8.40}$$

where A_0, ϕ_0, t_0, c_j, $j = 0, 1, 2, 3, \ldots$ are arbitrary real constants. For specific values of A_2, short-period solutions are obtained as summarized in table 8.1.

8.4 Ablowitz–Ladik Equation

Equation:

$$i\,\psi_{nt} + \psi_{n+1} + \psi_{n-1} - 2\,\psi_n + a_2\,(\psi_{n+1} + \psi_{n-1})\,|\psi_n|^2 = 0, \tag{8.41}$$

where a_2 is a real constant.

Solutions:

***Solution* 1. Constant Amplitude** *discrete CW, t- and n-dependent phase*

$$\psi(n, t) = A_0\,e^{i[A_1\,(n-n_0)+(A_2-2)\,(t-t_0)+\phi_0]}, \tag{8.42}$$

where

$A_2 = 2\cos(A_1)\,(1 + a_2\,A_0^2)$,
A_0, A_1, t_0, n_0, and ϕ_0 are arbitrary real constants.

Table 8.1. Short-period solutions to general, Kerr, and saturable nonlinearities of the discrete NLSE (8.37, 8.38, 8.39), where F is a general real function and, A_0, t_0, and ϕ_0 are arbitrary real constants.

Period	Nonlinearity*	Condition on A_2	Solution
1	GN	$A_2 = 0 - a_2\, F[A_0^2]$	$\psi(n, t) = A_0\, (\dots, 1, 1, 1, 1, \dots)\, e^{i\,[A_2\,(t-t_0)+\phi_0]}$
	KN	$A_2 = 0 - a_2 A_0^2$	
	SN	$A_2 = 0 - \frac{a_2\, A_0^2}{1+\mu\, A_0^2}$	
2	GN	$A_2 = 4 - a_2\, F[A_0^2]$	$\psi(n, t) = A_0\, (\dots, 1, -1, \dots)\, e^{i\,[A_2\,(t-t_0)+\phi_0]}$
	KN	$A_2 = 4 - a_2 A_0^2$	
	SN	$A_2 = 4 - \frac{a_2\, A_0^2}{1+\mu\, A_0^2}$	
3	GN	$A_2 = 3 - a_2\, F[A_0^2]$	$\psi(n, t) = A_0\, (\dots, 1, 0, -1, \dots)\, e^{i\,[A_2\,(t-t_0)+\phi_0]}$
	KN	$A_2 = 3 - a_2 A_0^2$	
	SN	$A_2 = 3 - \frac{a_2\, A_0^2}{1+\mu\, A_0^2}$	
4	GN	$A_2 = 2 - a_2\, F[A_0^2]$	$\psi(n, t) = A_0\, (\dots, 1, 1, -1, -1, \dots)\, e^{i\,[A_2\,(t-t_0)+\phi_0]}$
	KN	$A_2 = 2 - a_2 A_0^2$	
	SN	$A_2 = 2 - \frac{a_2\, A_0^2}{1+\mu\, A_0^2}$	$\psi(n, t) = A_0\, (\dots, 1, 0, -1, 0, \dots)\, e^{i\,[A_2\,(t-t_0)+\phi_0]}$
6	GN	$A_2 = 1 - a_2\, F[A_0^2]$	$\psi(n, t) = A_0\, (\dots, 1, 1, 0, -1, -1, 0, \dots)\, e^{i\,[A_2\,(t-t_0)+\phi_0]}$
	KN	$A_2 = 1 - a_2 A_0^2$	
	SN	$A_2 = 1 - \frac{a_2\, A_0^2}{1+\mu\, A_0^2}$	

* GN: General Nonlinearity, $a_2\, F[|\psi|^2]$, KN: Kerr Nonlinearity, $a_2\, |\psi|^2$, SN: Saturable Nonlinearity, $\frac{a_2\, |\psi|^2}{1+\mu\, |\psi|^2}$.

***Solution* 2. sech(n)** *discrete bright soliton*
(Figure 8.11)

$$\psi(n, t) = A_0\, \text{sech}[A_1\, (n - n_0)]\, e^{-i[(A_2+2)\,(t-t_0)+\phi_0]}, \tag{8.43}$$

where

$A_2 = -2\cosh(A_1)$,

$a_2 = \frac{\sinh^2(A_1)}{A_0^2}$,

$A_0 \neq 0$,

A_1, t_0, n_0, and ϕ_0 are arbitrary real constants.

- *Reference*: [1], *taken from the nonlocal case.*

***Solution* 3. tanh(n)** *discrete dark soliton*
(Figure 8.12)

$$\psi(n, t) = A_0\, \tanh[A_1\, (n - n_0)]\, e^{-i[(A_2+2)\,(t-t_0)+\phi_0]}, \tag{8.44}$$

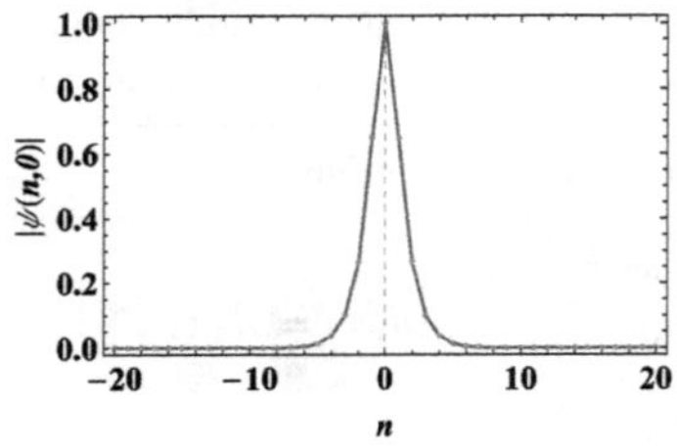

Figure 8.11. Discrete bright soliton (8.43) at $t = 0$ with $A_0 = A_1 = 1$, and $n_0 = t_0 = \phi_0 = 0$. The lines are guides for the eye.

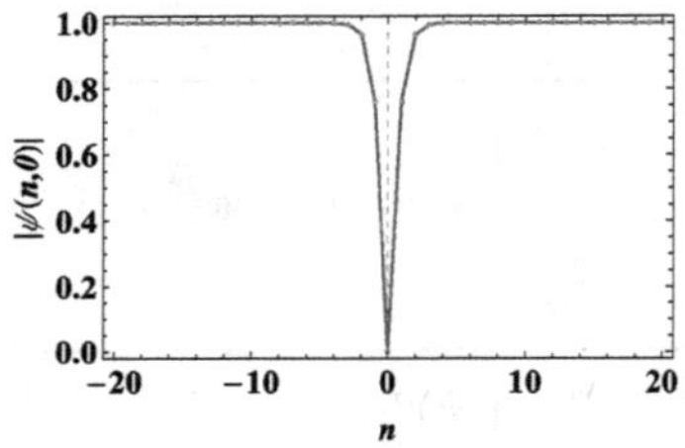

Figure 8.12. Discrete dark soliton (8.44) at $t = 0$ with $A_0 = A_1 = 1$, and $n_0 = t_0 = \phi_0 = 0$. The lines are guides for the eye.

where

$A_2 = -2\,\text{sech}^2(A_1)$,

$a_2 = \frac{-\tanh^2(A_1)}{A_0^2}$,

$A_0 \neq 0$,

A_1, t_0, n_0, and ϕ_0 are arbitrary real constants.

- *Reference*: [1], *taken from the nonlocal case.*

***Solution* 4. sn(*n*,*m*)** *discrete SW*

(Figure 8.13)

$$\psi(n,\, t) = A_0\, \sqrt{m}\, \text{sn}[A_1\, (n - n_0),\, m]\, e^{-i[(A_2+2)\,(t-t_0)+\phi_0]}, \tag{8.45}$$

where

$A_2 = -2\,\text{cn}(A_1,\, m)\,\text{dn}(A_1,\, m)$,

$a_2 = \frac{-1}{A_0^2\, \text{ns}^2(A_1,\, m)}$,

$0 \leqslant m \leqslant 1$,

$A_0 \neq 0$,

A_1, t_0, n_0, and ϕ_0 are arbitrary real constants.

- *Reference*: [1], *taken from the nonlocal case.*

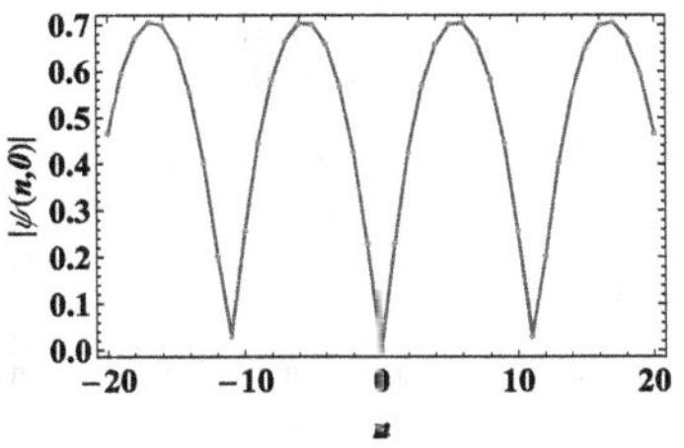

Figure 8.13. Discrete solitary wave (8.45) at $t = 0$ with $A_0 = 1$, $A_1 = 1/3$, $m = 1/2$, and $n_0 = t_0 = \phi_0 = 0$. The lines are guides for the eye.

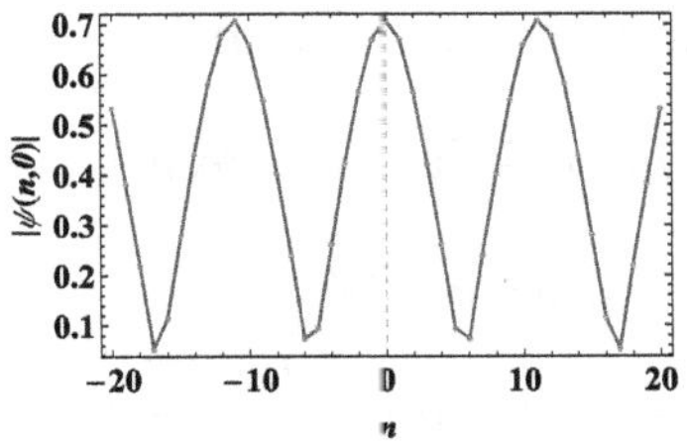

Figure 8.14. Discrete solitary wave (8.46) at $t = 0$ with $A_0 = 1$, $A_1 = 1/3$, $m = 1/2$, and $n_0 = t_0 = \phi_0 = 0$. The lines are guides for the eye.

***Solution* 5. cn(*n*,*m*)** *discrete SW*
(Figure 8.14)

$$\psi(n,\, t) = A_0\, \sqrt{m}\, \text{cn}[A_1\, (n - n_0),\, m]\, e^{-i[(A_2+2)\,(t-t_0)+\phi_0]}, \tag{8.46}$$

where

$A_2 = \dfrac{-2\,\text{cn}(A_1, m)}{\text{dn}^2(A_1, m)}$,

$a_2 = \dfrac{1}{A_0^2\, \text{ds}^2(A_1, m)}$,

$0 \leqslant m \leqslant 1$,

$A_0 \neq 0$,

A_1, t_0, n_0, and ϕ_0 are arbitrary real constants.

- *Reference*: [1], *taken from the nonlocal case.*

***Solution* 6. dn(*n*,*m*)** *discrete SW*
(Figure 8.15)

$$\psi(n,\, t) = A_0\, \text{dn}[A_1\, (n - n_0),\, m]\, e^{-i[(A_2+2)\,(t-t_0)+\phi_0]}, \tag{8.47}$$

where

$A_2 = \dfrac{-2\,\text{dn}(A_1, m)}{\text{cn}^2(A_1, m)}$,

$a_2 = \dfrac{1}{A_0^2\, \text{cs}^2(A_1, m)}$,

$0 \leqslant m \leqslant 1$,

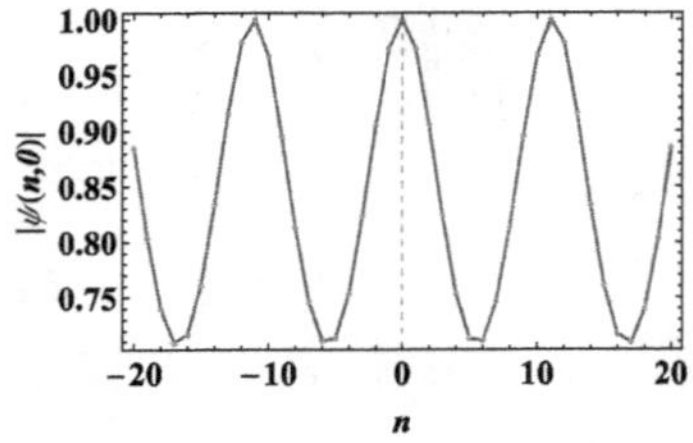

Figure 8.15. Discrete solitary wave (8.47) at $t = 0$ with $A_0 = 1$, $A_1 = 1/3$, $m = 1/2$, and $n_0 = t_0 = \phi_0 = 0$. The lines are guides for the eye.

$A_0 \neq 0$,
A_1, t_0, n_0, and ϕ_0 are arbitrary real constants.

- *Reference*: [1], *taken from the nonlocal case.*

***Solution* 7. dn(*n*,*m*) + cn(*n*,*m*)** *discrete SW*

$$\psi(n,\, t) = \left\{\frac{A_0}{2}\,\mathrm{dn}[A_1\,(n - n_0),\, m] + \frac{B_0\,\sqrt{m}}{2}\,\mathrm{cn}[A_1\,(n - n_0),\, m]\right\} \times e^{-i[(A_2+2)\,(t-t_0)+\phi_0]}, \tag{8.48}$$

where

$$A_2 = \frac{-4}{\mathrm{cn}(A_1, m) + \mathrm{dn}(A_1, m)},$$
$$a_2 = \frac{4}{A_0^2\,[\mathrm{ds}(A_1, m) + \mathrm{cs}(A_1, m)]^2},$$
$B_0 = \pm A_0$,
$0 \leqslant m \leqslant 1$,
$A_0 \neq 0$,
A_1, t_0, n_0, and ϕ_0 are arbitrary real constants.

- *Reference*: [1], *taken from the nonlocal case.*

***Solution* 8. cd(*n*,*m*)** *discrete SW*

$$\psi(n,\, t) = A_0\,\sqrt{m}\,\mathrm{cd}[A_1\,(n - n_0),\, m]\,e^{-i[(A_2+2)\,(t-t_0)+\phi_0]}, \tag{8.49}$$

where

$$A_2 = \frac{2\,\mathrm{ns}(A_1, m)\,[\mathrm{cs}^2(A_1, m) - \mathrm{ds}^2(A_1, m)]\,[\mathrm{cs}(2\,A_1, m) + \mathrm{ds}(2\,A_1, m)]}{\mathrm{ds}(A_1, m)\,\mathrm{cs}(A_1, m)\,[\mathrm{cs}(A_1, m)\mathrm{ds}(A_1, m) - 2\,\mathrm{cs}(A_1, m)\,\mathrm{ns}(A_1, m)]},$$
$$a_2 = \frac{2\,\mathrm{cs}(2\,A_1, m)\,\mathrm{ds}(A_1, m)\,\mathrm{ns}(A_1, m) - \mathrm{cs}^3(A_1, m)}{A_0^2\,[\mathrm{cs}(A_1, m)\,\mathrm{ds}^2(A_1, m) - 2\,\mathrm{cs}(2\,A_1, m)\,\mathrm{ds}(A_1, m)\,\mathrm{ns}(A_1, m)]},$$
$0 \leqslant m \leqslant 1$,
$A_0 \neq 0$,
A_1, t_0, n_0, and ϕ_0 are arbitrary real constants.

- *Reference*: [1], *taken from the nonlocal case.*

Solution* 9. Periodicity in *n* and Localization in *t *discrete Akhmediev breather* (Figure 8.16)

$$\psi(n,\, t) = \kappa \left\{ \frac{\cos[\alpha\,(n - n_0)] + i\,\sqrt{\dfrac{2 + \kappa^2}{1 + \kappa^2}}\,\sinh[2\,\kappa^2\,(t - t_0)]}{\sqrt{\dfrac{2 + \kappa^2}{1 + \kappa^2}}\,\cosh[2\,\kappa^2\,(t - t_0)] - \cos[\alpha\,(n - n_0)]} \right\} \times e^{i\,[2\,\kappa^2\,(t - t_0) + \phi_0]}, \tag{8.50}$$

where

$\alpha = \cos^{-1}\left(\frac{1}{1+\kappa^2}\right)$,

t_0, κ, n_0, and ϕ_0 are arbitrary real constants.

- *Reference*: [5].

Solution* 10. Localization in *n* and *t *discrete Peregrine soliton* (Figure 8.17)

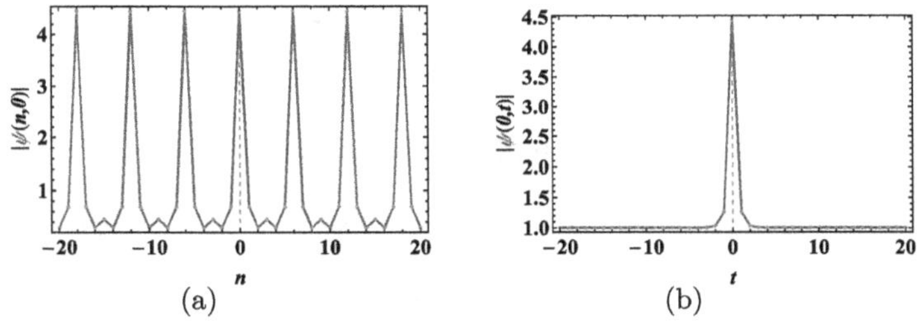

Figure 8.16. Discrete Akhmediev breather (8.50). (a) $t = 0$, (b) $n = 0$, with $\kappa = 1$ and $n_0 = t_0 = \phi_0 = 0$. The lines are guides for the eye. Animation available online at https://iopscience.iop.org/book/978-0-7503-2428-1.

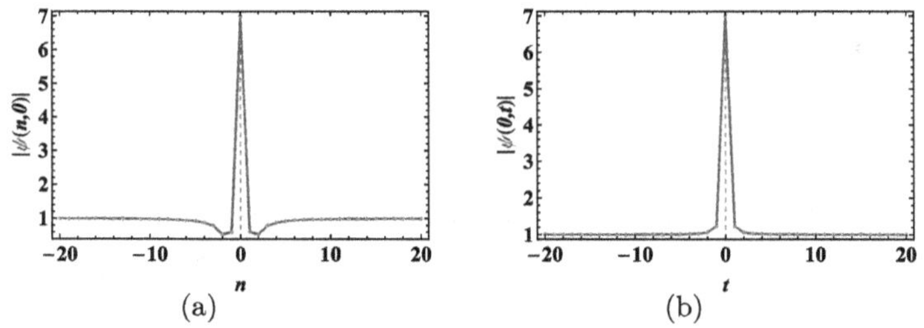

Figure 8.17. Discrete Peregrine soliton (8.51). (a) $t = 0$, (b) $n = 0$, with $n_0 = t_0 = \phi_0 = 0$. The lines are guides for the eye. Animation available online at https://iopscience.iop.org/book/978-0-7503-2428-1.

$$\psi(n,\,t) = \left\{\frac{8[1 + 4i(t - t_0)]}{1 + 4(n - n_0)^2 + 32(t - t_0)^2} - 1\right\}e^{i[(t-t_0)+\phi_0]}, \tag{8.51}$$

where t_0, n_0, and ϕ_0 are arbitrary real constants.

- *Reference*: [5].

Solution* 11. Periodicity in *n* and *t

(Figure 8.18)

$$\psi(n,\,t) = \kappa \left\{\frac{\sqrt{m}\ \mathrm{dn}[2\,\kappa^2\,(t - t_0),\, \sin^2(\theta)]\ \mathrm{cn}[A_0\,(n - n_0),\, m^2] + i\,A_1\,\sqrt{\sin(\theta)}\ \mathrm{sn}[2\,\kappa^2\,(t - t_0),\, \sin^2(\theta)]}{A_1 - \sqrt{m\,\sin(\theta)}\ \mathrm{cn}[2\,\kappa^2\,(t - t_0),\, \sin^2(\theta)]\ \mathrm{cn}[A_0\,(n - n_0),\, m^2]}\right\} \times e^{i\,[2\,\kappa^2\,\sin(\theta)\,(t-t_0)+\phi_0]}, \tag{8.52}$$

where

$A_1 = \{(1 - m^2)\,[1 - \sin^2(\theta)]\}^{1/4}$,

$\kappa = \sqrt{\frac{m\,\sqrt{1-m^2}\ \mathrm{sn}^2(A_0, m^2)}{\sqrt{1-\sin^2(\theta)}\ \mathrm{cn}(A_0, m^2)}}$,

$\theta = \tan^{-1}\{\frac{1}{m\,\sqrt{1-m^2}}\,[\frac{1}{1+\mathrm{cn}^2(A_0, m^2)} - m^2]\}$,

$0 < m < 1$,

A_0, t_0, n_0, and ϕ_0 are arbitrary real constants.

- *Reference*: [5].

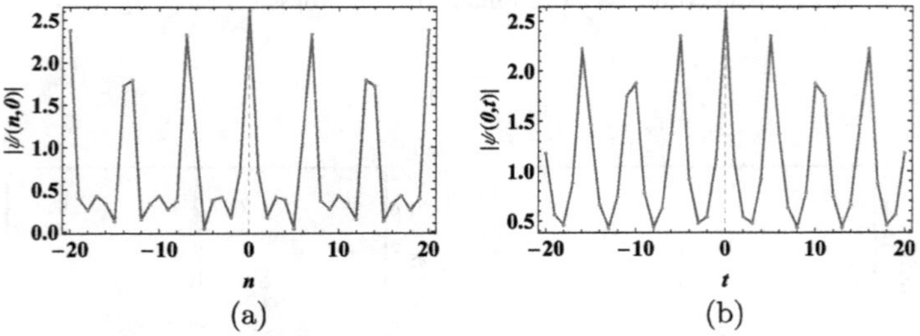

Figure 8.18. Plot of solution (8.52). (a) $t = 0$, (b) $n = 0$, with $A_0 = 1$, $m = 1/2$, and $n_0 = t_0 = \phi_0 = 0$. The lines are guides for the eye.

Solution* 12. Periodicity in *t* and Localization in *n* *discrete Kuznetsov–Ma breather (Figure 8.19)

$$\psi(n, t) = \left\{ \frac{\begin{array}{c}(1 + \kappa\,\alpha_1)\cos[\alpha_4\,(t - t_0)] + \kappa\,\alpha_1\,\alpha_2 \\ \cosh[\alpha_5\,(n - n_0)] + i\,\alpha_3\sin[\alpha_4\,(t - t_0)]\end{array}}{-\alpha_1\cos[\alpha_4\,(t - t_0)] - \alpha_1\,\alpha_2\cosh[\alpha_5\,(n - n_0)]} \right\} \quad (8.53)$$

$$\times\, e^{i\,[2\,\kappa^2\,(t - t_0) + \phi_0]},$$

where

$\alpha_1 = \frac{\kappa}{(1 + \kappa^2)\,[\cosh(\alpha_5) - 1]}$,

$\alpha_2 = \sqrt{1 + \frac{\tanh^2\left(\frac{\alpha_5}{2}\right)}{\kappa^2}}$,

$\alpha_3 = \sqrt{1 + 2\,\kappa\,\alpha_1}$,

$\alpha_4 = 4\,(1 + \kappa^2)\sinh\left(\frac{\alpha_5}{2}\right)\sqrt{\frac{\kappa^2}{1 + \kappa^2} + \sinh^2\left(\frac{\alpha_5}{2}\right)}$,

$\alpha_5 > 0$,

κ, t_0, n_0, and ϕ_0 are arbitrary real constants.

- *Reference*: [5].

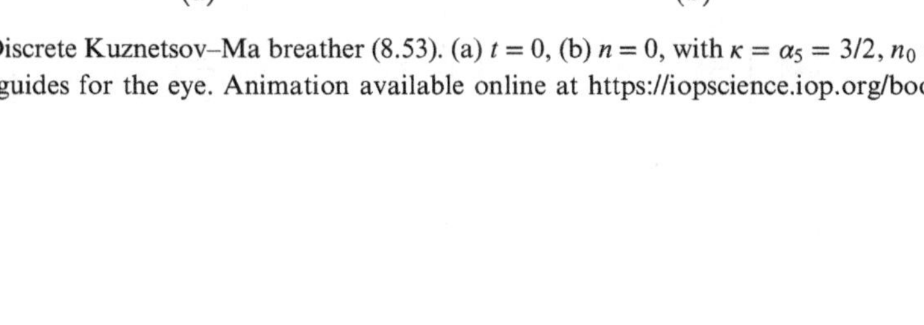

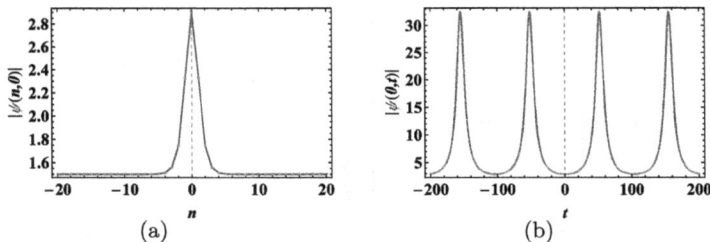

Figure 8.19. Discrete Kuznetsov–Ma breather (8.53). (a) $t = 0$, (b) $n = 0$, with $\kappa = \alpha_5 = 3/2$, $n_0 = t_0 = \phi_0 = 0$. The lines are guides for the eye. Animation available online at https://iopscience.iop.org/book/978-0-7503-2428-1.

8.5 Summary of Section 8.4

Note: For lengthy conditions, the reader is referred to the solutions in section 8.4.

Equation

$$i\,\psi_{nt} + \psi_{n+1} + \psi_{n-1} - 2\,\psi_n + a_2\,(\psi_{n+1} + \psi_{n-1})\,|\psi_n|^2 = 0$$

#	Solution	Conditions	Name	Eq. #
1.	$\psi(n, t) = A_0\, e^{i[A_1\,(n-n_0)+(A_2-2)\,(t-t_0)+\phi_0]}$	$A_2 = 2\cos(A_1)\,(1 + a_2\,A_0^2)$, A_0, A_1, t_0, and ϕ_0 are arbitrary real constants, n_0 is an arbitrary real integer	discrete continuous wave, t- and n-dependent phase	(8.42)
2.	$\psi(n, t) = A_0\,\mathrm{sech}[A_1\,(n - n_0)]\, e^{-i[(A_2+2)\,(t-t_0)+\phi_0]}$	$A_2 = -2\cosh(A_1)$, $a_2 = \frac{\sinh^2(A_1)}{A_0^2}$, $A_0 \neq 0$, A_1, t_0, and ϕ_0 are arbitrary real constants, n_0 is an arbitrary real integer	discrete bright soliton	(8.43)
3.	$\psi(n, t) = A_0\,\tanh[A_1\,(n - n_0)]\, e^{-i[(A_2+2)\,(t-t_0)+\phi_0]}$	$A_2 = -2\,\mathrm{sech}^2(A_1)$, $a_2 = \frac{-\tanh^2(A_1)}{A_0^2}$, $A_0 \neq 0$, A_1, t_0, and ϕ_0 are arbitrary real constants, n_0 is an arbitrary real integer	discrete dark soliton	(8.44)
4.	$\psi(n, t) = A_0\,\sqrt{m}\,\mathrm{sn}[A_1\,(n - n_0), m]\, e^{-i[(A_2+2)\,(t-t_0)+\phi_0]}$	$A_2 = -2\,\mathrm{cn}(A_1, m)\,\mathrm{dn}(A_1, m)$, $a_2 = \frac{-1}{A_0^2\,\mathrm{ns}^2(A_1, m)}$, $0 \leqslant m \leqslant 1$, $A_0 \neq 0$, A_1, t_0, and ϕ_0 are arbitrary real constants, n_0 is an arbitrary real integer	discrete solitary wave	(8.45)

5.	$\psi(n, t) = A_0\sqrt{m}\,\mathrm{cn}[A_1(n-n_0), m]\, e^{-i[(A_2+2)(t-t_0)+\phi_0]}$	$A_2 = \frac{-2\,\mathrm{cn}(A_1, m)}{\mathrm{dn}^2(A_1, m)}$, $a_2 = \frac{1}{A_0^2\,\mathrm{ds}^2(A_1, m)}$, $0 \leqslant m \leqslant 1$, $A_0 \neq 0$, A_1, t_0, and ϕ_0 are arbitrary real constants, n_0 is an arbitrary real integer	discrete solitary wave	(8.46)
6.	$\psi(n, t) = A_0\,\mathrm{dn}[A_1(n-n_0), m]\, e^{-i[(A_2+2)(t-t_0)+\phi_0]}$	$A_2 = \frac{-2\,\mathrm{dn}(A_1, m)}{\mathrm{cn}^2(A_1, m)}$, $a_2 = \frac{1}{A_0^2\,\mathrm{cs}^2(A_1, m)}$, $0 \leqslant m \leqslant 1$, $A_0 \neq 0$, A_1, t_0, and ϕ_0 are arbitrary real constants, n_0 is an arbitrary real integer	discrete solitary wave	(8.47)
7.	$\psi(n, t) = \left\{\frac{A_0}{2}\,\mathrm{dn}[A_1(n-n_0), m] + \frac{B_0\sqrt{m}}{2}\,\mathrm{cn}[A_1(n-n_0), m]\right\} \times e^{-i[(A_2+2)(t-t_0)+\phi_0]}$	$A_2 = \frac{-4}{\mathrm{cn}(A_1, m) + \mathrm{dn}(A_1, m)}$, $a_2 = \frac{4}{A_0^2\,[\mathrm{ds}(A_1, m) + \mathrm{cs}(A_1, m)]^2}$, $B_0 = \pm A_0$, $0 \leqslant m \leqslant 1$, $A_0 \neq 0$, A_1, t_0, and ϕ_0 are arbitrary real constants, n_0 is an arbitrary real integer	discrete solitary wave	(8.48)
8.	$\psi(n, t) = A_0\sqrt{m}\,\mathrm{cd}[A_1(n-n_0), m]\, e^{-i[(A_2+2)(t-t_0)+\phi_0]}$	See text	discrete solitary wave	(8.49)
9.	$\psi(n, t) = \kappa \left\{\frac{\cos[\alpha(n-n_0)] + i\sqrt{\frac{2+\kappa^2}{1+\kappa^2}}\sinh[2\kappa^2(t-t_0)]}{\sqrt{\frac{2+\kappa^2}{1+\kappa^2}}\cosh[2\kappa^2(t-t_0)] - \cos[\alpha(n-n_0)]}\right\} e^{i[2\kappa^2(t-t_0)+\phi_0]}$	$\alpha = \cos^{-1}\left(\frac{1}{1+\kappa^2}\right)$, t_0, κ, and ϕ_0 are arbitrary real constants, n_0 is an arbitrary real integer	discrete Akhmediev breather	(8.50)
10.	$\psi(n, t) = \left\{\frac{8[1 + 4i(t-t_0)]}{1 + 4(n-n_0)^2 + 32(t-t_0)^2} - 1\right\} e^{i[2(t-t_0)+\phi_0]}$	t_0 and ϕ_0 are arbitrary real constants, n_0 is an arbitrary real integer	discrete Peregrine soliton	(8.51)

(Continued)

11. $\psi(n,t)=\kappa\left\{\frac{\sqrt{m}\,\mathrm{dn}[2\kappa^2(t-t_0),\sin^2(\theta)]\,\mathrm{cn}[A_0(n-n_0),m^2]+i\,A_1\sqrt{\sin(\theta)}\,\mathrm{sn}[2\kappa^2(t-t_0),\sin^2(\theta)]}{A_1-\sqrt{m\,\sin(\theta)}\,\mathrm{cn}[2\kappa^2(t-t_0),\sin^2(\theta)]\,\mathrm{cn}[A_0(n-n_0),m^2]}\right\}$ $\times e^{i\,[2\kappa^2\sin(\theta)(t-t_0)+\phi_0]}$	$A_1=\{(1-m^2)[1-\sin^2(\theta)]\}^{1/4}$, $\kappa=\sqrt{\frac{m\sqrt{1-m^2}\,\mathrm{sn}^2(A_0,m^2)}{\sqrt{1-\sin^2(\theta)}\,\mathrm{cn}(A_0,m^2)}}$, $0<m<1$, $\theta=\tan^{-1}\left\{\frac{1}{m\sqrt{1-m^2}}\left[\frac{1}{1+\mathrm{cn}^2(A_0,m^2)}-m^2\right]\right\}$, A_0, t_0, and ϕ_0 are arbitrary real constants, n_0 is an arbitrary real integer	–	(8.52)
12. $\psi(n,t)=\left\{\frac{(1+\kappa\,\alpha_1)\cos[\alpha_4(t-t_0)]+\kappa\,\alpha_1\,\alpha_2\cosh[\alpha_5(n-n_0)]+i\,\alpha_3\sin[\alpha_4(t-t_0)]}{-\alpha_1\cos[\alpha_4(t-t_0)]-\alpha_1\,\alpha_2\cosh[\alpha_5(n-n_0)]}\right\}$ $\times e^{i\,[2\kappa^2(t-t_0)+\phi_0]}$	$\alpha_1=\frac{\kappa}{(1+\kappa^2)[\cosh(\alpha_5)-1]}$, $\alpha_2=\sqrt{1+\frac{\tanh^2\left(\frac{\alpha_5}{2}\right)}{\kappa^2}}$, $\alpha_3=\sqrt{1+2\,\kappa\,\alpha_1}$, $\alpha_4=4(1+\kappa^2)\sinh\left(\frac{\alpha_5}{2}\right)\times\sqrt{\frac{\kappa^2}{1+\kappa^2}+\sinh^2\left(\frac{\alpha_5}{2}\right)}$, $\alpha_5>0$, κ, t_0, and ϕ_0 are arbitrary real constants, n_0 is an arbitrary real integer	discrete Kuznetsov–Ma breather	(8.53)

8.6 Cubic-quintic Discrete NLSE

Equation:

$$i\,\psi_{nt} + a_1\,(\psi_{n+1} + \psi_{n-1} - 2\,\psi_n) + a_2\,|\psi_n|^2\,\psi_n + (a_3\,|\psi_n|^2 + a_4\,|\psi_n|^4)(\psi_{n+1} + \psi_{n-1}) = 0, \tag{8.54}$$

where

$\psi_n = \psi(n, t)$ is the complex function profile,
the integer site index, n, and t are its two independent variables,
a_1, a_2, a_3, and a_4 are real constants.

Solutions:

***Solution* 1. Constant Amplitude** *discrete CW, t- and n-dependent phase*

$$\psi(n, t) = A_0\,e^{i\,[A_1\,(n-n_0)+A_2\,(t-t_0)+\phi_0]}, \tag{8.55}$$

where

$A_2 = -2\,a_1 + A_0^2\,a_2 + 2\cos(A_1)\,(a_1 + A_0^2\,a_3 + A_0^4\,a_4)$,
A_0, A_1, t_0, n_0, and ϕ_0 are arbitrary real constants.

***Solution* 2. sech(*n*) I** *discrete bright soliton*
(Figure 8.20)

$$\psi(n, t) = A_0\,\mathrm{sech}[A_1\,(n - n_0)]\,e^{i\,[A_2\,(t-t_0)+\phi_0]}, \tag{8.56}$$

where

$$A_0 = \pm\frac{\sqrt{a_3\,(a_2^2 - 4\,a_1\,a_4) + \sqrt{a_3^2 - 4\,a_1\,a_4}\;(a_2^2 + 4\,a_1\,a_4)}}{2\,a_4\,\sqrt{2\,a_1}},$$

$$A_1 = \mathrm{sech}^{-1}\left(\frac{-a_3 + \sqrt{a_3^2 - 4\,a_1\,a_4}}{a_2}\right),$$

$$A_2 = -2\,a_1 - \frac{a_2\,(a_3 + \sqrt{a_3^2 - 4\,a_1\,a_4})}{2\,a_4},$$

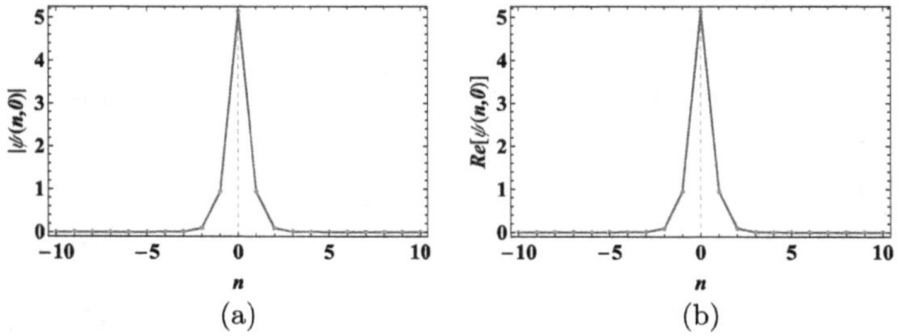

Figure 8.20. Discrete bright soliton (8.56) $t = 0$. (a) Absolute value, (b) real part, with $n_0 = \phi_0 = 0$, $a_1 = a_2 = a_3 = 1$, and $a_4 = -1/10$. The lines are guides for the eye.

$a_3^2 - 4\,a_4\,a_1 \geqslant 0,$
$a_1 > 0,$
$a_3\,(a_2^2 - 4\,a_1\,a_4) + \sqrt{a_3^2 - 4\,a_1\,a_4}\,(a_2^2 + 4\,a_1\,a_4) > 0,$
t_0, n_0, and ϕ_0 are arbitrary real constants.

- *Reference*: [6].

***Solution* 3. sech(*n*) II** *staggered discrete bright soliton* (Figure 8.21)

$$\psi(n,\,t) = (-1)^n\,A_0\,\text{sech}[A_1\,(n - n_0)]\,e^{i\,[A_2\,(t-t_0)+\phi_0]}, \tag{8.57}$$

where

$$A_0 = \pm\frac{\sqrt{a_3\,(a_2^2 - 4\,a_1\,a_4) + \sqrt{a_3^2 - 4\,a_1\,a_4}\,(a_2^2 + 4\,a_1\,a_4)}}{2\,a_4\,\sqrt{2\,a_1}},$$

$$A_1 = \text{sech}^{-1}\left(\frac{a_3 - \sqrt{a_3^2 - 4\,a_1\,a_4}}{a_2}\right),$$

$$A_2 = -2\,a_1 - \frac{a_2\,(a_3 + \sqrt{a_3^2 - 4\,a_1\,a_4})}{2\,a_4},$$

$a_3^2 - 4\,a_4\,a_1 \geqslant 0,$
$a_1 > 0,$
$a_3\,(a_2^2 - 4\,a_1\,a_4) + \sqrt{a_3^2 - 4\,a_1\,a_4}\,(a_2^2 + 4\,a_1\,a_4) > 0,$
t_0, n_0, and ϕ_0 are arbitrary real constants.

- *Reference*: [6].

***Solution* 4. tanh(*n*) I** *discrete dark soliton* (Figure 8.22)

$$\psi(n,\,t) = A_0\,\tanh[A_1\,(n - n_0)]\,e^{i\,[A_2\,(t-t_0)+\phi_0]}, \tag{8.58}$$

where

$$A_0 = \pm\sqrt{\frac{a_2 + a_3 + \sqrt{a_3^2 - 4\,a_1\,a_4}}{-2\,a_4}},$$

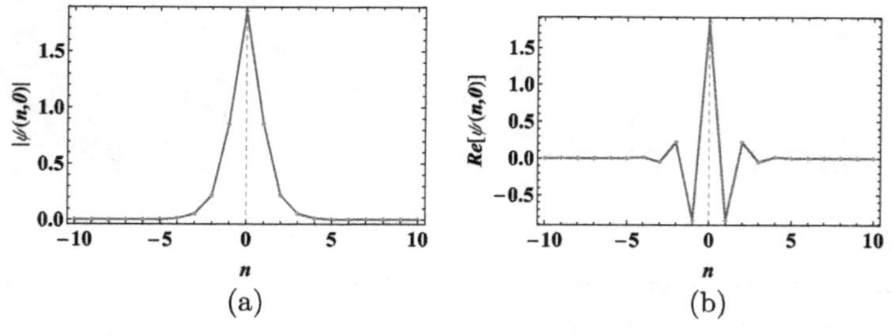

Figure 8.21. Staggered discrete bright soliton (8.57) $t = 0$. (a) Absolute value, (b) real part, with $n_0 = \phi_0 = 0$, $a_1 = a_3 = 1$, $a_2 = -4/10$, and $a_4 = -1/10$. The lines are guides for the eye.

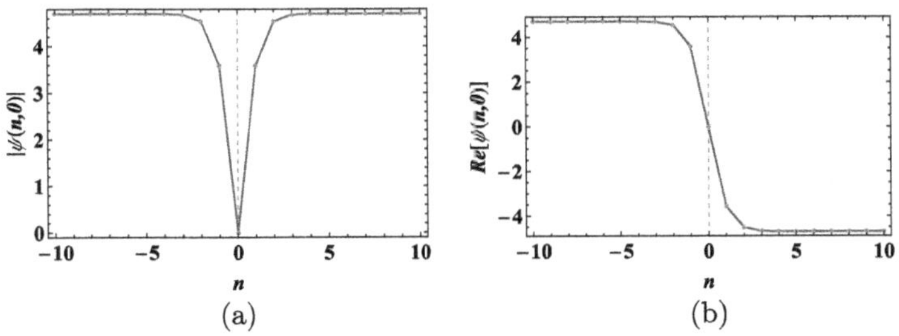

Figure 8.22. Discrete dark soliton (8.58) $t = 0$. (a) Absolute value, (b) real part, with $n_0 = \phi_0 = 0$, $a_1 = -18/10$, $a_2 = -8/10$, $a_3 = 1$, and $a_4 = -18/10$. The lines are guides for the eye.

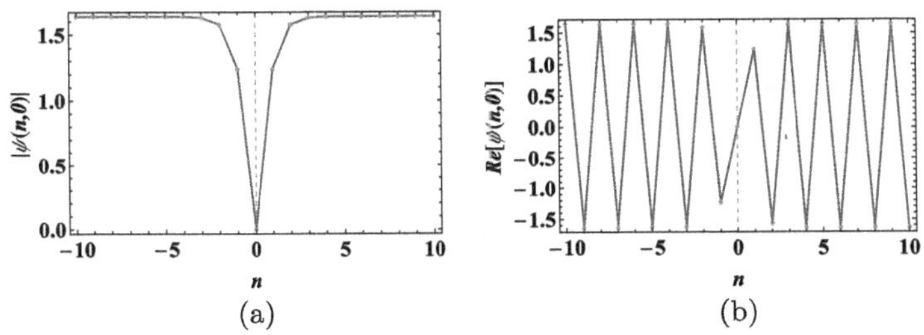

Figure 8.23. Staggered discrete dark soliton (8.59) at $t = 0$. (a) Absolute value, (b) real part, with $n_0 = \phi_0 = 0$, $a_1 = 8/10$, $a_2 = a_3 = 1$, and $a_4 = -25/100$. The lines are guides for the eye.

$$A_1 = \cosh^{-1}\left(\sqrt{\frac{a_3 + \sqrt{a_3^2 - 4\, a_1\, a_4}}{-a_2}}\right),$$

$$A_2 = -2\, a_1 - \frac{a_2\, (a_3 - \sqrt{a_3^2 - 4\, a_1\, a_4})}{2\, a_4},$$

$$a_3^2 - 4\, a_4\, a_1 \geqslant 0,$$

$$a_4\, (a_2 + a_3 + \sqrt{a_3^2 - 4\, a_1\, a_4}) < 0,$$

$$a_2\, (a_3 + \sqrt{a_3^2 - 4\, a_1\, a_4}) < 0,$$

t_0, n_0, and ϕ_0 are arbitrary real constants.

- *Reference*: [6].

Solution* 5. tanh(*n*) II *staggered discrete dark soliton
(Figure 8.23)

$$\psi(n,\, t) = (-1)^n\, A_0\, \tanh[A_1\, (n - n_0)]\, e^{i\, [A_2\, (t - t_0) + \phi_0]}, \tag{8.59}$$

where

$$A_0 = \pm\sqrt{\frac{a_2 - a_3 - \sqrt{a_3^2 - 4\,a_1\,a_4}}{2\,a_4}},$$

$$A_1 = \cosh^{-1}\left(\sqrt{\frac{a_3 + \sqrt{a_3^2 - 4\,a_1\,a_4}}{a_2}}\right),$$

$$A_2 = -2\,a_1 - \frac{a_2\,(a_3 - \sqrt{a_3^2 - 4\,a_1\,a_4})}{2\,a_4},$$

$a_3^2 - 4\,a_4\,a_1 \geqslant 0$,

$(a_2 - a_3 - \sqrt{a_3^2 - 4\,a_1\,a_4}) > 0$,

$a_2\,(a_3 + \sqrt{a_3^2 - 4\,a_1\,a_4}) > 0$,

t_0, n_0, and ϕ_0 are arbitrary real constants.

- *Reference*: [6].

8.7 Summary of Section 8.6

Equation

$$i\,\psi_{n_t} + a_1\,(\psi_{n+1} + \psi_{n-1} - 2\,\psi_n) + a_2\,|\psi_n|^2\,\psi_n + (a_3\,|\psi_n|^2 + a_4\,|\psi_n|^4)(\psi_{n+1} + \psi_{n-1}) = 0$$

# Solution	Conditions	Name	Eq. #
1. $\psi(n,\,t) = A_0\,e^{i\,[A_1\,(n-n_0)+A_2\,(t-t_0)+\phi_0]}$	$A_2 = -2\,a_1 + A_0^2\,a_2 + 2\cos(A_1)\,(a_1 + A_0^2\,a_3 + A_0^4\,a_4)$, A_0, A_1, t_0, and ϕ_0 are arbitrary real constants, n_0 is an arbitrary real integer	discrete continuous wave, t- and n-dependent phase	(8.55)
2. $\psi(n,\,t) = A_0\,\mathrm{sech}[A_1\,(n-n_0)]\,e^{i\,[A_2\,(t-t_0)+\phi_0]}$	$A_0 = \pm\frac{\sqrt{a_3\,(a_2^2-4\,a_1\,a_4)+\sqrt{a_3^2-4\,a_1\,a_4}\,(a_2^2+4\,a_1\,a_4)}}{2\,a_4\,\sqrt{2\,a_1}}$, $A_1 = \mathrm{sech}^{-1}\left(\frac{-a_3+\sqrt{a_3^2-4\,a_1\,a_4}}{a_2}\right)$, $A_2 = -2\,a_1 - \frac{a_2\,(a_3+\sqrt{a_3^2-4\,a_1\,a_4})}{2\,a_4}$, $a_3^2 - 4\,a_4\,a_1 \geqslant 0,\ a_1 > 0$, $a_3\,(a_2^2 - 4\,a_1\,a_4) + \sqrt{a_3^2 - 4\,a_1\,a_4}\,(a_2^2 + 4\,a_1\,a_4) > 0$, t_0 and ϕ_0 are arbitrary real constants, n_0 is an arbitrary real integer	discrete bright soliton	(8.56)
3. $\psi(n,\,t) = (-1)^n\,A_0\,\mathrm{sech}[A_1\,(n-n_0)]\,e^{i\,[A_2\,(t-t_0)+\phi_0]}$	$A_0 = \pm\frac{\sqrt{a_3\,(a_2^2-4\,a_1\,a_4)+\sqrt{a_3^2-4\,a_1\,a_4}\,(a_2^2+4\,a_1\,a_4)}}{2\,a_4\,\sqrt{2\,a_1}}$, $A_1 = \mathrm{sech}^{-1}\left(\frac{a_3-\sqrt{a_3^2-4\,a_1\,a_4}}{a_2}\right)$, $A_2 = -2\,a_1 - \frac{a_2\,(a_3+\sqrt{a_3^2-4\,a_1\,a_4})}{2\,a_4}$, $a_3^2 - 4\,a_4\,a_1 \geqslant 0,\ a_1 > 0$, $a_3\,(a_2^2 - 4\,a_1\,a_4) + \sqrt{a_3^2 - 4\,a_1\,a_4}\,(a_2^2 + 4\,a_1\,a_4) > 0$, t_0 and ϕ_0 are arbitrary real constants, n_0 is an arbitrary real integer	staggered discrete bright soliton	(8.57)

(Continued)

4. $\psi(n, t) = A_0 \tanh[A_1 (n - n_0)]\, e^{i\,[A_2 (t-t_0)+\phi_0]}$	$A_0 = \pm\sqrt{\frac{a_2+a_3+\sqrt{a_3^2-4\,a_1\,a_4}}{-2\,a_4}}$, $A_1 = \cosh^{-1}\left(\sqrt{\frac{a_3+\sqrt{a_3^2-4\,a_1\,a_4}}{-a_2}}\right)$, $A_2 = -2\,a_1 - \frac{a_2\,(a_3-\sqrt{a_3^2-4\,a_1\,a_4})}{2\,a_4}$, $a_3^2 - 4\,a_4\,a_1 \geqslant 0$, $a_4\,(a_2 + a_3 + \sqrt{a_3^2 - 4\,a_1\,a_4}) < 0$, $a_2\,(a_3 + \sqrt{a_3^2 - 4\,a_1\,a_4}) < 0$, t_0 and ϕ_0 are arbitrary real constants, n_0 is an arbitrary real integer	discrete dark soliton	(8.58)
5. $\psi(n, t) = (-1)^n A_0 \tanh[A_1 (n - n_0)]\, e^{i\,[A_2 (t-t_0)+\phi_0]}$	$A_0 = \pm\sqrt{\frac{a_2-a_3-\sqrt{a_3^2-4\,a_1\,a_4}}{2\,a_4}}$, $A_1 = \cosh^{-1}\left(\sqrt{\frac{a_3+\sqrt{a_3^2-4\,a_1\,a_4}}{a_2}}\right)$, $A_2 = -2\,a_1 - \frac{a_2\,(a_3-\sqrt{a_3^2-4\,a_1\,a_4})}{2\,a_4}$, $a_3^2 - 4\,a_4\,a_1 \geqslant 0$, $(a_2 - a_3 - \sqrt{a_3^2 - 4\,a_1\,a_4}) > 0$, $a_2\,(a_3 + \sqrt{a_3^2 - 4\,a_1\,a_4}) > 0$, t_0 and ϕ_0 are arbitrary real constants, n_0 is an arbitrary real integer	staggered discrete dark soliton	(8.59)

8.8 Generalized Discrete NLSE

Equation:

$$i\,\psi_{n_t} + a_1\,(\psi_{n+1} + \psi_{n-1} - 2\,\psi_n) + f[\psi_{n-1}, \psi_n, \psi_{n+1}] = 0, \qquad (8.60)$$

where

$$\begin{aligned} f[\psi_{n-1}, \psi_n, \psi_{n+1}] = {} & \alpha_1\,|\psi_n|^2\,\psi_n + \alpha_2\,|\psi_n|^2\,(\psi_{n+1} + \psi_{n-1}) + \alpha_3\,\psi_n^2\,(\psi_{n+1}^* + \psi_{n-1}^*) \\ & + \alpha_4\,\psi_n\,(|\psi_{n+1}|^2 + |\psi_{n-1}|^2) + \alpha_5\,\psi_n\,(\psi_{n+1}^*\,\psi_{n-1} + \psi_{n-1}^*\,\psi_{n+1}) \\ & + \alpha_6\,\psi_n^*\,(\psi_{n+1}^2 + \psi_{n-1}^2) + \alpha_7\,\psi_n^*\,\psi_{n+1}\,\psi_{n-1} \\ & + \alpha_8\,(|\psi_{n+1}|^2\,\psi_{n+1} + |\psi_{n-1}|^2\,\psi_{n-1}) + \alpha_9\,(\psi_{n-1}^*\,\psi_{n+1}^2 + \psi_{n+1}^*\,\psi_{n-1}^2) \\ & + \alpha_{10}\,(|\psi_{n+1}|^2\,\psi_{n-1} + |\psi_{n-1}|^2\,\psi_{n+1}) + \alpha_{11}\,(|\psi_{n-1}\,\psi_n| + |\psi_n\,\psi_{n+1}|)\,\psi_n \\ & + \alpha_{12}\,(\psi_{n+1}\,|\psi_{n+1}\,\psi_n| + \psi_{n-1}\,|\psi_n\,\psi_{n-1}|) \\ & + \alpha_{13}\,(\psi_{n+1}\,|\psi_{n-1}\,\psi_n| + \psi_{n-1}\,|\psi_n\,\psi_{n+1}|) \\ & + \alpha_{14}\,(\psi_{n+1}\,|\psi_{n-1}\,\psi_{n+1}| + \psi_{n-1}\,|\psi_{n-1}\,\psi_{n+1}|), \end{aligned}$$

$\psi_n = \psi(n, t)$ is the complex function profile,
the integer site index, n, and t are its two independent variables,
a_1 and $\alpha_1, \ldots, \alpha_{14}$ are real constants.

Solutions:

***Solution* 1. sech(*n*)** *discrete bright soliton*
(Figure 8.24)

$$\psi(n, t) = A_0\,\mathrm{sech}[\beta\,(n + \gamma)]\,e^{-i\,(A_2\,t+\phi_0)}, \qquad (8.61)$$

where

$\alpha_1 = \alpha_8 = 0,$

$A_2 = 2\,a_1\,[1 - \cosh(\beta)],$

$\alpha_4 = -\alpha_{12} + \alpha_{13} + \alpha_5 - \alpha_6 + \frac{\alpha_7}{2} + \cosh(\beta)\,[\alpha_2 + \alpha_3 + \alpha_{11} - \frac{a_1\,\sinh^2(\beta)}{A_0^2}],$

$\alpha_{10} = -\alpha_9 - \alpha_{14} - (2\,\alpha_{13} + 2\,\alpha_5 + \alpha_7)\cosh(\beta) - 2\,(\alpha_2 + \alpha_3 + \alpha_{11})\cosh^2(\beta),$
$+ \frac{2\,a_1\,\cosh^2(\beta)\,\sinh^2(\beta)}{A_0^2}$

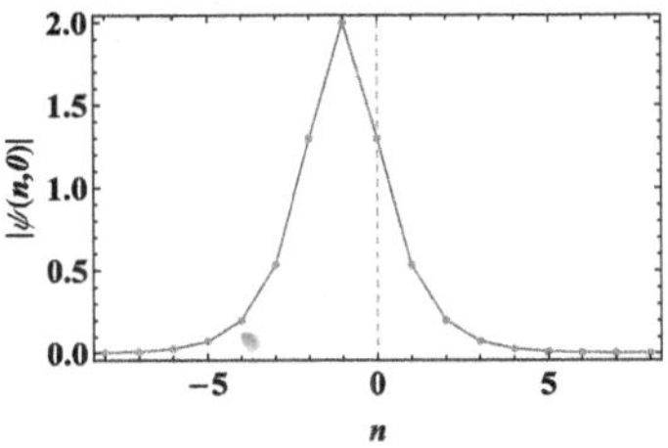

Figure 8.24. Discrete bright soliton (8.61) at $t = 0$ with $a_1 = A_0 = A_1 = 2$, $\alpha_2 = \alpha_3 = \alpha_5 = \alpha_7 = \alpha_9 = \alpha_{11} = \alpha_{12} = \alpha_{13} = \alpha_{14} = \beta = \gamma = 1$, and $\phi_0 = 0$. The lines are guides for the eye.

$\alpha_7 = -2\,\alpha_{13} - 2\,\alpha_5$,
$\alpha_{11} = -\alpha_2 - \alpha_3$,
a_1, A_0, A_1, α_2, α_3, α_5, α_6, α_9, α_{12}, α_{13}, α_{14}, γ, and ϕ_0 are arbitrary real constants.

- *Reference*: [7].

***Solution* 2. sech(*n*,*t*)** *moving discrete bright soliton* (Figure 8.25)

$$\psi(n,\,t) = A_0\,\mathrm{sech}[\beta\,(n - v\,t + \gamma)]\,e^{i\,(A_1\,n - A_2\,t + \phi_0)}, \tag{8.62}$$

where
$a_1 = 1$,
$\alpha_1 = \alpha_8 = 0$,
$v = \frac{2\,a_1\,\sin(A_1)\,\sinh(\beta)}{\beta}$,
$A_2 = 2\,a_1\,[1 - \cos(A_1)\cosh(\beta)]$,
$\alpha_2 = \alpha_3 + \frac{\sinh^2(\beta)}{A_0^2}$,
$\alpha_4 = \alpha_6 - (\alpha_{10} - \alpha_9)\cos(A_1)\,\mathrm{sech}(\beta)$,
$\alpha_6 = -\frac{1}{4}\sec(A_1)\,[2\,\alpha_{12} - \alpha_{10}\,\mathrm{sech}(\beta) + \alpha_{14}\,\mathrm{sech}(\beta) + \alpha_9\,\csc(A_1)\,\mathrm{sech}(\beta)\sin(3\,A_1)]$,
$$\alpha_7 = -2\,[\alpha_{13}\cos(A_1) + \alpha_5\cos(2\,A_1)] - [(\alpha_{10} + \alpha_{14})\cos(A_1) + \alpha_9\cos(3\,A_1)] \times \mathrm{sech}(\beta) + 2\cosh(\beta)[-\alpha_{11} - 2\,\alpha_3\cos(A_1) + \frac{(a_1 - 1)\cos(A_1)\sinh^2(\beta)}{A_0^2}],$$
A_0, A_1, α_3, α_5, α_9, α_{10}, α_{11}, α_{12}, α_{13}, α_{14}, γ, and ϕ_0 are arbitrary real constants.

- *Reference*: [7].

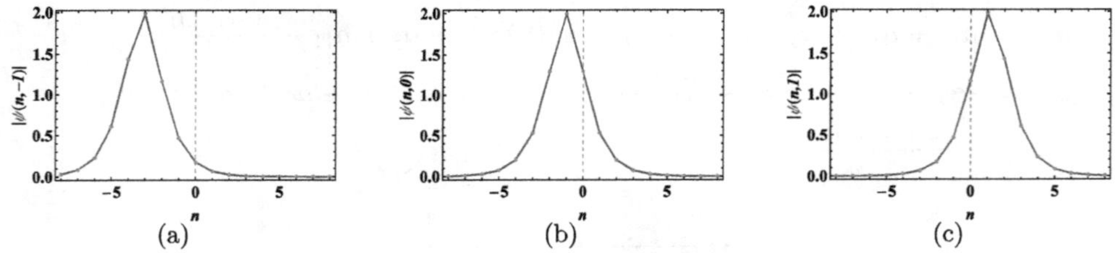

(a) (b) (c)

Figure 8.25. Moving discrete bright soliton (8.62) with $A_0 = A_1 = 2$, $a_1 = \alpha_3 = \alpha_5 = \alpha_9 = \alpha_{10} = \alpha_{11} = \alpha_{12} = \alpha_{13} = \alpha_{14} = \beta = \gamma = 1$, and $\phi_0 = 0$. (a) at $t = -1$, (b) at $t = 0$, and (c) at $t = 1$. The lines are guides for the eye. Animation available online at https://iopscience.iop.org/book/978-0-7503-2428-1.

***Solution* 3. tanh(*n*)** *discrete dark soliton*
(Figure 8.26)

$$\psi(n,\, t) = A_0 \tanh[\beta\, (n + \gamma)]\, e^{-i\, (A_2\, t + \phi_0)}, \tag{8.63}$$

where

$\alpha_1 = \alpha_8 = 0,$

$A_2 = \frac{2}{1 + \coth^2(\beta)} \left[A_0^2\, (\alpha_{10} - \alpha_{11} + \alpha_{14} - \alpha_2 - \alpha_3 + \alpha_9) + a_1\right],$

$\alpha_4 = \frac{-1}{1 + \cosh(2\,\beta)} \left[\alpha_6 + (\alpha_{10} + \alpha_{14} + \alpha_6 + \alpha_9) \cosh(2\,\beta) + 2\,\alpha_{12} \cosh^2(\beta)\right],$

$$\begin{aligned}\alpha_5 \;=\; & \frac{-1}{2\, A_0^2\, [1 + \tanh^2(\beta)]} \\ & \times\{A_0^2\, (\alpha_{10} + 2\,\alpha_{11} + 2\,\alpha_{13} + \alpha_{14} + 2\,\alpha_2 + 2\,\alpha_3 + \alpha_7 + \alpha_9) \\ & + [A_0^2\, (2\,\alpha_{10} + 2\,\alpha_{14} + 2\,\alpha_{13} + \alpha_7 + 2\,\alpha_9) + 2\, a_1] \\ & \times \tanh^2(\beta) - A_0^2\, (\alpha_{10} + \alpha_{14} + \alpha_9) \tanh^4(\beta)\},\end{aligned}$$

a_1, A_0, A_1, γ, α_2, α_3, α_6, α_7, α_9, α_{10}, α_{11}, α_{12}, α_{13}, α_{14}, and ϕ_0 are arbitrary real constants.

- *Reference*: [7].

***Solution* 4. tanh(*n*,*t*)** *moving discrete dark soliton*
(Figure 8.27)

$$\psi(n,\, t) = A_0 \tanh[\beta\, (n - v\, t + \gamma)]\, e^{i\, (A_1\, n - A_2\, t + \phi_0)}, \tag{8.64}$$

where

$\alpha_1 = \alpha_8 = \alpha_{14} = \gamma = 0,$

$v = \frac{2\, A_0^2\, (\alpha_2 - \alpha_3) \coth(\beta) \sin(A_1)}{\beta},$

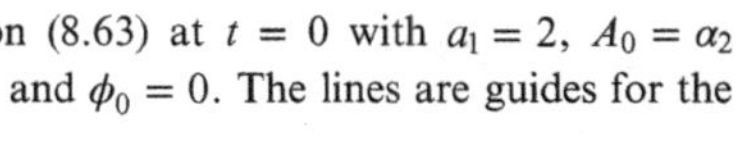

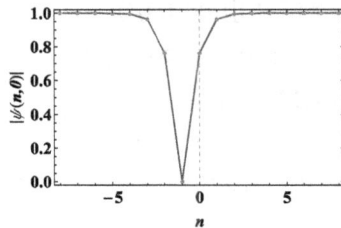

Figure 8.26. Discrete dark soliton (8.63) at $t = 0$ with $a_1 = 2$, $A_0 = \alpha_2 = \alpha_3 = \alpha_6 = \alpha_7 = \alpha_9 = \alpha_{10} = \alpha_{11} = \beta = \gamma = 1$, $\alpha_{12} = 7$, $\alpha_{13} = \alpha_{14} = 2$, and $\phi_0 = 0$. The lines are guides for the eye.

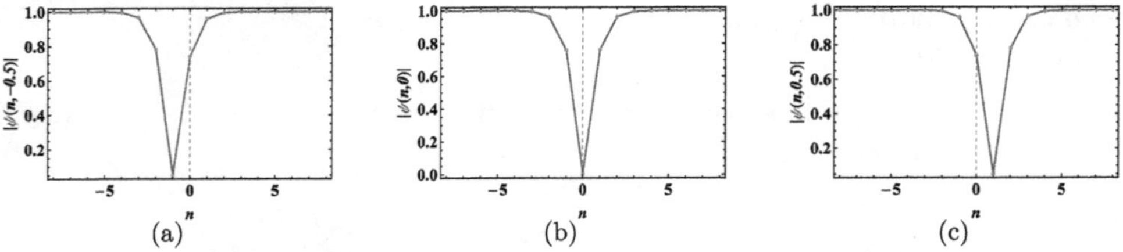

Figure 8.27. Moving discrete dark soliton (8.64) with $a_1 = A_0 = A_1 = \alpha_2 = \alpha_5 = \alpha_9 = \alpha_{10} = \alpha_{11} = \alpha_{12} = \alpha_{13} = \beta = 1$, and $\gamma = \phi_0 = 0$. (a) at $t = -0.5$, (b) at $t = 0$, and (c) at $t = 0.5$. The lines are guides for the eye. Animation available online at https://iopscience.iop.org/book/978-0-7503-2428-1.

$$A_2 = \frac{2}{1+\coth^2(\beta)}\,(a_1 + 2\,A_0^2\,\alpha_{10}\cos(A_1) - 2\,A_0^2\,\alpha_9\cos(A_1)$$
$$-A_0^2\,[\alpha_{11} + 2\,\alpha_2\cos(A_1)]\coth^4(\beta) + A_0^2\,[\alpha_{11} + (\alpha_2+\alpha_3)\cos(A_1)]\,\mathrm{csch}^2(\beta)$$
$$+\coth^2(\beta)\,\{a_1 + A_0^2\,[\alpha_{11} + (\alpha_2+\alpha_3)\cos(A_1)]\,\mathrm{csch}^2(\beta)\}),$$

$$\alpha_3 = \frac{1}{A_0^2}\,[-A_0^2\,\alpha_{10} + A_0^2\,\alpha_{14} - a_1\coth^2(\beta)$$
$$+A_0^2\,\alpha_2\coth^4(\beta) + A_0^2\,\alpha_9\csc(A_1)\sin(3\,A_1)]\tanh^4(\beta),$$

$$\alpha_4 = \frac{1}{2}\,\{\alpha_6\,\mathrm{sech}^2(\beta) + [\alpha_6 - 2\,(\alpha_{10}-\alpha_9)\cos(A_1)]\,[1+\tanh^2(\beta)]\},$$

$$\alpha_6 = \frac{1}{4\,A_0^2}\,[a_1 + 2\,A_0^2\,\alpha_{12} + a_1\coth^2(\beta) - A_0^2\,\alpha_2\coth^2(\beta)$$
$$+A_0^2\,\alpha_3\coth^2(\beta) - A_0^2\,\alpha_2\coth^4(\beta) + A_0^2\,\alpha_3\coth^4(\beta)]\sec(A_1),$$

$$\alpha_7 = \frac{1}{A_0^2\,(1+\coth^2(\beta))}$$
$$(-4\,A_0^2\,\alpha_{10}\cos(A_1) - 2\,A_0^2\,\alpha_{13}\cos(A_1) + 4\,A_0^2\,\alpha_9\cos(A_1)$$
$$-a_1\cos(A_1) - 2\,A_0^2\,\alpha_5\cos(2\,A_1)$$
$$-\{[A_0^2\,(2\,\alpha 10 + 2\,\alpha_{13} + 3\,\alpha_2 + \alpha_3 - 2\,\alpha_9) + 2\,a_1]\cos(A_1)$$
$$+2\,A_0^2\,[\alpha_{11} + \alpha_5\cos(2\,A_1)]\}\coth^2(\beta)$$
$$+[2\,A_0^2\,(\alpha_2-\alpha_3) - a_1]\cos(A_1)\coth^4(\beta)$$
$$+A_0^2\,(\alpha_2-\alpha_3)\cos(A_1)\coth^6(\beta)$$
$$+2\,A_0^2\,(\alpha_{10}-\alpha_9)\cos(A_1)\tanh^2(\beta)),$$

a_1, A_0, A_1, α_2, α_5, α_9, α_{10}, α_{11}, α_{12}, α_{13}, γ, and ϕ_0 are arbitrary real constants.

- *Reference*: [7].

***Solution* 5. sin(*n*,*t*)**

$$\psi(n,\ t) = A_0 \sin[\beta\ (n - v\ t + \gamma)]\ e^{i\ (A_1\ n - A_2\ t + \phi_0)}, \tag{8.65}$$

where

$\alpha_1 = \alpha_8 = 0,$

$v\ =\ \frac{2\sin(\beta)}{\beta}\{A_0^2\ \alpha_6 \sin(2\ A_1)\sin(\beta)\sin(2\ \beta) + \sin(A_1)$

$[a_1 + A_0^2\ (\alpha_2 - \alpha_3 + 4\ \alpha_8)\sin^2(\beta)$

$-\ 4\ A_0^2\ \alpha_8 \sin^4(\beta) + A_0^2\ \alpha_{12}\sin(\beta)\sin(2\ \beta)]\},$

$A_2\ =\ 2\ a_1 - A_0^2\cos(A_1)\cos(\beta)$

$\left[\alpha_2 + \alpha_3 + \alpha_8 + \frac{2\ a_1}{A_0^2} + \alpha_{12}\cos(\beta) - (\alpha_2 + \alpha_3)\cos(2\ \beta)\right.$

$-\ \alpha_{12}\cos(3\ \beta) - \alpha_8\cos(4\ \beta)]$

$-A_0^2\sin^2(\beta)\{\alpha_1 + 2\ \alpha_4 + 2\ \alpha_6\cos(2\ A_1)$

$+\ \alpha_6\cos[2\ (A_1 - \beta)] + 2\ \alpha_{11}\cos(\beta)\}$

$+2\ \alpha_4\cos(2\ \beta) + \alpha_6\cos[2\ (A_1 + \beta)],$

$\alpha_7\ =\ -\ \alpha_1 + \alpha_5 + \alpha_6 - 2\ (\alpha_5 - \alpha_6)\cos^2(A_1)$

$+\ (\alpha_6 - \alpha_5)\cos(2\ A_1) - 2\ \alpha_{11}\cos(\beta)$

$+2\cos(A_1)\ [\alpha_{12} - \alpha_{13} - 2\cos(\beta)\ (\alpha_{10} + \alpha_3 - \alpha_8 - \alpha_9)]$

$-2\alpha_4\cos(2\beta) + 2\alpha_6\cos(2\beta),$

$\alpha_{14} = \alpha_{10} - \alpha_2 + \alpha_3 - 3\ \alpha_8 - 2\cos(\beta)\ [\alpha_{12} + 2\ \alpha_6\cos(A_1)] - \alpha_9\csc(A_1)\sin(3\ A_1)$

$+\ 4\ \alpha_8\sin^2(\beta),$

$\gamma = \beta = 1,$

a_1, A_0, A_1, α_2, α_3, α_4, α_5, α_6, α_9, α_{10}, α_{11}, α_{12}, α_{13}, and ϕ_0 are arbitrary real constants.

- *Reference*: [7].

***Solution* 6. dn(*n*,*t*,*m*)** *moving discrete SW*
(Figure 8.28)

$$\psi(n,\ t) = A_0\ \mathrm{dn}[\beta\ (n - v\ t + \gamma),\ m]\ e^{i\ (A_1\ n - A_2\ t + \phi_0)}, \tag{8.66}$$

where

$a_1 = 1,$

$\alpha_1 = \alpha_8 = \alpha_{11} = \alpha_{12} = \alpha_{13} = \alpha_{14} = 0,$

$v = \frac{2\ A_0^2}{\beta}\ (\alpha_2 - \alpha_3)\sin(A_1)\ \mathrm{cs}(\beta,\ m),$

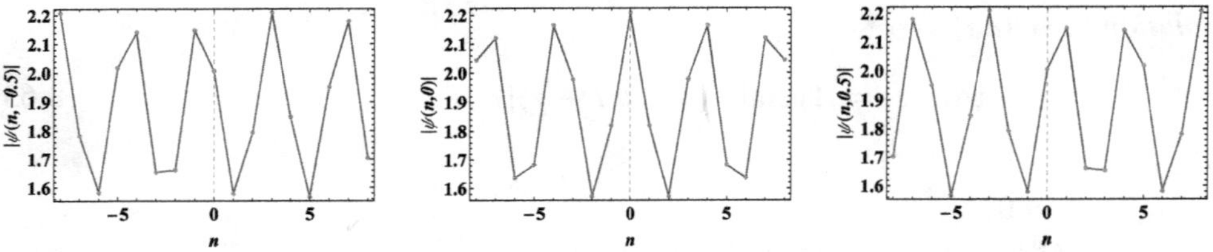

(a) (b) (c)

Figure 8.28. Moving discrete solitary wave (8.66) with $a_1 = A_1 = \alpha_2 = \alpha_9 = \alpha_{10} = \beta = 1$, $\alpha_3 = 2$, $m = 1/2$, and $\gamma = \phi_0 = 0$. (a) at $t = -0.5$, (b) at $t = 0$, and (c) at $t = 0.5$. The lines are guides for the eye.

$$A_0 = \sqrt{\frac{\sin(A_1)}{q_1 - q_2 - q_3\, q_4}},$$

$$q_1 = (\alpha_2 - \alpha_3)\sin(A_1)\,\mathrm{cs}^2(\beta, m),$$

$$q_2 = \alpha_6 \sin(2\,A_1)\,\mathrm{ds}(\beta, m)\,\mathrm{ns}(\beta, m),$$

$$q_3 = \alpha_9 \sin(3\,A_1) - \alpha_{10}\sin(A_1),$$

$$q_4 = \mathrm{cs}^2(2\,\beta, m) + \mathrm{ds}(2\,\beta, m)\,\mathrm{ns}(2\,\beta, m),$$

$$A_2 = A_0^2\,\{-2\,(\alpha_2 + \alpha_3)\cos(A_1)\,\mathrm{ds}(\beta, m)\,\mathrm{ns}(\beta, m) + 2\,[\alpha_4 + \alpha_6\cos(2\,A_1)]\,\mathrm{cs}^2(\beta, m) - [2\,\alpha_5\cos(2\,A_1) + \alpha_7]\,\mathrm{cs}^2(\beta, m) + \frac{2}{A_0^2}\},$$

$$\alpha_4 = -\alpha_6\cos(2\,A_1) - \frac{[\alpha_9\cos(3\,A_1) + \alpha_{10}\cos(A_1)]\,\mathrm{cs}(2\,\beta, m)}{\mathrm{cs}(\beta, m)},$$

$$\alpha_5 = -\frac{\alpha_7}{\cos(2\,A_1)} + \frac{1}{2\,\mathrm{cs}(\beta, m)\,\mathrm{cs}(2\,\beta, m)\cos(2\,A_1)}\left\{\frac{\cos(A_1)}{A_0^2} - (\alpha_2 + \alpha_3)\cos(A_1)\,\mathrm{cs}^2(\beta, m) + [\alpha_4 + \alpha_6\cos(2\,A_1)]\,\mathrm{ds}(\beta, m)\,\mathrm{ns}(\beta, m) + [\alpha_9\cos(3\,A_1) + \alpha_{10}\cos(A_1)]\,[\mathrm{ds}(2\,\beta, m)\,\mathrm{ns}(2\,\beta, m) - \mathrm{cs}^2(2\,\beta, m)],\right\},$$

$$\alpha_6 = -\frac{[\alpha_9\sin(3\,A_1) - \alpha_{10}\sin(A_1)]\,\mathrm{cs}(2\beta, m)}{\sin(2\,A_1)\,\mathrm{cs}(\beta, m)},$$

$$\alpha_7 = \frac{1}{\mathrm{cs}^2(\beta, m)\,\mathrm{cs}(2\,\beta, m)\,[\sec(2\,A_1) - 1]}(2\,(\alpha_2 + \alpha_3 + \alpha_9 + \alpha_{10} - 1)\,\mathrm{cs}^2(\beta, m)\,\mathrm{cs}(2\,\beta, m) - 2\,\alpha_3\cos(A_1)\,\mathrm{cs}^3(\beta, m)\sec(2\,A_1) + 2\,(\alpha_9 - \alpha_{10})\cos(A_1)\,\mathrm{cs}(2\,\beta, m)\,\mathrm{ds}(\beta, m)\,\mathrm{ns}(\beta, m)\sec(2\,A_1) + \mathrm{cs}(\beta, m)\,\{-2\,[\alpha_9 + (\alpha_{10} + 2\,\alpha_9)\cos(2\,A_1)]\,\mathrm{cs}^2(2\,\beta, m)\sec(A_1) + 2(\alpha_{10} - \alpha_9)\cos(A_1)\,\mathrm{ds}(2\,\beta, m)\,\mathrm{ns}(2\,\beta, m)\sec(2\,A_1)\}),$$

$0 \leqslant m \leqslant 1$,

$A_1, \beta, \gamma, \alpha_2, \alpha_3, \alpha_9, \alpha_{10}$, and ϕ_0 are arbitrary real constants.

- *Reference*: [8].

***Solution* 7. cn(*n*,*t*,*m*)** *moving discrete SW*
(Figure 8.29)

$$\psi(n,\, t) = A_0\, \sqrt{m}\, \mathrm{cn}[\beta\, (n - v\, t + \gamma),\, m]\, e^{i\,(A_1\, n - A_2\, t + \phi_0)}, \tag{8.67}$$

where

$a_1 = 1,$

$\alpha_1 = \alpha_8 = \alpha_{11} = \alpha_{12} = \alpha_{13} = \alpha_{14} = 0,$

$v = \frac{2\, A_0^2}{\beta}\, (\alpha_2 - \alpha_3)\, \sin(A_1)\, \mathrm{ds}(\beta,\, m),$

$A_0 = \sqrt{\frac{\sin(A_1)}{p_1 - p_2 - p_3\, p_4}},$

$p_1 = (\alpha_2 - \alpha_3)\, \sin(A_1)\, \mathrm{ds}^2(\beta,\, m),$

$p_2 = \alpha_6\, \sin(2\, A_1)\, \mathrm{cs}(\beta,\, m)\, \mathrm{ns}(\beta,\, m),$

$p_3 = \alpha_9\, \sin(3\, A_1) - \alpha_{10}\, \sin(A_1),$

$p_4 = \mathrm{ds}^2(2\, \beta,\, m) + \mathrm{cs}(2\, \beta,\, m)\, \mathrm{ns}(2\, \beta,\, m),$

$$\begin{aligned} A_2 = \; & A_0^2\, \{-2\, (\alpha_2 + \alpha_3)\, \cos(A_1)\, \mathrm{cs}(\beta,\, m)\, \mathrm{ns}(\beta,\, m) \\ & +2\, [\alpha_4 + \alpha_6\, \cos(2\, A_1)]\, \mathrm{ds}^2(\beta,\, m) \\ & -[2\, \alpha_5\, \cos(2\, A_1) + \alpha_7]\, \mathrm{ds}^2(\beta,\, m) + \tfrac{2}{A_0^2}\}, \end{aligned}$$

$\alpha_4 = -\alpha_6\, \cos(2\, A_1) - \frac{[\alpha_9\, \cos(3\, A_1) + \alpha_{10}\, \cos(A_1)]\, \mathrm{ds}(2\, \beta,\, m)}{\mathrm{ds}(\beta,\, m)},$

$$\begin{aligned} \alpha_5 = & -\frac{\alpha_7}{2\, \cos(2\, A_1)} + \frac{1}{2\, \mathrm{ds}(\beta,\, m)\, \mathrm{ds}(2\, \beta,\, m)\, \cos(2\, A_1)} \left\{ \frac{\cos(A_1)}{A_0^2} - (\alpha_2 + \alpha_3)\, \cos(A_1)\, \mathrm{ds}^2(\beta,\, m) \right. \\ & + [\alpha_4 + \alpha_6\, \cos(2\, A_1)]\, \mathrm{cs}(\beta,\, m)\, \mathrm{ns}(\beta,\, m) \\ & + [\alpha_9\, \cos(3\, A_1) + \alpha_{10}\, \cos(A_1)]\, [\mathrm{cs}(2\, \beta,\, m)\, \mathrm{ns}(2\, \beta,\, m) \\ & \left. - \, \mathrm{ds}^2(2\, \beta,\, m)] \right\}, \end{aligned}$$

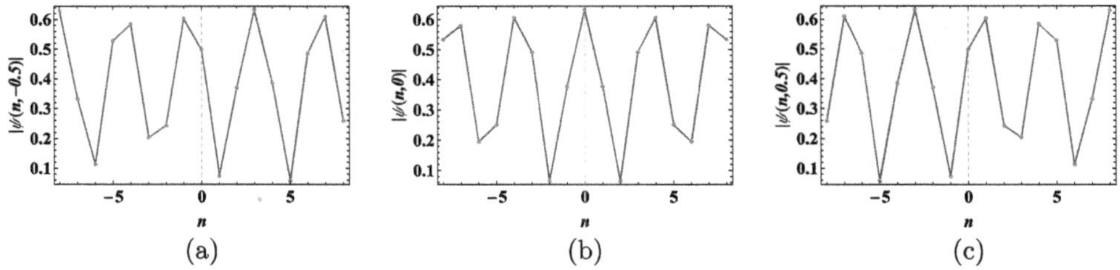

Figure 8.29. Moving discrete solitary wave (8.67) with $a_1 = A_1 = \alpha_2 = \alpha_9 = \alpha_{10} = \beta = 1$, $\alpha_3 = 2$, $m = 1/2$, and $\gamma = \phi_0 = 0$. (a) at $t = -0.5$, (b) at $t = 0$, and (c) at $t = 0.5$. The lines are guides for the eye.

$$\alpha_6 = -\frac{[\alpha_9 \sin(3\,A_1) - \alpha_{10} \sin(A_1)]\,\mathrm{ds}(2\beta, m)}{\sin(2\,A_1)\,\mathrm{ds}(\beta, m)},$$

$$\alpha_7 = \frac{1}{\mathrm{ds}^2(\beta, m)\,\mathrm{ds}(2\,\beta, m)\,[\sec(2\,A_1) - 1]}\,(2\,(\alpha_2 + \alpha_3 + \alpha_9 + \alpha_{10} - 1)\,\mathrm{ds}^2(\beta, m)\,\mathrm{ds}(2\,\beta, m)$$

$$- 2\,\alpha_3 \cos(A_1)\,\mathrm{ds}^3(\beta, m)\sec(2\,A_1)$$

$$+ 2\,(\alpha_9 - \alpha_{10})\cos(A_1)\,\mathrm{ds}(2\,\beta, m)\,\mathrm{cs}(\beta, m)\,\mathrm{ns}(\beta, m)\sec(2\,A_1)$$

$$+ \mathrm{ds}(\beta, m)\,\{-2\,[\alpha_9 + (\alpha_{10} + 2\,\alpha_9)\cos(2\,A_1)]\,\mathrm{ds}^2(2\,\beta, m)\sec(A_1)$$

$$+ 2(\alpha_{10} - \alpha_9)\cos(A_1)\,\mathrm{cs}(2\,\beta, m)\,\mathrm{ns}(2\,\beta, m)\sec(2\,A_1)\}),$$

$0 \leqslant m \leqslant 1$,

A_1, β, γ, α_2, α_3, α_9, α_{10}, and ϕ_0 are arbitrary real constants.

- *Reference*: [8].

8.9 Summary of Section 8.8

Note: For lengthy conditions, the reader is referred to the solutions in section 8.8.

Equation

$$i\,\psi_{n_t} + a_1\,(\psi_{n+1} + \psi_{n-1} - 2\,\psi_n) + f[\psi_{n-1}, \psi_n, \psi_{n+1}] = 0,$$

$$\begin{aligned} f[\psi_{n-1}, \psi_n, \psi_{n+1}] = {} & \alpha_1\,|\psi_n|^2\,\psi_n + \alpha_2\,|\psi_n|^2\,(\psi_{n+1} + \psi_{n-1}) + \alpha_3\,\psi_n^2\,(\psi_{n+1}^* + \psi_{n-1}^*) \\ & + \alpha_4\,\psi_n\,(|\psi_{n+1}|^2 + |\psi_{n-1}|^2) + \alpha_5\,\psi_n\,(\psi_{n+1}^*\,\psi_{n-1} + \psi_{n-1}^*\,\psi_{n+1}) \\ & + \alpha_6\,\psi_n^*\,(\psi_{n+1}^2 + \psi_{n-1}^2) + \alpha_7\,\psi_n^*\,\psi_{n+1}\,\psi_{n-1} \\ & + \alpha_8\,(|\psi_{n+1}|^2\,\psi_{n+1} + |\psi_{n-1}|^2\,\psi_{n-1}) + \alpha_9\,(\psi_{n-1}^*\,\psi_{n+1}^2 + \psi_{n+1}^*\,\psi_{n-1}^2) \\ & + \alpha_{10}\,(|\psi_{n+1}|^2\,\psi_{n-1} + |\psi_{n-1}|^2\,\psi_{n+1}) + \alpha_{11}\,(|\psi_{n-1}\,\psi_n| + |\psi_n\,\psi_{n+1}|)\,\psi_n \\ & + \alpha_{12}\,(\psi_{n+1}\,|\psi_{n+1}\,\psi_n| + \psi_{n-1}\,|\psi_n\,\psi_{n-1}|) \\ & + \alpha_{13}\,(\psi_{n+1}\,|\psi_{n-1}\,\psi_n| + \psi_{n-1}\,|\psi_n\,\psi_{n+1}|) \\ & + \alpha_{14}\,(\psi_{n+1}\,|\psi_{n-1}\,\psi_{n+1}| + \psi_{n-1}\,|\psi_{n-1}\,\psi_{n+1}|) \end{aligned}$$

#	Solution	Conditions	Name	Eq. #
1.	$\psi(n, t) = A_0\,\text{sech}[\beta\,(n + \gamma)]\,e^{-i\,(A_2\,t+\phi_0)}$	See text	discrete bright soliton	(8.61)
2.	$\psi(n, t) = A_0\,\text{sech}[\beta\,(n - v\,t + \gamma)]\,e^{i\,(A_1\,n-A_2\,t+\phi_0)}$	See text	moving discrete bright soliton	(8.62)
3.	$\psi(n, t) = A_0\,\tanh[\beta\,(n + \gamma)]\,e^{-i\,(A_2\,t+\phi_0)}$	See text	discrete dark soliton	(8.63)
4.	$\psi(n, t) = A_0\,\tanh[\beta\,(n - v\,t + \gamma)]\,e^{i\,(A_1\,n-A_2\,t+\phi_0)}$	See text	moving discrete dark soliton	(8.64)
5.	$\psi(n, t) = A_0\,\sin[\beta\,(n - v\,t + \gamma)]\,e^{i\,(A_1\,n-A_2\,t+\phi_0)}$	See text	–	(8.65)
6.	$\psi(n, t) = A_0\,\text{dn}[\beta\,(n - v\,t + \gamma), m]\,e^{i\,(A_1\,n-A_2\,t+\phi_0)}$	See text	moving discrete solitary wave	(8.66)
7.	$\psi(n, t) = A_0\,\sqrt{m}\,\text{cn}[\beta\,(n - v\,t + \gamma), m]\,e^{i\,(A_1\,n-A_2\,t+\phi_0)}$	See text	moving discrete solitary wave	(8.67)

8.10 Coupled Salerno Equations

Equation:

$$\begin{aligned} &i\,\psi_{1nt} + \psi_{1n+1} + \psi_{1n-1} - 2\,\psi_{1n} \\ &\quad + (\mu_1\,|\psi_{1n}|^2 + \mu_2\,|\psi_{2n}|^2)\left(\psi_{1n+1} + \psi_{1n-1} + \frac{\nu_1 - 2\,\mu_1}{\mu_1}\,\psi_{1n}\right) = 0, \\ &i\,\psi_{2nt} + \left[\psi_{2n+1} + \psi_{2n-1} - \left(2 + \frac{\nu_1\,\mu_2}{\mu_1^2} - \frac{\nu_2}{\mu_2}\right)\psi_{2n}\right] \\ &\quad + (\mu_1\,|\psi_{1n}|^2 + \mu_2\,|\psi_{2n}|^2)\left[\psi_{2n+1} + \psi_{2n-1} + \left(\frac{\nu_2 - 2\,\mu_2}{\mu_2}\right)\psi_{2n}\right] = 0, \end{aligned} \tag{8.68}$$

where

$\psi_j = \psi_j(n,\, t)$ is the complex function profile, $j = 1,\, 2$,
the integer site index, n, and t are its two independent variables,
μ_1, μ_2, ν_1, and ν_2 are real constants.

Solutions:

***Solution* 1. dn(*n*,*m*)-sn(*n*,*m*)** *discrete SW*
(Figure 8.30)

$$\begin{aligned} \psi_1(n,\, t) &= A_0\,\mathrm{dn}[A_1\,(n + n_0),\, m]\,e^{-i\,(\omega_1\,t + \phi_1)}, \\ \psi_2(n,\, t) &= B_0\,\sqrt{m}\,\mathrm{sn}[A_1\,(n + n_0),\, m]\,e^{-i\,(\omega_2\,t + \phi_2)}, \end{aligned} \tag{8.69}$$

where

$\omega_1 = \frac{\nu_1}{\mu_1}$,
$\omega_2 = \frac{\nu_1\,\mu_2}{\mu_1^2}$,
$\mu_1 = \frac{-1}{A_0^2}$,
$\mu_2 = \frac{\mu_1\,A_0^2}{B_0^2}$,
$0 \leqslant m \leqslant 1$,

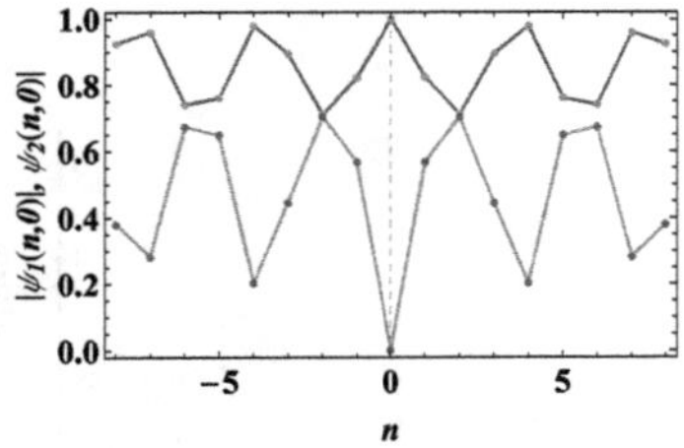

Figure 8.30. Discrete solitary wave (8.69) at $t = 0$. Blue is ψ_1 and red is ψ_2 with $A_0 = B_0 = A_1 = 1$, $\nu_1 = \nu_2 = 2$, $m = 1/2$, and $n_0 = t_0 = \phi_1 = \phi_2 = 0$. The lines are guides for the eye.

A_0, A_1, B_0, n_0, ϕ_1, and ϕ_2 are arbitrary real constants.

- *Reference*: [9].

***Solution* 2. cn(*n*,*m*)-sn(*n*,*m*)** *discrete SW*
(Figure 8.31)

$$\psi_1(n,\, t) = A_0\, \sqrt{m}\, \mathrm{cn}[A_1\, (n + n_0),\, m]\, e^{-i\,(\omega_1\, t+\phi_1)},$$
$$\psi_2(n,\, t) = B_0\, \sqrt{m}\, \mathrm{sn}[A_1\, (n + n_0),\, m]\, e^{-i\,(\omega_2\, t+\phi_2)}, \tag{8.70}$$

where
$\omega_1 = \frac{\nu_1}{\mu_1}$,
$\omega_2 = \frac{\nu_1\, \mu_2}{\mu_1^2}$,
$\mu_1 = \frac{-1}{m\, A_0^2}$,
$\mu_2 = \frac{\mu_1\, A_0^2}{B_0^2}$,
$0 \leqslant m \leqslant 1$,
A_0, A_1, B_0, n_0, ϕ_1, and ϕ_2 are arbitrary real constants.

- *Reference*: [9].

***Solution* 3. sech(*n*)-tanh(*n*)** *discrete bright-dark soliton*
(Figure 8.32)

$$\psi_1(n,\, t) = A_0\, \mathrm{sech}[A_1\, (n + n_0)]\, e^{-i\,(\omega_1\, t+\phi_1)},$$
$$\psi_2(n,\, t) = B_0\, \tanh[A_1\, (n + n_0)]\, e^{-i\,(\omega_2\, t+\phi_2)}, \tag{8.71}$$

where
$\omega_1 = \frac{\nu_1}{\mu_1}$,
$\omega_2 = \frac{\nu_1\, \mu_2}{\mu_1^2}$,
$\mu_1 = \frac{-1}{A_0^2}$,
$\mu_2 = \frac{\mu_1\, A_0^2}{B_0^2}$,

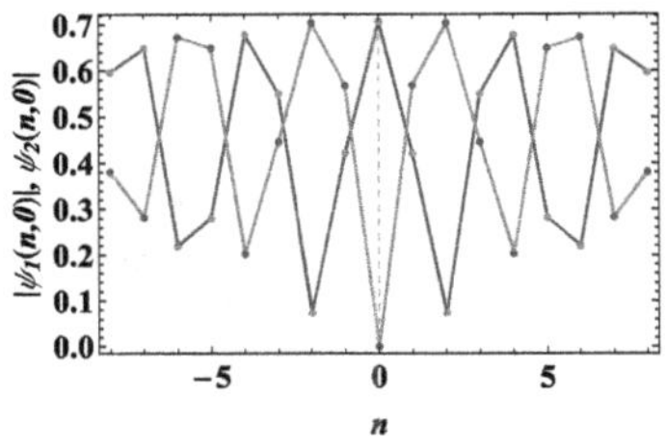

Figure 8.31. Discrete solitary wave (8.70) at $t = 0$. Blue is ψ_1 and red is ψ_2 with $A_0 = B_0 = A_1 = 1$, $\nu_1 = \nu_2 = 2$, $m = 1/2$, and $n_0 = t_0 = \phi_1 = \phi_2 = 0$. The lines are guides for the eye.

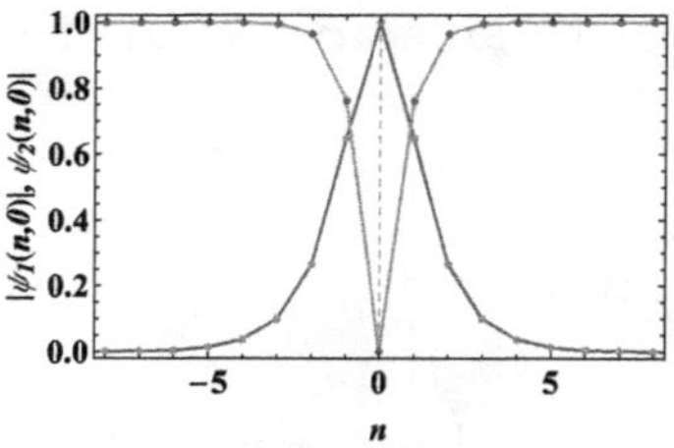

Figure 8.32. Discrete bright-dark soliton (8.71) at $t = 0$. Blue is ψ_1 and red is ψ_2 with $A_0 = B_0 = A_1 = 1$, $\nu_1 = \nu_2 = 2$, and $n_0 = t_0 = \phi_1 = \phi_2 = 0$. The lines are guides for the eye.

A_0, A_1, B_0, n_0, ϕ_1, and ϕ_2 are arbitrary real constants.

- *Reference*: [9].

Solution* 4. dn²(*n*,*m*)-sn(*n*,*m*) dn(*n*,*m*) *discrete SW

$$\begin{aligned} \psi_1(n, t) &= \left\{A_0 \,\mathrm{dn}^2[A_1 (n + n_0), m] + A_2\right\} e^{-i (\omega_1 t+\phi_1)}, \\ \psi_2(n, t) &= B_0 \sqrt{m}\, \mathrm{sn}[A_1 (n + n_0), m]\, \mathrm{dn}[A_1 (n + n_0), m]\, e^{-i (\omega_2 t+\phi_2)}, \end{aligned} \tag{8.72}$$

where

$A_0 = -2 A_2$,
$\omega_1 = \frac{\nu_1}{\mu_1}$,
$\omega_2 = \frac{\nu_1 \mu_2}{\mu_1^2}$,
$\mu_1 = \frac{-4}{A_0^2}$,
$\mu_2 = \frac{\mu_1 A_0^2}{B_0^2}$,
$0 \leqslant m \leqslant 1$,
A_1, A_2, B_0, n_0, ϕ_1, and ϕ_2 are arbitrary real constants.

- *Reference*: [9].

Solution* 5. dn²(*n*,*m*)-sn(*n*,*m*) cn(*n*,*m*) *discrete SW

$$\begin{aligned} \psi_1(n, t) &= \{A_0 \,\mathrm{dn}^2[A_1 (n + n_0), m] + A_2\} e^{-i (\omega_1 t+\phi_1)}, \\ \psi_2(n, t) &= B_0\, m\, \mathrm{sn}[A_1 (n + n_0), m]\, \mathrm{cn}[A_1 (n + n_0), m]\, e^{-i (\omega_2 t+\phi_2)}, \end{aligned} \tag{8.73}$$

where

$A_0 = \frac{-2 A_2}{2 - m}$,
$\omega_1 = \frac{\nu_1}{\mu_1}$,
$\omega_2 = \frac{\nu_1 \mu_2}{\mu_1^2}$,
$\mu_1 = \frac{-4}{m^2 A_0^2}$,
$\mu_2 = \frac{\mu_1 A_0^2}{B_0^2}$,

$0 \leqslant m \leqslant 1$,
A_1, A_2, B_0, n_0, ϕ_1, and ϕ_2 are arbitrary real constants.

- *Reference*: [9].

***Solution* 6. sech^2(*n*)-sech(*n*) tanh(*n*)**
(Figure 8.33)

$$\begin{aligned}\psi_1(n, t) &= \{A_0 \operatorname{sech}^2[A_1 (n + n_0)] + A_2\} e^{-i (\omega_1 t+\phi_1)},\\ \psi_2(n, t) &= B_0 \operatorname{sech}[A_1 (n + n_0)] \tanh[A_1 (n + n_0)] e^{-i (\omega_2 t+\phi_2)},\end{aligned} \tag{8.74}$$

where
$A_0 = -2 A_2$,
$\omega_1 = \frac{\nu_1}{\mu_1}$,
$\omega_2 = \frac{\nu_1 \mu_2}{\mu_1^2}$,
$\mu_1 = \frac{-4}{A_0^2}$,
$\mu_2 = \frac{\mu_1 A_0^2}{B_0^2}$,
A_1, A_2, B_0, n_0, ϕ_1, and ϕ_2 are arbitrary real constants.

- *Reference*: [9].

***Solution* 7. Rational Solution I**

$$\begin{aligned}\psi_1(n, t) &= \frac{A_0}{\sqrt{1 + n^2}} e^{-i (\omega_1 t+\phi_1)},\\ \psi_2(n, t) &= \frac{B_0 n}{\sqrt{1 + n^2}} e^{-i (\omega_2 t+\phi_2)},\end{aligned} \tag{8.75}$$

where
$\omega_1 = \frac{\nu_1}{\mu_1}$,
$\omega_2 = \frac{\nu_1 \mu_2}{\mu_1^2}$,
$\mu_1 = \frac{-1}{A_0^2}$,
$\mu_2 = \frac{\mu_1 A_0^2}{B_0^2}$,

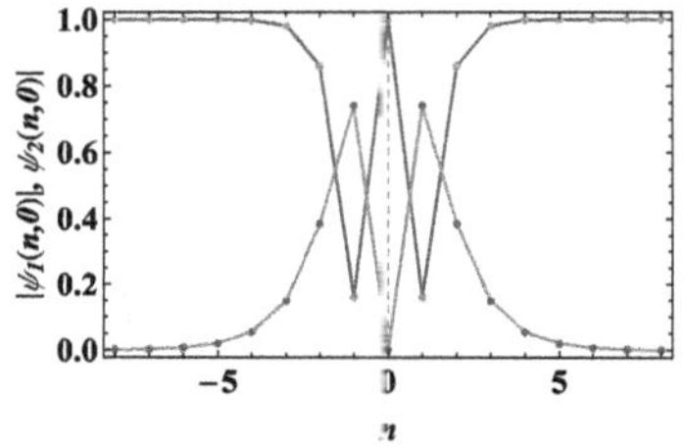

Figure 8.33. Plot of solution (8.74) at $t = 0$. Blue is ψ_1 and red is ψ_2 with $A_1 = A_2 = 1$, $B_0 = 3/2$, $\nu_1 = \nu_2 = 2$, $m = 1/2$, and $n_0 = t_0 = \phi_1 = \phi_2 = 0$. The lines are guides for the eye.

A_0, B_0, ϕ_1, and ϕ_2 are arbitrary real constants.

- *Reference*: [9].

***Solution* 8. Rational Solution II**

$$\begin{aligned}\psi_1(n,\,t) &= A_0\sqrt{\frac{1+n^2}{1+n^2+n^4}}\;e^{-i\,(\omega_1\,t+\phi_1)},\\ \psi_2(n,\,t) &= \frac{B_0\,n^2}{\sqrt{1+n^2+n^4}}\;e^{-i\,(\omega_2\,t+\phi_2)},\end{aligned}\tag{8.76}$$

where

$\omega_1 = \frac{\nu_1}{\mu_1}$,
$\omega_2 = \frac{\nu_1\,\mu_2}{\mu_1^2}$,
$\mu_1 = \frac{-1}{A_0^2}$,
$\mu_2 = \frac{\mu_1\,A_0^2}{B_0^2}$,
A_0, B_0, ϕ_1, and ϕ_2 are arbitrary real constants.

- *Reference*: [9].

***Solution* 9. Rational Solution III**

$$\begin{aligned}\psi_1(n,\,t) &= A_0\sqrt{\frac{2+n^2}{1+n^2}}\;e^{-i\,(\omega_1\,t+\phi_1)},\\ \psi_2(n,\,t) &= \frac{B_0}{\sqrt{1+n^2}}\;e^{-i\,(\omega_2\,t+\phi_2)},\end{aligned}\tag{8.77}$$

where

$\omega_1 = \frac{\nu_1}{\mu_1}$,
$\omega_2 = \frac{\nu_1\,\mu_2}{\mu_1^2}$,
$\mu_1 = \frac{-1}{A_0^2}$,
$\mu_2 = 1$,
A_0, B_0, ϕ_1, and ϕ_2 are arbitrary real constants.

- *Reference*: [9].

***Solution* 10. cos(*n*)-sin (*n*)**

$$\begin{aligned}\psi_1(n,\,t) &= A_0\cos[A_1\,(n+n_0)]\;e^{-i\,(\omega_1\,t+\phi_1)},\\ \psi_2(n,\,t) &= B_0\sin[A_1\,(n+n_0)]\;e^{-i\,(\omega_2\,t+\phi_2)},\end{aligned}\tag{8.78}$$

where

$\omega_1 = \frac{\nu_1}{\mu_1}$,

$\omega_2 = \frac{\nu_1\,\mu_2}{\mu_1^2}$,

$\mu_1 = \frac{-1}{A_0^2}$,

$\mu_2 = \frac{\mu_1\,A_0^2}{B_0^2}$,

A_0, B_0, n_0, ϕ_1, and ϕ_2 are arbitrary real constants.

- *Reference*: [9].

***Solution* 11. nd(*n*,*m*)-sd(*n*,*m*)** *discrete SW*

$$\begin{aligned}\psi_1(n,\,t) &= A_0\ \mathrm{nd}[A_1\,(n+n_0),\,m]\ e^{-i\,(\omega_1\,t+\phi_1)},\\ \psi_2(n,\,t) &= B_0\ \sqrt{m}\ \mathrm{sd}[A_1\,(n+n_0),\,m]\ e^{-i\,(\omega_2\,t+\phi_2)},\end{aligned} \tag{8.79}$$

where

$A_0 = \frac{\mu_2\,B_0^2}{|\,\mu_1\,|}$,

$\omega_1 = \frac{\nu_1}{\mu_1}$,

$\omega_2 = \frac{\nu_1\,\mu_2}{\mu_1^2}$,

$\mu_1 = -1$,

$\mu_2 = 1$,

$0 \leqslant m \leqslant 1$,

A_1, A_2, B_0, n_0, ϕ_1, and ϕ_2 are arbitrary real constants.

- *Reference*: [9].

***Solution* 12. cosh(*n*)-sinh(*n*)**

$$\begin{aligned}\psi_1(n,\,t) &= A_0\ \cosh[A_1\,(n+n_0)]\ e^{-i\,(\omega_1\,t+\phi_1)},\\ \psi_2(n,\,t) &= B_0\ \sinh[A_1\,(n+n_0)]\ e^{-i\,(\omega_2\,t+\phi_2)},\end{aligned} \tag{8.80}$$

where

$A_0 = \frac{\mu_2\,B_0^2}{|\,\mu_1\,|}$,

$\omega_1 = \frac{\nu_1}{\mu_1}$,

$\omega_2 = \frac{\nu_1\,\mu_2}{\mu_1^2}$,

$\mu_1 = -1$,

$\mu_2 = 1$,

A_1, B_0, n_0, ϕ_1, and ϕ_2 are arbitrary real constants.

- *Reference*: [9].

***Solution* 13.** *discrete SW*

$$\psi_1(n,\,t) = \{A_0\,\mathrm{nd}^2[A_1\,(n+n_0),\,m] + A_2\}\,e^{-i\,(\omega_1\,t+\phi_1)},$$
$$\psi_2(n,\,t) = \frac{B_0\,\sqrt{m}\,\mathrm{sn}[A_1\,(n+n_0),\,m]}{\mathrm{dn}^2[A_1\,(n+n_0),\,m]}\,e^{-i\,(\omega_2\,t+\phi_2)}, \tag{8.81}$$

where

$A_0 = -2\,A_2,$
$B_0 = \sqrt{\frac{|\mu_1|\,A_0^2}{\mu_2}},$
$\omega_1 = \frac{\nu_1}{\mu_1},$
$\omega_2 = \frac{\nu_1\,\mu_2}{\mu_1^2},$
$\mu_1 = \frac{-4}{A_0^2},$
$\mu_2 > 0,$
$0 \leqslant m \leqslant 1,$
A_1, A_2, n_0, ϕ_1, and ϕ_2 are arbitrary real constants.

- *Reference*: [9].

***Solution* 14.**

$$\psi_1(n,\,t) = \{A_0\,\cosh^2[A_1\,(n+n_0)] + A_2\}\,e^{-i\,(\omega_1\,t+\phi_1)},$$
$$\psi_2(n,\,t) = B_0\,\sinh[A_1\,(n+n_0)]\,\cosh[A_1\,(n+n_0)]\,e^{-i\,(\omega_2\,t+\phi_2)}, \tag{8.82}$$

where

$A_0 = -2\,A_2,$
$B_0 = \sqrt{\frac{|\mu_1|\,A_0^2}{\mu_2}},$
$\omega_1 = \frac{\nu_1}{\mu_1},$
$\omega_2 = \frac{\nu_1\,\mu_2}{\mu_1^2},$
$\mu_1 = \frac{-4}{A_0^2},$
$\mu_2 > 0,$
A_1, A_2, n_0, ϕ_1, and ϕ_2 are arbitrary real constants.

- *Reference*: [9].

8.11 Summary of Section 8.10

Equation

$$i\,\psi_{1n_t} + (\psi_{1n+1} + \psi_{1n-1} - 2\,\psi_{1n}) + (\mu_1\,|\psi_{1n}|^2 + \mu_2\,|\psi_{2n}|^2)\,(\psi_{1n+1} + \psi_{1n-1} + \tfrac{\nu_1 - 2\,\mu_1}{\mu_1}\,\psi_{1n}) = 0,$$

$$i\,\psi_{2n_t} + [\psi_{2n+1} + \psi_{2n-1} - (2 + \tfrac{\nu_1\,\mu_2}{\mu_1^2} - \tfrac{\nu_2}{\mu_2})\,\psi_{2n}] + (\mu_1\,|\psi_{1n}|^2 + \mu_2\,|\psi_{2n}|^2)\,(\psi_{2n+1} + \psi_{2n-1} + (\tfrac{\nu_2 - 2\,\mu_2}{\mu_2})\,\psi_{2n}) = 0$$

#	Solution	Conditions	Name	Eq. #
1.	$\psi_1(n,\,t) = A_0\,\mathrm{dn}[A_1\,(n + n_0),\,m]\,e^{-i\,(\omega_1\,t+\phi_1)}$, $\psi_2(n,\,t) = B_0\,\sqrt{m}\,\mathrm{sn}[A_1\,(n + n_0),\,m]\,e^{-i\,(\omega_2\,t+\phi_2)}$	$\omega_1 = \frac{\nu_1}{\mu_1}$, $\omega_2 = \frac{\nu_1\,\mu_2}{\mu_1^2}$, $\mu_1 = \frac{-1}{A_0^2}$, $\mu_2 = \frac{\mu_1\,A_0^2}{B_0^2}$, $0 \leqslant m \leqslant 1$, A_0, A_1, B_0, ϕ_1, and ϕ_2 are arbitrary real constants, n_0 is an arbitrary real integer	discrete solitary wave	(8.69)
2.	$\psi_1(n,\,t) = A_0\,\sqrt{m}\,\mathrm{cn}[A_1\,(n + n_0),\,m]\,e^{-i\,(\omega_1\,t+\phi_1)}$, $\psi_2(n,\,t) = B_0\,\sqrt{m}\,\mathrm{sn}[A_1\,(n + n_0),\,m]\,e^{-i\,(\omega_2\,t+\phi_2)}$	$\omega_1 = \frac{\nu_1}{\mu_1}$, $\omega_2 = \frac{\nu_1\,\mu_2}{\mu_1^2}$, $\mu_1 = \frac{-1}{m\,A_0^2}$, $\mu_2 = \frac{\mu_1\,A_0^2}{B_0^2}$, $0 \leqslant m \leqslant 1$, A_0, A_1, B_0, ϕ_1, and ϕ_2 are arbitrary real constants, n_0 is an arbitrary real integer	discrete solitary wave	(8.70)
3.	$\psi_1(n,\,t) = A_0\,\mathrm{sech}[A_1\,(n + n_0)]\,e^{-i\,(\omega_1\,t+\phi_1)}$, $\psi_2(n,\,t) = B_0\,\tanh[A_1\,(n + n_0)]\,e^{-i\,(\omega_2\,t+\phi_2)}$	$\omega_1 = \frac{\nu_1}{\mu_1}$, $\omega_2 = \frac{\nu_1\,\mu_2}{\mu_1^2}$, $\mu_1 = \frac{-1}{A_0^2}$, $\mu_2 = \frac{\mu_1\,A_0^2}{B_0^2}$, A_0, A_1, B_0, ϕ_1, and ϕ_2 are arbitrary real constants, n_0 is an arbitrary real integer	discrete bright-dark soliton	(8.71)
4.	$\psi_1(n,\,t) = \{A_0\,\mathrm{dn}^2[A_1\,(n + n_0),\,m] + A_2\}\,e^{-i\,(\omega_1\,t+\phi_1)}$, $\psi_2(n,\,t) = B_0\,\sqrt{m}\,\mathrm{sn}[A_1\,(n + n_0),\,m]\,\mathrm{dn}[A_1\,(n + n_0),\,m]\,e^{-i\,(\omega_2\,t+\phi_2)}$	$A_0 = -2\,A_2$, $\omega_1 = \frac{\nu_1}{\mu_1}$, $\omega_2 = \frac{\nu_1\,\mu_2}{\mu_1^2}$, $\mu_1 = \frac{-4}{A_0^2}$, $\mu_2 = \frac{\mu_1\,A_0^2}{B_0^2}$, $0 \leqslant m \leqslant 1$, A_1, A_2, B_0, ϕ_1, and ϕ_2 are arbitrary real constants, n_0 is an arbitrary real integer	discrete solitary wave	(8.72)

(*Continued*)

5. $\psi_1(n, t) = \{A_0\, \mathrm{dn}^2[A_1\,(n + n_0), m] + A_2\}\, e^{-i\,(\omega_1 t + \phi_1)}$, $\psi_2(n, t) = B_0\, m\, \mathrm{sn}[A_1\,(n + n_0), m]\, \mathrm{cn}[A_1\,(n + n_0), m]\, e^{-i\,(\omega_2 t + \phi_2)}$	$A_0 = \frac{-2\,A_2}{2 - m}$, $\omega_1 = \frac{\nu_1}{\mu_1}$, $\omega_2 = \frac{\nu_1\,\mu_2}{\mu_1^2}$, $\mu_1 = \frac{-4}{m^2\,A_0^2}$, $\mu_2 = \frac{\mu_1\,A_0^2}{B_0^2}$, $0 \leqslant m \leqslant 1$, A_1, A_2, B_0, ϕ_1, and ϕ_2 are arbitrary real constants, n_0 is an arbitrary real integer	discrete solitary wave	(8.73)
6. $\psi_1(n, t) = \{A_0\, \mathrm{sech}^2[A_1\,(n + n_0)] + A_2\}\, e^{-i\,(\omega_1 t + \phi_1)}$, $\psi_2(n, t) = B_0\, \mathrm{sech}[A_1\,(n + n_0)]\, \tanh[A_1\,(n + n_0)]\, e^{-i\,(\omega_2 t + \phi_2)}$	$A_0 = -2\,A_2$, $\omega_1 = \frac{\nu_1}{\mu_1}$, $\omega_2 = \frac{\nu_1\,\mu_2}{\mu_1^2}$, $\mu_1 = \frac{-4}{A_0^2}$, $\mu_2 = \frac{\mu_1\,A_0^2}{B_0^2}$, A_1, A_2, B_0, ϕ_1, and ϕ_2 are arbitrary real constants, n_0 is an arbitrary real integer	—	(8.74)
7. $\psi_1(n, t) = \frac{A_0}{\sqrt{1 + n^2}}\, e^{-i\,(\omega_1 t + \phi_1)}$, $\psi_2(n, t) = \frac{B_0\, n}{\sqrt{1 + n^2}}\, e^{-i\,(\omega_2 t + \phi_2)}$	$\omega_1 = \frac{\nu_1}{\mu_1}$, $\omega_2 = \frac{\nu_1\,\mu_2}{\mu_1^2}$, $\mu_1 = \frac{-1}{A_0^2}$, $\mu_2 = \frac{\mu_1\,A_0^2}{B_0^2}$, A_0, B_0, ϕ_1, and ϕ_2 are arbitrary real constants	—	(8.75)
8. $\psi_1(n, t) = A_0\,\sqrt{\frac{1 + n^2}{1 + n^2 + n^4}}\, e^{-i\,(\omega_1 t + \phi_1)}$, $\psi_2(n, t) = \frac{B_0\, n^2}{\sqrt{1 + n^2 + n^4}}\, e^{-i\,(\omega_2 t + \phi_2)}$	$\omega_1 = \frac{\nu_1}{\mu_1}$, $\omega_2 = \frac{\nu_1\,\mu_2}{\mu_1^2}$, $\mu_1 = \frac{-1}{A_0^2}$, $\mu_2 = \frac{\mu_1\,A_0^2}{B_0^2}$, A_0, B_0, ϕ_1, and ϕ_2 are arbitrary real constants	—	(8.76)
9. $\psi_1(n, t) = A_0\,\sqrt{\frac{2 + n^2}{1 + n^2}}\, e^{-i\,(\omega_1 t + \phi_1)}$, $\psi_2(n, t) = \frac{B_0}{\sqrt{1 + n^2}}\, e^{-i\,(\omega_2 t + \phi_2)}$	$\omega_1 = \frac{\nu_1}{\mu_1}$, $\omega_2 = \frac{\nu_1\,\mu_2}{\mu_1^2}$, $\mu_1 = \frac{-1}{A_0^2}$, $\mu_2 = 1$, A_0, B_0, ϕ_1, and ϕ_2 are arbitrary real constants	—	(8.77)
10. $\psi_1(n, t) = A_0\, \cos[A_1\,(n + n_0)]\, e^{-i\,(\omega_1 t + \phi_1)}$, $\psi_2(n, t) = B_0\, \sin[A_1\,(n + n_0)]\, e^{-i\,(\omega_2 t + \phi_2)}$	$\omega_1 = \frac{\nu_1}{\mu_1}$, $\omega_2 = \frac{\nu_1\,\mu_2}{\mu_1^2}$, $\mu_1 = \frac{-1}{A_0^2}$, $\mu_2 = \frac{\mu_1\,A_0^2}{B_0^2}$, A_0, B_0, ϕ_1, and ϕ_2 are arbitrary real constants, n_0 is an arbitrary real integer	—	(8.78)

11. $\psi_1(n,t) = A_0\,\mathrm{nd}[A_1\,(n+n_0),\,m]\,e^{-i\,(\omega_1 t+\phi_1)}$, $\psi_2(n,t) = B_0\,\sqrt{m}\,\mathrm{sd}[A_1\,(n+n_0),\,m]\,e^{-i\,(\omega_2 t+\phi_2)}$	$A_0 = \frac{\mu_2 B_0^2}{\lvert\mu_1\rvert}$, $\omega_1 = \frac{\nu_1}{\mu_1}$, $\omega_2 = \frac{\nu_1\mu_2}{\mu_1^2}$, $\mu_1 = -1$, $\mu_2 = 1$, $0 \leqslant m \leqslant 1$, A_1, A_2, B_0, ϕ_1, and ϕ_2 are arbitrary real constants, n_0 is an arbitrary real integer	discrete solitary wave	(8.79)
12. $\psi_1(n,t) = A_0\cosh[A_1\,(n+n_0)]\,e^{-i\,(\omega_1 t+\phi_1)}$, $\psi_2(n,t) = B_0\sinh[A_1\,(n+n_0)]\,e^{-i\,(\omega_2 t+\phi_2)}$	$A_0 = \frac{\mu_2 B_0^2}{\lvert\mu_1\rvert}$, $\omega_1 = \frac{\nu_1}{\mu_1}$, $\omega_2 = \frac{\nu_1\mu_2}{\mu_1^2}$, $\mu_1 = -1$, $\mu_2 = 1$, A_1, B_0, ϕ_1, and ϕ_2 are arbitrary real constants, n_0 is an arbitrary real integer	—	(8.80)
13. $\psi_1(n,t) = \{A_0\,\mathrm{nd}^2[A_1\,(n+n_0),\,m] + A_2\}\,e^{-i\,(\omega_1 t+\phi_1)}$, $\psi_2(n,t) = \frac{B_0\sqrt{m}\,\mathrm{sn}[A_1\,(n+n_0),\,m]}{\mathrm{dn}^2[A_1\,(n+n_0),\,m]}\,e^{-i\,(\omega_2 t+\phi_2)}$	$A_0 = -2\,A_2$, $B_0 = \sqrt{\frac{\lvert\mu_1\rvert\,A_0^2}{\mu_2}}$, $\omega_1 = \frac{\nu_1}{\mu_1}$, $\omega_2 = \frac{\nu_1\mu_2}{\mu_1^2}$, $\mu_1 = \frac{-4}{A_0^2}$, $\mu_2 > 0$, $0 \leqslant m \leqslant 1$, A_1, A_2, ϕ_1, and ϕ_2 are arbitrary real constants, n_0 is an arbitrary real integer	discrete solitary wave	(8.81)
14. $\psi_1(n,t) = \{A_0\cosh^2[A_1\,(n+n_0)] + A_2\}\,e^{-i\,(\omega_1 t+\phi_1)}$, $\psi_2(n,t) = B_0\sinh[A_1\,(n+n_0)]\cosh[A_1\,(n+n_0)]\,e^{-i\,(\omega_2 t+\phi_2)}$	$A_0 = -2\,A_2$, $B_0 = \sqrt{\frac{\lvert\mu_1\rvert\,A_0^2}{\mu_2}}$, $\omega_1 = \frac{\nu_1}{\mu_1}$, $\omega_2 = \frac{\nu_1\mu_2}{\mu_1^2}$, $\mu_1 = \frac{-4}{A_0^2}$, $\mu_2 > 0$, A_1, A_2, ϕ_1, and ϕ_2 are arbitrary real constants, n_0 is an arbitrary real integer	—	(8.82)

8.12 Coupled Ablowitz–Ladik Equation

Equation:

$$\begin{aligned} &i\,\psi_{1nt} + \psi_{1n+1} + \psi_{1n-1} - 2\,\psi_{1n} + (\mu_1\,|\psi_{1n}|^2 + \mu_2\,|\psi_{2n}|^2)\,(\psi_{1n+1} + \psi_{1n-1}) = 0,\\ &i\,\psi_{2nt} + \psi_{2n+1} + \psi_{2n-1} - \frac{2\,\mu_2}{\mu_1}\,\psi_{2n} + (\mu_1\,|\psi_{1n}|^2 + \mu_2\,|\psi_{2n}|^2)\,(\psi_{2n+1} + \psi_{2n-1}) = 0, \end{aligned} \tag{8.83}$$

where

μ_1 and μ_2 are real constants,
$\psi_j = \psi_j(n,\,t)$ is the complex function profile, $j = 1,\,2$,
the integer site index, n, and t are its two independent variables.

Solutions:

***Solution* 1. dn(*n*,*t*,*m*)-sn(*n*,*t*,*m*)** *moving discrete SW*
(Figure 8.34)

$$\begin{aligned} \psi_1(n,\,t) &= A_0\,\mathrm{dn}[A_1\,(n - v\,t + n_0),\,m]\,e^{-i\,(\omega_1\,t - \kappa_1\,n + \phi_1)},\\ \psi_2(n,\,t) &= B_0\,\sqrt{m}\,\mathrm{sn}[A_1\,(n - v\,t + n_0),\,m]\,e^{-i\,(\omega_2\,t - \kappa_2\,n + \phi_2)}, \end{aligned} \tag{8.84}$$

where

$A_0 = \sqrt{\dfrac{1 + \mu_2\,B_0^2\,\mathrm{ns}^2(A_1,m)}{\mu_1\,\mathrm{cs}^2(A_1,m)}}$,
$\kappa_2 = \sin^{-1}[\mathrm{cn}(A_1,\,m)\,\sin(\kappa_1)]$,
$v = \dfrac{2\,\sin(\kappa_1)\,(1 + \mu_2\,B_0^2)}{A_1\,\mathrm{cs}(A_1,m)}$,
$\omega_1 = 2 - \dfrac{2\,(1 + \mu_2\,B_0^2)\cos(\kappa_1)\,\mathrm{dn}(A_1,m)}{\mathrm{cn}^2(A_1,m)}$,
$\omega_2 = \dfrac{2\,\mu_2}{\mu_1} - \dfrac{2\,(1 + \mu_2\,B_0^2)\cos(\kappa_2)\,\mathrm{dn}(A_1,m)}{\mathrm{cn}(A_1,m)}$,

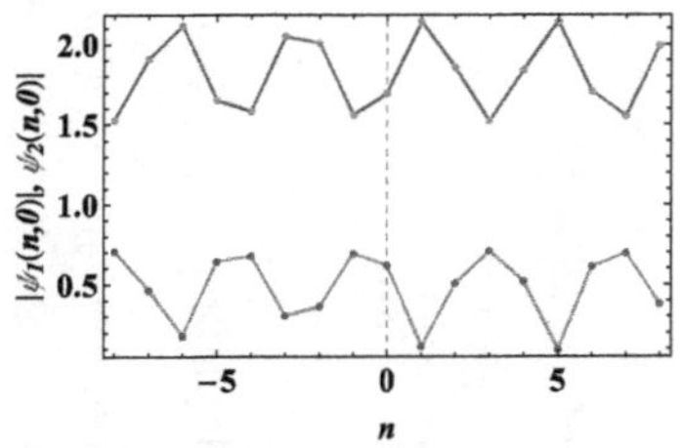

Figure 8.34. Moving discrete solitary wave (8.84) at $t = 0$. Blue is ψ_1 and red is ψ_2 with $B_0 = A_1 = 1$, $\mu_1 = -1$, $\mu_2 = \kappa_1 = 1$, $m = 1/2$, $n_0 = -1/3$, and $\phi_1 = \phi_2 = 0$. Lines are guide to the eye. Animation available online at https://iopscience.iop.org/book/978-0-7503-2428-1.

$\frac{1+\mu_2\, B_0^2\, \mathrm{ns}^2(A_1, m)}{\mu_1\, \mathrm{cs}^2(A_1, m)} > 0,$
$0 \leqslant m \leqslant 1,$
A_1, B_0, κ_1, ϕ_1, ϕ_2, μ_1, μ_2, and n_0 are arbitrary real constants.

- *Reference*: [9].

***Solution* 2. cn(*n*,*t*,*m*)-sn(*n*,*t*,*m*)** *moving discrete SW*

$$\psi_1(n,\, t) = A_0\, \sqrt{m}\, \mathrm{cn}[A_1\,(n - v\,t + n_0),\, m]\, e^{-i\,(\omega_1\,t - \kappa_1\,n + \phi_1)},$$
$$\psi_2(n,\, t) = B_0\, \sqrt{m}\, \mathrm{sn}[A_1\,(n - v\,t + n_0),\, m]\, e^{-i\,(\omega_2\,t - \kappa_2\,n + \phi_2)}, \tag{8.85}$$

where

$A_0 = \sqrt{\frac{1+\mu_2\, B_0^2\, \mathrm{ns}^2(A_1, m)}{\mu_1\, \mathrm{ds}^2(A_1, m)}},$
$\kappa_2 = \sin^{-1}[\mathrm{dn}(A_1,\, m)\, \sin(\kappa_1)],$
$v = \frac{2\sin(\kappa_1)\,(1 + m\,\mu_2\, B_0^2)}{A_1\, \mathrm{ds}(A_1, m)},$
$\omega_1 = 2 - \frac{2\,(1 + m\,\mu_2\, B_0^2)\cos(\kappa_1)\,\mathrm{cn}(A_1, m)}{\mathrm{dn}^2(A_1, m)},$
$\omega_2 = \frac{2\,\mu_2}{\mu_1} - \frac{2\,(1 + m\,\mu_2\, B_0^2)\cos(\kappa_2)\,\mathrm{cn}(A_1, m)}{\mathrm{dn}(A_1, m)},$
$\frac{1+\mu_2\, B_0^2\, \mathrm{ns}^2(A_1, m)}{\mu_1\, \mathrm{ds}^2(A_1, m)} > 0,$
$0 \leqslant m \leqslant 1,$
A_1, B_0, κ_1, ϕ_1, ϕ_2, μ_1, μ_2, and n_0 are arbitrary real constants.

- *Reference*: [9].

***Solution* 3. sech(*n*,*t*)-tanh(*n*,*t*)** *moving discrete bright-dark soliton* (Figure 8.35)

$$\psi_1(n,\, t) = A_0\, \mathrm{sech}[A_1\,(n - v\,t + n_0)]\, e^{-i\,(\omega_1\,t - \kappa_1\,n + \phi_1)},$$
$$\psi_2(n,\, t) = B_0\, \tanh[A_1\,(n - v\,t + n_0)]\, e^{-i\,(\omega_2\,t - \kappa_2\,n + \phi_2)}, \tag{8.86}$$

where

$A_0 = \sqrt{\frac{\sinh^2(A_1) + \mu_2\, B_0^2\, \cosh^2(A_1)}{\mu_1}},$
$\kappa_2 = \sin^{-1}[\mathrm{sech}(A_1)\, \sin(\kappa_1)],$
$v = \frac{2\sin(\kappa_1)\sinh(A_1)\,(1 + \mu_2\, B_0^2)}{A_1},$
$\omega_1 = 2 - 2\,(1 + \mu_2\, B_0^2)\cos(\kappa_1)\cosh(A_1),$
$\omega_2 = \frac{2\,\mu_2}{\mu_1} - 2\,(1 + \mu_2\, B_0^2)\cos(\kappa_2),$
$\frac{\sinh^2(A_1) + \mu_2\, B_0^2\, \cosh^2(A_1)}{\mu_1} > 0,$

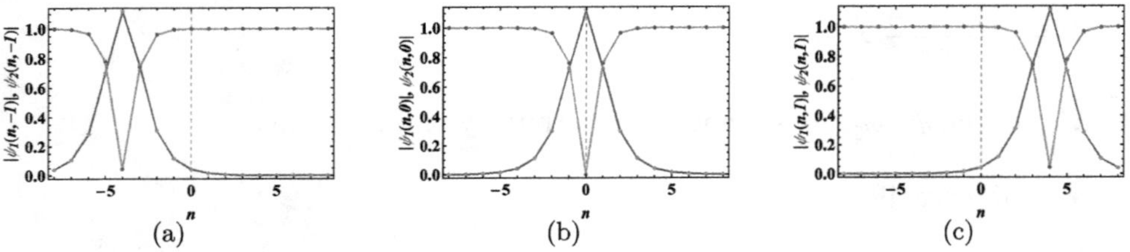

Figure 8.35. Moving discrete bright-dark soliton (8.86). Blue is ψ_1 and red is ψ_2 with $B_0 = A_1 = \mu_2 = \kappa_1 = 1$, $\mu_1 = 3$, and $n_0 = \phi_1 = \phi_2 = 0$. (a) at $t = -1$, (b) at $t = 0$, and (c) at $t = 1$. The lines are guides for the eye. Animation available online at https://iopscience.iop.org/book/978-0-7503-2428-1.

A_1, B_0, κ_1, ϕ_1, ϕ_2, μ_1, μ_2, and n_0 are arbitrary real constants.

- *Reference*: [9].

***Solution* 4. dn(*n*,*t*,*m*)-cn(*n*,*t*,*m*)** *moving discrete SW*

$$\begin{aligned}\psi_1(n,\,t) &= A_0\ \mathrm{dn}[A_1\,(n - v\,t + n_0),\,m]\,e^{-i\,(\omega_1 t - \kappa_1 n + \phi_1)},\\ \psi_2(n,\,t) &= B_0\,\sqrt{m}\ \mathrm{cn}[A_1\,(n - v\,t + n_0),\,m]\,e^{-i\,(\omega_2 t - \kappa_2 n + \phi_2)},\end{aligned} \tag{8.87}$$

where

$$A_0 = \sqrt{\frac{1 - \mu_2\,B_0^2\,\mathrm{ds}^2(A_1,m)}{\mu_1\,\mathrm{cs}^2(A_1,m)}},$$

$$\kappa_2 = \sin^{-1}[\frac{\sin(\kappa_1)\,\mathrm{cn}(A_1,m)}{\mathrm{dn}(A_1,m)}],$$

$$v = \frac{2\sin(\kappa_1)\,[1 - (1-m)\,\mu_2\,B_0^2]}{A_1\,\mathrm{cs}(A_1,m)},$$

$$\omega_1 = 2 - 2\,[1 - (1-m)\,\mu_2\,B_0^2]\left[\frac{\cos(\kappa_1)\,\mathrm{dn}(A_1,m)}{\mathrm{cn}^2(A_1,m)}\right],$$

$$\omega_2 = \frac{2\,\mu_2}{\mu_1} - 2\,[1 - (1-m)\,\mu_2\,B_0^2]\left[\frac{\cos(\kappa_2)}{\mathrm{cn}(A_1,m)}\right],$$

$$\frac{1 - \mu_2\,B_0^2\,\mathrm{ds}^2(A_1,m)}{\mu_1\,\mathrm{cs}^2(A_1,m)} > 0,$$

$$0 < m \leqslant 1,$$

A_1, B_0, κ_1, ϕ_1, ϕ_2, μ_1, μ_2, and n_0 are arbitrary real constants.

- *Reference*: [9].

Note: Solutions (5–18) below can be obtained from solutions (1–14) in section 8.10 with the replacements: $\omega_1 = 2$ and $\omega_2 = \frac{2\,\mu_2}{\mu_1}$.

***Solution* 5. dn(*n*,*m*)-sn(*n*,*m*)** *discrete SW*
(Figure 8.36)

$$\begin{aligned}\psi_1(n,\,t) &= A_0\ \mathrm{dn}[A_1\,(n + n_0),\,m]\,e^{-i\,(\omega_1 t + \phi_1)},\\ \psi_2(n,\,t) &= B_0\,\sqrt{m}\ \mathrm{sn}[A_1\,(n + n_0),\,m]\,e^{-i\,(\omega_2 t + \phi_2)},\end{aligned} \tag{8.88}$$

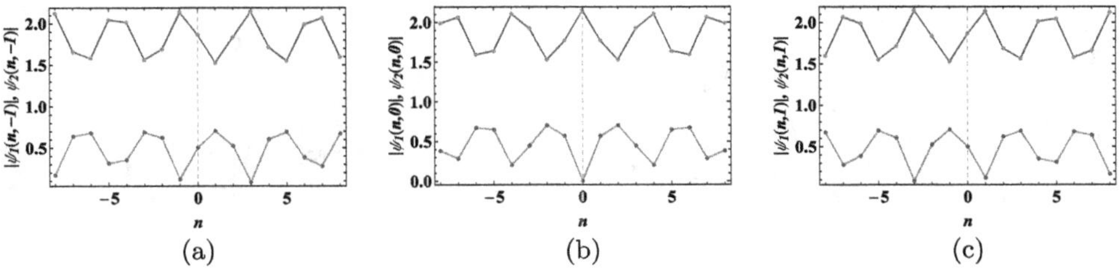

Figure 8.36. Discrete solitary wave (8.88). Blue is ψ_1 and red is ψ_2 with $B_0 = A_1 = \mu_1 = \mu_2 = \kappa_1 = 1$, $m = 1/2$, and $n_0 = \phi_1 = \phi_2 = 0$. (a) at $t = -1$, (b) at $t = 0$, and (c) at $t = 1$. The lines are guides for the eye.

where

$\omega_1 = 2,$

$\omega_2 = \frac{2\,\mu_2}{\mu_1},$

$\mu_1 = \frac{-1}{A_0^2},$

$\mu_2 = \frac{\mu_1\,A_0^2}{B_0^2},$

$0 \leqslant m \leqslant 1,$

A_0, A_1, B_0, ϕ_1, and ϕ_2 are arbitrary real constants,
n_0 is an arbitrary real integer.

- *Reference*: [9].

***Solution* 6. cn(*n*,*m*)-sn(*n*,*m*)** *discrete SW*
(Figure 8.37)

$$\begin{aligned}\psi_1(n,\,t) &= A_0\,\sqrt{m}\,\operatorname{cn}[A_1\,(n+n_0),\,m]\,e^{-i\,(\omega_1\,t+\phi_1)},\\ \psi_2(n,\,t) &= B_0\,\sqrt{m}\,\operatorname{sn}[A_1\,(n+n_0),\,m]\,e^{-i\,(\omega_2\,t+\phi_2)},\end{aligned} \tag{8.89}$$

where

$\omega_1 = 2,$

$\omega_2 = \frac{2\,\mu_2}{\mu_1},$

$\mu_1 = \frac{-1}{m\,A_0^2},$

$\mu_2 = \frac{\mu_1\,A_0^2}{B_0^2},$

$0 \leqslant m \leqslant 1,$

A_0, A_1, B_0, n_0, ϕ_1, and ϕ_2 are arbitrary real constants.

- *Reference*: [9].

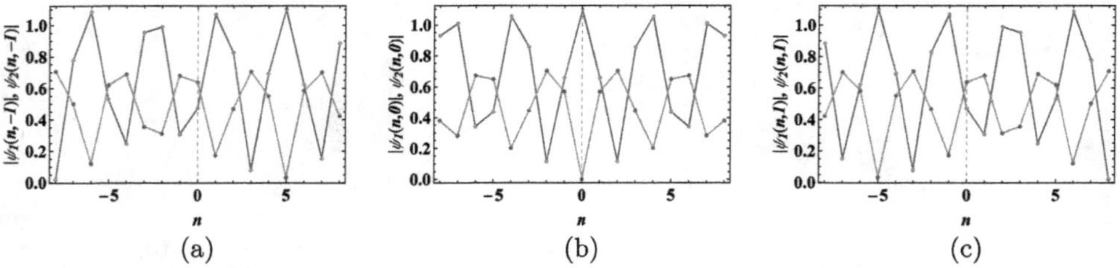

Figure 8.37. Discrete solitary wave (8.89). Blue is ψ_1 and red is ψ_2 with $B_0 = A_1 = \mu_1 = \mu_2 = \kappa_1 = 1$, $m = 1/2$, and $n_0 = \phi_1 = \phi_2 = 0$. (a) at $t = -1$, (b) at $t = 0$, and (c) at $t = 1$. The lines are guides for the eye.

***Solution* 7. sech(n)-tanh(n)** *discrete bright-dark soliton*

$$\begin{aligned} \psi_1(n,\, t) &= A_0\ \mathrm{sech}[A_1\,(n + n_0)]\ e^{-i\,(\omega_1\, t+\phi_1)}, \\ \psi_2(n,\, t) &= B_0\ \tanh[A_1\,(n + n_0)]\ e^{-i\,(\omega_2\, t+\phi_2)}, \end{aligned} \tag{8.90}$$

where

$\omega_1 = 2,$

$\omega_2 = \frac{2\,\mu_2}{\mu_1},$

$\mu_1 = \frac{-1}{A_0^2},$

$\mu_2 = \frac{\mu_1\, A_0^2}{B_0^2},$

A_0, A_1, B_0, n_0, ϕ_1, and ϕ_2 are arbitrary real constants.

- *Reference*: [9].

***Solution* 8. dn^2(n,m)-sn(n,m) dn(n,m)** *discrete SW*

$$\begin{aligned} \psi_1(n,\, t) &= \{A_0\ \mathrm{dn}^2[A_1\,(n + n_0),\, m] + A_2\}\ e^{-i\,(\omega_1\, t+\phi_1)}, \\ \psi_2(n,\, t) &= B_0\ \sqrt{m}\ \mathrm{sn}[A_1\,(n + n_0),\, m]\ \mathrm{dn}[A_1\,(n + n_0),\, m]\ e^{-i\,(\omega_2\, t+\phi_2)}, \end{aligned} \tag{8.91}$$

where

$A_0 = -2\,A_2,$

$\omega_1 = 2,$

$\omega_2 = \frac{2\,\mu_2}{\mu_1},$

$\mu_1 = \frac{-4}{A_0^2},$

$\mu_2 = \frac{\mu_1\, A_0^2}{B_0^2},$

$0 \leqslant m \leqslant 1,$

A_1, A_2, B_0, n_0, ϕ_1, and ϕ_2 are arbitrary real constants.

- *Reference*: [9].

Solution* 9. $\mathbf{dn^2(n,m)}$-$\mathbf{sn(n,m)cn(n,m)}$ *discrete SW

$$\psi_1(n,\, t) = \{A_0\, \mathrm{dn}^2[A_1\,(n + n_0),\, m] + A_2\}\; e^{-i\,(\omega_1\, t+\phi_1)},$$
$$\psi_2(n,\, t) = B_0\, m\, \mathrm{sn}[A_1\,(n + n_0),\, m]\, \mathrm{cn}[A_1\,(n + n_0),\, m]\; e^{-i\,(\omega_2\, t+\phi_2)}, \tag{8.92}$$

where

$A_0 = \frac{-2\,A_2}{2-m}$,
$\omega_1 = 2$,
$\omega_2 = \frac{2\,\mu_2}{\mu_1}$,
$\mu_1 = \frac{-4}{m^2\,A_0^2}$,
$\mu_2 = \frac{\mu_1\,A_0^2}{B_0^2}$,
$0 \leqslant m \leqslant 1$,
A_1, A_2, B_0, n_0, ϕ_1, and ϕ_2 are arbitrary real constants.

- *Reference*: [9].

***Solution* 10. $\mathbf{sech^2(n)}$-$\mathbf{sech(n)\,tanh(n)}$**

$$\psi_1(n,\, t) = \{A_0\, \mathrm{sech}^2[A_1\,(n + n_0)] - A_2\}\; e^{-i\,(\omega_1\, t+\phi_1)},$$
$$\psi_2(n,\, t) = B_0\, \mathrm{sech}[A_1\,(n + n_0)]\, \tanh[A_1\,(n + n_0)]\; e^{-i\,(\omega_2\, t+\phi_2)}, \tag{8.93}$$

where

$A_0 = -2\,A_2$,
$\omega_1 = 2$,
$\omega_2 = \frac{2\,\mu_2}{\mu_1}$,
$\mu_1 = \frac{-4}{A_0^2}$,
$\mu_2 = \frac{\mu_1\,A_0^2}{B_0^2}$,
A_1, A_2, B_0, n_0, ϕ_1, and ϕ_2 are arbitrary real constants.

- *Reference*: [9].

***Solution* 11. Rational Solution I**

$$\psi_1(n,\, t) = \frac{A_0}{\sqrt{1 + n^2}}\; e^{-i\,(\omega_1\, t+\phi_1)},$$
$$\psi_2(n,\, t) = \frac{B_0\, n}{\sqrt{1 + n^2}}\; e^{-i\,(\omega_2\, t+\phi_2)}, \tag{8.94}$$

where

$\omega_1 = 2,$

$\omega_2 = \frac{2\,\mu_2}{\mu_1},$

$\mu_1 = \frac{-1}{A_0^2},$

$\mu_2 = \frac{\mu_1\,A_0^2}{B_0^2},$

A_0, B_0, ϕ_1, and ϕ_2 are arbitrary real constants.

- *Reference*: [9].

***Solution* 12. Rational Solution II**

$$\begin{aligned} \psi_1(n,\,t) &= A_0\,\sqrt{\frac{1+n^2}{1+n^2+n^4}}\;e^{-i\,(\omega_1\,t+\phi_1)}, \\ \psi_2(n,\,t) &= \frac{B_0\,n^2}{\sqrt{1+n^2+n^4}}\;e^{-i\,(\omega_2\,t+\phi_2)}, \end{aligned} \tag{8.95}$$

where

$\omega_1 = 2,$

$\omega_2 = \frac{2\,\mu_2}{\mu_1},$

$\mu_1 = \frac{-1}{A_0^2},$

$\mu_2 = \frac{\mu_1\,A_0^2}{B_0^2},$

A_0, B_0, ϕ_1, and ϕ_2 are arbitrary real constants.

- *Reference*: [9].

***Solution* 13. Rational Solution III**

$$\begin{aligned} \psi_1(n,\,t) &= A_0\,\sqrt{\frac{2+n^2}{1+n^2}}\;e^{-i\,(\omega_1\,t+\phi_1)}, \\ \psi_2(n,\,t) &= \frac{B_0}{\sqrt{1+n^2}}\;e^{-i\,(\omega_2\,t+\phi_2)}, \end{aligned} \tag{8.96}$$

where

$\omega_1 = 2,$

$\omega_2 = \frac{2\,\mu_2}{\mu_1},$

$\mu_1 = \frac{-1}{A_0^2},$

$\mu_2 = 1,$

A_0, B_0, ϕ_1, and ϕ_2 are arbitrary real constants.

- *Reference*: [9].

***Solution* 14. cos(*n*)-sin(*n*)**

$$\psi_1(n,\, t) = A_0 \cos[A_1\,(n + n_0)]\, e^{-i\,(\omega_1\, t+\phi_1)},$$
$$\psi_2(n,\, t) = B_0 \sin[A_1\,(n + n_0)]\, e^{-i\,(\omega_2\, t+\phi_2)}, \tag{8.97}$$

where

$\omega_1 = 2,$
$\omega_2 = \frac{2\,\mu_2}{\mu_1},$
$\mu_1 = \frac{-1}{A_0^2},$
$\mu_2 = \frac{\mu_1\, A_0^2}{B_0^2},$
A_0, B_0, n_0, ϕ_1, and ϕ_2 are arbitrary real constants.

- *Reference*: [9].

***Solution* 15. nd(*n*,*m*)-sd(*n*,*m*)** *discrete SW*

$$\psi_1(n,\, t) = A_0\, \mathrm{nd}[A_1\,(n + n_0),\, m]\, e^{-i\,(\omega_1\, t+\phi_1)},$$
$$\psi_2(n,\, t) = B_0\, \sqrt{m}\, \mathrm{sd}[A_1\,(n + n_0),\, m]\, e^{-i\,(\omega_2\, t+\phi_2)}, \tag{8.98}$$

where

$A_0 = \frac{\mu_2\, B_0^2}{|\,\mu_1\,|},$
$\omega_1 = 2,$
$\omega_2 = \frac{2\,\mu_2}{\mu_1},$
$\mu_1 = -1,$
$\mu_2 = 1,$
$0 \leqslant m \leqslant 1,$
A_1, A_2, B_0, n_0, ϕ_1, and ϕ_2 are arbitrary real constants.

- *Reference*: [9].

***Solution* 16. cosh(*n*)-sinh(*n*)**

$$\psi_1(n,\, t) = A_0 \cosh[A_1\,(n + n_0)]\, e^{-i\,(\omega_1\, t+\phi_1)},$$
$$\psi_2(n,\, t) = B_0 \sinh[A_1\,(n + n_0)]\, e^{-i\,(\omega_2\, t+\phi_2)}, \tag{8.99}$$

where

$A_0 = \frac{\mu_2\, B_0^2}{|\,\mu_1\,|},$
$\omega_1 = 2,$
$\omega_2 = \frac{2\,\mu_2}{\mu_1},$

$\mu_1 = -1,$
$\mu_2 = 1$, A_1, B_0, n_0, ϕ_1, and ϕ_2 are arbitrary real constants.

- *Reference*: [9].

***Solution* 17.** *discrete SW*

$$\psi_1(n, t) = \{A_0 \operatorname{nd}^2[A_1 (n + n_0), m] + A_2\} e^{-i (\omega_1 t + \phi_1)},$$
$$\psi_2(n, t) = \frac{B_0 \sqrt{m} \operatorname{sn}[A_1 (n + n_0), m]}{\operatorname{dn}^2[A_1 (n + n_0), m]} e^{-i (\omega_2 t + \phi_2)}, \tag{8.100}$$

where
$A_0 = -2 A_2,$
$B_0 = \sqrt{\frac{|\mu_1| A_0^2}{\mu_2}},$
$\omega_1 = 2,$
$\omega_2 = \frac{2 \mu_2}{\mu_1},$
$\mu_1 = \frac{-4}{A_0^2},$
$\mu_2 > 0,$
$0 \leqslant m \leqslant 1,$
A_1, A_2, n_0, ϕ_1, and ϕ_2 are arbitrary real constants.

- *Reference*: [9].

***Solution* 18.**

$$\psi_1(n, t) = \{A_0 \cosh^2[A_1 (n + n_0)] + A_2\} e^{-i (\omega_1 t + \phi_1)},$$
$$\psi_2(n, t) = B_0 \sinh[A_1 (n + n_0)] \cosh[A_1 (n + n_0)] e^{-i (\omega_2 t + \phi_2)}, \tag{8.101}$$

where
$A_0 = -2 A_2,$
$B_0 = \sqrt{\frac{|\mu_1| A_0^2}{\mu_2}},$
$\omega_1 = 2,$
$\omega_2 = \frac{2 \mu_2}{\mu_1},$
$\mu_1 = \frac{-4}{A_0^2},$
$\mu_2 > 0,$
A_1, A_2, n_0, ϕ_1, and ϕ_2 are arbitrary real constants.

- *Reference*: [9].

8.13 Summary of Section 8.12

Equation

$$i\,\psi_{1n_t} + \psi_{1n+1} + \psi_{1n-1} - 2\,\psi_{1n} + (\mu_1\,|\psi_{1n}|^2 + \mu_2\,|\psi_{2n}|^2)\,(\psi_{1n+1} + \psi_{1n-1}) = 0,$$
$$i\,\psi_{2n_t} + \psi_{2n+1} + \psi_{2n-1} - \frac{2\,\mu_2}{\mu_1}\,\psi_{2n} + (\mu_1\,|\psi_{1n}|^2 + \mu_2\,|\psi_{2n}|^2)\,(\psi_{2n+1} + \psi_{2n-1}) = 0$$

#	Solution	Conditions	Name	Eq. #
1.	$\psi_1(n,\,t) = A_0\,\mathrm{dn}[A_1\,(n - v\,t + n_0),\,m]\,e^{-i\,(\omega_1\,t-\kappa_1\,n+\phi_1)}$, $\psi_2(n,\,t) = B_0\,\sqrt{m}\,\mathrm{sn}[A_1\,(n - v\,t + n_0),\,m]\,e^{-i\,(\omega_2\,t-\kappa_2\,n+\phi_2)}$	$A_0 = \sqrt{\frac{1+\mu_2\,B_0^2\,\mathrm{ns}^2(A_1,m)}{\mu_1\,\mathrm{cs}^2(A_1,m)}}$, $v = \frac{2\,\sin(\kappa_1)\,(1+\mu_2\,B_0^2)}{A_1\,\mathrm{cs}(A_1,m)}$, $\kappa_2 = \sin^{-1}[\mathrm{cn}(A_1,\,m)\,\sin(\kappa_1)]$, $\frac{1+\mu_2\,B_0^2\,\mathrm{ns}^2(A_1,m)}{\mu_1\,\mathrm{cs}^2(A_1,m)} > 0$, $\omega_1 = 2 - \frac{2\,(1+\mu_2\,B_0^2)\cos(\kappa_1)\,\mathrm{dn}(A_1,m)}{\mathrm{cn}^2(A_1,m)}$, $\omega_2 = \frac{2\,\mu_2}{\mu_1} - \frac{2\,(1+\mu_2\,B_0^2)\cos(\kappa_2)\,\mathrm{dn}(A_1,m)}{\mathrm{cn}(A_1,m)}$, $0 \leqslant m \leqslant 1$, A_1, B_0, κ_1, ϕ_1, ϕ_2, μ_1, μ_2, and n_0 are arbitrary real constants	moving discrete solitary wave	(8.84)
2.	$\psi_1(n,\,t) = A_0\,\sqrt{m}\,\mathrm{cn}[A_1\,(n - v\,t + n_0),\,m]\,e^{-i\,(\omega_1\,t-\kappa_1\,n+\phi_1)}$, $\psi_2(n,\,t) = B_0\,\sqrt{m}\,\mathrm{sn}[A_1\,(n - v\,t + n_0),\,m]\,e^{-i\,(\omega_2\,t-\kappa_2\,n+\phi_2)}$	$A_0 = \sqrt{\frac{1+\mu_2\,B_0^2\,\mathrm{ns}^2(A_1,m)}{\mu_1\,\mathrm{ds}^2(A_1,m)}}$, $v = \frac{2\,\sin(\kappa_1)\,(1+m\,\mu_2\,B_0^2)}{A_1\,\mathrm{ds}(A_1,m)}$, $\kappa_2 = \sin^{-1}[\mathrm{dn}(A_1,\,m)\,\sin(\kappa_1)]$, $\frac{1+\mu_2\,B_0^2\,\mathrm{ns}^2(A_1,m)}{\mu_1\,\mathrm{ds}^2(A_1,m)} > 0$, $\omega_1 = 2 - \frac{2\,(1+m\,\mu_2\,B_0^2)\cos(\kappa_1)\,\mathrm{cn}(A_1,m)}{\mathrm{dn}^2(A_1,m)}$, $\omega_2 = \frac{2\,\mu_2}{\mu_1} - \frac{2\,(1+m\,\mu_2\,B_0^2)\cos(\kappa_2)\,\mathrm{cn}(A_1,m)}{\mathrm{dn}(A_1,m)}$, $0 \leqslant m \leqslant 1$, A_1, B_0, κ_1, ϕ_1, ϕ_2, μ_1, μ_2, and n_0 are arbitrary real constants	moving discrete solitary wave	(8.85)

(*Continued*)

3. $\psi_1(n, t) = A_0 \operatorname{sech}[A_1 (n - v t + n_0)]\, e^{-i (\omega_1 t - \kappa_1 n + \phi_1)}$, $\psi_2(n, t) = B_0 \tanh[A_1 (n - v t + n_0)]\, e^{-i (\omega_2 t - \kappa_2 n + \phi_2)}$	$A_0 = \sqrt{\frac{\sinh^2(A_1) + \mu_2 B_0^2 \cosh^2(A_1)}{\mu_1}}$, $\kappa_2 = \sin^{-1}[\operatorname{sech}(A_1) \sin(\kappa_1)]$, $v = \frac{2 \sin(\kappa_1) \sinh(A_1) (1 + \mu_2 B_0^2)}{A_1}$, $\omega_1 = 2 - 2 (1 + \mu_2 B_0^2) \cos(\kappa_1) \cosh(A_1)$, $\omega_2 = \frac{2 \mu_2}{\mu_1} - 2 (1 + \mu_2 B_0^2) \cos(\kappa_2)$, $\frac{\sinh^2(A_1) + \mu_2 B_0^2 \cosh^2(A_1)}{\mu_1} > 0$, A_1, B_0, κ_1, ϕ_1, ϕ_2, μ_1, μ_2, and n_0 are arbitrary real constants	moving discrete bright-dark soliton	(8.86)
4. $\psi_1(n, t) = A_0 \operatorname{dn}[A_1 (n - v t + n_0), m]\, e^{-i (\omega_1 t - \kappa_1 n + \phi_1)}$, $\psi_2(n, t) = B_0 \sqrt{m} \operatorname{cn}[A_1 (n - v t + n_0), m]\, e^{-i (\omega_2 t - \kappa_2 n + \phi_2)}$	$A_0 = \sqrt{\frac{1 - \mu_2 B_0^2 \operatorname{ds}^2(A_1, m)}{\mu_1 \operatorname{cs}^2(A_1, m)}}$, $\kappa_2 = \sin^{-1}[\frac{\sin(\kappa_1) \operatorname{cn}(A_1, m)}{\operatorname{dn}(A_1, m)}]$, $\omega_1 = 2 - 2 [1 - (1 - m) \mu_2 B_0^2]\left[\frac{\cos(\kappa_1) \operatorname{dn}(A_1, m)}{\operatorname{cn}^2(A_1, m)}\right]$, $\omega_2 = \frac{2 \mu_2}{\mu_1} - 2 [1 - (1 - m) \mu_2 B_0^2]\left[\frac{\cos(\kappa_2)}{\operatorname{cn}(A_1, m)}\right]$, $0 < m \leqslant 1$, $v = \frac{2 \sin(\kappa_1) [1 - (1 - m) \mu_2 B_0^2]}{A_1 \operatorname{cs}(A_1, m)}$, $\frac{1 - \mu_2 B_0^2 \operatorname{ds}^2(A_1, m)}{\mu_1 \operatorname{cs}^2(A_1, m)} > 0$, A_1, B_0, κ_1, ϕ_1, ϕ_2, μ_1, μ_2, and n_0 are arbitrary real constants	moving discrete solitary wave	(8.87)
5. $\psi_1(n, t) = A_0 \operatorname{dn}[A_1 (n + n_0), m]\, e^{-i (\omega_1 t + \phi_1)}$, $\psi_2(n, t) = B_0 \sqrt{m} \operatorname{sn}[A_1 (n + n_0), m]\, e^{-i (\omega_2 t + \phi_2)}$	$\omega_1 = 2$, $\omega_2 = \frac{2 \mu_2}{\mu_1}$, $\mu_1 = \frac{-1}{A_0^2}$, $\mu_2 = \frac{\mu_1 A_0^2}{B_0^2}$, $0 \leqslant m \leqslant 1$, A_0, A_1, B_0, ϕ_1, and ϕ_2 are arbitrary real constants, n_0 is an arbitrary real integer	discrete solitary wave	(8.88)
6. $\psi_1(n, t) = A_0 \sqrt{m} \operatorname{cn}[A_1 (n + n_0), m]\, e^{-i (\omega_1 t + \phi_1)}$, $\psi_2(n, t) = B_0 \sqrt{m} \operatorname{sn}[A_1 (n + n_0), m]\, e^{-i (\omega_2 t + \phi_2)}$	$\omega_1 = 2$, $\omega_2 = \frac{2 \mu_2}{\mu_1}$, $\mu_1 = \frac{-1}{m A_0^2}$, $\mu_2 = \frac{\mu_1 A_0^2}{B_0^2}$, $0 \leqslant m \leqslant 1$, A_0, A_1, B_0, ϕ_1, and ϕ_2 are arbitrary real constants, n_0 is an arbitrary real integer	discrete solitary wave	(8.89)

7. $\psi_1(n, t) = A_0 \operatorname{sech}[A_1 (n + n_0)]\, e^{-i(\omega_1 t + \phi_1)}$, $\psi_2(n, t) = B_0 \tanh[A_1 (n + n_0)]\, e^{-i(\omega_2 t + \phi_2)}$	$\omega_1 = 2$, $\omega_2 = \frac{2\mu_2}{\mu_1}$, $\mu_1 = \frac{-1}{A_0^2}$, $\mu_2 = \frac{\mu_1 A_0^2}{B_0^2}$, A_0, A_1, B_0, ϕ_1, and ϕ_2 are arbitrary real constants, n_0 is an arbitrary real integer	discrete bright-dark soliton	(8.90)
8. $\psi_1(n, t) = \{A_0 \operatorname{dn}^2[A_1 (n + n_0), m] + A_2\}\, e^{-i(\omega_1 t + \phi_1)}$, $\psi_2(n, t) = B_0 \sqrt{m} \operatorname{sn}[A_1 (n + n_0), m] \operatorname{dn}[A_1 (n + n_0), m]\, e^{-i(\omega_2 t + \phi_2)}$	$A_0 = -2A_2$, $\omega_1 = 2$, $\omega_2 = \frac{2\mu_2}{\mu_1}$, $\mu_1 = \frac{-4}{A_0^2}$, $\mu_2 = \frac{\mu_1 A_0^2}{B_0^2}$, $0 \leqslant m \leqslant 1$, A_1, A_2, B_0, ϕ_1, and ϕ_2 are arbitrary real constants, n_0 is an arbitrary real integer	discrete solitary wave	(8.91)
9. $\psi_1(n, t) = \{A_0 \operatorname{dn}^2[A_1 (n + n_0), m] + A_2\}\, e^{-i(\omega_1 t + \phi_1)}$, $\psi_2(n, t) = B_0\, m \operatorname{sn}[A_1 (n + n_0), m] \operatorname{cn}[A_1 (n + n_0), m]\, e^{-i(\omega_2 t + \phi_2)}$	$A_0 = \frac{-2A_2}{2-m}$, $\omega_1 = 2$, $\omega_2 = \frac{2\mu_2}{\mu_1}$, $\mu_1 = \frac{-4}{m^2 A_0^2}$, $\mu_2 = \frac{\mu_1 A_0^2}{B_0^2}$, $0 \leqslant m \leqslant 1$, A_1, A_2, B_0, ϕ_1, and ϕ_2 are arbitrary real constants, n_0 is an arbitrary real integer	discrete solitary wave	(8.92)
10. $\psi_1(n, t) = \{A_0 \operatorname{sech}^2[A_1 (n + n_0)] + A_2\}\, e^{-i(\omega_1 t + \phi_1)}$, $\psi_2(n, t) = B_0 \operatorname{sech}[A_1 (n + n_0)] \tanh[A_1 (n + n_0)]\, e^{-i(\omega_2 t + \phi_2)}$	$A_0 = -2A_2$, $\omega_1 = 2$, $\omega_2 = \frac{2\mu_2}{\mu_1}$, $\mu_1 = \frac{-4}{A_0^2}$, $\mu_2 = \frac{\mu_1 A_0^2}{B_0^2}$, A_1, A_2, B_0, ϕ_1, and ϕ_2 are arbitrary real constants, n_0 is an arbitrary real integer	—	(8.93)
11. $\psi_1(n, t) = \frac{A_0}{\sqrt{1+n^2}}\, e^{-i(\omega_1 t + \phi_1)}$, $\psi_2(n, t) = \frac{B_0\, n}{\sqrt{1+n^2}}\, e^{-i(\omega_2 t + \phi_2)}$	$\omega_1 = 2$, $\omega_2 = \frac{2\mu_2}{\mu_1}$, $\mu_1 = \frac{-1}{A_0^2}$, $\mu_2 = \frac{\mu_1 A_0^2}{B_0^2}$, A_0, B_0, ϕ_1, and ϕ_2 are arbitrary real constants	—	(8.94)
12. $\psi_1(n, t) = A_0 \sqrt{\frac{1+n^2}{1+n^2+n^4}}\, e^{-i(\omega_1 t + \phi_1)}$, $\psi_2(n, t) = \frac{B_0\, n^2}{\sqrt{1+n^2+n^4}}\, e^{-i(\omega_2 t + \phi_2)}$	$\omega_1 = 2$, $\omega_2 = \frac{2\mu_2}{\mu_1}$, $\mu_1 = \frac{-1}{A_0^2}$, $\mu_2 = \frac{\mu_1 A_0^2}{B_0^2}$, A_0, B_0, ϕ_1, and ϕ_2 are arbitrary real constants	—	(8.95)
13. $\psi_1(n, t) = A_0 \sqrt{\frac{2+n^2}{1+n^2}}\, e^{-i(\omega_1 t + \phi_1)}$, $\psi_2(n, t) = \frac{B_0}{\sqrt{1+n^2}}\, e^{-i(\omega_2 t + \phi_2)}$	$\omega_1 = 2$, $\omega_2 = \frac{2\mu_2}{\mu_1}$, $\mu_1 = \frac{-1}{A_0^2}$, $\mu_2 = 1$, A_0, B_0, ϕ_1, and ϕ_2 are arbitrary real constants	—	(8.96)

(Continued)

14. $\psi_1(n,t) = A_0 \cos[A_1 (n+n_0)]\, e^{-i(\omega_1 t+\phi_1)}$, $\psi_2(n,t) = B_0 \sin[A_1 (n+n_0)]\, e^{-i(\omega_2 t+\phi_2)}$	$\omega_1 = 2$, $\omega_2 = \frac{2\mu_2}{\mu_1}$, $\mu_1 = \frac{-1}{A_0^2}$, $\mu_2 = \frac{\mu_1 A_0^2}{B_0^2}$, A_0, B_0, ϕ_1, and ϕ_2 are arbitrary real constants,		(8.97)
15. $\psi_1(n,t) = A_0\, \mathrm{nd}[A_1 (n+n_0), m]\, e^{-i(\omega_1 t+\phi_1)}$, $\psi_2(n,t) = B_0 \sqrt{m}\, \mathrm{sd}[A_1 (n+n_0), m]\, e^{-i(\omega_2 t+\phi_2)}$	n_0 is an arbitrary real integer $A_0 = \frac{\mu_2 B_0^2}{\lvert\mu_1\rvert}$, $\omega_1 = 2$, $\omega_2 = \frac{2\mu_2}{\mu_1}$, $\mu_1 = -1$, $\mu_2 = 1$, $0 \leqslant m \leqslant 1$, A_1, A_2, B_0, ϕ_1, and ϕ_2 are arbitrary real constants,	discrete solitary wave	(8.98)
16. $\psi_1(n,t) = A_0 \cosh[A_1 (n+n_0)]\, e^{-i(\omega_1 t+\phi_1)}$, $\psi_2(n,t) = B_0 \sinh[A_1 (n+n_0)]\, e^{-i(\omega_2 t+\phi_2)}$	n_0 is an arbitrary real integer $A_0 = \frac{\mu_2 B_0^2}{\lvert\mu_1\rvert}$, $\omega_1 = 2$, $\omega_2 = \frac{2\mu_2}{\mu_1}$, $\mu_1 = -1$, $\mu_2 = 1$, A_1, B_0, ϕ_1, and ϕ_2 are arbitrary real constants, n_0 is an arbitrary real integer	–	(8.99)
17. $\psi_1(n,t) = \{A_0\, \mathrm{nd}^2[A_1 (n+n_0), m] + A_2\}\, e^{-i(\omega_1 t+\phi_1)}$, $\psi_2(n,t) = \frac{B_0 \sqrt{m}\, \mathrm{sn}[A_1 (n+n_0), m]}{\mathrm{dn}^2[A_1 (n+n_0), m]}\, e^{-i(\omega_2 t+\phi_2)}$	$A_0 = -2A_2$, $B_0 = \sqrt{\frac{\lvert\mu_1\rvert A_0^2}{\mu_2}}$, $\omega_1 = 2$, $\omega_2 = \frac{2\mu_2}{\mu_1}$, $\mu_1 = \frac{-4}{A_0^2}$, $\mu_2 > 0$, $0 \leqslant m \leqslant 1$, A_1, A_2, ϕ_1, and ϕ_2 are arbitrary real constants, n_0 is an arbitrary real integer	discrete solitary wave	(8.100)
18. $\psi_1(n,t) = \{A_0 \cosh^2[A_1 (n+n_0)] + A_2\}\, e^{-i(\omega_1 t+\phi_1)}$, $\psi_2(n,t) = B_0 \sinh[A_1 (n+n_0)] \cosh[A_1 (n+n_0)]\, e^{-i(\omega_2 t+\phi_2)}$	$A_0 = -2A_2$, $B_0 = \sqrt{\frac{\lvert\mu_1\rvert A_0^2}{\mu_2}}$, $\omega_1 = 2$, $\omega_2 = \frac{2\mu_2}{\mu_1}$, $\mu_1 = \frac{-4}{A_0^2}$, $\mu_2 > 0$, A_1, A_2, ϕ_1, and ϕ_2 are arbitrary real constants, n_0 is an arbitrary real integer	–	(8.101)

8.14 Coupled Saturable Discrete NLSE

Equation:

$$i\,\psi_{1n_t} + \psi_{1n+1} + \psi_{1n-1} - 2\,\psi_{1n} + \frac{\nu_1\,(\mu_1\,|\psi_{1n}|^2 + \mu_2\,|\psi_{2n}|^2)\,\psi_{1n}}{\mu_1\,(1 + \mu_1\,|\psi_{1n}|^2 + \mu_2\,|\psi_{2n}|^2)} = 0,$$

$$i\,\psi_{2n_t} + \psi_{2n+1} + \psi_{2n-1} - 2\,\psi_{2n} + \frac{\left[\nu_2 - \frac{\nu_1\,\mu_2^2}{\mu_1^2} + \nu_2\,(\mu_1\,|\psi_{1n}|^2 + \mu_2\,|\psi_{2n}|^2)\right]\psi_{2n}}{\mu_2\,(1 + \mu_1\,|\psi_{1n}|^2 + \mu_2\,|\psi_{2n}|^2)} = 0, \quad (8.102)$$

where

$\psi_j = \psi_j(n, t)$ is the complex function profile, $j = 1, 2$,
the integer site index, n, and t are its two independent variables,
μ_1, μ_2, ν_1, and ν_2 are real constants.

Solutions:

***Solution* 1. dn(*n*,*m*)-sn(*n*,*m*)** *discrete SW*

$$\psi_1(n, t) = A_0\,\mathrm{dn}[A_1\,(n + n_0), m]\,e^{-i\,(\omega_1 t + \phi_1)},$$
$$\psi_2(n, t) = B_0\,\sqrt{m}\,\mathrm{sn}[A_1\,(n + n_0), m]\,e^{-i\,(\omega_2 t + \phi_2)}, \quad (8.103)$$

where

$A_0 = \sqrt{\frac{\nu_1}{2\,\mu_1^2\,\mathrm{dn}(A_1, m)} - \frac{1}{\mu_1}}$,
$B_0 = \sqrt{\frac{\nu_1\,\mathrm{cn}^2(A_1, m)}{2\,\mu_1\,\mu_2\,\mathrm{dn}(A_1, m)} - \frac{1}{\mu_2}}$,
$A_1 = \mu_1$,
$\omega_1 = 2 - \frac{\nu_1}{\mu_1}$,
$\omega_2 = 2 - \frac{\nu_2}{\mu_2}$,
$\mu_2 = \mu_1\,\mathrm{cn}(A_1, m)$,
$\frac{\nu_1}{2\,\mu_1^2\,\mathrm{dn}(A_1, m)} > \frac{1}{\mu_1}$,
$\frac{\nu_1\,\mathrm{cn}^2(A_1, m)}{2\,\mu_1\,\mu_2\,\mathrm{dn}(A_1, m)} > \frac{1}{\mu_2}$,
$0 \leqslant m \leqslant 1$,
ν_1, ν_2, μ_1, n_0, ϕ_1, and ϕ_2 are arbitrary real constants.

- *Reference*: [9].

***Solution* 2. cn(*n*,*m*)-sn(*n*,*m*)** *discrete SW*

$$\begin{aligned}\psi_1(n,\, t) &= A_0\,\sqrt{m}\,\operatorname{cn}[A_1\,(n + n_0),\, m]\, e^{-i\,(\omega_1\, t+\phi_1)},\\ \psi_2(n,\, t) &= B_0\,\sqrt{m}\,\operatorname{sn}[A_1\,(n + n_0),\, m]\, e^{-i\,(\omega_2\, t+\phi_2)},\end{aligned} \tag{8.104}$$

where

$A_0 = \sqrt{\frac{\nu_1}{2\, m\, \mu_1^2\, \operatorname{cn}(A_1, m)} - \frac{1}{m\, \mu_1}}$,

$B_0 = \sqrt{\frac{\nu_1\, \operatorname{dn}^2(A_1, m)}{2\, m\, \mu_1\, \mu_2\, \operatorname{cn}(A_1, m)} - \frac{1}{m\, \mu_2}}$,

$A_1 = \mu_1$,

$\omega_1 = 2 - \frac{\nu_1}{\mu_1}$,

$\omega_2 = 2 - \frac{\nu_2}{\mu_2}$,

$\mu_2 = \mu_1\, \operatorname{dn}(A_1,\, m)$,

$\frac{\nu_1}{2\, m\, \mu_1^2\, \operatorname{cn}(A_1, m)} > \frac{1}{m\, \mu_1}$,

$\frac{\nu_1\, \operatorname{dn}^2(A_1, m)}{2\, m\, \mu_1\, \mu_2\, \operatorname{cn}(A_1, m)} > \frac{1}{m\, \mu_2}$,

$0 \leqslant m \leqslant 1$,

ν_1, ν_2, μ_1, n_0, ϕ_1, and ϕ_2 are arbitrary real constants.

- *Reference*: [9].

***Solution* 3. sech(*n*)-tanh(*n*)** *discrete bright-dark soliton*

$$\begin{aligned}\psi_1(n,\, t) &= A_0\, \operatorname{sech}[A_1\,(n + n_0)]\, e^{-i\,(\omega_1\, t+\phi_1)},\\ \psi_2(n,\, t) &= B_0\, \tanh[A_1\,(n + n_0)]\, e^{-i\,(\omega_2\, t+\phi_2)},\end{aligned} \tag{8.105}$$

where

$A_0 = \sqrt{\frac{\nu_1 \cosh(A_1)}{2\, \mu_1^2} - \frac{1}{\mu_1}}$,

$B_0 = \sqrt{\frac{\nu_1}{2\, \mu_1\, \mu_2\, \cosh(A_1)} - \frac{1}{\mu_2}}$,

$A_1 = \mu_1$,

$\omega_1 = 2 - \frac{\nu_1}{\mu_1}$,

$\omega_2 = 2 - \frac{\nu_2}{\mu_2}$,

$\mu_2 = \mu_1\, \operatorname{sech}(A_1)$,

$\frac{\nu_1 \cosh(A_1)}{2\, \mu_1^2} > \frac{1}{\mu_1}$,

$\frac{\nu_1}{2\, \mu_1\, \mu_2\, \cosh(A_1)} > \frac{1}{\mu_2}$,

ν_1, ν_2, μ_1, n_0, ϕ_1, and ϕ_2 are arbitrary real constants.

- *Reference*: [9].

8.15 Summary of Section 8.14

Equation

$$i\,\psi_{1n_t} + \psi_{1n+1} + \psi_{1n-1} - 2\,\psi_{1n} + \frac{\nu_1\,(\mu_1\,|\,\psi_{1n}\,|^2 + \mu_2\,|\,\psi_{2n}\,|^2)\,\psi_{1n}}{\mu_1\,(1 + \mu_1\,|\,\psi_{1n}\,|^2 + \mu_2\,|\,\psi_{2n}\,|^2)} = 0,$$

$$i\,\psi_{2n_t} + \psi_{2n+1} + \psi_{2n-1} - 2\,\psi_{2n} + \frac{\left[\nu_2 - \frac{\nu_1\,\mu_2^2}{\mu_1^2} + \nu_2\,(\mu_1\,|\,\psi_{1n}\,|^2 + \mu_2\,|\,\psi_{2n}\,|^2)\right]\psi_{2n}}{\mu_2\,(1 + \mu_1\,|\,\psi_{1n}\,|^2 + \mu_2\,|\,\psi_{2n}\,|^2)} = 0$$

#	Solution	Conditions	Name	Eq. #
1.	$\psi_1(n,\,t) = A_0\,\mathrm{dn}[A_1\,(n + n_0),\,m]\,e^{-i\,(\omega_1\,t + \phi_1)}$, $\psi_2(n,\,t) = B_0\,\sqrt{m}\,\mathrm{sn}[A_1\,(n + n_0),\,m]\,e^{-i\,(\omega_2\,t + \phi_2)}$	$A_0 = \sqrt{\frac{\nu_1}{2\,\mu_1^2\,\mathrm{dn}(A_1,m)} - \frac{1}{\mu_1}}$, $B_0 = \sqrt{\frac{\nu_1\,\mathrm{cn}^2(A_1,m)}{2\,\mu_1\,\mu_2\,\mathrm{dn}(A_1,m)} - \frac{1}{\mu_2}}$, $A_1 = \mu_1$, $\omega_1 = 2 - \frac{\nu_1}{\mu_1}$, $\omega_2 = 2 - \frac{\nu_2}{\mu_2}$, $\mu_2 = \mu_1\,\mathrm{cn}(A_1,\,m)$, $\frac{\nu_1}{2\,\mu_1^2\,\mathrm{dn}(A_1,m)} > \frac{1}{\mu_1}$, $\frac{\nu_1\,\mathrm{cn}^2(A_1,m)}{2\,\mu_1\,\mu_2\,\mathrm{dn}(A_1,m)} > \frac{1}{\mu_2}$, $0 \leqslant m \leqslant 1$, ν_1, ν_2, μ_1, ϕ_1, and ϕ_2 are arbitrary real constants, n_0 is an arbitrary real integer	discrete solitary wave	(8.103)
2.	$\psi_1(n,\,t) = A_0\,\sqrt{m}\,\mathrm{cn}[A_1\,(n + n_0),\,m]\,e^{-i\,(\omega_1\,t + \phi_1)}$, $\psi_2(n,\,t) = B_0\,\sqrt{m}\,\mathrm{sn}[A_1\,(n + n_0),\,m]\,e^{-i\,(\omega_2\,t + \phi_2)}$	$A_0 = \sqrt{\frac{\nu_1}{2\,m\,\mu_1^2\,\mathrm{cn}(A_1,m)} - \frac{1}{m\,\mu_1}}$, $B_0 = \sqrt{\frac{\nu_1\,\mathrm{dn}^2(A_1,m)}{2\,m\,\mu_1\,\mu_2\,\mathrm{cn}(A_1,m)} - \frac{1}{m\,\mu_2}}$, $A_1 = \mu_1$, $\omega_1 = 2 - \frac{\nu_1}{\mu_1}$, $\omega_2 = 2 - \frac{\nu_2}{\mu_2}$, $\mu_2 = \mu_1\,\mathrm{dn}(A_1,\,m)$, $\frac{\nu_1}{2\,m\,\mu_1^2\,\mathrm{cn}(A_1,m)} > \frac{1}{m\,\mu_1}$, $\frac{\nu_1\,\mathrm{dn}^2(A_1,m)}{2\,m\,\mu_1\,\mu_2\,\mathrm{cn}(A_1,m)} > \frac{1}{m\,\mu_2}$, $0 \leqslant m \leqslant 1$, ν_1, ν_2, μ_1, ϕ_1, and ϕ_2 are arbitrary real constants, n_0 is an arbitrary real integer	discrete solitary wave	(8.104)
3.	$\psi_1(n,\,t) = A_0\,\mathrm{sech}[A_1\,(n + n_0)]\,e^{-i\,(\omega_1\,t + \phi_1)}$, $\psi_2(n,\,t) = B_0\,\tanh[A_1\,(n + n_0)]\,e^{-i\,(\omega_2\,t + \phi_2)}$	$A_0 = \sqrt{\frac{\nu_1\,\cosh(A_1)}{2\,\mu_1^2} - \frac{1}{\mu_1}}$, $B_0 = \sqrt{\frac{\nu_1}{2\,\mu_1\,\mu_2\,\cosh(A_1)} - \frac{1}{\mu_2}}$, $A_1 = \mu_1$, $\omega_1 = 2 - \frac{\nu_1}{\mu_1}$, $\omega_2 = 2 - \frac{\nu_2}{\mu_2}$, $\mu_2 = \mu_1\,\mathrm{sech}(A_1)$, $\frac{\nu_1\,\cosh(A_1)}{2\,\mu_1^2} > \frac{1}{\mu_1}$, $\frac{\nu_1}{2\,\mu_1\,\mu_2\,\cosh(A_1)} > \frac{1}{\mu_2}$, ν_1, ν_2, μ_1, ϕ_1, and ϕ_2 are arbitrary real constants, n_0 is an arbitrary real integer	discrete bright-dark soliton	(8.105)

References

[1] Khare A and Saxena A 2015 Periodic and hyperbolic soliton solutions of a number of nonlocal nonlinear equations *J. Math. Phys.* **56** 032104–27

[2] Khare A, Rasmussen K Ø, Samuelsen M R and Saxena A 2005 Exact solutions of the saturable discrete nonlinear Schrödinger equation *J. Phys. A: Math. Gen.* **38** 807–14

[3] Yan Z 2009 Envelope solution profiles of the discrete nonlinear Schrödinger equation with a saturable nonlinearity *Appl. Math. Lett.* **22** 448–52

[4] Khare A and Rasmussen K Ø 2009 Staggered and short-period solutions of the saturable discrete nonlinear Schrödinger equation *J. Phys. A: Math. Theor.* **42** 085002-6

[5] Ankiewicz A, Akhmediev N and Lederer F 2011 Approach to first-order exact solutions of the Ablowitz-Ladik equation *Phys. Rev.* E **83** 056602-6

[6] Hua-Mei L and Feng-Min W 2005 Exact discrete soliton solutions of quintic discrete nonlinear Schrödinger equation *Chin. Phys.* **14** 1069-7

[7] Kevrekidis P G 2009 *The Discrete Nonlinear Schrödinger Equation: Mathematical Analysis, Numerical Computations and Physical Perspectives* (Springer Tracts in Modern Physics vol 232) (Berlin: Springer)

[8] Khare A, Dmitriev S V and Saxena A 2007 Exact moving and stationary solutions of a generalized discrete nonlinear Schrödinger equation *J. Phys. A: Math. Theor.* **40** 11301–17

[9] Khare A and Saxena A 2012 Solutions of several coupled discrete models in terms of Lamé polynomials of order one and two *Pramana* **78** 187–213

IOP Publishing

Handbook of Exact Solutions to the Nonlinear Schrödinger Equations

Usama Al Khawaja and Laila Al Sakkaf

Chapter 9

Nonlocal Nonlinear Schrödinger Equation

A Glance at Chapter 9

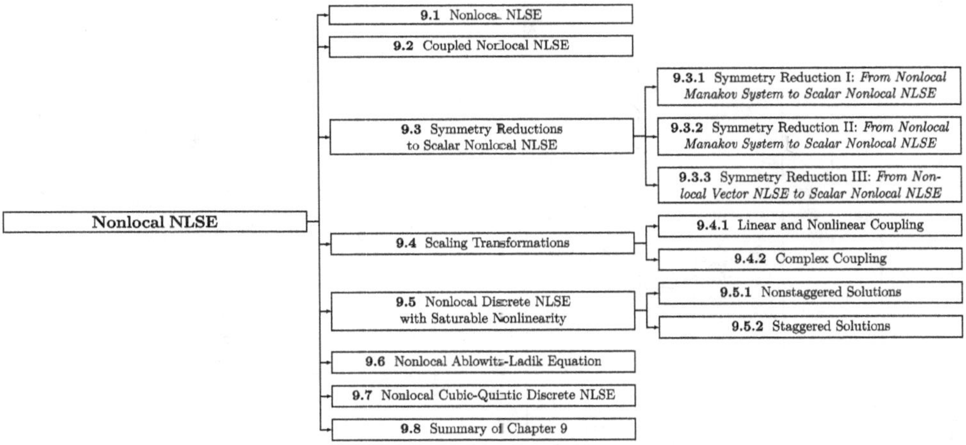

A Statistical View of Chapter 9

	Equation	Solutions
1	$i\,\Phi_t + a_1\,\Phi_{xx} + a_2\,\Phi^2\,\bar{\Phi} = 0$	2
2	$i\,\Phi_{1t} + \Phi_{1xx} + (a_1\,\Phi_1\,\bar{\Phi}_1 + a_2\,\Phi_2\,\bar{\Phi}_2)\,\Phi_1 = 0,$ $i\,\Phi_{2t} + \Phi_{2xx} + (b_1\,\Phi_1\,\bar{\Phi}_1 + b_2\,\Phi_2\,\bar{\Phi}_2)\,\Phi_2 = 0$	4
3	$i\,\Phi_{1t} + b_0\,\Phi_{1xx} + (c_1 + c_2\lvert\sigma\rvert^2)\,\Phi_1^2\,\bar{\Phi}_1 = 0$	0
4	$i\,\Phi_{1t} + b_0\,\Phi_{1xx} - (c_1 + c_2)\,\Phi_1^2\,\bar{\Phi}_1 = 0$	0
5	$i\,\Phi_{1t} + a_1\,\Phi_{1xx} + \sum_{k=1}^{N} b_{1k}\lvert\sigma_k\rvert^2\,\Phi_1^2\,\bar{\Phi}_1 = 0$	0
6	$i\,\Phi_{1t} + \Phi_{1xx} + (g_1\,\Phi_1\,\bar{\Phi}_1 - g_2\,\Phi_2\,\bar{\Phi}_2)\,\Phi_1 + g_0\,(g_1 + g_2)\,\Phi_1 - 2\,g_0\,g_2\,\Phi_2 = 0,$ $i\,\Phi_{2t} + \Phi_{2xx} + (g_1\,\Phi_1\,\bar{\Phi}_1 - g_2\,\Phi_2\,\bar{\Phi}_2)\,\Phi_2 - g_0\,(g_1 + g_2)\,\Phi_2 + 2\,g_0\,g_1\,\Phi_1 = 0$	0
7	$i\,\Phi_{1t} + \Phi_{1xx} + (g_1\,\Phi_1\,\bar{\Phi}_1 - g_2\,\Phi_2\,\bar{\Phi}_2)\,\Phi_1 = 0,$ $i\,\Phi_{2t} + \Phi_{2xx} + (g_1\,\Phi_1\,\bar{\Phi}_1 - g_2\,\Phi_2\,\bar{\Phi}_2)\,\Phi_2 = 0$	0
8	$i\,\Phi_{1t} + \Phi_{1xx} + 2\,(a_{11}\,\Phi_1\,\bar{\Phi}_1 + a_{12}\,\Phi_2\,\bar{\Phi}_2)\,\Phi_1 + 2\,(b_{11}\,\Phi_1\,\bar{\Phi}_2 + b_{12}\,\Phi_2\,\bar{\Phi}_1)\,\Phi_1 = 0,$ $i\,\Phi_{2t} + \Phi_{2xx} + 2\,(a_{21}\,\Phi_1\,\bar{\Phi}_1 + a_{22}\,\Phi_2\,\bar{\Phi}_2)\,\Phi_2 + 2\,(b_{21}\,\Phi_1\,\bar{\Phi}_2 + b_{22}\,\Phi_2\,\bar{\Phi}_1)\,\Phi_2 = 0,$	0
9	$i\,\Phi_{1t} + \Phi_{1xx} - 2(a + b)\,(\Phi_1\,\bar{\Phi}_1 + \Phi_2\,\bar{\Phi}_2)\,\Phi_1 + 2\,((a + i\,b)\,\Phi_1\,\bar{\Phi}_2 + (a - i\,b)\,\Phi_2\,\bar{\Phi}_1)\,\Phi_1 = 0,$ $i\,\Phi_{2t} + \Phi_{2xx} - 2(a + b)\,(\Phi_1\,\bar{\Phi}_1 + \Phi_2\,\bar{\Phi}_2)\,\Phi_2 + 2\,((a + i\,b)\,\Phi_1\,\bar{\Phi}_2 + (a - i\,b)\,\Phi_2\bar{\Phi}_1)\,\Phi_2 = 0$	0
10	$i\,\Phi_{nt} + \Phi_{n+1} + \Phi_{n-1} - 2\,\Phi_n + \frac{a_2\,\Phi_n^2\,\bar{\Phi}_n}{1 + \mu\,\Phi_n\,\bar{\Phi}_n} = 0$	3
11	$i\,\Phi_{nt} + \Phi_{n+1} + \Phi_{n-1} - 2\,\Phi_n + a_2\,(\Phi_{n+1} + \Phi_{n-1})\,\Phi_n\,\bar{\Phi}_n = 0$	2
12	$i\,\Phi_{nt} + a_1\,(\Phi_{n+1} + \Phi_{n-1} - 2\,\Phi_n) + a_2\,\Phi_n^2\,\bar{\Phi}_n + \left(a_3\,\Phi_n\,\bar{\Phi}_n + a_4\,\Phi_n^2\,\bar{\Phi}_n^2\right)(\Phi_{n+1} + \Phi_{n-1}) = 0$	4
Total	12	15

9.1 Nonlocal NLSE

If $\psi(x, t)$ is a solution of the fundamental NLSE, (2.1),

$$i\,\psi_t + a_1\,\psi_{xx} + a_2\,|\psi|^2\,\psi = 0,$$

then

$$\Phi(x, t; a_1, a_2) = \begin{cases} \psi(x, t, a_1, a_2), & \psi \text{ is an even function in } x, \\ \psi(x, t, a_1, -a_2), & \psi \text{ is an odd function in } x \end{cases}$$

is a solution of

$$i\,\Phi_t + a_1\,\Phi_{xx} + a_2\,\Phi^2\,\bar{\Phi} = 0, \tag{9.1}$$

where
- $\psi = \psi(x, t; a_1, a_2)$,
- $\Phi = \Phi(x, t; a_1, a_2)$,
- $\bar{\Phi} = \Phi^*(-x, t; a_1, a_2)$,
- a_1 and a_2 are arbitrary real constants.

***Example* 1. Even function: sech(x)** *bright soliton*

Given

$$\psi(\zeta, t) = A_0\,\sqrt{\frac{2\,a_1}{a_2}}\,\text{sech}[A_0\,\zeta]\,e^{i\left[a_1\,A_0^2\,(t-t_0)+\phi_0\right]}$$

is a solution of (2.1), then

$$\Phi(\zeta, t) = A_0\,\sqrt{\frac{2\,a_1}{a_2}}\,\text{sech}[A_0\,\zeta]\,e^{i\left[a_1\,A_0^2\,(t-t_0)+\phi_0\right]} \tag{9.2}$$

is a solution of (9.1), where
- $\zeta = x - x_0$,
- $a_1\,a_2 > 0$,
- A_0, x_0, t_0, and ϕ_0 are arbitrary real constants.

***Example* 2. Odd function: tanh(x)** *dark soliton*

Given

$$\psi(\zeta, t) = A_0\,\tanh\left[\frac{A_1}{\sqrt{a_1}}\,\zeta\right]\,e^{-i\,[A_2\,(t-t_0)+\phi_0]}$$

is a solution of (2.1), where
- $A_2 = 2\,A_1^2$,
- $a_2 = -\frac{2\,A_1^2}{A_0^2}$,
- $a_1 > 0$,
- $\zeta = x - x_0$,

A_0, A_1, x_0, t_0, and ϕ_0 are arbitrary real constants, then

$$\Phi(\zeta,\, t) = A_0 \tanh\left[\frac{A_1}{\sqrt{a_1}}\,\zeta\right] e^{-i\,[A_2\,(t-t_0)+\phi_0]} \tag{9.3}$$

is a solution of (9.1), where
$A_2 = 2\,A_1^2$,
$a_2 = \frac{2\,A_1^2}{A_0^2}$,
(*notice the change of sign compared to the local case*),
$a_1 > 0$,
$\zeta = x - x_0$,
A_0, A_1, x_0, t_0, and ϕ_0 are arbitrary.

9.2 Nonlocal Coupled NLSE

If $(\psi_1,\, \psi_2)$ is a solution of

$$\begin{aligned} i\,\psi_{1_t} + \psi_{1xx} + \left(a_1\,\psi_1\,\psi_1^* + a_2\,\psi_2\,\psi_2^*\right)\psi_1 = 0,\\ i\,\psi_{2_t} + \psi_{2xx} + \left(b_1\,\psi_1\,\psi_1^* + b_2\,\psi_2\,\psi_2^*\right)\psi_2 = 0, \end{aligned} \tag{9.4}$$

then

$$\Phi_1(x,\, t;\, a_1,\, a_2,\, b_1,\, b_2) = \begin{cases} \psi_1(x,\, t,\, a_1,\, a_2,\, b_1,\, b_2), & \psi_1 \text{ is an even function in } x,\\ \psi_1(x,\, t,\, -a_1,\, a_2,\, -b_1,\, b_2), & \psi_1 \text{ is an odd function in } x, \end{cases}$$

$$\Phi_2(x,\, t;\, a_1,\, a_2,\, b_1,\, b_2) = \begin{cases} \psi_2(x,\, t,\, a_1,\, a_2,\, b_1,\, b_2), & \psi_2 \text{ is an even function in } x,\\ \psi_2(x,\, t,\, a_1,\, -a_2,\, b_1,\, -b_2), & \psi_2 \text{ is an odd function in } x \end{cases}$$

is a solution of

$$\begin{aligned} i\,\Phi_{1t} + \Phi_{1xx} + (a_1\,\Phi_1\,\bar{\Phi}_1 + a_2\,\Phi_2\,\bar{\Phi}_2)\,\Phi_1 = 0,\\ i\,\Phi_{2t} + \Phi_{2xx} + (b_1\,\Phi_1\,\bar{\Phi}_1 + b_2\,\Phi_2\,\bar{\Phi}_2)\,\Phi_2 = 0, \end{aligned} \tag{9.5}$$

where
$\psi_{1,2} = \psi_{1,2}(x,\, t;\, a_1,\, a_2,\, b_1,\, b_2)$,
$\Phi_{1,2} = \Phi_{1,2}(x,\, t;\, a_1,\, a_2,\, b_1,\, b_2)$,
$\bar{\Phi}_{1,2} = \Phi_{1,2}^*(-x,\, t;\, a_1,\, a_2,\, b_1,\, b_2)$,
a_1, a_2, b_1, and b_2 are arbitrary real constants.

***Example* 1. Even–Odd function: sech(*x*)-tanh(*x*)** *bright-dark soliton*
Given

$$\begin{aligned} \psi_1(\zeta,\, t) = A_0\,\text{sech}[A_1\,\zeta]\,e^{-i\,[\omega_1\,(t-t_0)+\phi_1]},\\ \psi_2(\zeta,\, t) = B_0\,\tanh[A_1\,\zeta]\,e^{-i\,[\omega_2\,(t-t_0)+\phi_2]} \end{aligned}$$

is a solution of (9.4), where

$\omega_1 = A_1^2 - a_1 A_0^2$,
$\omega_2 = 2 A_1^2 - b_1 A_0^2$,
$a_1 = \frac{2 A_1^2 + a_2 B_0^2}{A_0^2}$,
$b_1 = \frac{2 A_1^2 + b_2 B_0^2}{A_0^2}$,
$\zeta = x - x_0$,
$A_0 \neq 0$,
A_1, B_0, x_0, t_0, ϕ_1, and ϕ_2 are arbitrary real constants, then

$$\Phi_1(\zeta, t) = A_0 \operatorname{sech}[A_1 \zeta] e^{-i [\omega_1 (t-t_0)+\phi_1]}, \quad \Phi_2(\zeta, t) = B_0 \tanh[A_1 \zeta] e^{-i [\omega_2 (t-t_0)+\phi_2]} \tag{9.6}$$

is a solution of (9.5), where

$\omega_1 = A_1^2 - a_1 A_0^2$,
$\omega_2 = 2 A_1^2 - b_1 A_0^2$,
$a_1 = \frac{2 A_1^2 - a_2 B_0^2}{A_0^2}$, *(changed the sign of* a_2*)*,
$b_1 = \frac{2 A_1^2 - b_2 B_0^2}{A_0^2}$, *(changed the sign of* b_2*)*,
$\zeta = x - x_0$,
$A_0 \neq 0$,
A_1, B_0, x_0, t_0, ϕ_1, and ϕ_2 are arbitrary real constants.

***Example* 2. Odd–Odd function: cn(*x*,*m*)sn(*x*,*m*)-dn(*x*,*m*)sn(*x*,*m*)** *solitary wave (SW)*
Given

$$\psi_1(\zeta, t) = A_0\, m \operatorname{cn}[A_1 \zeta, m] \operatorname{sn}[A_1 \zeta, m] e^{-i [\omega_1 (t-t_0)+\phi_1]}, \quad \psi_2(\zeta, t) = B_0 \sqrt{m} \operatorname{dn}[A_1 \zeta, m] \operatorname{sn}[A_1 \zeta, m] e^{-i [\omega_2 (t-t_0)+\phi_2]}$$

is a solution of (9.4), where

$\omega_1 = (4 + m) A_1^2$,
$\omega_2 = (1 + 4 m) A_1^2$,
$a_1 = \frac{6 A_1^2}{A_0^2 (1 - m)}$,
$a_2 = \frac{-6 A_1^2}{B_0^2 (1 - m)}$,
$b_1 = a_1$,
$b_2 = a_2$,
$0 \leqslant m < 1$,
$\zeta = x - x_0$,
$A_0 \neq 0$,
$B_0 \neq 0$,

A_1, x_0, t_0, ϕ_1, and ϕ_2 are arbitrary real constants, then

$$\begin{aligned}\Phi_1(\zeta, t) &= A_0\, m\, \mathrm{cn}[A_1\, \zeta, m]\, \mathrm{sn}[A_1\, \zeta, m]\, e^{-i\,[\omega_1\,(t-t_0)+\phi_1]},\\ \Phi_2(\zeta, t) &= B_0\, \sqrt{m}\, \mathrm{dn}[A_1\, \zeta, m]\, \mathrm{sn}[A_1\, \zeta, m]\, e^{-i\,[\omega_2\,(t-t_0)+\phi_2]}\end{aligned} \tag{9.7}$$

is a solution of (9.5), where
$\omega_1 = (4 + m)\, A_1^2$,
$\omega_2 = (1 + 4\, m)\, A_1^2$,
$a_1 = \frac{-6\, A_1^2}{A_0^2\, (1-m)}$, *(changed the sign of* a_1*)*,
$a_2 = \frac{6\, A_1^2}{B_0^2\, (1-m)}$, *(changed the sign of* a_2*)*,
$b_1 = a_1$,
$b_2 = a_2$,
$0 \leqslant m < 1$,
$\zeta = x - x_0$,
$A_0 \neq 0$,
$B_0 \neq 0$,
A_1, x_0, t_0, ϕ_1, and ϕ_2 are arbitrary real constants.

***Example* 3. Odd–Even function: sd(*x*,*m*)-nd(*x*,*m*)** *SW.*
Given

$$\begin{aligned}\psi_1(\zeta, t) &= A_0\, \sqrt{m\,(1-m)}\, \mathrm{sd}[A_1\, \zeta, m]\, e^{-i\,[\omega_1\,(t-t_0)+\phi_1]},\\ \psi_2(\zeta, t) &= B_0\, \sqrt{1-m}\, \mathrm{nd}[A_1\, \zeta, m]\, e^{-i\,[\omega_2\,(t-t_0)+\phi_2]}\end{aligned}$$

is a solution of (9.4), where
$\omega_1 = -(m-1)\, a_1\, A_0^2 - A_1^2$,
$\omega_2 = (m-2)\, A_1^2 - (m-1)\, b_1\, A_0^2$,
$a_1 = -\frac{-2\, A_1^2 + a_2\, B_0^2}{A_0^2}$,
$b_1 = -\frac{-2\, A_1^2 + b_2\, B_0^2}{A_0^2}$,
$0 \leqslant m \leqslant 1$,
$\zeta = x - x_0$,
$A_0 \neq 0$,
A_1, B_0, x_0, t_0, ϕ_1, and ϕ_2 are arbitrary real constants, then

$$\begin{aligned}\Phi_1(\zeta, t) &= A_0\, \sqrt{m\,(1-m)}\, \mathrm{sd}[A_1\, \zeta, m]\, e^{-i\,[\omega_1\,(t-t_0)+\phi_1]},\\ \Phi_2(\zeta, t) &= B_0\, \sqrt{1-m}\, \mathrm{nd}[A_1\, \zeta, m]\, e^{-i\,[\omega_2\,(t-t_0)+\phi_2]}\end{aligned} \tag{9.8}$$

is a solution of (9.5), where
$\omega_1 = (m-1)\, a_1\, A_0^2 - A_1^2$, *(changed the sign of* a_1*)*,
$\omega_2 = (m-2)\, A_1^2 + (m-1)\, b_1\, A_0^2$, *(changed the sign of* b_1*)*,

$a_1 = \frac{-2\,A_1^2 + a_2\,B_0^2}{A_0^2}$, *(changed the sign of* a_1*)*,
$b_1 = \frac{-2\,A_1^2 + b_2\,B_0^2}{A_0^2}$, *(changed the sign of* b_1*)*,
$0 \leqslant m \leqslant 1$,
$\zeta = x - x_0$,
$A_0 \neq 0$,
A_1, B_0, x_0, t_0, ϕ_1, and ϕ_2 are arbitrary real constants.

***Example* 4. Even–Even function: sech²(*x*)-sech²(*x*).**
Given

$$\psi_1(\zeta,\,t) = [A_0\,\mathrm{sech}^2(A_1\,\zeta) + A_3]\,e^{-i\,[\omega_1\,(t-t_0)+\phi_1]},$$
$$\psi_2(\zeta,\,t) = B_0\,\mathrm{sech}^2(A_1\,\zeta)\,e^{-i\,[\omega_2\,(t-t_0)+\phi_2]}$$

is a solution of (9.4), then

$$\begin{aligned}\Phi_1(\zeta,\,t) &= \{A_0\,\mathrm{sech}^2[A_1\,\zeta] + A_3\}\,e^{-i\,[\omega_1\,(t-t_0)+\phi_1]},\\ \Phi_2(\zeta,\,t) &= B_0\,\mathrm{sech}^2[A_1\,\zeta]\,e^{-i\,[\omega_2\,(t-t_0)+\phi_2]}\end{aligned} \tag{9.9}$$

is a solution of (9.5), where
$A_3 = \frac{-2\,A_0}{3}$,
$\omega_1 = 2\,A_1^2$,
$\omega_2 = -2\,A_1^2$,
$a_1 = \frac{-9\,A_1^2}{2\,A_0^2}$,
$a_2 = \frac{9\,A_1^2}{2\,B_0^2}$,
$b_1 = a_1$,
$b_2 = a_2$,
$\zeta = x - x_0$,
$A_0 \neq 0$,
$B_0 \neq 0$,
A_1, x_0, t_0, ϕ_1, and ϕ_2 are arbitrary real constants.

9.3 Symmetry Reductions to Scalar Nonlocal NLSE

9.3.1 Symmetry Reduction I

From Nonlocal Manakov System to Scalar Nonlocal NLSE
The nonlocal CNLSE, (9.5),

$$i\,\Phi_{1_t} + b_0\,\Phi_{1_x x} + (b_1\,\Phi_1\,\bar{\Phi}_1 + b_2\,\Phi_2\,\bar{\Phi}_2)\,\Phi_1 = 0,$$
$$i\,\Phi_{2_t} + b_0\,\Phi_{2xx} + (c_1\,\Phi_1\,\bar{\Phi}_1 + c_2\,\Phi_2\,\bar{\Phi}_2)\,\Phi_2 = 0,$$

transforms to the scalar nonlocal NLSE

$$i\,\Phi_{1t} + b_0\,\Phi_{1xx} + (c_1 + c_2|\sigma|^2)\,\Phi_1^2\,\bar{\Phi}_1 = 0, \tag{9.10}$$

with the replacements:

1. $\Phi_2(x, t) = \sigma\, \Phi_1(x, t)$,
2. $b_1 = c_1 + (c_2 - b_2)\, |\sigma|^2$,

where σ is an arbitrary complex constant.

Conclusion:

If $\Phi_1(x, t)$ is a solution of the nonlocal NLSE

$$i\, \Phi_{1t} + a_1\, \Phi_{1xx} + a_2\, \Phi_1^2\, \bar{\Phi}_1 = 0,$$

then

$$(\Phi_1, \Phi_2) = (\Phi_1, \sigma\, \Phi_1) \tag{9.11}$$

is a solution of the nonlocal CNLSE

$$i\, \Phi_{1t} + b_0\, \Phi_{1xx} + (b_1\, \Phi_1\, \bar{\Phi}_1 + b_2\, \Phi_2\, \bar{\Phi}_2)\, \Phi_1 = 0,$$
$$i\, \Phi_{2t} + b_0\, \Phi_{2xx} + (c_1\, \Phi_1\, \bar{\Phi}_1 + c_2\, \Phi_2\, \bar{\Phi}_2)\, \Phi_2 = 0,$$

with

$a_1 = b_0$,
$a_2 = c_1 + c_2|\sigma|^2$,
$b_1 = c_1 + (c_2 - b_2)\, |\sigma|^2$.

9.3.2 Symmetry Reduction II

From Nonlocal Manakov System to Scalar Nonlocal NLSE

The nonlocal CNLSE, (9.5),

$$i\, \Phi_{1t} + b_0\, \Phi_{1_x x} + (b_1\, \Phi_1\, \bar{\Phi}_1 + b_2\, \Phi_2\, \bar{\Phi}_2)\, \Phi_1 = 0,$$
$$i\, \Phi_{2t} - b_0\, \Phi_{2_x x} + (c_1\, \Phi_1\, \bar{\Phi}_1 + c_2\, \Phi_2\, \bar{\Phi}_2)\, \Phi_2 = 0,$$

transforms to the scalar nonlocal NLSE

$$i\, \Phi_{1t} + b_0\, \Phi_{1xx} - (c_1 + c_2)\, \Phi_1^2\, \bar{\Phi}_1 = 0, \tag{9.12}$$

with the replacements:

1. $\Phi_2(x, t; b_1, b_2, c_1, c_2)$
$$= e^{i\,\phi}\begin{cases} \bar{\Phi}_1(x, t, b_1, b_2, c_1, c_2) & \text{if } \Phi_1 \text{ is an even function in } x, \\ \bar{\Phi}_1(x, t, b_1, b_2, -c_1, -c_2) & \text{if } \Phi_1 \text{ is an odd function in } x, \end{cases}$$
2. $b_1 = -(c_1 + c_2 + b_2)$,

where ϕ is an arbitrary real constant.

Conclusion:

I. If $\Phi_1(x, t)$ is an even solution of the nonlocal NLSE

$$i\,\Phi_{1_t} + b_0\,\Phi_{1xx} + a_2\,\Phi_1^2\,\bar{\Phi}_1 = 0,$$

then

$$(\Phi_1, \Phi_2) = \left[\Phi_1(x, t, b_1, b_2, c_1, c_2), e^{i\,\phi}\,\Phi_1^*(x, t, b_1, b_2, c_1, c_2)\right]$$

is a solution of the nonlocal CNLSE

$$\begin{aligned} i\,\Phi_{1_t} + b_0\,\Phi_{1xx} + (b_1\,\Phi_1\,\bar{\Phi}_1 + b_2\,\Phi_2\,\bar{\Phi}_2)\,\Phi_1 = 0,\\ i\,\Phi_{2_t} - b_0\,\Phi_{2xx} + (c_1\,\Phi_1\,\bar{\Phi}_1 + c_2\,\Phi_2\,\bar{\Phi}_2)\,\Phi_2 = 0, \end{aligned}$$

with $a_2 = -(c_1 + c_2)$.

II. If $\Phi_1(x, t)$ is an odd solution of the nonlocal NLSE

$$i\,\Phi_{1_t} + b_0\,\Phi_{1xx} + a_2\,\Phi_1^2\,\bar{\Phi}_1 = 0,$$

then

$$(\Phi_1, \Phi_2) = \left[\Phi_1(x, t, b_1, b_2, c_1, c_2), e^{i\,\phi}\,\Phi_1^*(-x, t, b_1, b_2, -c_1, -c_2)\right] \tag{9.13}$$

is a solution of the nonlocal CNLSE

$$\begin{aligned} i\,\Phi_{1_t} + b_0\,\Phi_{1xx} + (b_1\,\Phi_1\,\bar{\Phi}_1 + b_2\,\Phi_2\,\bar{\Phi}_2)\,\Phi_1 = 0,\\ i\,\Phi_{2_t} - b_0\,\Phi_{2xx} + (c_1\,\Phi_1\,\bar{\Phi}_1 + c_2\,\Phi_2\,\bar{\Phi}_2)\,\Phi_2 = 0, \end{aligned}$$

with $a_2 = -(c_1 + c_2)$.

9.3.3 Symmetry Reduction III

From Nonlocal Vector NLSE to Scalar Nonlocal NLSE

The generalized nonlocal CNLSE

$$i\,\Phi_{j_t} + a_1\,\Phi_{j_{xx}} + \left(\sum_{k=1}^{N} b_{1k}\,\Phi_k\,\bar{\Phi}_k\right)\Phi_j = 0, \quad j = 1, 2, \ldots, N, \tag{9.14}$$

transforms to the scalar nonlocal NLSE

$$i\,\Phi_{1_t} + a_1\,\Phi_{1xx} + \sum_{k=1}^{N} b_{1k}|\sigma_k|^2\,\Phi_1^2\,\bar{\Phi}_1 = 0,$$

with the replacement:

$$\Phi_j(x, t) = \sigma_j \Phi_1(x, t),$$

where σ_j are arbitrary complex constants and $\sigma_1 = 1$.

Conclusion:

If $\Phi_1(x, t)$ is a solution of the nonlocal NLSE

$$i \Phi_{1t} + a_1 \Phi_{1xx} + a_2 \Phi_1^2 \Phi_1 = 0,$$

then

$$(\Phi_1, \Phi_2, \Phi_3, \dots, \Phi_N) = (\Phi_1, \sigma_2 \Phi_1, \sigma_3 \Phi_1, \dots, \sigma_N \Phi_N) \tag{9.15}$$

is a solution of the generalized nonlocal CNLSE

$$i \Phi_{j_t} + a_1 \Phi_{j_{xx}} + \left(\sum_{k=1}^{N} b_{1k} \Phi_k \bar{\Phi}_k \right) \Phi_j = 0, \quad j = 1, 2, \dots, N,$$

with $a_2 = \sum_{k=1}^{N} b_{1k} |\sigma_k|^2$.

9.4 Scaling Transformations

9.4.1 Linear and Nonlinear Coupling

9.4.1.1 General Case

If (ψ_1, ψ_2) is a solution of

$$\begin{aligned} i \psi_{1_t} + \psi_{1xx} + (b_1 \psi_1 \bar{\psi}_1 + b_2 \psi_2 \bar{\psi}_2) \psi_1 = 0, \\ i \psi_{2_t} + \psi_{2xx} + (b_1 \psi_1 \bar{\psi}_1 + b_2 \psi_2 \bar{\psi}_2) \psi_2 = 0, \end{aligned} \tag{9.16}$$

then

$$\begin{aligned} \Phi_1(x, t) &= \sqrt{\frac{b_1}{g_1 - g_2}} \psi_1(x, t) e^{i g_0 (g_1 - g_2) t} + \sqrt{\frac{g_2 b_2}{g_1 (g_2 - g_1)}} \psi_2(x, t) \\ &\quad e^{-i g_0 (g_1 - g_2) t}, \\ \Phi_2(x, t) &= \sqrt{\frac{b_1}{g_1 - g_2}} \psi_1(x, t) e^{i g_0 (g_1 - g_2) t} + \sqrt{\frac{g_1 b_2}{g_2 (g_2 - g_1)}} \psi_2(x, t) \\ &\quad e^{-i g_0 (g_1 - g_2) t} \end{aligned} \tag{9.17}$$

is a solution of

$$
\begin{aligned}
&i\,\Phi_{1t} + \Phi_{1xx} + (g_1\,\Phi_1\,\bar{\Phi}_1 - g_2\,\Phi_2\,\bar{\Phi}_2)\,\Phi_1 + g_0\,(g_1 + g_2)\,\Phi_1 - 2\,g_0\,g_2\,\Phi_2 = 0,\\
&i\,\Phi_{2t} + \Phi_{2xx} + (g_1\,\Phi_1\,\bar{\Phi}_1 - g_2\,\Phi_2\,\bar{\Phi}_2)\,\Phi_2 - g_0\,(g_1 + g_2)\,\Phi_2 + 2\,g_0\,g_1\,\Phi_1 = 0,
\end{aligned}
\tag{9.18}
$$

where

$b_1\,(g_1 - g_2) > 0$,
$b_2\,g_1\,g_2\,(g_2 - g_1) > 0$,
b_1, b_2, g_0, g_1, and g_2 are real constants.

9.4.1.2 Specific Case I: Nonlocal Manakov System to Another Nonlocal Manakov System

If (ψ_1, ψ_2) is a solution of

$$
\begin{aligned}
&i\,\psi_{1t} + \psi_{1xx} + (b_1\,\psi_1\,\bar{\psi}_1 + b_2\,\psi_2\,\bar{\psi}_2)\,\psi_1 = 0,\\
&i\,\psi_{2t} + \psi_{2xx} + (b_1\,\psi_1\,\bar{\psi}_1 + b_2\,\psi_2\,\bar{\psi}_2)\,\psi_2 = 0,
\end{aligned}
$$

then

$$
\begin{aligned}
\Phi_1(x,\,t) &= \sqrt{\frac{b_1}{g_1 - g_2}}\,\psi_1(x,\,t) + \sqrt{\frac{g_2\,b_2}{g_1\,(g_2 - g_1)}}\,\psi_2(x,\,t),\\
\Phi_2(x,\,t) &= \sqrt{\frac{b_1}{g_1 - g_2}}\,\psi_1(x,\,t) + \sqrt{\frac{g_1\,b_2}{g_2\,(g_2 - g_1)}}\,\psi_2(x,\,t)
\end{aligned}
\tag{9.19}
$$

is a solution of

$$
\begin{aligned}
&i\,\Phi_{1t} + \Phi_{1xx} + (g_1\,\Phi_1\,\bar{\Phi}_1 - g_2\,\Phi_2\,\bar{\Phi}_2)\,\Phi_1 = 0,\\
&i\,\Phi_{2t} + \Phi_{2xx} + (g_1\,\Phi_1\,\bar{\Phi}_1 - g_2\,\Phi_2\,\bar{\Phi}_2)\,\Phi_2 = 0,
\end{aligned}
\tag{9.20}
$$

where

$b_1\,(g_1 - g_2) > 0$,
$b_2\,g_1\,g_2\,(g_2 - g_1) > 0$,
b_1, b_2, g_1 and g_2 are real constants.

9.4.1.3 Specific Case II: Nonlocal Manakov System to the Same Nonlocal Manakov System

Superposition Principle for a Nonlocal Nonlinear System

If (ψ_1, ψ_2) is a solution of (9.16) then

$$
\begin{aligned}
\Phi_1(x,\,t) &= \sqrt{\frac{b_1}{b_1 + b_2}}\left[\psi_1(x,\,t) - \frac{b_2}{b_1}\,\psi_2(x,\,t)\right],\\
\Phi_2(x,\,t) &= \sqrt{\frac{b_1}{b_1 + b_2}}\,[\psi_1(x,\,t) + \psi_2(x,\,t)]
\end{aligned}
\tag{9.21}
$$

is also a solution of (9.16), where $b_1 + b_2 \neq 0$.

9.4.2 Complex Coupling

9.4.2.1 General Case

If (ψ_1, ψ_2) is a solution of

$$\begin{aligned} i\,\psi_{1_t} + \psi_{1xx} + q_1\,(q_2\,\psi_1\,\bar{\psi}_1 + q_3\,\psi_2\,\bar{\psi}_2)\,\psi_1 = 0, \\ i\,\psi_{2_t} + \psi_{2xx} + q_1\,(q_2\,\psi_1\,\bar{\psi}_1 + q_3\,\psi_2\,\bar{\psi}_2)\,\psi_2 = 0, \end{aligned} \tag{9.22}$$

then

$$\begin{aligned} \Phi_1(x,\,t) = c_1\,\psi_1(x,\,t) + c_2\,\psi_2(x,\,t), \\ \Phi_2(x,\,t) = c_3\,\psi_1(x,\,t) + c_4\,\psi_2(x,\,t) \end{aligned} \tag{9.23}$$

is a solution of

$$\begin{aligned} i\,\Phi_{1_t} + \Phi_{1xx} + 2\,(a_{11}\,\Phi_1\,\bar{\Phi}_1 + a_{12}\,\Phi_2\,\bar{\Phi}_2)\,\Phi_1 + 2\,(b_{11}\,\Phi_1\,\bar{\Phi}_2 + b_{12}\,\Phi_2\,\bar{\Phi}_1)\,\Phi_1 = 0, \\ i\,\Phi_{2_t} + \Phi_{2xx} + 2\,(a_{21}\,\Phi_1\,\bar{\Phi}_1 + a_{22}\,\Phi_2\,\bar{\Phi}_2)\,\Phi_2 + 2\,(b_{21}\,\Phi_1\,\bar{\Phi}_2 + b_{22}\,\Phi_2\,\bar{\Phi}_1)\,\Phi_2 = 0, \end{aligned} \tag{9.24}$$

where

$$q_1 = \frac{2\,(c_1\,c_4 - c_2\,c_3)\left(c_2^*\,c_3^* - c_1^*\,c_4^*\right)}{c_1^*\,c_2\,c_3\,c_4^* - c_1\,c_2^*\,c_3^*\,c_4},$$

$$q_2 = (a - i\,b)\,c_1^*\,c_3 - (a + i\,b)\,c_1\,c_3^*,$$

$$q_3 = a\,(c_2\,c_4^* - c_2^*\,c_4) + i\,b\,(c_2^*\,c_4 + c_2\,c_4^*),$$

$$a_{12} = \frac{b_{12}\,c_1^*\,c_2^*\,(c_2\,c_3 - c_1\,c_4) + b_{11}\,c_1\,c_2\left(c_1^*\,c_4^* - c_2^*\,c_3^*\right)}{c_1\,c_2^*\,c_3^*\,c_4 - c_1^*\,c_2\,c_3\,c_4^*},$$

$$a_{11} = \frac{b_{11}\,c_3^*\,c_4^*\,(c_2\,c_3 - c_1\,c_4) + b_{12}\,c_3\,c_4\left(c_1^*\,c_4^* - c_2^*\,c_3^*\right)}{c_1\,c_2^*\,c_3^*\,c_4 - c_1^*\,c_2\,c_3\,c_4^*},$$

$a_{22} = a_{12}$,
$b_{12} = a - i\,b$,
$b_{21} = a + i\,b$,
$b_{11} = b_{21}$,
$b_{22} = b_{12}$,
$c_1\,c_2^*\,c_3^*\,c_4$ should not be pure real or pure imaginary,
$c_2\,c_3 - c_1\,c_4 \neq 0$,
c_j, $j = 1, 2, 3, 4$ are complex constants,
a and b are real constants.

9.4.2.2 Specific Case

If (ψ_1, ψ_2) is a solution of

$$\begin{aligned} i\,\psi_{1_t} + \psi_{1xx} - 4\,(b\,\psi_1\,\bar{\psi}_1 + a\,\psi_2\,\bar{\psi}_2)\,\psi_1 = 0, \\ i\,\psi_{2_t} + \psi_{2xx} - 4\,(b\,\psi_1\,\bar{\psi}_1 + a\,\psi_2\,\bar{\psi}_2)\,\psi_2 = 0, \end{aligned} \tag{9.25}$$

then

$$\begin{aligned} \Phi_1(x, t) &= \psi_1(x, t) + \psi_2(x, t), \\ \Phi_2(x, t) &= \psi_1(x, t) + i\,\psi_2(x, t) \end{aligned} \tag{9.26}$$

is a solution of

$$\begin{aligned} & i\,\Phi_{1t} + \Phi_{1xx} - 2(a + b)\,(\Phi_1\,\bar{\Phi}_1 + \Phi_2\,\bar{\Phi}_2)\,\Phi_1 \\ &\quad + 2\,[(a + i\,b)\,\Phi_1\,\bar{\Phi}_2 + (a - i\,b)\,\Phi_2\,\bar{\Phi}_1]\,\Phi_1 = 0, \\ & i\,\Phi_{2t} + \Phi_{2xx} - 2(a + b)\,(\Phi_1\,\bar{\Phi}_1 + \Phi_2\,\bar{\Phi}_2)\,\Phi_2 \\ &\quad +2\,[(a + i\,b)\,\Phi_1\,\bar{\Phi}_2 + (a - i\,b)\,\Phi_2\bar{\Phi}_1]\,\Phi_2 = 0, \end{aligned} \tag{9.27}$$

where a and b are real constants.

9.5 Nonlocal Discrete NLSE with Saturable Nonlinearity

If $\psi(x, t)$ is a solution of the NLSE with saturable nonlinearity, (8.1),

$$i\,\psi_{nt} + \psi_{n+1} + \psi_{n-1} - 2\,\psi_n + \frac{a_2\,|\psi_n|^2\,\psi_n}{1 + \mu\,|\psi_n|^2} = 0,$$

then

$$\Phi_n(n,\, t;\, a_2,\, \mu) = \begin{cases} \psi_n(n,\, t,\, a_2,\, \mu) & \text{if } \psi_n \text{ is an even function in } n, \\ \psi_n(n,\, t,\, -a_2,\, -\mu) & \text{if } \psi_n \text{ is an odd function in } n \end{cases}$$

is a solution of

$$i\,\Phi_{nt} + \Phi_{n+1} + \Phi_{n-1} - 2\,\Phi_n + \frac{a_2\,\Phi_n^2\,\bar{\Phi}_n}{1 + \mu\,\Phi_n\,\bar{\Phi}_n} = 0, \tag{9.28}$$

where

$\psi_n = \psi_n(n,\, t;\, a_2,\, \mu)$,
$\Phi_n = \Phi_n(n,\, t;\, a_2,\, \mu)$,
$\bar{\Phi}_n = \Phi_n^*(-n,\, t;\, a_2,\, \mu)$,
a_2 and μ are real constants.

9.5.1 Nonstaggered Solutions

***Example* 1. Even function: sech(*n*)** *discrete bright soliton*

Given

$$\psi(\zeta,\, t) = A_0\,\text{sech}[A_1\,\zeta]\,e^{-i\,[A_2\,(t-t_0)+\phi_0]}$$

is a solution of (8.1), then

$$\Phi(\zeta, t) = A_0 \operatorname{sech}[A_1 \zeta]\, e^{-i\,[A_2\,(t-t_0)+\phi_0]} \tag{9.29}$$

is a solution of (9.28), where
$A_0 = \frac{\sinh(A_1)}{\sqrt{\mu}}$,
$A_2 = \frac{2\,\mu - a_2}{\mu}$,
$\mu = \frac{a_2 \operatorname{sech}(A_1)}{2} > 0$,
$\zeta = n - n_0$,
A_1, t_0, n_0, and ϕ_0 are arbitrary real constants.

***Example* 2. Odd function: tanh(*n*)** *discrete dark soliton*
Given

$$\psi(\zeta, t) = A_0 \tanh[A_1 \zeta]\, e^{-i\,[A_2\,(t-t_0)+\phi_0]}$$

is a solution of (8.1), where
$A_0 = \frac{\tanh(A_1)}{\sqrt{-\mu}}$,
$A_2 = \frac{-2\,\mu + a_2}{-\mu}$,
$a_2 = 2\,\mu \operatorname{sech}^2(A_1)$,
$\zeta = n - n_0$,
$\mu < 0$,
A_1, n_0, and ϕ_0 are arbitrary real constants, then

$$\Phi(\zeta, t) = A_0 \tanh[A_1 \zeta]\, e^{-i\,[A_2\,(t-t_0)+\phi_0]} \tag{9.30}$$

is a solution of (9.28), where
$A_0 = \frac{\tanh(A_1)}{\sqrt{\mu}}$, *(changed the sign of* μ*)*,
$A_2 = \frac{2\,\mu - a_2}{\mu}$, *(changed the sign of* μ *and* a_2*)*,
$a_2 = 2\,\mu \operatorname{sech}^2(A_1)$,
$\zeta = n - n_0$,
$\mu > 0$,
A_1, n_0, and ϕ_0 are arbitrary real constants.

9.5.2 Staggered Solutions

If $\Phi_n(n, t; a_2)$ is a nonstaggered solution of (9.28), then

$$\Phi_{ns}(n, t, a_2) = (-1)^n \, \Phi_n^*(n, t, -a_2) \, e^{-4\,i\,(t-t_0)}, \tag{9.31}$$

is a staggered solution of the same equation.

***Example* 1. sech(*n*)** *staggered discrete bright soliton.*
Given

$$\Phi(\zeta, t) = A_0 \operatorname{sech}[A_1 \zeta] \, e^{i\,[A_2\,(t-t_0)+\phi_0]}$$

is a nonstaggered solution of (9.28), where

$A_0 = \frac{\sinh(A_1)}{\sqrt{\mu}}$,

$A_2 = \frac{2\mu - a_2}{\mu}$, *(changed the sign of* a_2*)*,

$\mu = \frac{a_2 \operatorname{sech}(A_1)}{2} > 0$, *(changed the sign of* a_2*)*,

$\zeta = n - n_0$,

A_1, n_0, t_0, and ϕ_0 are arbitrary real constants, then

$$\Phi_s(\zeta, t) = (-1)^{\zeta+n_0} A_0 \operatorname{sech}(A_1 \zeta) \, e^{i\,[A_2\,(t-t_0)-4\,(t-t_0)+\phi_0]} \tag{9.32}$$

is a staggered solution of (9.28), where

$A_0 = \frac{\sinh(A_1)}{\sqrt{\mu}}$,

$A_2 = \frac{2\mu + a_2}{\mu}$, *(changed the sign of* a_2*)*,

$\mu = -\frac{a_2 \operatorname{sech}(A_1)}{2} > 0$, *(changed the sign of* a_2*)*,

$\zeta = n - n_0$,

A_1, n_0, t_0, and ϕ_0 are arbitrary real constants.

9.6 Nonlocal Ablowitz–Ladik Equation

If $\psi(x, t)$ is a solution of the Ablowitz–Ladik equation, (8.41),

$$i\,\psi_{nt} + \psi_{n+1} + \psi_{n-1} - 2\,\psi_n + a_2\,(\psi_{n+1} + \psi_{n-1})\,|\psi_n|^2 = 0,$$

then

$$\Phi_n(n, t; a_2) = \begin{cases} \psi_n(n, t; a_2) & \text{if } \psi_n \text{ is an even function in } n, \\ \psi_n(n, t; -a_2) & \text{if } \psi_n \text{ is an odd function in } n \end{cases}$$

is a solution of

$$i\,\Phi_{nt} + \Phi_{n+1} + \Phi_{n-1} - 2\,\Phi_n + a_2\,(\Phi_{n+1} + \Phi_{n-1})\,\Phi_n\,\bar{\Phi}_n = 0, \tag{9.33}$$

where
$\psi_n = \psi_n(n,\, t;\, a_2),$
$\Phi_n = \Phi_n(n,\, t;\, a_2),$
$\bar{\Phi}_n = \Phi_n^*(-n,\, t;\, a_2),$
a_2 is an arbitrary real constant.

***Example* 1. Even function: sech(*n*)** *discrete bright soliton*
Given

$$\psi(\zeta,\, t) = A_0\, \text{sech}[A_1\, \zeta]\, e^{-i\, [(A_2+2)\, (t-t_0)+\phi_0]}$$

is a solution of (8.41), then

$$\Phi(\zeta,\, t) = A_0\, \text{sech}[A_1\, \zeta]\, e^{-i\, [(A_2+2)\, (t-t_0)+\phi_0]} \tag{9.34}$$

is a solution of (9.33), where
$A_2 = -2\cosh(A_1),$
$a_2 = \frac{\sinh^2(A_1)}{A_0^2},$
$\zeta = n - n_0,$
$A_0 \neq 0,$
A_1, n_0, t_0, and ϕ_0 are arbitrary real constants.

***Example* 2. Odd function: tanh(*n*)** *discrete dark soliton*
Given

$$\psi(\zeta,\, t) = A_0\, \tanh[A_1\, \zeta]\, e^{-i\, [(A_2+2)\, (t-t_0)+\phi_0]}$$

is a solution of (8.41), where
$A_2 = -2\,\text{sech}(A_1),$
$a_2 = -\frac{\tanh^2(A_1)}{A_0^2},$
$\zeta = n - n_0,$
$A_0 \neq 0,$
A_1, n_0, t_0, and ϕ_0 are arbitrary real constants, then

$$\Phi(\zeta,\, t) = A_0\, \tanh[A_1\, \zeta]\, e^{-i\, [(A_2+2)\, (t-t_0)+\phi_0]} \tag{9.35}$$

is a solution of (9.33), where
$A_2 = -2\,\text{sech}(A_1),$
$a_2 = \frac{\tanh^2(A_1)}{A_0^2},$ *(changed the sign of* a_2*)*,
$\zeta = n - n_0,$
$A_0 \neq 0,$
A_1, n_0, t_0, and ϕ_0 are arbitrary real constants.

9.7 Nonlocal Cubic-Quintic Discrete NLSE

If $\psi(x, t)$ is a solution of the cubic-quintic discrete NLSE, (8.54),

$$i\,\psi_{nt} + a_1\,(\psi_{n+1} + \psi_{n-1} - 2\,\psi_n) + a_2\,|\psi_n|^2\,\psi_n + (a_3\,|\psi_n|^2 + a_4\,|\psi_n|^4)(\psi_{n+1} + \psi_{n-1}) = 0,$$

then

$$\Phi_n(n, t; a_1, a_2, a_3, a_4) = \begin{cases} \psi_n(n, t; a_1, a_2, a_3, a_4) & \text{if } \psi_n \text{ is an even function in } n, \\ \psi_n(n, t; a_1, -a_2, -a_3, a_4) & \text{if } \psi_n \text{ is an odd function in } n \end{cases}$$

is a solution of

$$i\,\Phi_{nt} + a_1\,(\Phi_{n+1} + \Phi_{n-1} - 2\,\Phi_n) + a_2\,\Phi_n^2\,\bar{\Phi}_n + (a_3\,\Phi_n\,\bar{\Phi}_n + a_4\,\Phi_n^2\,\bar{\Phi}_n^2)(\Phi_{n+1} + \Phi_{n-1}) = 0, \tag{9.36}$$

where

$\psi_n = \psi_n(n, t; a_1, a_2, a_3, a_4)$,
$\Phi_n = \Phi_n(n, t; a_1, a_2, a_3, a_4)$,
$\bar{\Phi}_n = \Phi_n^*(-n, t; a_1, a_2, a_3, a_4)$,
a_1, a_2, a_3, and a_4 are arbitrary real constants.

***Example* 1. Even function: sech(n) I** *discrete bright soliton*

Given

$$\psi(\zeta, t) = A_0\,\text{sech}[A_1\,\zeta]\,e^{i\,[A_2\,(t-t_0)+\phi_0]}$$

is a solution of (8.54), then

$$\Phi(\zeta, t) = A_0\,\text{sech}[A_1\,\zeta]\,e^{i\,[A_2\,(t-t_0)+\phi_0]} \tag{9.37}$$

is a solution of (9.33), where

$A_0 = \pm\dfrac{\sqrt{a_3\,(a_2^2 - 4\,a_1\,a_4) + \sqrt{a_3^2 - 4\,a_1\,a_4}\,(a_2^2 + 4\,a_1\,a_4)}}{2\,a_4\,\sqrt{2\,a_1}}$,

$A_1 = \text{sech}^{-1}\left(\dfrac{-a_3 + \sqrt{a_3^2 - 4\,a_1\,a_4}}{a_2}\right)$,

$A_2 = -2\,a_1 - \dfrac{a_2\left(a_3 + \sqrt{a_3^2 - 4\,a_1\,a_4}\right)}{2\,a_4}$,

$a_3^2 - 4\,a_4\,a_1 \geqslant 0$,
$a_1 > 0$,
$a_3\,(a_2^2 - 4\,a_1\,a_4) + \sqrt{a_3^2 - 4\,a_1\,a_4}\,(a_2^2 + 4\,a_1\,a_4) > 0$,
$\zeta = n - n_0$,
t_0, n_0, and ϕ_0 are arbitrary real constants.

***Example* 2. Even function: sech(*n*) II** *staggered discrete bright soliton*

Given

$$\psi(\zeta, t) = (-1)^{\zeta+n_0} A_0 \operatorname{sech}[A_1 \zeta]\, e^{i\,[A_2 (t-t_0)+\phi_0]}$$

is a solution of (8.54), then

$$\Phi(\zeta, t) = (-1)^{\zeta+n_0} A_0 \operatorname{sech}[A_1 \zeta]\, e^{i\,[A_2 (t-t_0)+\phi_0]} \tag{9.38}$$

is a solution of (9.33), where

$$A_0 = \pm\frac{\sqrt{a_3\,(a_2^2 - 4\,a_1\,a_4) + \sqrt{a_3^2 - 4\,a_1\,a_4}\;(a_2^2 + 4\,a_1\,a_4)}}{2\,a_4\,\sqrt{2\,a_1}},$$

$$A_1 = \operatorname{sech}^{-1}\left(\frac{a_3 - \sqrt{a_3^2 - 4\,a_1\,a_4}}{a_2}\right),$$

$$A_2 = -2\,a_1 - \frac{a_2\left(a_3 + \sqrt{a_3^2 - 4\,a_1\,a_4}\right)}{2\,a_4},$$

$a_3^2 - 4\,a_4\,a_1 \geqslant 0$,

$a_1 > 0$,

$a_3\,(a_2^2 - 4\,a_1\,a_4) + \sqrt{a_3^2 - 4\,a_1\,a_4}\;(a_2^2 + 4\,a_1\,a_4) > 0$,

$\zeta = n - n_0$,

t_0, n_0, and ϕ_0 are arbitrary real constants.

***Example* 3. Odd function: tanh(*n*) I** *discrete dark soliton*

Given

$$\psi(\zeta, t) = A_0 \tanh[A_1 \zeta]\, e^{i\,[A_2 (t-t_0)+\phi_0]}$$

is a solution of (8.54), where

$$A_0 = \pm\sqrt{\frac{a_2 + a_3 + \sqrt{a_3^2 - 4\,a_1\,a_4}}{-2\,a_4}},$$

$$A_1 = \cosh^{-1}\left(\sqrt{\frac{a_3 + \sqrt{a_3^2 - 4\,a_1\,a_4}}{-a_2}}\right),$$

$$A_2 = -2\,a_1 - \frac{a_2\left(a_3 - \sqrt{a_3^2 - 4\,a_1\,a_4}\right)}{2\,a_4},$$

$a_3^2 - 4\,a_4\,a_1 \geqslant 0$,

$a_4\left(a_2 + a_3 + \sqrt{a_3^2 - 4\,a_1\,a_4}\right) < 0$,

$a_2\left(a_3 + \sqrt{a_3^2 - 4\,a_1\,a_4}\right) < 0$,

$\zeta = n - n_0$,

t_0, n_0, and ϕ_0 are arbitrary real constants, then

$$\Phi(\zeta, t) = A_0 \tanh[A_1 \zeta]\, e^{i\,[A_2\,(t-t_0)+\phi_0]} \tag{9.39}$$

is a solution of (9.33), where

$A_0 = \pm\sqrt{\dfrac{-a_2 - a_3 + \sqrt{a_3^2 - 4\,a_1\,a_4}}{-2\,a_4}}$, *(changed the sign of* a_2 *and* a_3*)*,

$A_1 = \cosh^{-1}\left(\sqrt{\dfrac{-a_3 + \sqrt{a_3^2 - 4\,a_1\,a_4}}{a_2}}\right)$, *(changed the sign of* a_2 *and* a_3*)*,

$A_2 = -2\,a_1 - \dfrac{-a_2\left(-a_3 - \sqrt{a_3^2 - 4\,a_1\,a_4}\right)}{2\,a_4}$, *(changed the sign of* a_2 *and* a_3*)*,

$a_3^2 - 4\,a_4\,a_1 \geqslant 0$,

$a_4\left(-a_2 - a_3 + \sqrt{a_3^2 - 4\,a_1\,a_4}\right) < 0$,

$a_2\left(-a_3 + \sqrt{a_3^2 - 4\,a_1\,a_4}\right) > 0$,

$\zeta = n - n_0$,

t_0, n_0, and ϕ_0 are arbitrary real constants.

***Example* 4. Odd function: tanh(*n*) II** *staggered discrete dark soliton*

Given

$$\psi(\zeta, t) = (-1)^{\zeta+n_0} A_0 \tanh[A_1 \zeta]\, e^{i\,[A_2\,(t-t_0)+\phi_0]}$$

is a solution of (8.54), where

$A_0 = \pm\sqrt{\dfrac{a_2 - a_3 - \sqrt{a_3^2 - 4\,a_1\,a_4}}{2\,a_4}}$,

$A_1 = \cosh^{-1}\left(\sqrt{\dfrac{a_3 + \sqrt{a_3^2 - 4\,a_1\,a_4}}{a_2}}\right)$,

$A_2 = -2\,a_1 - \dfrac{a_2\left(a_3 - \sqrt{a_3^2 - 4\,a_1\,a_4}\right)}{2\,a_4}$,

$a_3^2 - 4\,a_4\,a_1 \geqslant 0$,

$\left(a_2 - a_3 - \sqrt{a_3^2 - 4\,a_1\,a_4}\right) > 0$,

$a_2\left(a_3 + \sqrt{a_3^2 - 4\,a_1\,a_4}\right) > 0$,

$\zeta = n - n_0$,

t_0, n_0, and ϕ_0 are arbitrary real constants, then

$$\Phi(\zeta, t) = (-1)^{\zeta+n_0} A_0 \tanh[A_1 \zeta]\, e^{i\,[A_2\,(t-t_0)+\phi_0]} \tag{9.40}$$

is a solution of (9.33), where

$A_0 = \pm\sqrt{\frac{-a_2 + a_3 - \sqrt{a_3^2 - 4\,a_1\,a_4}}{2\,a_4}}$, *(changed the sign of a_2 and a_3)*,

$A_1 = \cosh^{-1}\left(\sqrt{\frac{-a_3 + \sqrt{a_3^2 - 4\,a_1\,a_4}}{-a_2}}\right)$, *(changed the sign of a_2 and a_3)*,

$A_2 = -2\,a_1 - \frac{a_2\left(a_3 + \sqrt{a_3^2 - 4\,a_1\,a_4}\right)}{2\,a_4}$, *(changed the sign of a_2 and a_3)*,

$a_3^2 - 4\,a_4\,a_1 \geqslant 0$,

$a_4\left(-a_2 + a_3 - \sqrt{a_3^2 - 4\,a_1\,a_4}\right) > 0$,

$a_2\left(-a_3 + \sqrt{a_3^2 - 4\,a_1\,a_4}\right) < 0$,

$\zeta = n - n_0$,

t_0, n_0, and ϕ_0 are arbitrary real constants.

9.8 Summary of Chapter 9

Nonlocal NLSE

Transformation: $\Phi(x, t; a_1, a_2) = \begin{cases} \psi(x, t, a_1, a_2) & \text{if } \psi \text{ is an even function in } x, \\ \psi(x, t, a_1, -a_2) & \text{if } \psi \text{ is an odd function in } x \end{cases}$ — ψ is a solution of the fundamental NLSE (2.1)

Equation: $i\,\Phi_t + a_1\,\Phi_{xx} + a_2\,|\Phi^2\,\bar{\Phi} = 0$

# Example	Conditions	Name	Eq. #
1. $\Phi(\zeta, t) = A_0\sqrt{\frac{2\,a_1}{a_2}}\,\text{sech}[A_0\,\zeta]\,e^{i\left[a_1\,A_0^2\,(t-t_0)+\phi_0\right]}$	$\zeta = x - x_0$, $a_1\,a_2 > 0$, A_0, x_0, t_0, and ϕ_0 are arbitrary real constants	bright soliton	(9.2)
2. $\Phi(\zeta, t) = A_0\tanh\left[\frac{A_1}{\sqrt{a_1}}\,\zeta\right]e^{-i\,[A_2\,(t-t_0)+\phi_0]}$	$A_2 = 2\,A_1^2$, $a_2 = \frac{2\,A_1^2}{A_0^2}$, $a_1 > 0$, $\zeta = x - x_0$, A_0, A_1, x_0, t_0, and ϕ_0 are arbitrary	dark soliton	(9.3)

Nonlocal Coupled NLSE

Transformation:

$$\Phi_1(x, t; a_1, a_2, b_1, b_2) = \begin{cases} \psi_1(x, t, a_1, a_2, b_1, b_2) & \text{if } \psi_1 \text{ is an even function in } x, \\ \psi_1(x, t, -a_1, a_2, -b_1, b_2) & \text{if } \psi_1 \text{ is an odd function in } x, \end{cases}$$

$$\Phi_2(x, t; a_1, a_2, b_1, b_2) = \begin{cases} \psi_2(x, t, a_1, a_2, b_1, b_2) & \text{if } \psi_2 \text{ is an even function in } x, \\ \psi_2(x, t, a_1, -a_2, b_1, -b_2) & \text{if } \psi_2 \text{ is an odd function in } x \end{cases}$$

Equation: $i\,\Phi_{1t} + \Phi_{1xx} + (a_1\,\Phi_1\,\bar{\Phi}_1 + a_2\,\Phi_2\,\bar{\Phi}_2)\,\Phi_1 = 0$, $i\,\Phi_{2t} + \Phi_{2xx} + (b_1\,\Phi_1\,\bar{\Phi}_1 + b_2\,\Phi_2\,\bar{\Phi}_2)\,\Phi_2 = 0$

# Example	Conditions	Name	Eq. #
1. $\Phi_1(\zeta, t) = A_0\,\text{sech}[A_1\,\zeta]\,e^{-i\,[\omega_1\,(t-t_0)+\phi_1]}$,	$\omega_1 = A_1^2 - a_1\,A_0^2$, $\omega_2 = 2\,A_1^2 - b_1\,A_0^2$, $a_1 = \frac{2\,A_1^2 - a_2\,B_0^2}{A_0^2}$, $b_1 = \frac{2\,A_1^2 - b_2\,B_0^2}{A_0^2}$, $\zeta = x - x_0$, $A_0 \neq 0$,	bright-dark soliton	(9.6)

(*Continued*)

(Continued)

Nonlocal NLSE		
$\Phi_2(\zeta, t) = B_0 \tanh[A_1 \zeta]\, e^{-i\,[\omega_2 (t-t_0)+\phi_2]}$	A_1, B_0, x_0, t_0, ϕ_1, and ϕ_2 are arbitrary real constants	
2. $\Phi_1(\zeta, t) = A_0\, m\, \mathrm{cn}[A_1 \zeta, m]\, \mathrm{sn}[A_1 \zeta, m]\, e^{-i\,[\omega_1 (t-t_0)+\phi_1]}$, $\Phi_2(\zeta, t) = B_0 \sqrt{m}\, \mathrm{dn}[A_1 \zeta, m]\, \mathrm{sn}[A_1 \zeta, m]\, e^{-i\,[\omega_2 (t-t_0)+\phi_2]}$	$\omega_1 = (4+m) A_1^2$, $\omega_2 = (1+4m) A_1^2$, $a_1 = \frac{-6 A_1^2}{A_0^2 (1-m)}$, $a_2 = \frac{6 A_1^2}{B_0^2 (1-m)}$, $b_1 = a_1$, $b_2 = a_2$, $0 \leqslant m < 1$, $\zeta = x - x_0$, $A_0 \neq 0$, $B_0 \neq 0$, A_1, x_0, t_0, ϕ_1, and ϕ_2 are arbitrary real constants	solitary wave (9.7)
3. $\Phi_1(\zeta, t) = A_0 \sqrt{m\,(1-m)}\, \mathrm{sd}[A_1 \zeta, m]\, e^{-i\,[\omega_1 (t-t_0)+\phi_1]}$, $\Phi_2(\zeta, t) = B_0 \sqrt{1-m}\, \mathrm{nd}[A_1 \zeta, m]\, e^{-i\,[\omega_2 (t-t_0)+\phi_2]}$	$\omega_1 = (m-1)\, a_1 A_0^2 - A_1^2$, $\omega_2 = (m-2) A_1^2 + (m-1)\, b_1 A_0^2$, $a_1 = \frac{-2 A_1^2 + a_2 B_0^2}{A_0^2}$, $b_1 = \frac{-2 A_1^2 + b_2 B_0^2}{A_0^2}$, $0 \leqslant m \leqslant 1$, $\zeta = x - x_0$, $A_0 \neq 0$, A_1, B_0, x_0, t_0, ϕ_1, and ϕ_2 are arbitrary real constants	solitary wave (9.8)
4. $\Phi_1(\zeta, t) = \{A_0 \,\mathrm{sech}^2[A_1 \zeta] + A_3\}\, e^{-i\,[\omega_1 (t-t_0)+\phi_1]}$, $\Phi_2(\zeta, t) = B_0 \,\mathrm{sech}^2[A_1 \zeta]\, e^{-i\,[\omega_2 (t-t_0)+\phi_2]}$	$A_3 = \frac{-2 A_0}{3}$, $\omega_1 = 2 A_1^2$, $\omega_2 = -2 A_1^2$, $a_1 = \frac{-9 A_1^2}{2 A_0^2}$, $a_2 = \frac{9 A_1^2}{2 B_0^2}$, $b_1 = a_1$, $b_2 = a_2$, $\zeta = x - x_0$, $A_0 \neq 0$, $B_0 \neq 0$, A_1, x_0, t_0, ϕ_1, and ϕ_2 are arbitrary real constants	— (9.9)
Symmetry Reductions to Scalar Nonlocal NLSE		
Symmetry Reductions I: ***From Nonlocal Manakov System to Scalar Nonlocal NLSE***		
Transformation: $\Phi_2(x, t) = \sigma\, \Phi_1(x, t)$, $\quad b_1 = c_1 + (c_2 - b_2)\, \lvert\sigma\rvert^2$		
Equation: $i\, \Phi_{1t} + b_0\, \Phi_{1xx} + (c_1 + c_2 \lvert\sigma\rvert^2)\, \Phi_1^2\, \bar{\Phi}_1 = 0$		
Symmetry Reductions II: *From Nonlocal Manakov System to Scalar Nonlocal NLSE*		
Transformation: $\Phi_2(x, t; b_1, b_2, c_1, c_2) = e^{i\,\phi} \begin{cases} \Phi_1^*(-x, t, b_1, b_2, c_1, c_2) & \text{if } \Phi_1 \text{ is an even function in } x, \\ \Phi_1^*(-x, t, b_1, b_2, -c_1, -c_2) & \text{if } \Phi_1 \text{ is an odd function in } x, \end{cases}$ $b_1 = -(c_1 + c_2 + b_2)$		

Equation: $i\,\Phi_{1t} + b_0\,\Phi_{1xx} - (c_1 + c_2)\,\Phi_1^2\,\bar{\Phi}_1 = 0$

Symmetry Reductions III: *From Nonlocal Vector NLSE to Scalar Nonlocal NLSE*

Transformation: $\Phi_j(x, t) = \sigma_j\,\Phi_1(x, t)$

Equation: $i\,\Phi_{1t} + a_1\,\Phi_{1xx} + \sum_{k=1}^{N} b_{1k}|\sigma_k|^2\,\Phi_1^2\,\bar{\Phi}_1 = 0$

Scaling Transformations

Linear and Nonlinear Coupling

General Case

Transformation:

$$\Phi_1(x, t) = \sqrt{\frac{b_1}{g_1 - g_2}}\,\psi_1(x, t)\,e^{i\,g_0\,(g_1 - g_2)\,t} + \sqrt{\frac{g_2\,b_2}{g_1\,(g_2 - g_1)}}\,\psi_2(x, t)\,e^{-i\,g_0\,(g_1 - g_2)\,t},$$

$$\Phi_2(x, t) = \sqrt{\frac{b_1}{g_1 - g_2}}\,\psi_1(x, t)\,e^{i\,g_0\,(g_1 - g_2)\,t} + \sqrt{\frac{g_1\,b_2}{g_2\,(g_2 - g_1)}}\,\psi_2(x, t)\,e^{-i\,g_0\,(g_1 - g_2)\,t}$$

Equation:

$$i\,\Phi_{1t} + \Phi_{1xx} + (g_1\,\Phi_1\,\bar{\Phi}_1 - g_2\,\Phi_2\,\bar{\Phi}_2)\,\Phi_1 + g_0\,(g_1 + g_2)\,\Phi_1 - 2\,g_0\,g_2\,\Phi_2 = 0,\quad i\,\Phi_{2t} + \Phi_{2xx} + (g_1\,\Phi_1\,\bar{\Phi}_1 - g_2\,\Phi_2\,\bar{\Phi}_2)\,\Phi_2 - g_0\,(g_1 + g_2)\,\Phi_2 + 2\,g_0\,g_1\,\Phi_1 = 0$$

Specific Case I: Nonlocal Manakov System to Another Nonlocal Manakov System

Transformation:

$$\Phi_1(x, t) = \sqrt{\frac{b_1}{g_1 - g_2}}\,\psi_1(x, t) + \sqrt{\frac{g_2\,b_2}{g_1\,(g_2 - g_1)}}\,\psi_2(x, t),$$

$$\Phi_2(x, t) = \sqrt{\frac{b_1}{g_1 - g_2}}\,\psi_1(x, t) + \sqrt{\frac{g_1\,b_2}{g_2\,(g_2 - g_1)}}\,\psi_2(x, t)$$

Equation:

$$i\,\Phi_{1t} + \Phi_{1xx} + (g_1\,\Phi_1\,\bar{\Phi}_1 - g_2\,\Phi_2\,\bar{\Phi}_2)\,\Phi_1 = 0,$$

$$i\,\Phi_{2t} + \Phi_{2xx} + (g_1\,\Phi_1\,\bar{\Phi}_1 - g_2\,\Phi_2\,\bar{\Phi}_2)\,\Phi_2 = 0$$

(Continued)

(Continued)

Nonlocal NLSE

Specific Case II: Nonlocal Manakov System to the Same System
Superposition Principle for a Nonlocal Nonlinear System

Transformation:
$$\Phi_1(x,\,t) = \sqrt{\frac{b_1}{b_1+b_2}}\left[\psi_1(x,\,t) - \frac{b_2}{b_1}\,\psi_2(x,\,t)\right],$$
$$\Phi_2(x,\,t) = \sqrt{\frac{b_1}{b_1+b_2}}\,[\psi_1(x,\,t) + \psi_2(x,\,t)]$$

Complex Coupling

General Case

Transformation:
$$\Phi_1(x,\,t) = c_1\,\psi_1(x,\,t) + c_2\,\psi_2(x,\,t),$$
$$\Phi_2(x,\,t) = c_3\,\psi_1(x,\,t) + c_4\,\psi_2(x,\,t)$$

Equation:
$$i\,\Phi_{1t} + \Phi_{1xx} + 2\,(a_{11}\,\Phi_1\,\bar{\Phi}_1 + a_{12}\,\Phi_2\,\bar{\Phi}_2)\,\Phi_1 + 2\,(b_{11}\,\Phi_1\,\bar{\Phi}_2 + b_{12}\,\Phi_2\,\bar{\Phi}_1)\,\Phi_1 = 0,$$
$$i\,\Phi_{2t} + \Phi_{2xx} + 2\,(a_{21}\,\Phi_1\,\bar{\Phi}_1 + a_{22}\,\Phi_2\,\bar{\Phi}_2)\,\Phi_2 + 2\,(b_{21}\,\Phi_1\,\bar{\Phi}_2 + b_{22}\,\Phi_2\,\bar{\Phi}_1)\,\Phi_2 = 0$$

Specific Case

Transformation: $\Phi_1(x,\,t) = \psi_1(x,\,t) + \psi_2(x,\,t), \quad \Phi_2(x,\,t) = \psi_1(x,\,t) + i\,\psi_2(x,\,t)$

Equation:
$$i\,\Phi_{1t} + \Phi_{1xx} - 2(a+b)\,(\Phi_1\,\bar{\Phi}_1 + \Phi_2\,\bar{\Phi}_2)\,\Phi_1 + 2\,[(a+i\,b)\,\Phi_1\,\bar{\Phi}_2 + (a-i\,b)\,\Phi_2\,\bar{\Phi}_1]\,\Phi_1 = 0,$$
$$i\,\Phi_{2t} + \Phi_{2xx} - 2(a+b)\,(\Phi_1\,\bar{\Phi}_1 + \Phi_2\,\bar{\Phi}_2)\,\Phi_2 + 2\,[(a+i\,b)\,\Phi_1\,\bar{\Phi}_2 + (a-i\,b)\,\Phi_2\bar{\Phi}_1]\,\Phi_2 = 0$$

Nonlocal Discrete NLSE with Saturable Nonlinearity

Transformation:
$$\Phi_n(n,\,t;\,a_2,\,\mu) = \begin{cases} \psi_n(n,\,t,\,a_2,\,\mu) & \text{if } \psi_n \text{ is an even function in } n, \\ \psi_n(n,\,t,\,-a_2,\,-\mu) & \text{if } \psi_n \text{ is an odd function in } n \end{cases}$$

Equation: $i\,\Phi_{nt} + \Phi_{n+1} + \Phi_{n-1} - 2\,\Phi_n + \frac{a_2\,\Phi_n^2\,\bar{\Phi}_n}{1+\mu\,\Phi_n\,\bar{\Phi}_n} = 0$

# Example	Conditions	Name	Eq. #
1. $\Phi(\zeta, t) = A_0\,\text{sech}[A_1\,\zeta]\,e^{-i\,[A_2\,(t-t_0)+\phi_0]}$	$A_0 = \frac{\sinh(A_1)}{\sqrt{\mu}}$, $A_2 = \frac{2\,\mu - a_2}{\mu}$, $\mu = \frac{a_2\,\text{sech}(A_1)}{2} > 0$, $\zeta = n - n_0$, A_1, n_0, t_0, and ϕ_0 are arbitrary real constants	bright soliton	(9.29)
2. $\Phi(\zeta, t) = A_0\,\tanh[A_1\,\zeta]\,e^{-i\,[A_2\,(t-t_0)+\phi_0]}$	$A_0 = \frac{\tanh(A_1)}{\sqrt{\mu}}$, $A_2 = \frac{2\,\mu - a_2}{\mu}$, $a_2 = 2\,\mu\,\text{sech}^2(A_1)$, $\mu > 0$, $\zeta = n - n_0$, A_1, n_0, and ϕ_0 are arbitrary real constants	dark soliton	(9.30)
3. $\Phi_s(\zeta, t) = (-1)^{\zeta+n_0}\,A_0\,\text{sech}[A_1\,\zeta]\,e^{i\,[A_2\,(t-t_0)-4\,(t-t_0)+\phi_0]}$	$A_0 = \frac{\sinh(A_1)}{\sqrt{\mu}}$, $A_2 = \frac{2\,\mu + a_2}{\mu}$, $\mu = -\frac{a_2\,\text{sech}(A_1)}{2} > 0$, $\zeta = n - n_0$, A_1, n_0, t_0, and ϕ_0 are arbitrary real constants	staggered bright soliton	(9.32)

Nonlocal Ablowitz–Ladik equation

Transformation: $\Phi_n(n, t; a_2) = \begin{cases} \psi_n(n, t; a_2) & \text{if } \psi_n \text{ is an even function in } n, \\ \psi_n(n, t; -a_2) & \text{if } \psi_n \text{ is an odd function in } n \end{cases}$

Equation: $i\,\Phi_{nt} + \Phi_{n+1} + \Phi_{n-1} - 2\,\Phi_n + a_2\,(\Phi_{n+1} + \Phi_{n-1})\,\Phi_n\,\bar{\Phi}_n = 0$

# Example	Conditions	Name	Eq. #
1. $\Phi(\zeta, t) = A_0\,\text{sech}[A_1\,\zeta]\,e^{-i\,[(A_2+2)\,(t-t_0)+\phi_0]}$	$A_2 = -2\cosh(A_1)$, $a_2 = \frac{\sinh^2(A_1)}{A_0^2}$, $\zeta = n - n_0$, $A_0 \neq 0$, A_0, A_1, n_0, t_0, and ϕ_0 are arbitrary real constants	discrete bright soliton	(9.34)

(Continued)

(Continued)

# Example	Conditions	Name	Eq. #
Nonlocal NLSE			
2. $\Phi(\zeta, t) = A_0 \tanh[A_1 \zeta]\, e^{-i\,[(A_2+2)\,(t-t_0)+\phi_0]}$	$A_2 = -2\,\text{sech}(A_1)$, $a_2 = \frac{\tanh^2(A_1)}{A_0^2}$, $\zeta = n - n_0$, $A_0 \neq 0$, A_0, A_1, n_0, t_0, and ϕ_0 are arbitrary real constants	discrete dark soliton	(9.35)

Nonlocal Cubic-Quintic Discrete NLSE

Transformation: $\Phi_n(n, t;\, a_1, a_2, a_3, a_4) = \begin{cases} \psi_n(n, t;\, a_1, a_2, a_3, a_4) & \text{if } \psi_n \text{ is an even function in } n, \\ \psi_n(n, t;\, a_1, -a_2, -a_3, a_4) & \text{if } \psi_n \text{ is an odd function in } n \end{cases}$

Equation: $i\,\Phi_{nt} + a_1\,(\Phi_{n+1} + \Phi_{n-1} - 2\,\Phi_n) + a_2\,\Phi_n^2\,\bar{\Phi}_n + \left(a_3\,\Phi_n\,\bar{\Phi}_n + a_4\,\Phi_n^2\,\bar{\Phi}_n^2\right)(\Phi_{n+1} + \Phi_{n-1}) = 0$

# Example	Conditions	Name	Eq. #
1. $\Phi(\zeta, t) = A_0\,\text{sech}[A_1\,\zeta]\, e^{i\,[A_2\,(t-t_0)+\phi_0]}$	$A_0 = \pm\frac{\sqrt{a_3\,(a_2^2 - 4\,a_1\,a_4) + \sqrt{a_3^2 - 4\,a_1\,a_4}\,(a_2^2 + 4\,a_1\,a_4)}}{2\,a_4\,\sqrt{2\,a_1}}$, $A_1 = \text{sech}^{-1}\left(\frac{-a_3 + \sqrt{a_3^2 - 4\,a_1\,a_4}}{a_2}\right)$, $\zeta = n - n_0$, $A_2 = -2\,a_1 - \frac{a_2\,(a_3 + \sqrt{a_3^2 - 4\,a_1\,a_4})}{2\,a_4}$, $a_3^2 - 4\,a_4\,a_1 \geqslant 0$, $a_3\,(a_2^2 - 4\,a_1\,a_4) + \sqrt{a_3^2 - 4\,a_1\,a_4}\,(a_2^2 + 4\,a_1\,a_4) > 0$, $a_1 > 0$, t_0, n_0, and ϕ_0 are arbitrary real constants	discrete bright soliton	(9.37)
2. $\Phi(\zeta, t) = (-1)^{\zeta+n_0}\,A_0\,\text{sech}[A_1\,\zeta]\, e^{i\,[A_2\,(t-t_0)+\phi_0]}$	$A_0 = \pm\frac{\sqrt{a_3\,(a_2^2 - 4\,a_1\,a_4) + \sqrt{a_3^2 - 4\,a_1\,a_4}\,(a_2^2 + 4\,a_1\,a_4)}}{2\,a_4\,\sqrt{2\,a_1}}$, $A_1 = \text{sech}^{-1}\left(\frac{a_3 - \sqrt{a_3^2 - 4\,a_1\,a_4}}{a_2}\right)$, $\zeta = n - n_0$, $A_2 = -2\,a_1 - \frac{a_2\left(a_3 + \sqrt{a_3^2 - 4\,a_1\,a_4}\right)}{2\,a_4}$, $a_3^2 - 4\,a_4\,a_1 \geqslant 0$, $a_3\,(a_2^2 - 4\,a_1\,a_4) + \sqrt{a_3^2 - 4\,a_1\,a_4}\,(a_2^2 + 4\,a_1\,a_4) > 0$, $a_1 > 0$, t_0, n_0, and ϕ_0 are arbitrary real constants	staggered discrete bright soliton	(9.38)

3. $\Phi(\zeta, t) = A_0 \tanh[A_1 \zeta]\, e^{i\,[A_2 (t-t_0)+\phi_0]}$	$A_0 = \pm\sqrt{\frac{-a_2 - a_3 + \sqrt{a_3^2 - 4 a_1 a_4}}{-2 a_4}}$, $A_1 = \cosh^{-1}\left(\sqrt{\frac{-a_3 + \sqrt{a_3^2 - 4 a_1 a_4}}{a_2}}\right)$, $\zeta = n - n_0$, $A_2 = -2 a_1 - \frac{-a_2\left(-a_3 - \sqrt{a_3^2 - 4 a_1 a_4}\right)}{2 a_4}$, $a_3^2 - 4 a_4 a_1 \geqslant 0$, $a_4\left(-a_2 - a_3 + \sqrt{a_3^2 - 4 a_1 a_4}\right) < 0$, $a_2\left(-a_3 + \sqrt{a_3^2 - 4 a_1 a_4}\right) > 0$, t_0, n_0, and ϕ_0 are arbitrary real constants	discrete dark soliton	(9.39)
4. $\Phi(\zeta, t) = (-1)^{\zeta+n_0} A_0 \tanh[A_1 \zeta]\, e^{i\,[A_2 (t-t_0)+\phi_0]}$	$A_0 = \pm\sqrt{\frac{-a_2 + a_3 - \sqrt{a_3^2 - 4 a_1 a_4}}{2 a_4}}$, $A_1 = \cosh^{-1}\left(\sqrt{\frac{-a_3 + \sqrt{a_3^2 - 4 a_1 a_4}}{-a_2}}\right)$, $\zeta = n - n_0$, $A_2 = -2 a_1 - \frac{a_2\left(a_3 + \sqrt{a_3^2 - 4 a_1 a_4}\right)}{2 a_4}$, $a_3^2 - 4 a_4 a_1 \geqslant 0$, $a_4\left(-a_2 + a_3 - \sqrt{a_3^2 - 4 a_1 a_4}\right) > 0$, $a_2\left(-a_3 + \sqrt{a_3^2 - 4 a_1 a_4}\right) < 0$, t_0, n_0, and ϕ_0 are arbitrary real constants	staggered discrete dark soliton	(9.40)

IOP Publishing

Handbook of Exact Solutions to the Nonlinear Schrödinger Equations

Usama Al Khawaja and Laila Al Sakkaf

Appendix A

Derivation of Some Solutions of Chapters 2 and 3

Remark: Throughout this appendix, x_0, t_0, ϕ_0, A_0, A_1, and A_2 are arbitrary real constants.

A.1 Derivation of Some Solutions of Section 2.1

A.1.1 Schematic Representation

Equation

$$i\,\psi_t + a_1\,\psi_{xx} + a_2\,|\psi|^2\,\psi = 0$$

Solutions with

real a_1 and a_2	**complex a_1 and a_2**
Case A1: $\psi(x,\,t) = A_0\,e^{i\,\phi(t)}$	**Case A1′:** $\psi(x,\,t) = A_0\,e^{i\,\phi(t)}$
Solution (2.2)	Solution (2.52)
Case A2: $\psi(x,\,t) = A_0\,e^{i\,\phi(x)}$	**Case A2′:** $\psi(x,\,t) = A_0\,e^{i\,\phi(x)}$
Solution (2.3)	Solution (2.53)
Case A3: $\psi(x,\,t) = A_0\,e^{i\,\phi(x,t)}$	**Case A3′:** $\psi(x,\,t) = A(x)\,e^{i\,\phi_0}$
	Solution (5.59)
Solution (2.4)	Solution (2.60)
	Solution (2.61)
Case A4: $\psi(x,\,t) = A(x)\,e^{i\,\phi_0}$	**Case A4′:** $\psi(x,\,t) = A(x)\,e^{i\,\phi(t)}$
$c \neq 0$[a] $\quad$ $c = 0$	

(Continued)

Solution (2.24)		
Solution (2.26)	Solution (2.6)	Solution (2.54)
Solution (2.28)		
Case A5: $\psi(x, t) = A(x)\, e^{i\,\phi(t)}$		**Case A5′:** $\psi(x, t) = A(t)\, e^{i\,\phi(x,t)}$
$c \neq 0$	$c = 0$	
Solution (2.10)		Solution (2.55)
Solution (2.23)	Solution (2.8)	
Solution (2.14)		Solution (2.56)
Solution (2.25)		
Solution (2.12)	Solution (2.13)	Solution (2.57)
Solution (2.27)		
Case A6: $\psi(x, t) = A(x)\, e^{i\,\phi(x,t)}$		Solution (2.58)
Solution (2.32)		
Case A7: $\psi(x, t) = A(t)\, e^{i\,\phi(x,t)}$		
Solution (2.5)		

[a] c is an arbitrary constant of integration resulting from integrating the NLSE, as detailed below.

A.1.2 Detailed Derivations

A.1.2.1 Real Coefficients

The general solution to (2.1) can be written in the polar form:

$$\psi(x, t) = Z\, e^{i\,\phi}, \tag{A.1}$$

where $Z = Z(x, t)$ and $\phi = \phi(x, t)$ are real functions. Substituting (A.1) in the fundamental NLSE (2.1) generates its real and imaginary parts:

$$\text{Re[Equation (2.1)]:} \quad a_2\, Z^3 - Z\, \phi_t - a_1\, Z\, \phi_x^2 + a_1\, Z_{xx} = 0, \tag{A.2}$$

and

$$\text{Im[Equation (2.1)]:} \quad Z_t + 2\, a_1\, Z_x\, \phi_x + a_1\, Z\, \phi_{xx} = 0, \tag{A.3}$$

where the subscripts indicate differentiation with respect to x and t. In the following, we take cases for Z and ϕ.

Case A1:

$$Z(x, t) = A_0, \tag{A.4}$$

$$\phi(x, t) = \phi(t). \tag{A.5}$$

Substituting back into (A.2) and (A.3) leads to the following Re[Equation (2.1)] and Im[Equation (2.1)]:

$$\text{Re[Equation (2.1)]:} \quad A_0^3\, a_2 - A_0\, \phi'(t) = 0, \tag{A.6}$$

$$\text{Im[Equation (2.1)]:} \quad 0 = 0. \tag{A.7}$$

Solving (A.6) for $\phi(t)$ gives

$$\phi(t) = A_0^2\, a_2\, (t - t_0) + \phi_0. \tag{A.8}$$

Substituting (A.4) and (A.8) back into (A.1) leads to solution (2.2).

Case A2:

$$Z(x,\, t) = A_0, \tag{A.9}$$

$$\phi(x,\, t) = \phi(x). \tag{A.10}$$

Substituting back into (A.2) and (A.3) leads to the updated forms of Re[Equation (2.1)] and Im[Equation (2.1)]:

$$\text{Re[Equation (2.1)]:} \quad A_0^3\, a_2 - A_0\, a_1\, \phi'^2(x) = 0, \tag{A.11}$$

$$\text{Im[Equation (2.1)]:} \quad A_0\, a_1\, \phi''(x) = 0. \tag{A.12}$$

Solving (A.12) for $\phi(x)$ gives

$$\phi(x) = A_0 \sqrt{\frac{a_2}{a_1}}\, (x - x_0) + \phi_0. \tag{A.13}$$

Substituting (A.9) and (A.13) back into (A.1) leads to solution (2.3).

Case A3:

$$Z(x,\, t) = A_0, \tag{A.14}$$

$$\phi(x,\, t) = \phi(x,\, t), \tag{A.15}$$

where A_0 is an arbitrary real constant. Substituting back into (A.2) and (A.3) leads to the following Re[Equation (2.1)] and Im[Equation (2.1)]:

$$\text{Re[Equation (2.1)]:} \quad A_0^3\, a_2 - A_0\, \phi_t(x,\, t) - A_0\, a_1\, \phi_x^2(x,\, t) = 0, \tag{A.16}$$

$$\text{Im[Equation (2.1)]:} \quad A_0\, a_1\, \phi_{xx}(x,\, t) = 0. \tag{A.17}$$

Solving (A.17) for $\phi(x,\, t)$ leads to

$$\phi(x,\, t) = s_1(t) + (x - x_0)\, s_2(t), \tag{A.18}$$

where $s_1(t)$ and $s_2(t)$ are real functions of t. Substituting back into (A.16), collecting coefficients of $(x - x_0)^0$ and $(x - x_0)$, separately, and equating to zero, we obtain

$$(x - x_0)^0: \quad A_0^3\, a_2 - A_0\, a_1\, s_2^2(t) - A_0\, s_1'(t) = 0, \tag{A.19}$$

$$(x - x_0): -A_0\, s_2'(t) = 0. \tag{A.20}$$

Solving (A.20) for $s_2(t)$ reads

$$s_2(t) = A_1. \tag{A.21}$$

Substituting this result in (A.19) and solving for $s_1(t)$, we get

$$s_1(t) = \left(A_0^2\, a_2 - a_1\, A_1^2\right)(t - t_0) + \phi_0. \tag{A.22}$$

The form of $\phi(x, t)$ in (A.18) will then be

$$\phi(x, t) = A_1\,(x - x_0) + \left(A_0^2\, a_2 - a_1\, A_1^2\right)(t - t_0) + \phi_0. \tag{A.23}$$

Using (A.23) back into (A.1) leads to solution (2.4).

Case A4:

$$Z(x, t) = A(x), \tag{A.24}$$

$$\phi(x, t) = \phi_0, \tag{A.25}$$

where $A(x)$ is a real function to be determined. Substituting back into (A.2) and (A.3) leads to the following new forms of Re[Equation (2.1)] and Im[Equation (2.1)]:

$$\text{Re[Equation (2.1)]:} \quad a_2\, A^3(x) + a_1\, A''(x) = 0, \tag{A.26}$$

$$\text{Im[Equation (2.1)]:} \quad 0 = 0. \tag{A.27}$$

Employing the separation of variables method using the chain rule $A''(x) = \frac{A'(x)\, dA'(x)}{dA(x)}$ in (A.26), we get the following integral of the independent variable x:

$$x - x_0 = \int \frac{1}{\sqrt{c - \dfrac{a_2}{2\, a_1}\, A^4(x)}}\, dA(x), \tag{A.28}$$

where c is a real constant of integration. In the following, we take two categories of c.
If $c \neq 0$:

The integration above can be written as

$$x - x_0 = \frac{1}{\sqrt{c}} \int \frac{1}{\sqrt{1 - b\, A^2(x)}\, \sqrt{1 + b\, A^2(x)}}\, dA(x), \tag{A.29}$$

where $b = \sqrt{\frac{a_2}{2\,c\,a_1}}$. In the following, we take two options for $A(x)$.

1. $A(x) = \frac{1}{\sqrt{b}}\sin(\theta)$, $0 > \theta > \frac{\pi}{2}$.

The integral in (A.29) becomes

$$x - x_0 = \frac{1}{\sqrt{c\,b}} \int \frac{\cos(\theta)}{\sqrt{1 - \sin^2(\theta)}\,\sqrt{1 + \sin^2(\theta)}}\, d\theta. \tag{A.30}$$

This integration gives

$$x - x_0 = \frac{1}{\sqrt{c\,b}}\,\mathrm{F}(\theta, -1), \tag{A.31}$$

where F gives the elliptic integral of the first kind. Resubstituting $\theta = \sin^{-1}[\sqrt{b}\,A(x)]$ and $b = \sqrt{\frac{a_2}{2\,c\,a_1}}$ in (A.31) and solving for $A(x)$ leads to

$$A(x) = \left(\frac{2\,c\,a_1}{a_2}\right)^{\frac{1}{4}} \mathrm{sn}\left[\left(\frac{c\,a_2}{2\,a_1}\right)^{\frac{1}{4}} (x - x_0), -1\right]. \tag{A.32}$$

From (A.32) and (A.25), (A.1) will lead to (2.24), where $\left(\frac{c\,a_2}{2\,a_1}\right)^{\frac{1}{4}} \to A_0$.

2. $A(x) = \frac{1}{\sqrt{b}}\cos(\theta)$, $0 > \theta > \frac{\pi}{2}$.

The integral in (A.29) becomes

$$x - x_0 = \frac{1}{\sqrt{c\,b}} \int \frac{-\sin(\theta)}{\sqrt{1 - \cos^2(\theta)}\,\sqrt{1 + \cos^2(\theta)}}\, d\theta. \tag{A.33}$$

This integration leads to

$$x - x_0 = \frac{-1}{\sqrt{2\,c\,b}}\,\mathrm{F}\left(\theta, \frac{1}{2}\right), \tag{A.34}$$

where F gives the elliptic integral of the first kind. Resubstituting $\theta = \cos^{-1}[\sqrt{b}\,A(x)]$ and $b = \sqrt{\frac{a_2}{2\,c\,a_1}}$ in (A.34) and solving for $A(x)$ lead to

$$A(x) = \left(\frac{2\,c\,a_1}{a_2}\right)^{\frac{1}{4}} \mathrm{cn}\left[\left(\frac{2\,c\,a_2}{a_1}\right)^{\frac{1}{4}} (x - x_0), \frac{1}{2}\right]. \tag{A.35}$$

From (A.25) and (A.35), (A.1) will lead to (2.26), where $\left(\frac{2\,c\,a_2}{a_1}\right)^{\frac{1}{4}} \to A_0$.

3.

$$A(x) = c\,\sqrt{\frac{2\,a_1}{a_2}}\,\mathrm{dn}\,[c\,(x - x_0), 2]. \tag{A.36}$$

From (A.25) and (A.36), (A.1) will lead to (2.28), where $c \to A_0$.

The procedure of how we get this solution will be shown in the next case, item 4 in Case A5.

If $c = 0$:

Solving (A.28) for $A(x)$, we get

$$A(x) = \sqrt{\frac{-2\,a_1}{a_2}}\,\frac{1}{x - x_0}. \tag{A.37}$$

Substituting (A.25) and (A.37) back into (A.1) leads to solution (2.6).

Case A5:

$$Z(x, t) = A(x), \tag{A.38}$$

$$\phi(x, t) = e^{-i\,[\lambda\,(t - t_0) + \phi_0]}, \tag{A.39}$$

where $A(x)$ is a real function to be determined and λ is an arbitrary real constant. (2.1) takes the form

$$\lambda\,A(x) + a_1\,A''(x) + a_2\,A^3(x) = 0. \tag{A.40}$$

Using the chain rule $A''(x) = \frac{A'(x)\,dA'(x)}{dA(x)}$, we get the following integral of the independent variable x:

$$x - x_0 = \int \frac{1}{\sqrt{c - \frac{\lambda}{a_1}\,A^2(x) - \frac{a_2}{2\,a_1}\,A^4(x)}}\,dA(x), \tag{A.41}$$

where c is a real constant of integration. In the following, we take two categories of c.

If $c \neq 0$:

1. From (A.41), with $c = \frac{-\lambda^2}{2\,a_1\,a_2}$, $x - x_0$ will be given by

$$x - x_0 = \sqrt{\frac{-2\,a_1}{\lambda}}\,\tan^{-1}\left[\sqrt{\frac{a_2}{\lambda}}\,A(x)\right]. \tag{A.42}$$

Solving (A.42) for $A(x)$, we get

$$A(x) = \sqrt{\frac{\lambda}{a_2}}\,\tan\left[\sqrt{\frac{-\lambda}{2\,a_1}}\,(x - x_0)\right]. \tag{A.43}$$

Substituting (A.39) and (A.43) back into (A.1) leads to (2.10), where $\lambda \to -2\,a_1\,A_0^2$.

2. The integral definition of the Jacobi $\mathrm{sn}(x, m)$ elliptic function with modulus m is given by:

$$x - x_0 = \int_0^{\mathrm{sn}(x,m)} \frac{1}{\sqrt{[1 - A^2(x)]\,[1 - m\,A^2(x)]}}\,dA(x). \tag{A.44}$$

Equating this integration with (A.41) gives

$$c - \frac{\lambda}{a_1} A^2(x) - \frac{a_2}{2\,a_1} A^4(x) - c\,[1 - c_1\,A^2(x)]\,[1 - c_1\,m\,A^2(x)] = 0, \tag{A.45}$$

where c_1 is a required real constant to be determined. Equating the coefficients of $A^2(x)$ and $A^4(x)$ to zero, separately, and solving for c and c_1, we get

$$c_1 = \frac{-a_2\,(1+m)}{2\,m\,\lambda}, \qquad c = \frac{-2\,m\,\lambda^2}{a_1\,a_2\,(1+m)^2}. \tag{A.46}$$

Solving

$$x - x_0 = \int_0^{\mathrm{sn}(x,m)} \frac{1}{\sqrt{c\,[1 - c_1\,A^2(x)]\,[1 - c_1\,m\,A^2(x)]}}\,dA(x), \tag{A.47}$$

for $A(x)$ with the expressions of c an c_1 in (A.46), we get

$$A(x) = \sqrt{\frac{2\,m\,\lambda}{-a_2\,(1+m)}}\,\mathrm{sn}\left[\sqrt{\frac{\lambda}{a_1\,(1+m)}}\,(x - x_0),\,m\right]. \tag{A.48}$$

Substituting (A.39) and (A.48) back into (A.1) leads to solution (2.23), where $\lambda \to A_0^2$.

If we take the limit of (A.48) when $\lambda \to c\,a_1\,(1+m)$ and replace $m \to -1$, we return back to (A.32).

For the special case of $m = 1$, equation (A.48) reads

$$A(x) = \sqrt{\frac{-\lambda}{a_2}}\,\tanh\left[\sqrt{\frac{\lambda}{2\,a_1}}\,(x - x_0)\right], \tag{A.49}$$

which leads to solution (2.14).

3. The integral definition of the Jacobi elliptic function, $\mathrm{cn}(x, m)$, with modulus m is given by:

$$x - x_0 = \int_{\mathrm{cn}(x,m)}^{1} \frac{1}{\sqrt{[1 - A^2(x)]\,[1 - m\,A^2(x)]}}\,dA(x). \tag{A.50}$$

By pulling out a factor of $\frac{1}{1-m}$ and replacing $m \to \frac{m}{m-1}$ and $x - x_0 \to \frac{x - x_0}{\sqrt{1-m}}$, this integral definition can be transformed into

$$x - x_0 = \int_{\mathrm{cn}(x,m)}^{1} \frac{1}{\sqrt{[1 - A^2(x)]\,[1 - m + m\,A^2(x)]}}\,dA(x). \tag{A.51}$$

Equating the above integral with (A.41), we get

$$c - \frac{\lambda}{a_1} A^2(x) - \frac{a_2}{2\,a_1} A^4(x) - c\,[1 - c_1\,A^2(x)]\,[1 - m + c_1\,m\,A^2(x)] = 0, \tag{A.52}$$

where c_1 is a real constant to be determined. Equating the coefficients of $A^2(x)$ and $A^4(x)$ to zero, separately, and solving for c and c_1, we obtain

$$c_1 = \frac{a_2\,(1-2\,m)}{2\,m\,\lambda}, \qquad c = \frac{2\,m\,\lambda^2}{a_1\,a_2\,(1-2\,m)^2}. \tag{A.53}$$

Solving

$$x - x_0 = \int_{\mathrm{cn}(x,\,m)}^{1} \frac{1}{\sqrt{c\,[1 - c_1\,A^2(x)]\,[1 - m + c_1\,m\,A^2(x)]}}\,dA(x), \tag{A.54}$$

for $A(x)$ with the help of (A.53), we find

$$A(x) = \sqrt{\frac{2\,m\,\lambda}{a_2\,(1-2\,m)}}\;\mathrm{cd}\left[\sqrt{\frac{\lambda\,(1-m)}{a_1\,(1-2\,m)}}\,(x - x_0),\ \frac{m}{m-1}\right], \tag{A.55}$$

where $\mathrm{cd}(x, m)$ is a Jacobi elliptic function with modulus m. This is equivalent to

$$A(x) = \sqrt{\frac{2\,m\,\lambda}{a_2\,(1-2\,m)}}\;\mathrm{cn}\left[\sqrt{\frac{\lambda}{a_1\,(1-2\,m)}}\,(x - x_0),\ m\right]. \tag{A.56}$$

Substituting (A.39) and (A.56) back into (A.1) leads to the solution in (2.25), where $\lambda \to -A_0^2$.

If we take the limit of (A.56) when $\lambda \to c^2\,a_1\,(1-2\,m)$ and replace $m \to \frac{1}{2}$, we return back to (A.35).

For the special case of $m = 1$, equation (A.56) reads

$$A(x) = \sqrt{\frac{-2\,\lambda}{a_2}}\;\mathrm{sech}\left[\sqrt{\frac{-\lambda}{a_1}}\,(x - x_0)\right], \tag{A.57}$$

which leads to (2.12).

4. The integral definition of the Jacobi elliptic function, $\mathrm{dn}(x, m)$, with modulus m is given by:

$$x - x_0 = \int_{\mathrm{dn}(x,m)}^{1} \frac{1}{\sqrt{[1 - A^2(x)]\,[A^2(x) - 1 + m]}}\,dA(x). \tag{A.58}$$

Equating the above integration with the integral in (A.41)

$$c - \frac{\lambda}{a_1}\,A^2(x) - \frac{a_2}{2\,a_1}\,A^4(x) - c\,[1 - c_1\,A^2(x)]\,[c_1\,A^2(x) - 1 + m] = 0, \tag{A.59}$$

where c_1 is a real constant to be determined. Equating the coefficients of $A^2(x)$ and $A^4(x)$ to zero, and solving for c and c_1, we get

$$c_1 = \frac{a_2\,(m-2)}{2\,\lambda}, \qquad c = \frac{2\,\lambda^2}{a_1\,a_2\,(m-2)^2}. \tag{A.60}$$

Solving

$$x - x_0 = \int_{\mathrm{dn}(x,m)}^{1} \frac{1}{\sqrt{c\,[1 - c_1\,A^2(x)]\,[c_1\,A^2(x) - 1 + m]}}\,dA(x), \tag{A.61}$$

for $A(x)$ with the help of (A.60), we obtain

$$A(x) = \sqrt{\frac{2\,\lambda}{a_2\,(m-2)}}\;\mathrm{cd}\left[\sqrt{\frac{\lambda\,(m-1)}{a_1\,(m-2)}}\,(x - x_0),\,\frac{1}{1-m}\right], \tag{A.62}$$

where cd(x, m) is a Jacobi elliptic function with modulus m, which is equivalent to

$$A(x) = \sqrt{\frac{2\,\lambda}{a_2\,(m-2)}}\;\mathrm{dn}\left[\sqrt{\frac{\lambda}{a_1\,(m-2)}}\,(x - x_0),\,m\right]. \tag{A.63}$$

Substituting (A.39) and (A.63) back into (A.1) leads to (2.27), where $\lambda \to -A_0^2$.

If we take the limit of (A.63) when $\lambda \to c^2\,a_1\,(m-2)$ and replace $m \to 2$, we get (A.36).

For the special case of $m = 1$, equation (A.63) reads

$$A(x) = \sqrt{\frac{-2\,\lambda}{a_2}}\;\mathrm{sech}\left[\sqrt{\frac{-\lambda}{a_1}}\,(x - x_0)\right], \tag{A.64}$$

which again leads to (2.12).

If $c = 0$:

From (A.41), $x - x_0$ is given by

$$x - x_0 = \sqrt{\frac{-a_1}{\lambda}}\left(\ln[A(x)] - \ln\left\{2\,\lambda + \sqrt{2\,\lambda}\,\sqrt{[a_2\,A^2(x) + 2\,\lambda]}\right\}\right). \tag{A.65}$$

Solving (A.65) for $A(x)$, we get

$$A(x) = \frac{-4\,\lambda\,e^{\sqrt{\frac{-\lambda\,x^2}{a_1}}}}{2\,\lambda\,a_2\,e^{\sqrt{\frac{4\,\lambda\,x^2}{a_1}}} - 1}. \tag{A.66}$$

In the following, we represent the different possible cases of $A(x)$ depending upon the sign of each of a_1, a_2, and λ.

1. $a_1 > 0$, $a_2 < 0$, and $\lambda > 0$:

$$A(x) = 2\,\lambda\,\sec\left[\sqrt{\frac{\lambda}{a_1}}\,(x - x_0)\right]. \tag{A.67}$$

Substituting (A.39) and (A.67) back into (A.1) leads to (2.8), where $\lambda \to a_1\,A_0^2$.

2. $a_1 < 0$, $a_2 > 0$, and $\lambda > 0$:

$$A(x) = 2\,\lambda\,\mathrm{csch}[\sqrt{\frac{\lambda}{a_1}}\,(x - x_0)]. \tag{A.68}$$

Substituting (A.39) and (A.68) back into (A.1) leads to (2.13), where $\lambda \to -a_1\,A_0^2$. Other cases produce either imaginary or repeated solutions.

Case A6:

$$Z(x, t) = A(x), \tag{A.69}$$

$$\phi(x, t) = \phi(x, t), \tag{A.70}$$

where $A(x)$ is a real function to be determined. Substituting back into (A.2) and (A.3) leads to the following forms of Re[Equation (2.1)] and Im[Equation (2.1)]:

$$\text{Re[Equation (2.1)]:} \quad a_2\, A^3(x) + a_1\, A''(x) - A(x)\left[\phi_t(x, t) + a_1\, \phi_x^2(x, t)\right] = 0, \tag{A.71}$$

$$\text{Im[Equation (2.1)]:} \quad a_1\, [2\, A'(x)\, \phi_x(x, t) + A(x)\, \phi_{xx}(x, t)] = 0. \tag{A.72}$$

Solving (A.72) for $\phi(x, t)$ will give

$$\phi(x, t) = \int \frac{s_1(t)}{A^2(x)}\, dx + s_2(t), \tag{A.73}$$

where $s_1(t)$ and $s_2(t)$ are two real functions of t to be determined. Taking a special case of these two functions, $s_1(t) = \lambda_0$ and $s_2(t) = \lambda_2\,(t - t_0)$, equation (A.71) becomes

$$\text{Re[Equation (2.1)]:} \quad a_2\, A^3(x) - \frac{a_1\, \lambda_0^2}{A^3(x)} - \lambda_2\, A(x) + a_1\, A''(x) = 0, \tag{A.74}$$

where λ_0 and λ_2 are arbitrary real constants. Solving (A.74) for $A(x)$, we get

$$A(x) = \sqrt{R_3 + m_1\, \text{sn}^2\left[\sqrt{\frac{-a_2\, m_2}{2\, a_1}}\,(x - x_0),\, m\right]}, \tag{A.75}$$

where sn is a Jacobi elliptic function of the modulus $m = \frac{R_2 - R_3}{R_1 - R_3}$, R_j, $j = 1, 2, 3$, are the three roots of $Y(x) = 2\, a_1\, \lambda_0^2 - 2\, a_1\, \lambda_1\, x - 2\, \lambda_2\, x^2 + a_2\, x^3$, $m_1 = R_1 - R_3$, $m_2 = R_1 - R_3$, and λ_1 is an arbitrary real constant. From (A.73), $\phi(x, t)$ takes the form

$$\phi(x, t) = \frac{\sqrt{2}\, \lambda_0\, \Pi\left\{\frac{R_3 - R_2}{R_3},\, \text{am}\left[\sqrt{\frac{-a_2\, m_2}{2\, a_1}}\,(x - x_0),\, m\right],\, m\right\} \text{dn}\left[\sqrt{\frac{-a_2\, m_2}{2\, a_1}}\,(x - x_0),\, m\right]}{R_3 \sqrt{\frac{-a_2\, m_2}{a_1}}\, \sqrt{1 - m\, \text{sn}^2\left[\sqrt{\frac{-a_2\, m_2}{2\, a_1}}\,(x - x_0),\, m\right]}} + \lambda_2\,(t - t_0), \tag{A.76}$$

where dn is the Jacobi elliptic function of the modulus m, Π is the incomplete elliptic integral, and am is the amplitude for Jacobi elliptic functions. Substituting (A.75) and (A.76) back into (A.1) leads to (2.32).

Case A7:

$$Z(x, t) = A(t), \tag{A.77}$$

$$\phi(x, t) = \phi(x, t), \tag{A.78}$$

where $A(t)$ is a real function to be determined. Substituting back into (A.2) and (A.3) leads to the following new forms of Re[Equation (2.1)] and Im[Equation (2.1)]:

$$\text{Re[Equation (2.1)]:} \quad A(t)\left[a_2\, A^2(t) - \phi_t(x, t) - a_1\, \phi_x^2(x, t)\right] = 0, \tag{A.79}$$

$$\text{Im[Equation (2.1)]:} \quad A'(t) + a_1\, A(t)\, \phi_{xx}(x, t) = 0. \tag{A.80}$$

Solving (A.80) for $\phi(x, t)$ gives

$$\phi(x, t) = s_1(t) + s_2(t)\,(x - x_0) - \frac{(x - x_0)^2\, A'(t)}{2\, a_1\, A(t)}, \tag{A.81}$$

where $s_1(t)$ and $s_2(t)$ are real functions to be determined. Using (A.81) in (A.79), we get

$$\begin{aligned} &\text{Re[Equation (2.1)]:} \\ &A(t)\left[a_2\, A^2(t) - a_1\, s_2^2(t) - s_1'(t)\right] + \left[2\, s_2(t)\, A'(t) - s_2'(t)\, A(t)\right](x - x_0) \\ &+\left[\frac{A(t)\, A''(t) - 3\, A'^2(t)}{2\, a_1\, A(t)}\right](x - x_0)^2 = 0. \end{aligned} \tag{A.82}$$

Collecting coefficients of $(x - x_0)^0$, $(x - x_0)$, and $(x - x_0)^2$, separately, and equating to zero, we obtain

$$(x - x_0)^0: \quad a_2\, A^2(t) - a_1\, s_2^2(t) - s_1'(t) = 0, \tag{A.83}$$

$$(x - x_0)^1: \quad 2\, s_2(t)\, A'(t) - s_2'(t)\, A(t) = 0, \tag{A.84}$$

$$(x - x_0)^2: \quad \frac{A(t)\, A''(t) - 3\, A'^2(t)}{2\, a_1\, A(t)} = 0. \tag{A.85}$$

Solving for $A(t)$, $s_1(t)$, and $s_2(t)$, we get

$$A(t) = \frac{A_0}{\sqrt{A_1 + 2\,(t - t_0)}}, \tag{A.86}$$

$$s_1(t) = \frac{a_1 A_2^2}{2 A_1 + 4 (t - t_0)} + \frac{1}{2} a_2 A_0^2 \ln[A_1 + 2 (t - t_0)] + \phi_0, \tag{A.87}$$

$$s_2(t) = \frac{A_2}{A_1 + 2 (t - t_0)}. \tag{A.88}$$

Substituting (A.86), (A.87), and (A.88) in (A.81) and then back into (A.1) leads to (2.5).

A.1.2.2 Complex Coefficients

For complex parameters, we define $a_1 = a_{1r} + i\, a_{1i}$ and $a_2 = a_{2r} + i\, a_{2i}$, where a_{1r}, a_{1i}, a_{2r}, and a_{1i} are real constants. The real and imaginary parts of (2.1) are given by

$$\begin{aligned}&\text{Re[Equation (2.1)]:}\\ &a_{2r}\, Z^3 - Z\, \phi_t - 2\, a_{1i}\, Z_x\, \phi_x - a_{1r}\, Z\, \phi_x^2 + a_{1r}\, Z_{xx} - a_{1i}\, Z\, \phi_{xx} = 0,\end{aligned} \tag{A.89}$$

and

$$\begin{aligned}&\text{Im[Equation (2.1)]:}\\ &a_{2i}\, Z^3 + Z_t + 2\, a_{1r}\, Z_x\, \phi_x - a_{1i}\, Z\, \phi_x^2 + a_{1i}\, Z_{xx} + a_{1r}\, Z\, \phi_{xx} = 0,\end{aligned} \tag{A.90}$$

respectively. In the following, we take cases for Z and ϕ.

Case A1′:

$$Z(x, t) = A_0, \tag{A.91}$$

$$\phi(x, t) = \phi(t). \tag{A.92}$$

Substituting back into (A.89) and (A.90) leads to the following Re[Equation (2.1)] and Im[Equation (2.1)]:

$$\text{Re[Equation (2.1)]:} \quad A_0^3\, a_{2r} - A_0\, \phi'(t) = 0, \tag{A.93}$$

$$\text{Im[Equation (2.1)]:} \quad A_0^3\, a_{2i} = 0. \tag{A.94}$$

Solving (A.93) for $\phi(t)$ gives

$$\phi(t) = A_0^2\, a_{2r}\, (t - t_0) + \phi_0. \tag{A.95}$$

Using (A.91) and (A.95) in (A.1) leads to solution (2.52).

Case A2′:

$$Z(x, t) = A_0, \tag{A.96}$$

$$\phi(x, t) = \phi(x). \tag{A.97}$$

Substituting back into (A.89) and (A.90) leads to the updated forms of Re[Equation (2.1)] and Im[Equation (2.1)]:

$$\text{Re[Equation (2.1)]:} \quad A_0^3\, a_{2r} - A_0\, a_{1r}\, \phi'^2(x) - A_0\, a_{1i}\, \phi''(x) = 0, \tag{A.98}$$

$$\text{Im[Equation (2.1)]:} \quad A_0^3\, a_{2i} - A_0\, a_{1i}\, \phi'^2(x) + A_0\, a_{1r}\, \phi''(x) = 0. \tag{A.99}$$

Multiplying (A.98) by a_{1r} and (A.99) by a_{1i} and taking the summation of the resulting equations leads to

$$A_0^3\,(a_{1i}\, a_{2i} + a_{1r}\, a_{2r}) - A_0\left(a_{1i}^2 + a_{1r}^2\right)\phi'^2(x) = 0. \tag{A.100}$$

Solving (A.100) for $\phi(x)$ gives

$$\phi(x) = \pm\sqrt{\frac{A_0^2\,(a_{1i}\, a_{2i} + a_{1r}\, a_{2r})}{a_{1i}^2 + a_{1r}^2}}\,(x - x_0) + \phi_0. \tag{A.101}$$

Resubstituting (A.101) into (A.98) and (A.99), equating the two resulting equations, and solving for a_{1i}, we find

$$a_{1i} = \frac{a_{1r}\, a_{2i}}{a_{2r}}. \tag{A.102}$$

Substituting (A.102) back into (A.101) and then into (A.1) with (A.96) leads to solution (2.53).

Case A3′:

$$Z(x, t) = A(x), \tag{A.103}$$

$$\phi(x, t) = \phi_0, \tag{A.104}$$

where $A(x)$ is a real function to be determined. Substituting back into (A.89) and (A.90) leads to the following Re[Equation (2.1)] and Im[Equation (2.1)]:

$$\text{Re[Equation (2.1)]:} \quad a_{2r}\, A^3(x) + a_{1r}\, A''(x) = 0, \tag{A.105}$$

$$\text{Im[Equation (2.1)]:} \quad a_{2i}\, A^3(x) + a_{1i}\, A''(x) = 0. \tag{A.106}$$

Solving the above two equations separately by following the steps used with real a_1 and a_2 in Case A4, we get two expressions for $A(x)$. For each of the two solutions to satisfy (2.1), a condition arises, namely $a_{1i} = \frac{a_{1r}\, a_{2i}}{a_{2r}}$. The resulting solutions are (2.59), (2.60), and (2.61).

Case A4′:

$$Z(x, t) = A(t), \tag{A.107}$$

$$\phi(x, t) = \phi_0, \tag{A.108}$$

where $A(t)$ is a real function to be determined. Substituting back into (A.89) and (A.90) leads to the following new forms of Re[Equation (2.1)] and Im[Equation (2.1)]:

$$\text{Re[Equation (2.1)]:} \quad a_{2r}\, A^3(t) = 0, \tag{A.109}$$

$$\text{Im[Equation (2.1)]:} \quad a_{2i}\, A^3(t) + A'(t) = 0. \tag{A.110}$$

Solving (A.110) for $A(t)$

$$A(t) = \frac{1}{\sqrt{2\, a_{2i}\, (t - t_0)}}. \tag{A.111}$$

Substituting (A.108) and (A.111) back into (A.1) leads to solution (2.54).

Case A5′:

$$Z(x, t) = A(t), \tag{A.112}$$

$$\phi(x, t) = \phi(x, t), \tag{A.113}$$

where $A(t)$ is a real function to be determined. Substituting back into (A.89) and (A.90) leads to the following Re[Equation (2.1)] and Im[Equation (2.1)]:

$$\begin{gathered}\text{Re[Equation (2.1)]:}\\ A(t)\left[a_{2r}\, A^2(t) - \phi_t(x, t) - a_{1r}\, \phi_x^2(x, t) - a_{1i}\, \phi_{xx}(x, t)\right] = 0,\end{gathered} \tag{A.114}$$

$$\begin{gathered}\text{Im[Equation (2.1)]:}\\ a_{2i}\, A^3(t) + A'(t) + A(t)\left[a_{1r}\, \phi_{xx}(x, t) - a_{1i}\, \phi_x^2(x, t)\right] = 0.\end{gathered} \tag{A.115}$$

Multiplying (A.114) by a_{1r} and (A.115) by a_{1i} and taking the summation of the resulting equations, we get

$$(a_{1i}\, a_{2i} + a_{1r}\, a_{2r})\, A^3(t) + a_{1i}\, A'(t) - A(t)\, [a_{1r}\, \phi_t(x, t) + (a_{1i}^2 + a_{1r}^2)\, \phi_x^2(x, t)] = 0. \tag{A.116}$$

Solving (A.116) for $\phi(x, t)$

$$\begin{aligned}\phi(x, t) = {} & A_0\, (x - x_0)\\ & + \frac{1}{a_{1r}} \int \left[(a_{1i}\, a_{2i} + a_{1r}\, a_{2r})\, A^2(t) - A_0^2\left(a_{1i}^2 + a_{1r}^2\right) + \frac{a_{1i}\, A'(t)}{A(t)}\right] dt + \phi_0\end{aligned} \tag{A.117}$$

and resubstituting in both (A.114) and (A.115), equating the two resulting equations, and solving for a_{1i}, we get

$$a_{1i} = \frac{a_{2i}\, A^3(t) + A'(t)}{A_0^2\, A(t)}. \tag{A.118}$$

In the following, we take four cases for a_{1i} and a_{2i}, simultaneously.

1. $a_{1i} \neq 0$ and $a_{2i} \neq 0$:
Solving (A.118) for $A(t)$

$$A(t) = \frac{\sqrt{a_{1i}\, A_0^2}\; e^{a_{1i}\, A_0^2\,(t-t_0)}}{\sqrt{-1 + a_{2i}\, e^{2\, a_{1i}\, A_0^2\,(t-t_0)}}}, \tag{A.119}$$

2. $a_{1i} = 0$ and $a_{2i} \neq 0$:
Solving (A.118) for $A(t)$

$$A(t) = \frac{1}{\sqrt{2\, a_{2i}\,(t - t_0)}}, \tag{A.120}$$

3. $a_{1i} \neq 0$ and $a_{2i} = -a_{1r}$:
Solving (A.118) for $A(t)$

$$A(t) = \sqrt{\frac{a_{1r}\, A_0^2}{-a_{2i} + e^{2\, a_{1r}\, A_0^2\,(t-t_0)}}}, \tag{A.121}$$

4. $a_{1i} = -a_{1r}$ and $a_{2i} = 0$:
Solving (A.118) for $A(t)$

$$A(t) = A_0\, e^{-a_{1r}\, A_1^2\,(t-t_0)}. \tag{A.122}$$

We then use the resulting four expressions of $A(t)$ individually in (A.117) to find the corresponding $\phi(x, t)$ for each case. Substituting $A(t)$ and $\phi(x, t)$ for each case into (A.1) leads to solutions (2.57), (2.55), (2.56), and (2.58), respectively.

A.2 Derivation of Some Solutions of Section 3.1

A.2.1 Schematic Representation

	Equation
	$i\, \psi_t + a_1\, \psi_{xx} + a_2\, \lvert\psi\rvert^n\, \psi = 0$
	Solutions
	Case B1: $\psi(x, t) = A_0\, e^{i\, \phi(t)}$
Solution (3.6)	
	Case B2: $\psi(x, t) = A_0\, e^{i\, \phi(x)}$
Solution (3.7)	

(Continued)

	Case B3: $\psi(x, t) = A_0\, e^{i\,\phi(x,t)}$	
Solution (3.8)		
	Case B4: $\psi(x, t) = A(x)\, e^{i\,\phi_0}$	
$c \neq 0$		$c = 0$
Solution (3.15)		Solution (3.10)
	Case B5: $\psi(x, t) = A(x)\, e^{i\,\phi(t)}$	
$c \neq 0$		$c = 0$
Solution (3.16)		
Solution (3.17)		Solution (3.13)
	Case B6: $\psi(x, t) = A(x)\, e^{i\,\phi(x,t)}$	
Solution (3.18)		
	Case B7: $\psi(x, t) = A(t)\, e^{i\,\phi(x,t)}$	
Solution (3.9)		

A.2.2 Detailed Derivations

The general solution to (3.1) can be written in the polar form:

$$\psi(x, t) = Z\, e^{i\,\phi}, \tag{A.123}$$

where $Z = Z(x, t)$ and $\phi = \phi(x, t)$ are real functions. Substituting (A.123) in (3.1) generates its real and imaginary parts:

$$\text{Re[Equation (3.1)]:} \quad a_2\, Z^{n+1} - Z\, \phi_t - a_1\, Z\, \phi_x^2 + a_1\, Z_{xx} = 0, \tag{A.124}$$

and

$$\text{Im[Equation (3.1)]:} \quad Z_t + 2\, a_1\, Z_x\, \phi_x + a_1\, Z\, \phi_{xx} = 0. \tag{A.125}$$

In the following, we take cases for Z and ϕ.

Case B1:

$$Z(x, t) = A_0, \tag{A.126}$$

$$\phi(x, t) = \phi(t). \tag{A.127}$$

Substituting back into (A.124) and (A.125) leads to the following Re[Equation (3.1)] and Im[Equation (3.1)]:

$$\text{Re[Equation (3.1)]:} \quad A_0 \left[A_0^n\, a_2 - \phi'(t) \right] = 0, \tag{A.128}$$

$$\text{Im[Equation (3.1)]:} \quad 0 = 0. \tag{A.129}$$

Solving (A.128) for $\phi(t)$ gives

$$\phi(t) = A_0^n\, a_2\, (t - t_0) + \phi_0. \tag{A.130}$$

Substituting (A.126) and (A.130) back into (A.123) leads to solution (3.6).

Case B2:

$$Z(x, t) = A_0, \tag{A.131}$$

$$\phi(x, t) = \phi(x). \tag{A.132}$$

Substituting back into (A.124) and (A.125) leads to the following Re[Equation (3.1)] and Im[Equation (3.1)]:

$$\text{Re[Equation (3.1)]:} \quad A_0 \left[A_0^n\, a_2 - a_1\, \phi'^2(x)\right] = 0, \tag{A.133}$$

$$\text{Im[Equation (3.1)]:} \quad A_0\, a_1\, \phi''(x) = 0. \tag{A.134}$$

Solving (A.134) for $\phi(x)$ gives

$$\phi(x) = A_0^{n/2} \sqrt{\frac{a_2}{a_1}}\, (x - x_0) + \phi_0. \tag{A.135}$$

Substituting (A.131) and (A.135) back into (A.123) leads to the continuous wave solution (3.7).

Case B3:

$$Z(x, t) = A_0, \tag{A.136}$$

$$\phi(x, t) = \phi(x, t). \tag{A.137}$$

Substituting back into (A.124) and (A.125) leads to the following Re[Equation (3.1)] and Im[Equation (3.1)]:

$$\text{Re[Equation (3.1)]:} \quad A_0 \left[A_0^n\, a_2 - \phi_t(x, t) - a_1\, \phi_x^2(x, t)\right] = 0, \tag{A.138}$$

$$\text{Im[Equation (3.1)]:} \quad A_0\, a_1\, \phi_{xx}(x, t) = 0. \tag{A.139}$$

Solving (A.139) for $\phi(x, t)$ gives

$$\phi(x, t) = s_1(t) + (x - x_0)\, s_2(t), \tag{A.140}$$

where $s_1(t)$ and $s_2(t)$ are real functions of t. Substituting back into (A.138), collecting coefficients of $(x - x_0)^0$ and $(x - x_0)$, separately, and equating to zero, we obtain:

$$(x - x_0)^0: \quad A_0 \left[A_0^n\, a_2 - a_1\, s_2^2(t) - s_1'(t)\right] = 0, \tag{A.141}$$

$$(x - x_0): \; -A_0\, s_2'(t) = 0. \tag{A.142}$$

Solving (A.142) for $s_2(t)$ reads

$$s_2(t) = A_1. \tag{A.143}$$

Substituting this result in (A.141) and solving for $s_1(t)$, we get

$$s_1(t) = \left(A_0^n\, a_2 - A_1^2\, a_1\right)(t - t_0) + \phi_0. \tag{A.144}$$

Then, $\phi(x, t)$ in (A.140) becomes

$$\phi(x, t) = A_1\,(x - x_0) + \left(A_0^n\, a_2 - A_1^2\, a_1\right)(t - t_0) + \phi_0. \tag{A.145}$$

Using (A.145) back into (A.123) leads to the continuous wave solution (3.8).

Case B4:

$$Z(x, t) = A(x), \tag{A.146}$$

$$\phi(x, t) = \phi_0, \tag{A.147}$$

where $A(x)$ is a real function to be determined. Substituting back into (A.124) and (A.125) leads to the following Re[Equation (3.1)] and Im[Equation (3.1)]:

$$\text{Re[Equation (3.1)]:} \quad a_2\, A^{1+n}(x) + a_1\, A''(x) = 0, \tag{A.148}$$

$$\text{Im[Equation (3.1)]:} \quad 0 = 0. \tag{A.149}$$

By employing the chain rule $A''(x) = \frac{A'(x)\, dA'(x)}{dA(x)}$ in (A.148), we get the following integral of the independent variable x:

$$x - x_0 = \int \frac{1}{\sqrt{c - \dfrac{2\, a_2}{(n+2)\, a_1}\, A^{n+2}(x)}}\, dA(x), \tag{A.150}$$

where c is the real constant of integration. In the following, we take two categories of c.
If $c \neq 0$:

The integration above reads

$$x - x_0 = \frac{A(x)}{\sqrt{A_0}}\, {}_2F_1\!\left[\frac{1}{2}, \frac{1}{n+2}, \frac{n+3}{n+2}, \frac{2\, a_2\, A^{n+2}(x)}{a_1\, A_0\,(n+2)}\right] = Y[A(x)] \tag{A.151}$$

$$\psi(x, t) = A(x)\, e^{i\,\phi_0}, \tag{A.152}$$

where

$$Y[A(x)] = \frac{A(x)}{\sqrt{A_0}}\, {}_2F_1\!\left[\frac{1}{2}, \frac{1}{n+2}, \frac{n+3}{n+2}, \frac{2\, a_2\, A^{n+2}(x)}{a_1\, A_0\,(n+2)}\right], \tag{A.153}$$

which is formally solved as

$$A(x) = Y^{-1}(x - x_0). \tag{A.154}$$

Here, ${}_2F_1$ is the hypergeometric function and Y^{-1} indicates the inverse operator of the function $Y[A(x)]$, and hence, we infer the solution expressed in (3.15).

If $c = 0$:

Solving (A.150) for $A(x)$

$$A(x) = \left[\frac{1}{\sqrt{\frac{-a_2\, n^2}{2\, a_1\, (2+n)}}\,(x - x_0)}\right]^{\frac{2}{n}}. \tag{A.155}$$

Substituting (A.147) and (A.155) back into (A.123) leads to (3.10).

Case B5:

$$Z(x, t) = A(x), \tag{A.156}$$

$$\phi(x, t) = e^{i\,[\lambda\,(t - t_0) + \phi_0]}, \tag{A.157}$$

where $A(x)$ is a real function to be determined and λ, t_0, and ϕ_0 are arbitrary real constants. Equation(3.1) becomes

$$-\lambda\, A(x) + a_1\, A''(x) + a_2\, A^{n+1}(x) = 0. \tag{A.158}$$

Using the chain rule $A''(x) = \frac{A'(x)\, dA'(x)}{d\, A(x)}$, we get the following integral of the independent variable x:

$$x - x_0 = \int \frac{1}{\sqrt{c + \frac{\lambda}{a_1} A^2(x) - \frac{2\, a_2}{(n+2)\, a_1} A^{n+2}(x)}}\, dA(x), \tag{A.159}$$

where c is real constant of integration. In the following, we take two categories of c.

If $c \neq 0$:

1. From (A.159), with $n = 1$, $x - x_0$ reads

$$x - x_0 = \frac{2\,(R_3 - R_2)\sqrt{\frac{3\, a_1\, [R_1 - A(x)]\, [R_3 - A(x)]\, [A(x) - R_2]}{(R_1 - R_3)\, (R_2 - R_3)^2}}\; \mathrm{F}\left\{\sin^{-1}\left[\sqrt{\frac{R_3 - A(x)}{R_3 - R_2}}\right], \frac{R_2 - R_3}{R_1 - R_3}\right\}}{\sqrt{-2\, a_2\, A^3(x) + 3\, a_1\, c + 3\, \lambda\, A^2(x)}}, \tag{A.160}$$

where R_j, $j = 1, 2, 3$, are the three roots of $Y(x) = 3\, c + \frac{3\, \lambda\, x}{a_1} - \frac{a_2\, x^3}{a_1}$, and F is the elliptic integral of the first kind. By solving (A.160) for $A(x)$, we obtain

$$A(x) = R_3 + (R_2 - R_3)\,\mathrm{sn}^2\left[\sqrt{\frac{-a_2\,(R_1 - R_3)}{6\,a_1}}\,(x - x_0),\ \frac{R_2 - R_3}{R_1 - R_3}\right], \tag{A.161}$$

where sn is the Jacobi elliptic function. Substituting (A.157) and (A.161) back into (A.123), we get (3.16), where $\frac{R_2 - R_3}{R_1 - R_3} \to m$, $c \to A_1$, and $\lambda \to A_0$.

2. From (A.159), with $n = 4$, $x - x_0$ is given by

$$x - x_0 = -\sqrt{\frac{9\,a_1}{a_2\,R_2\,(R_1 - R_3)}}\,\mathrm{F}\left\{\sin^{-1}\left[\sqrt{\frac{A^2(x)\,(R_3 - R_1)}{R_3\,[A^2(x) - R_1]}}\right],\ \frac{R_3\,(R_1 - R_2)}{R_2\,(R_1 - R_3)}\right\}, \tag{A.162}$$

where R_j, $j = 1, 2, 3$, are the three roots of $m_0 = 3\,c + \frac{3\,\lambda\,x^2}{a_1} - \frac{2\,a_2\,x^3}{a_1}$, and F gives the elliptic integral of the first kind. By solving (A.162) for $A(x)$, we obtain

$$A(x) = -\frac{\sqrt{R_1}\,\mathrm{sn}\left[\sqrt{\frac{a_2\,R_2\,(R_1 - R - 3)}{3\,a_1}}\,(x - x_0),\ \frac{R_3\,(R_1 - R_2)}{R_2\,(R_1 - R_3)}\right]}{\sqrt{\frac{R_1 - R_3}{R_3} + \mathrm{sn}^2\left[\sqrt{\frac{a_2\,R_2\,(R_1 - R - 3)}{3\,a_1}}\,(x - x_0),\ \frac{R_3\,(R_1 - R_2)}{R_2\,(R_1 - R_3)}\right]}}, \tag{A.163}$$

where sn is the Jacobi elliptic function. Substituting (A.157) and (A.163) back into (A.123), we get (3.17), where $\frac{R_3\,(R_1 - R_2)}{R_2\,(R_1 - R_3)} \to m$, $c \to A_1$, and $\lambda \to A_0$.

If $c = 0$:

Solving (A.159) for $A(x)$ with $c = 0$, we get

$$A(x) = \left\{\frac{\lambda\,(2 + n)}{2\,a_2}\,\mathrm{sech}^2\left[\sqrt{\frac{n^2\,\lambda}{4\,a_1}}\,(x - x_0)\right]\right\}^{\frac{1}{n}}. \tag{A.164}$$

Substituting (A.157) and (A.163) back into (A.123), we get (3.13), where $\lambda \to \frac{4\,a_1\,A_0^2}{n^2}$.

Case B6:

$$Z(x, t) = A(x), \tag{A.165}$$

$$\phi(x, t) = \phi(x, t), \tag{A.166}$$

where $A(x)$ is a real function to be determined. Substituting back into (A.124) and (A.125) leads to the following Re[Equation (3.1)] and Im[Equation (3.1)]:

Re[Equation (3.1)]:

$$a_2\,A^{1+n}(x) + a_1\,A''(x) - A(x)\left[\phi_t(x, t) + a_1\,\phi_x^2(x, t)\right] = 0, \tag{A.167}$$

Im[Equation (3.1)]: $\quad a_1\,[2\,A'(x)\,\phi_x(x, t) + A(x)\,\phi_{xx}(x, t)] = 0.$ (A.168)

Solving (A.168) for $\phi(x, t)$

$$\phi(x, t) = \int \frac{s_1(t)}{A^2(x)}\, dx + s_2(t), \tag{A.169}$$

where $s_1(t)$ and $s_2(t)$ are two real functions to be determined. Taking a special case of these two functions, $s_1(t) = \lambda_0$ and $s_2(t) = \lambda_2\,(t - t_0)$, the equation (A.167) becomes

$$\text{Re[Equation (3.1)]:}\quad a_2\, A^{1+n}(x) - \frac{a_1\,\lambda_0^2}{A^3(x)} - \lambda_2\, A(x) + a_1\, A''(x) = 0, \tag{A.170}$$

where λ_0 and λ_2 are arbitrary real constants. Using the chain rule $A''(x) = \frac{A'(x)\, dA'(x)}{dA(x)}$, we get the following integral of the independent variable x:

$$x - x_0 = \int \frac{1}{\sqrt{c + \frac{\lambda}{a_1}\, A^2(x) - \frac{2\, a_2}{(n+2)\, a_1}\, A^{n+2}(x) - \frac{a_1\,\lambda_0^2}{A^2(x)}}}\, dA(x). \tag{A.171}$$

Solving (A.171) for $A(x)$ with $n = 4$, we get

$$A(x) = \frac{\sqrt{R_1\,(R_2 - R_4) + R_2\,(R_4 - R_1)\,\text{sn}^2\left[\sqrt{\frac{-a_2\,(R_1 - R_3)\,(R_2 - R_4)}{3\,a_1}}\,(x - x_0), \frac{(R_2 - R_3)\,(R_1 - R_4)}{(R_1 - R_3)\,(R_2 - R_4)}\right]}}{\sqrt{-R_2 + R_4 + (R_1 - R_4)\,\text{sn}^2\left[\sqrt{\frac{-a_2\,(R_1 - R_3)\,(R_2 - R_4)}{3\,a_1}}\,(x - x_0), \frac{(R_2 - R_3)\,(R_1 - R_4)}{(R_1 - R_3)\,(R_2 - R_4)}\right]}}, \tag{A.172}$$

and hence, from (A.169), $\phi(x, t)$ will read

$$\phi(x, t) = \frac{A_0\left(\sqrt{\frac{-3\,a_1\,(R_1 - R_2)^2}{a_2\, m_2}}\;\Pi\left\{\frac{R_2\,(R_1 - R_4)}{R_1\,(R_2 - R_4)}, \text{am}\left[\sqrt{\frac{-a_2\, m_2}{3\,a_1}}\,(x - x_0), m\right], m\right\}\text{dn}\left[\sqrt{\frac{-a_2\, m_2}{3\,a_1}}\,(x - x_0), m\right]\right)}{R_1\, R_2\sqrt{\frac{m_2 + (R_3 - R_2)\,(R_1 - R_4)\,\text{sn}^2\left[\sqrt{\frac{-a_2\, m_2}{3\,a_1}}\,(x - x_0), m\right]}{m_2}}} + \frac{A_0}{R_2}\,(x - x_0) + A_2\,(t - t_0) + \phi_0, \tag{A.173}$$

where $m = \frac{m_1}{m_2}$, R_j, $j = 1, 2, 3, 4$, are the four roots of $Y(x) = 3\,a_1^2\,\lambda_0^2 - 3\,a_1\,c\,x - 3\,\lambda_2\,x^2 + a_2\,x^4$, $m_1 = (R_2 - R_3)\,(R_1 - R_4)$, and $m_2 = (R_1 - R_3)\,(R_2 - R_4)$. Here, sn and dn are Jacobi elliptic functions, Π is the incomplete elliptic integral, am is the amplitude for Jacobi elliptic functions. Substituting (A.172) and (A.173) back into (A.123) leads to (3.18), where $\lambda_0 \to A_0$, $c \to A_1$ and $\lambda_2 \to A_2$.

Case B7:

$$Z(x, t) = A(t), \tag{A.174}$$

$$\phi(x, t) = \phi(x, t), \tag{A.175}$$

where $A(t)$ is a real function to be determined. Substituting back into (A.124) and (A.125) leads to the following forms of Re[Equation (3.1)] and Im[Equation (3.1)]:

$$\text{Re[Equation (3.1)]:} \quad A(t)\left[a_2\, A^n(t) - \phi_t(x, t) - a_1\, \phi_x^2(x, t)\right] = 0, \tag{A.176}$$

$$\text{Im[Equation (3.1)]:} \quad A'(t) + a_1\, A(t)\, \phi_{xx}(x, t) = 0. \tag{A.177}$$

Solving (A.177) for $\phi(x, t)$ reads

$$\phi(x, t) = s_1(t) + s_2(t)\,(x - x_0) - \frac{(x - x_0)^2\, A'(t)}{2\, a_1\, A(t)}, \tag{A.178}$$

where $s_1(t)$ and $s_2(t)$ are real functions to be determined. Using the later expression of $\phi(x, t)$ in (A.176) gives

$$\begin{aligned} &\text{Re[Equation (3.1)]:} \\ &A(t)[a_2\, A^n(t) - a_1\, s_2^2(t) - s_1'(t)] + [2\, s_2(t)\, A'(t) - s_2'(t)\, A(t)]\,(x - x_0) \\ &+ \left[\frac{A(t)\, A''(t) - 3\, A'^2(t)}{2\, a_1\, A(t)}\right](x - x_0)^2 = 0. \end{aligned} \tag{A.179}$$

Collecting coefficients of $(x - x_0)^0$, $(x - x_0)$, and $(x - x_0)^2$, separately, and equating to zero, we obtain:

$$(x - x_0)^0: \quad a_2\, A^n(t) - a_1\, s_2^2(t) - s_1'(t) = 0, \tag{A.180}$$

$$(x - x_0)^1: \quad 2\, s_2(t)\, A'(t) - s_2'(t)\, A(t) = 0, \tag{A.181}$$

$$(x - x_0)^2: \quad \frac{A(t)\, A''(t) - 3\, A'^2(t)}{2\, a_1\, A(t)} = 0. \tag{A.182}$$

Solving for $A(t)$, $s_1(t)$, and $s_2(t)$, we get

$$A(t) = \frac{A_0}{\sqrt{A_1 + 2\,(t - t_0)}}, \tag{A.183}$$

$$s_1(t) = \frac{a_1\, A_2^2}{2\, A_1 + 4\,(t - t_0)} + \frac{a_2\,(A_1 + 2\, t)}{(2 - n)}\left[\frac{A_0}{\sqrt{A_1 + 2\,(t - t_0)}}\right]^n + \phi_0, \tag{A.184}$$

$$s_2(t) = \frac{A_2}{A_1 + 2\,(t - t_0)}. \tag{A.185}$$

Substituting (A.183), (A.184), and (A.185) in (A.178) and then back into (A.123) leads to (3.9).

A.3 Derivation of Some Solutions of Section 3.3

A.3.1 Schematic Representation

	Equation
	$i\,\psi_t + a_1\,\psi_{xx} + a_2\,\lvert\psi\rvert^n\,\psi + a_3\,\lvert\psi\rvert^m\,\psi = 0$
	Solutions
	Case C1: $\psi(x,\,t) = A_0\,e^{i\,\phi(t)}$
Solution (3.20)	
	Case C2: $\psi(x,\,t) = A_0\,e^{i\,\phi(x)}$
Solution (3.21)	
	Case C3: $\psi(x,\,t) = A_0\,e^{i\,\phi(x,t)}$
Solution (3.22)	
	Case C4: $\psi(x,\,t) = A(t)\,e^{i\,\phi(x,t)}$
Solution (3.23)	

A.3.2 Detailed Derivations

The general solution to equation (3.19) can be written in the polar form:

$$\psi(x,\,t) = Z\,e^{i\,\phi}, \tag{A.186}$$

where $Z = Z(x,\,t)$ and $\phi = \phi(x,\,t)$ are real functions. Substituting (A.186) in (3.19) generates its real and imaginary parts:

$$\text{Re[Equation (3.19)]:} \quad a_3\,Z^{m+1} + a_2\,Z^{n+1} - Z\,\phi_t - a_1\,Z\,\phi_x^2 + a_1\,Z_{xx} = 0, \tag{A.187}$$

and

$$\text{Im[Equation (3.19)]:} \quad Z_t + 2\,a_1\,Z_x\,\phi_x + a_1\,Z\,\phi_{xx} = 0. \tag{A.188}$$

In the following, we take cases for Z and ϕ:

Case C1:

$$Z(x,\,t) = A_0, \tag{A.189}$$

$$\phi(x,\,t) = \phi(t). \tag{A.190}$$

Substituting back into (A.187) and (A.188) leads to the following Re[Equation (3.19)] and Im[Equation (3.19)]:

$$\text{Re[Equation (3.19)]:} \quad A_0\left[A_0^n\,a_2 + A_0^m\,a_3 - \phi'(t)\right] = 0, \tag{A.191}$$

$$\text{Im[Equation (3.19)]:} \quad 0 = 0. \tag{A.192}$$

Solving (A.191) for $\phi(t)$ gives

$$\phi(t) = \left[A_0^n\, a_2 + A_0^m\, a_3\right](t - t_0) + \phi_0. \tag{A.193}$$

Substituting (A.189) and (A.193) back into (A.186) leads to solution (3.20).

Case C2:

$$Z(x, t) = A_0, \tag{A.194}$$

$$\phi(x, t) = \phi(x). \tag{A.195}$$

Substituting (A.194) and (A.195) back into (A.187) and (A.188) leads to the following Re[Equation (3.19)] and Im[Equation (3.19)]:

$$\text{Re[Equation (3.19)]:} \quad A_0\left[A_0^n\, a_2 + A_0^m\, a_3 - a_1\, \phi'^2(x)\right] = 0, \tag{A.196}$$

$$\text{Im[Equation (3.19)]:} \quad A_0\, a_1\, \phi''(x) = 0. \tag{A.197}$$

Solving (A.197) for $\phi(x)$ gives

$$\phi(x) = \pm\left(A_0^{n/2}\sqrt{\frac{a_2}{a_1}} + A_0^{m/2}\sqrt{\frac{a_3}{a_1}}\right)(x - x_0) + \phi_0. \tag{A.198}$$

Substituting (A.194) and (A.198) back into (A.186) leads to solution (3.21).

Case C3:

$$Z(x, t) = A_0, \tag{A.199}$$

$$\phi(x, t) = \phi(x, t). \tag{A.200}$$

Substituting back into (A.199) and (A.188) leads to the following Re[Equation (3.19)] and Im[Equation (3.19)]:

$$\text{Re[Equation (3.19)]:} \quad A_0\left[A_0^n\, a_2 + A_0^m\, a_3 - \phi_t(x, t) - a_1\, \phi_x^2(x, t)\right] = 0, \tag{A.201}$$

$$\text{Im[Equation (3.19)]:} \quad A_0\, a_1\, \phi_{xx}(x, t) = 0. \tag{A.202}$$

Solving (A.201) for $\phi(x, t)$ gives

$$\phi(x, t) = s_1(t) + (x - x_0)\, s_2(t), \tag{A.203}$$

where $s_1(t)$ and $s_2(t)$ are real functions of t. Substituting back into (A.201), collecting coefficients of $(x - x_0)^0$ and $(x - x_0)$, separately, and equating to zero, we obtain:

$$(x - x_0)^0: \quad A_0\left[A_0^n\, a_2 + A_0^m\, a_3 - a_1\, s_2^2(t) - s_1'(t)\right] = 0, \tag{A.204}$$

$$(x - x_0): \; -A_0\, s_2'(t) = 0. \tag{A.205}$$

Solving (A.205) for $s_2(t)$ gives

$$s_2(t) = A_1. \tag{A.206}$$

Substituting this result in (A.204) and solving for $s_1(t)$, we get

$$s_1(t) = \left[A_0^n\, a_2 + A_0^m\, a_3 - A_1^2\, a_1\right](t - t_0) + \phi_0. \tag{A.207}$$

The final form of $\phi(x, t)$ in (A.203) will be

$$\phi(x, t) = A_1\,(x - x_0) + \left(A_0^n\, a_2 + A_0^m\, a_3 - a_1\, A_1^2\right)(t - t_0) + \phi_0. \tag{A.208}$$

Using (A.208) back into (A.186) leads to solution (3.22).

Case C4:

$$Z(x, t) = A(t), \tag{A.209}$$

$$\phi(x, t) = \phi(x, t), \tag{A.210}$$

where $A(t)$ is a real function to be determined. Substituting back into (A.209) and (A.188) leads to the following Re[Equation (3.19)] and Im[Equation (3.19)]:

$$\begin{aligned}&\text{Re[Equation (3.19)]:}\\ &A(t)\left[a_2\, A^n(t) + a_3\, A^m(t) - \phi_t(x, t) - a_1\, \phi_x^2(x, t)\right] = 0,\end{aligned} \tag{A.211}$$

$$\text{Im[Equation (3.19)]:} \quad A'(t) + a_1\, A(t)\, \phi_{xx}(x, t) = 0. \tag{A.212}$$

Solving (A.212) for $\phi(x, t)$

$$\phi(x, t) = s_1(t) + s_2(t)\,(x - x_0) - \frac{(x - x_0)^2\, A'(t)}{2\, a_1\, A(t)}, \tag{A.213}$$

where $s_1(t)$ and $s_2(t)$ are real functions to be determined. Using the expression of $\phi(x, t)$ in (A.211), we get

$$\begin{aligned}&\text{Re[Equation (3.19)]:}\\ &A(t)[a_2\, A^n(t) + a_3\, A^m(t) - a_1\, s_2^2(t) - s_1'(t)] + [2\, s_2(t)\, A'(t) - s_2'(t)\, A(t)]\,(x - x_0)\\ &+\left[\frac{A(t)\, A''(t) - 3\, A'^2(t)}{2\, a_1\, A(t)}\right](x - x_0)^2 = 0.\end{aligned} \tag{A.214}$$

Collecting coefficients of $(x - x_0)^0$, $(x - x_0)$, and $(x - x_0)^2$, separately, and equating to zero, we obtain:

$$(x - x_0)^0: \quad a_2\, A^n(t) + a_3\, A^m(t) - a_1\, s_2^2(t) - s_1'(t) = 0, \tag{A.215}$$

$$(x - x_0)^1: \quad 2\, s_2(t)\, A'(t) - s_2'(t)\, A(t) = 0, \tag{A.216}$$

$$(x - x_0)^2: \quad \frac{A(t)\, A''(t) - 3\, A'^2(t)}{2\, a_1\, A(t)} = 0. \tag{A.217}$$

Solving for $A(t)$, $s_1(t)$, and $s_2(t)$, we get

$$A(t) = \frac{A_0}{\sqrt{A_1 + 2\,(t - t_0)}}, \tag{A.218}$$

$$s_1(t) = \frac{a_1\, A_2^2}{2\, A_1 + 4\,(t - t_0)} + \frac{a_2\, [A_1 + 2\,(t - t_0)]}{(2 - n)} \left[\frac{A_0}{\sqrt{A_1 + 2\,(t - t_0)}}\right]^n \tag{A.219}$$

$$+ \frac{a_3\, [A_1 + 2\,(t - t_0)]}{(2 - m)} \left[\frac{A_0}{\sqrt{A_1 + 2\,(t - t_0)}}\right]^m + \phi_0, \tag{A.220}$$

$$s_2(t) = \frac{A_2}{A_1 + 2\,(t - t_0)}. \tag{A.221}$$

Substituting (A.218), (A.220), and (A.221) in (A.213) and then back into (A.186) leads to (3.23).

IOP Publishing

Handbook of Exact Solutions to the Nonlinear Schrödinger Equations

Usama Al Khawaja and Laila Al Sakkaf

Appendix B

Darboux Transformation: Single Soliton and Breather Solutions

B.1 Darboux Transformation

The fundamental nonlinear Schrödinger equation (NLSE) to be solved is

$$i\,u_t + \frac{1}{2}u_{xx} + |u|^2\,u = 0, \qquad u = u(x,\,t). \tag{B.1}$$

Darboux transformation applies only for linear differential equations. Therefore, this equation is associated with a linear system as follows.

Consider the field

$$\Phi = \begin{pmatrix} \psi_1 & \psi_2 \\ \phi_1 & \phi_2 \end{pmatrix}, \tag{B.2}$$

with all components being complex functions of x and t. Consider the linear (Zakharov–Shabat) system of differential equations in this field

$$\Phi_x = U\cdot\Phi + J\cdot\Phi\cdot\Lambda, \tag{B.3}$$

$$\Phi_t = V\cdot\Phi + i\,U\cdot\Phi\cdot\Lambda + i\,J\cdot\Phi\cdot\Lambda^2, \tag{B.4}$$

where

$$U = \begin{pmatrix} 0 & u \\ -u^* & 0 \end{pmatrix}, \tag{B.5}$$

doi:10.1088/978-0-7503-2428-1ch11

$$V = \frac{i}{2}\begin{pmatrix} |u|^2 & u_x \\ u_x^* & -|u|^2 \end{pmatrix}, \tag{B.6}$$

$$J = \begin{pmatrix} 1 & 0 \\ 0 & -1 \end{pmatrix}, \tag{B.7}$$

$$\Lambda = \begin{pmatrix} \lambda_1 & 0 \\ 0 & \lambda_2 \end{pmatrix}. \tag{B.8}$$

The arbitrary complex constants, $\lambda_{1,2}$, are called the *spectral parameters*. We refer to (B.3, B.4) as the *Lax Pair* (LP). It refers also sometimes to the matrices U and V.

Link between NLSE and LP

The *compatability condition*

$$\Phi_{xt} = \Phi_{tx}, \tag{B.9}$$

requires

$$U_t - V_x + [U,\ V] = 0, \tag{B.10}$$

where $[U,\ V]$ is the commutator of U and V. Substituting for U and V from (B.5, B.6), the compatibility condition reads

$$\begin{pmatrix} -i\,\phi_1\!\left(i\,u_t + \frac{1}{2}u_{xx} + |u|^2\,u\right) & -i\,\phi_2\!\left(i\,u_t + \frac{1}{2}u_{xx} + |u|^2\,u\right) \\ -i\,\psi_1\!\left(-i\,u_t^* + \frac{1}{2}u_{xx}^* + |u|^2\,u^*\right) & -i\,\psi_2\!\left(-i\,u_t^* + \frac{1}{2}u_{xx}^* + |u|^2\,u^*\right) \end{pmatrix} = 0. \tag{B.11}$$

Clearly, the compatibility condition is nothing but the NLSE and its complex conjugate; it requires that u is a solution to the NLSE and u^* is a solution to the complex conjugate of the NLSE. This is the link between the NLSE and the linear system.

Seed Solution

For a given (seed) solution of the NLSE, namely u_0, the linear system will have a solution Φ_0.

Darboux Transformation

The Darboux Transformation is defined as

$$\Phi[1] = \Phi \cdot \Lambda - \sigma\,\Phi, \tag{B.12}$$

where

$$\sigma = \Phi_0 \cdot \Lambda \cdot \Phi_0^{-1}. \tag{B.13}$$

Here, Φ_0 is a seed solution of the linear system for a given seed solution of the NLSE, Φ denotes any solution of the linear system, and $\Phi[1]$ is the transformed (new) solution of the linear system.

We request that the LP, (B.3, B.4), is *covariant* under the Darboux transformation

$$\Phi[1]_x = U[1] \cdot \Phi[1] + J \cdot \Phi[1] \cdot \Lambda, \tag{B.14}$$

$$\Phi[1]_t = V[1] \cdot \Phi[1] + i\, U[1] \cdot \Phi[1] \cdot \Lambda + i\, J \cdot \Phi[1] \cdot \Lambda^2. \tag{B.15}$$

By substituting in this system for $\Phi[1]$ from (B.12) and using (B.5, B.6), the transformed LP $U[1]$ and $V[1]$, must satisfy

$$U[1] = U_0 + [J, \sigma], \tag{B.16}$$

$$V[1] = V_0 + i\,[U_0, \sigma], \tag{B.17}$$

where U_0 and V_0 are the LP in terms of the seed solution. It should be noted that J and Λ are constant matrices and do not change under the Darboux transformation. The new solution, $u[1]$, is obtained from the last equation by noting that

$$\begin{aligned} U[1] &= \begin{pmatrix} 0 & u[1] \\ -u[1]^* & 0 \end{pmatrix} \\ &= \begin{pmatrix} 0 & u_0 \\ -u_0^* & 0 \end{pmatrix} + \begin{pmatrix} 0 & \dfrac{2(\lambda_1 - \lambda_2)\psi_1\psi_2}{\varphi_1\psi_2 - \varphi_2\psi_1} \\ \dfrac{2(\lambda_1 - \lambda_2)\varphi_1\varphi_2}{\varphi_1\psi_2 - \varphi_2\psi_1} & 0 \end{pmatrix}, \end{aligned} \tag{B.18}$$

$$V[1] = \frac{i}{2}\begin{pmatrix} |u[1]|^2 & u[1]_x \\ u[1]_x^* & -|u[1]|^2 \end{pmatrix}, \tag{B.19}$$

lead to a covariant compatibility condition

$$U[1]_t - V[1]_x + [U[1], V[1]] = 0. \tag{B.20}$$

This means that $u[1]$ is a solution of the NLSE

$$i\, u[1]_t + \frac{1}{2} u[1]_{xx} + |u[1]|^2 u[1] = 0. \tag{B.21}$$

Thus, from the solution u to the NLSE, we obtained a new solution $u[1]$ to the same NLSE. The new solution is extracted from (B.18) as

$$u[1] = u_0 + \frac{2(\lambda_1 - \lambda_2)\psi_1\,\psi_2}{\varphi_1\,\psi_2 - \varphi_2\,\psi_1}, \tag{B.22}$$

together with its complex conjugate

$$u[1]^* = u_0^* - \frac{2(\lambda_1 - \lambda_2)\varphi_1\,\varphi_2}{\varphi_1\,\psi_2 - \varphi_2\,\psi_1}. \tag{B.23}$$

The second term on the right hand side of the last two equations is called the *Darboux dressing*.

Symmetry Reduction

For a general seed the linear system reads
***x*-equations:**

$$\psi_{1x} - u\,\phi_1 - \lambda_1\,\psi_1 = 0, \tag{B.24}$$

$$\psi_{2x} - u\,\phi_2 - \lambda_2\,\psi_2 = 0, \tag{B.25}$$

$$\phi_{1x} + u^*\,\psi_1 + \lambda_1\,\phi_1 = 0, \tag{B.26}$$

$$\phi_{2x} + u^*\,\psi_2 + \lambda_2\,\phi_2 = 0. \tag{B.27}$$

***t*-equations:**

$$i\,\psi_{1_t} + \left(\frac{1}{2}|u|^2 + \lambda_1^2\right)\psi_1 + \left(\frac{1}{2}u_x + \lambda_1\,u\right)\phi_1 = 0, \tag{B.28}$$

$$i\,\psi_{2_t} + \left(\frac{1}{2}|u|^2 + \lambda_2^2\right)\psi_2 + \left(\frac{1}{2}u_x + \lambda_2\,u\right)\phi_2 = 0, \tag{B.29}$$

$$i\,\phi_{1_t} - \left(\frac{1}{2}|u|^2 + \lambda_1^2\right)\phi_1 + \left(\frac{1}{2}u_x^* - \lambda_1\,u^*\right)\psi_1 = 0, \tag{B.30}$$

$$i\,\phi_{2_t} - \left(\frac{1}{2}|u|^2 + \lambda_2^2\right)\phi_2 + \left(\frac{1}{2}u_x^* - \lambda_2\,u^*\right)\psi_2 = 0. \tag{B.31}$$

Symmetry reduction:
With the relations:

$$\phi_2^* = \psi_1, \tag{B.32}$$

$$\psi_2^* = -\phi_1, \tag{B.33}$$

$$\lambda_2^* = -\lambda_1, \tag{B.34}$$

the linear system of eight equations reduces to four equations

$$\psi_{1x} - u\,\phi_1 - \lambda_1\,\psi_1 = 0, \tag{B.35}$$

$$\phi_{1x} + u^*\,\psi_1 + \lambda_1\,\phi_1 = 0, \tag{B.36}$$

$$i\,\psi_{1_t} + \left(\frac{1}{2}|u|^2 + \lambda_1^2\right)\psi_1 + \left(\frac{1}{2}u_x + \lambda_1\,u\right)\phi_1 = 0, \tag{B.37}$$

$$i\,\phi_{1_t} - \left(\frac{1}{2}|u|^2 + \lambda_1^2\right)\phi_1 + \left(\frac{1}{2}u_x^* - \lambda_1\,u^*\right)\psi_1 = 0. \tag{B.38}$$

The new solution (B.22) then takes the form

$$u[1] = u + \frac{2(\lambda_1 + \lambda_1^*)\psi_1\,\phi_1^*}{|\phi_1|^2 + |\psi_1|^2}, \tag{B.39}$$

and

$$u[1]^* = u^* + \frac{2(\lambda_1 + \lambda_1^*)\phi_1\,\psi_1^*}{|\phi_1|^2 + |\psi_1|^2}. \tag{B.40}$$

B.1.1 Bright Soliton Solution: Zero Seed

For $u = 0$, the linear system simplifies to

$$\psi_{1x} - \lambda_1\,\psi_1 = 0, \tag{B.41}$$

$$\phi_{1x} + \lambda_1\,\phi_1 = 0, \tag{B.42}$$

$$i\,\psi_{1_t} + \lambda_1^2\,\psi_1 = 0, \tag{B.43}$$

$$i\,\phi_{1_t} - \lambda_1^2\,\phi_1 = 0, \tag{B.44}$$

with a general solution

$$\psi_1 = c_1\,e^{\lambda_1 x + i\,\lambda_1^2 t}, \tag{B.45}$$

$$\phi_1 = c_2\,e^{-\lambda_1 x - i\,\lambda_1^2 t}. \tag{B.46}$$

And upon employing the symmetry reduction, we get

$$\psi_2 = -c_2^*\,e^{-\lambda_1^* x + i\,\lambda_1^{*2} t}, \tag{B.47}$$

$$\phi_2 = c_1^*\,e^{\lambda_1^* x - i\,\lambda_1^{*2} t}. \tag{B.48}$$

The new solution is then given by

$$u[1] = \frac{2(\lambda_1 + \lambda_1^*)c_1\,e^{\lambda_1 x + i\,\lambda_1^2 t}\,c_2^*\,e^{-\lambda_1^* x + i\,\lambda_1^{*2} t}}{|c_2|^2\,e^{-(\lambda_1^* + \lambda_1)x - i\,(\lambda_1^2 - \lambda_1^{*2})t} + |c_1|^2\,e^{(\lambda_1^* + \lambda_1)x + i(\lambda_1^2 - \lambda_1^{*2})t}}, \tag{B.49}$$

which, without loss of generality, simplifies to

$$u[1] = \alpha \operatorname{sech}[\alpha(x - x_0 - v\,t)]\, e^{i\left[\frac{1}{2}(\alpha^2 - v^2)t + v\,x + \phi_0\right]}, \tag{B.50}$$

where $\alpha = 2\lambda_{1r}$, $v = 2\lambda_{1i}$, $c_2/c_1 = e^{2\lambda_{1r} x_0 + i\,\phi_0}$, and ϕ_0 is an arbitrary real constant.

Remark: This is the bright soliton solution (2.12) characterized by four arbitrary parameters: initial position x_0, initial speed v, amplitude (or inverse width) α, and arbitrary global phase ϕ_0.

B.1.2 Generalized Breather Solution for Focusing and Defocusing Nonlinearity: CW Seed

Here, we derive the generalized breather solution of the fundamental NLSE

$$i\,u_t + \frac{1}{2}u_{xx} - c\,|u|^2\,u = 0, \tag{B.51}$$

where $c = 1(-1)$ corresponds to the defocusing (focusing) case.

The corresponding LP takes the form of (B.3, B.4) where

$$U = \sqrt{-c}\begin{pmatrix} 0 & u \\ -u^* & 0 \end{pmatrix}, \tag{B.52}$$

$$V = \frac{i}{2}\begin{pmatrix} -c\,|u|^2 & \sqrt{-c}\,u_x \\ \sqrt{-c}\,u_x^* & -c|u|^2 \end{pmatrix}, \tag{B.53}$$

and J and Λ are given by (B.7, B.8). The linear system (B.3, B.4) reads explicitly

***x*-equations:**

$$-i\,\sqrt{c}\,u\,\phi_1 - \lambda_1\,\psi_1 + \psi_{1x} = 0, \tag{B.54}$$

$$-i\,\sqrt{c}\,u\,\phi_2 - \lambda_2\,\psi_2 + \psi_{2x} = 0, \tag{B.55}$$

$$\lambda_1\,\phi_1 + i\,\sqrt{c}\,u^*\,\psi_1 + \phi_{1x} = 0, \tag{B.56}$$

$$\lambda_2\,\phi_2 + i\,\sqrt{c}\,u^*\,\psi_2 + \phi_{2x} = 0. \tag{B.57}$$

***t*-equations:**

$$-i\,\lambda_1^2\,\psi_1 + \left(\sqrt{c}\,\lambda_1\,\phi_1 + \frac{1}{2}\,i\,c\,u^*\,\psi_1\right)u + \psi_{1t} + \frac{1}{2}\,\sqrt{c}\,\phi_1\,u_x = 0, \tag{B.58}$$

$$-i\,\lambda_2^2\,\psi_1 + \left(\sqrt{c}\,\lambda_2\,\phi_2 + \frac{1}{2}\,i\,c\,u^*\,\psi_2\right)u + \psi_{2t} + \frac{1}{2}\,\sqrt{c}\,\phi_2\,u_x = 0, \tag{B.59}$$

$$\frac{1}{2}\, i(2\lambda_1^2 - c\, |u|^2)\phi_1 - \sqrt{c}\, \lambda_1\, u^*\, \psi_1 + \phi_{1_t} + \frac{1}{2}\, \sqrt{c}\, \psi_1\, u_x^* = 0, \tag{B.60}$$

$$\frac{1}{2}\, i(2\lambda_2^2 - c\, |u|^2)\phi_2 - \sqrt{c}\, \lambda_2\, u^*\, \psi_2 + \phi_{2_t} + \frac{1}{2}\, \sqrt{c}\, \psi_2\, u_x^* = 0. \tag{B.61}$$

Symmetry reduction:

Requiring the complex conjugate of (B.54) and (B.55) to be identical with (B.57) and (B.56), respectively, is possible with

$$\phi_2 = \psi_1^*, \tag{B.62}$$

$$\psi_2 = c\, \phi_1^*, \tag{B.63}$$

$$\lambda_2 = -\lambda_1^*, \tag{B.64}$$

where the system of eight equations reduces to four equations

$$-i\, \sqrt{c}\, u\, \phi_1 - \lambda_1\, \psi_1 + \psi_{1x} = 0, \tag{B.65}$$

$$-i\, \sqrt{c}\, u^*\, \psi_1 - \lambda_1\, \phi_1 - \phi_{1x} = 0, \tag{B.66}$$

$$-\frac{1}{2}\, i\left(2\lambda_1^2 - c\, |u|^2\right)\psi_1 + \sqrt{c}\, \lambda_1\, u\, \phi_1 + \psi_{1_t} + \frac{1}{2}\, \sqrt{c}\, \phi_1\, u_x = 0, \tag{B.67}$$

$$\frac{1}{2}\, i\left(2\lambda_1^2 - c\, |u|^2\right)\phi_1 - \sqrt{c}\, \lambda_1\, u^*\, \psi_1 + \phi_{1_t} + \frac{1}{2}\, \sqrt{c}\, \psi_1\, u_x^* = 0. \tag{B.68}$$

For the CW seed

$$u_0 = A\, e^{i\, A^2\, t} \tag{B.69}$$

with arbitrary real constant A, the general solution of the reduced linear system (B.65–B.68) is

$$\begin{aligned}\psi_1(x,\, t) = {}& [(A\sqrt{c}\; c_1\, i + c_2\lambda_1)\sinh((x + i\, \lambda_1\, t)\omega) + c_2\, \omega\, \cosh((x + i\, \lambda_1\, t)\omega)] \\ & \times \frac{1}{\omega}\, e^{-\frac{1}{2}\, i\, A^2\, c\, t},\end{aligned} \tag{B.70}$$

$$\begin{aligned}\phi_1(x,\, t) = {}& [-\, (A\sqrt{c}\; c_2\, i + c_1\lambda_1)\sinh((x + i\, \lambda_1\, t)\omega) + c_1\, \omega\, \cosh((x + i\, \lambda_1\, t)\omega)] \\ & \times \frac{1}{\omega}\, e^{-\frac{1}{2}\, i\, A^2\, c\, t},\end{aligned} \tag{B.71}$$

where

$$\omega = \sqrt{A^2\, c + \lambda_1^2}, \tag{B.72}$$

and c_1 and c_2 are arbitrary constants. The seed solution of the linear system thus reads

$$\Phi_0 = \begin{pmatrix} \psi_1 & c\,\phi_1^* \\ \phi_1 & \psi_1^* \end{pmatrix}, \tag{B.73}$$

where the symmetry reductions (B.62–B.64) have been taken into account and ψ_1 and ϕ_1 are given by (B.70) and (B.71), respectively.

The new solution, given formally by (B.16) and upon using the symmetry reduction conditions, now reads

$$u[1] = u_0 - 4i\,\sqrt{c}\,\lambda_{1r}\,\frac{\psi_1\,\phi_1^*}{c\,|\phi_1|^2 - |\psi_1|^2}, \tag{B.74}$$

where λ_{1r} is the real part of λ_1.

The breather solution can be viewed at as a combination of two *generalized* solitons where each soliton has a localization component and an oscillatory component, both in x and t. This is verified by rewriting the breather solution as

$$u[1] = \left[1 - i\,\frac{8\lambda_{1r}\,c^{3/2}}{A}\,\frac{\cos(\zeta_1 - i\,\chi_1)\cos(\zeta_2 + i\,\chi_2)}{\cos(2\zeta_1) - c\,\cos(2\zeta_2) + \cosh(2\chi_1) - c\,\cosh(2\chi_2)}\right] \times A\,e^{i\,A^2\,t}, \tag{B.75}$$

where

$$\zeta_{1,2} = \kappa\,X_{1,2} - \Omega\,T_{1,2}, \tag{B.76}$$

$$\chi_{1,2} = \frac{X_{1,2}}{\alpha} - \frac{T_{1,2}}{\tau}, \tag{B.77}$$

$$X_{1,2} = x - x_{01,02}, \tag{B.78}$$

$$T_{1,2} = t - T_{01,02}, \tag{B.79}$$

$$\kappa = \mathrm{Im}[\omega], \tag{B.80}$$

$$\Omega = -\mathrm{Re}[\lambda_1\,\omega], \tag{B.81}$$

$$\alpha = \frac{1}{\mathrm{Re}[\omega]}, \tag{B.82}$$

$$\tau = \frac{1}{\mathrm{Im}[\lambda_1\,\omega]}, \tag{B.83}$$

$$x_{02} = x_{01} + \frac{1}{2\lambda_{1r}}\left(\lambda_{1i}\,\mathrm{Im}\left[\frac{\log q}{\omega}\right] + \lambda_{1r}\,\mathrm{Re}\left[\frac{\log q}{\omega}\right]\right), \tag{B.84}$$

$$t_{02} = t_{01} + \frac{1}{2\lambda_{1r}}\,\mathrm{Im}\left[\frac{\log q}{\omega}\right]. \tag{B.85}$$

In addition to the arbitrary CW amplitude, A, there are four arbitrary parameters: x_{01}, t_{01}, λ_{1r}, and λ_{1r}:

x_{01}: sets the reference for x,
t_{01}: sets the reference for t,

λ_{1r} and λ_{1i} set:

α: width of localization in x,
τ: width of localization in t,
κ: frequency of oscillation in x,
Ω: frequency of oscillation in t.

Note that x_{02} and t_{02}, which correspond to the second generalized soliton, are not arbitrary as they are given in terms of the above four arbitrary parameters. Furthermore, since the four parameters α, τ, κ, and Ω, are given in terms of λ_{1r}, and λ_{1i}, only two out these four parameters are to be considered arbitrary while the other two are not. Any two of the four parameters can be chosen to be arbitrary.

IOP Publishing

Usama Al Khawaja and Laila Al Sakkaf

Appendix C

Derivation of the Similarity Transformations in Chapter 5

C.1 Function Coefficients

Given the generalized NLSE (5.64)

$$i\,\Phi_t + b_1(x,\,t)\,\Phi_{xx} + b_2(x,\,t)\,|\Phi|^2\,\Phi + [b_{3r}(x,\,t) + i\,b_{3i}(x,\,t)]\,\Phi = 0, \tag{C.1}$$

we aim at transforming it to the fundamental NLSE

$$p(x,\,t)(i\,\psi_T + a_1\,\psi_{XX} + a_2\,|\psi|^2\,\psi) = 0, \tag{C.2}$$

with the scaling (similarity) transformation

$$\Phi(x,\,t) = A(x,\,t)\,e^{i\,B(x,t)}\,\psi[X(x,\,t),\,T(x,\,t)], \tag{C.3}$$

where $a_{1,2}$ are arbitrary real constants and $b_{1,2}(x,\,t)$, $b_{3r,i}(x,\,t)$, and $p(x,\,t)$ are arbitrary real functions. The unknown functions $X(x,\,t)$, $T(x,\,t)$, $A(x,\,t)$, and $B(x,\,t)$ are assumed to be real and need to be determined in terms of the function coefficients of the generalized NLSE and the constant coefficients of the fundamental NLSE. It is essential to have the function $p(x,\,t)$ since otherwise the transformation will restrict the coefficients $b_{1,2}(x,\,t)$ to only time-dependent ones and consequently the potential $b_{3r,i}(x,\,t)$ will be real and quadratic in x.

Substituting (C.3) in (C.1) and requesting the result to take the form of the fundamental NLSE (C.2) gives

$$A\,b_1\,p\,T_x^2 = 0, \quad (\text{from } \psi_{TT}), \tag{C.4}$$

$$e^{i\,B}p[2b_1\,A_x\,T_x + A\,(i\,T_t + b_1\,(2i\,B_x\,T_x + T_{xx}))] = i, \quad (\text{from } \psi_T), \tag{C.5}$$

$$e^{i\,B}p\,b_1\,A\,X_x^2 = a_1, \quad (\text{from } \psi_{XX}), \tag{C.6}$$

$$e^{i\,B} p\, b_2\, A^3 = a_2, \quad (\text{from } |\psi|^2\psi), \tag{C.7}$$

$$e^{i\,B} p\, [2b_1\, A_x\, X_x + A\,(i\, X_t + b_1(2i\, B_x\, X_x + X_{xx}))] = 0, \quad (\text{from } \psi_X), \tag{C.8}$$

$$\begin{aligned} &e^{i\,B} p \left[i\, A_t + b_1\,(2i\, A_x\, B_x + A_{xx}) \right. \\ &\quad \left. + A\left(i\, b_{3i} + b_{3r} - B_t - b_1\left(B_x^2 - i\, B_{xx}\right)\right)\right] = 0, \quad (\text{from } \psi). \end{aligned} \tag{C.9}$$

The solution of this system is given by (6.43–6.50), and

$$p(r,\, t) = \frac{1}{e^{i\,B}\, A\, g_1'}. \tag{C.10}$$

As a hint on the procedure of solving this system, one starts with solving (C.4) for T, then (C.5) for p, then (C.6) and (C.7) for b_1 and b_2, respectively, then the real part of (C.8) gives A while the imaginary part gives B, and finally, the real part of (C.9) is solved for b_{3r} and the imaginary part of (C.9) is solved for b_{3i}.

C.2 Solution-Dependent Transformation

This is very similar to the previous case. The only difference is in the procedure of solving (C.8) and (C.9). While in the previous section we set the coefficient of ψ and ψ_X to zero separately, here we require the sum of the two terms to be zero, namely

$$\begin{aligned} &[2b_1\, A_x\, X_x + A\,(i\, X_t + b_1(2i\, B_x\, X_x + X_{xx}))]\, \psi \\ &+ \left[i\, A_t + b_1\,(2i\, A_x\, B_x + A_{xx}) \right. \\ &\quad \left. + A\left(i\, b_{3i} + b_{3r} - B_t - b_1\left(B_x^2 - i\, B_{xx}\right)\right)\right] \psi_X = 0. \end{aligned} \tag{C.11}$$

The two procedures should both be valid since they lead to satisfying the generalized and fundamental NLSE. Solving the real part of the last equation for b_{3r} and the imaginary part for b_{3i} gives (5.83) and (5.84), respectively. The solutions of $b_{1,2}$, X, T, A, and B remain the same as in the previous case.

C.3 Similarity Transformation for the NLSE in (N + 1)-Dimensions

Given the generalized NLSE in N-spacial dimensions and polar coordinates

$$\begin{aligned} &i\, \Phi_t + b_1(r,\, t)\left(\Phi_{rr} + \frac{N-1}{r}\, \Phi_r\right) + b_2(x,\, t)\, |\Phi|^2\, \Phi \\ &+ [b_{3r}(r,\, t) + i\, b_{3i}(r,\, t)]\, \Phi = 0, \end{aligned} \tag{C.12}$$

we aim at transforming it to the fundamental NLSE

$$p(r,\, t)(i\, \psi_T + a_1\, \psi_{RR} + a_2\, |\psi|^2\, \psi) = 0, \tag{C.13}$$

with the similarity transformation

$$\Phi(r,\, t) = A(r,\, t)\, e^{i\, B(r,t)}\, \psi[R(r,\, t),\, T(r,\, t)], \tag{C.14}$$

where $a_{1,2}$ are arbitrary real constants and $b_{1,2}(r,\, t)$, $b_{3r}(r,\, t)$, $b_{3i}(r,\, t)$, and $p(r,\, t)$ are arbitrary real functions. The unknown functions $R(r,\, t)$, $T(r,\, t)$, $A(r,\, t)$, and $B(r,\, t)$ are assumed to be real and need to be determined in terms of the function coefficients of the generalized NLSE and the constant coefficients of the fundamental NLSE.

Substituting (C.14) in (C.12) and requesting the result to take the form of the fundamental NLSE (C.13) gives

$$A\, b_1\, p\, T_r^2 = 0, \quad (\text{from } \psi_{TT}), \tag{C.15}$$

$$\begin{aligned} &\frac{1}{r}\, e^{i\, B} p [2r\, b_1\, A_r\, T_r \\ &\quad + A\, (i\, r\, T_t + b_1\, ((N - 1 + 2i\, r\, B_r)\, T_r + r\, T_{rr}))] = i, \quad (\text{from } \psi_T), \end{aligned} \tag{C.16}$$

$$e^{i\, B} p\, b_1\, A\, R_r^2 = a_1, \quad (\text{from } \psi_{RR}), \tag{C.17}$$

$$e^{i\, B} p\, b_2\, A^3 = a_2, \quad (\text{from } |\psi|^2 \psi), \tag{C.18}$$

$$\begin{aligned} &\frac{1}{r}\, e^{i\, B} p\, [2r\, b_1\, A_r\, R_r \\ &\quad + A\, (i\, r\, R_t + b_1 ((N - 1 + 2ir\, B_r)\, R_r + r\, R_{rr}))] = 0, \quad (\text{from } \psi_R), \end{aligned} \tag{C.19}$$

$$\begin{aligned} &\frac{1}{r}\, e^{i\, B} p \Big[\, i\, r\, A_t + b_1\, ((N - 1 + 2i\, r\, B_r)\, A_r \\ &\quad + r\, A_{rr}) + i\, A\, (r\, b_{3i} + (N - 1)\, b_1\, B_r \\ &\quad + i\, r \left(-b_{3r} + B_t + b_1\, B_r^2\right) + r\, b_1\, B_{rr}\Big)\Big] = 0, \quad (\text{from } \psi). \end{aligned} \tag{C.20}$$

The solution of this system is given by (5.20–5.26). The procedure of solving the system is similar to that described in Section C.1.

www.ingramcontent.com/pod-product-compliance
Lightning Source LLC
Chambersburg PA
CBHW082129210326
41599CB00031B/5921
* 9 7 8 0 7 5 0 3 2 4 2 9 8 *